Some Fundamental Constants[a]

Quantity	Symbol	Value[b]
Atomic mass unit	u	$1.660\ 540\ 2(10) \times 10^{-27}$ kg
		$931.434\ 32(28)$ MeV/c^2
Avogadro's number	N_A	$6.022\ 136\ 7(36) \times 10^{23}$ (mol)$^{-1}$
Bohr magneton	$\mu_B = \dfrac{e\hbar}{2m_e}$	$9.274\ 015\ 4(31) \times 10^{-24}$ J/T
Bohr radius	$a_0 = \dfrac{\hbar^2}{m_e e^2 k}$	$0.529\ 177\ 249(24) \times 10^{-10}$ m
Boltzmann's constant	$k_b = R/N_A$	$1.380\ 658(12) \times 10^{-23}$ J/K
Compton wavelength	$\lambda_c = \dfrac{h}{m_e c}$	$2.426\ 310\ 58(22) \times 10^{-12}$ m
Deuteron mass	m_d	$3.343\ 586\ 0(20) \times 10^{-27}$ kg
		$2.013\ 553\ 214(24)$ u
Electron mass	m_e	$9.109\ 389\ 7(54) \times 10^{-31}$ kg
		$5.485\ 799\ 03(13) \times 10^{-4}$ u
		$0.510\ 999\ 06(15)$ MeV/c^2
Electron-volt	eV	$1.602\ 177\ 33(49) \times 10^{-19}$ J
Elementary charge	e	$1.602\ 177\ 33(49) \times 10^{-19}$ C
Gas constant	R	$8.314\ 510(70)$ J/K·mol
Gravitational constant	G	$6.672\ 59(85) \times 10^{-11}$ N·m^2/kg^2
Hydrogen ground state	$E_0 = \dfrac{m_e e^4 k_e^2}{2\hbar^2} = \dfrac{e^2 k_e}{2a_0}$	$13.605\ 698(40)$ eV
Josephson frequency-voltage ratio	$2e/h$	$4.835\ 976\ 7(14) \times 10^{14}$ Hz/V
Magnetic flux quantum	$\Phi_0 = \dfrac{h}{2e}$	$2.067\ 834\ 61(61) \times 10^{-15}$ Wb
Neutron mass	m_n	$1.674\ 928\ 6(10) \times 10^{-27}$ kg
		$1.008\ 664\ 904(14)$ u
		$939.565\ 63(28)$ MeV/c^2
Nuclear magneton	$\mu_n = \dfrac{e\hbar}{2m_p}$	$5.050\ 786\ 6(17) \times 10^{-27}$ J/T
Permeability of free space	μ_0	$4\pi \times 10^{-7}$ N/A^2 (exact)
Permittivity of free space	$\epsilon_0 = 1/\mu_0 c^2$	$8.854\ 187\ 817 \times 10^{-12}$ C^2/N·m^2 (exact)
Planck's constant	h	$6.626\ 075(40) \times 10^{-34}$ J·s
	$\hbar = h/2\pi$	$1.054\ 572\ 66(63) \times 10^{-34}$ J·s
Proton mass	m_p	$1.672\ 623(10) \times 10^{-27}$ kg
		$1.007\ 276\ 470(12)$ u
		$938.272\ 3(28)$ MeV/c^2
Quantized Hall resistance	h/e^2	$25812.805\ 6(12)$ Ω
Rydberg constant	R_H	$1.097\ 373\ 153\ 4(13) \times 10^7$ m^{-1}
Speed of light in vacuum	c	$2.997\ 924\ 58 \times 10^8$ m/s (exact)

[a] These constants are the values recommended in 1986 by CODATA, based on a least-squares adjustment of data from different measurements. For a more complete list, see Cohen, E. Richard, and Barry N. Taylor, *Rev. Mod. Phys.* **59**:1121, 1987.

[b] The numbers in parentheses for the values below represent the uncertainties in the last two digits.

Principles of Physics

SECOND EDITION

Volume 1

Principles
of Physics

SECOND EDITION
Volume 1

Raymond A. Serway
James Madison University

with contributions by
John W. Jewett, Jr.
California State Polytechnic University, Pomona

SAUNDERS COLLEGE PUBLISHING
Harcourt Brace College Publishers

Fort Worth   Philadelphia   San Diego   New York   Orlando   Austin
San Antonio   Toronto   Montreal   London   Sydney   Tokyo

Requests for permission to make copies of any part of the work should be mailed to:
Permissions Department, Harcourt Brace & Company, 6277 Sea Harbor Drive, Orlando, Florida 32887-6777

Publisher: Emily Barrosse
Publisher: John Vondeling
Product Manager: Angus McDonald
Developmental Editor: Susan Dust Pashos
Project Editor: Elizabeth Ahrens
Production Manager: Charlene Catlett Squibb
Art Director: Carol Bleistine

Cover Credit: Wolf Howling, Aurora Borealis. (© 1997 Michael DeYoung/Alaska Stock)
Frontispiece Credit: David Malin, Anglo-Australian Observatory

Printed in the United States of America

PRINCIPLES OF PHYSICS, Second Edition: Volume 1
0-03-024558-3

Library of Congress Catalog Card Number: 97-65256

7890123456   032   10 987654321

Preface

*P*rinciples of Physics is designed for a one-year introductory calculus-based physics course for engineering and science students. This second edition contains many new pedagogical features—most notably, an increased emphasis on physical concepts through the use of conceptual examples and conceptual problems. Based on comments from users of the first edition and reviewers' suggestions, a major effort was made to improve organization, clarity of presentation, precision of language, and accuracy throughout.

This project was conceived because of the well-known problem we continue to wrestle with in teaching the introductory calculus-based physics course. The course content (and hence the size of textbooks) continues to grow, while the number of contact hours with students has either dropped or remained unchanged. Furthermore, traditional one-year courses cover little if any 20th-century physics.

In preparing this book, I was motivated by the spreading interest in reforming this course, primarily through efforts of the Introductory University Physics Project (IUPP) sponsored by the American Association of Physics Teachers and American Institute of Physics. The primary goals and guidelines of this project are to:

Bruce Byers, FPG

- reduce course content following the "less may be more" theme
- incorporate contemporary physics naturally into the course
- organize the course in the context of one or more "story lines"
- treat all student constituents equitably

Recognizing a need for a textbook that could meet these guidelines several years ago, I studied the various proposed IUPP models and the many reports from IUPP committees. Eventually, I became actively involved in the review and planning of one specific model, initially developed at the U.S. Air Force Academy, entitled "A Particles Approach to Introductory Physics." I spent part of the summer of 1990 at the Academy working with Colonel

James Head and Lt. Col. Rolf Enger, the primary authors of the IUPP model, and other members of that department. This most useful collaboration was the starting point of this project.

In my opinion, the IUPP model developed at the U.S. Air Force Academy, and modified by a team of people over the last few years, was an excellent choice for designing a reformed curriculum for several reasons:

- It is an evolutionary approach (rather than a revolutionary approach), which should meet the current demands of the physics community.
- It deletes many topics in classical physics (such as alternating current circuits and optical instruments), and places less emphasis on rigid body motion, optics, and thermodynamics.
- Some topics in 20th century physics are introduced early in the textbook such as special relativity, energy and momentum quantization, and the Bohr model of the hydrogen atom.
- A deliberate attempt is made at showing the unity of physics.

CHANGES TO THE SECOND EDITION

A number of changes and improvements have been made in the second edition of this text. Many changes are in response to comments and suggestions provided by reviewers of the manuscript and instructors using the first edition. The following represent the major changes in the second edition:

Superstock

Organization   The organization of the textbook is slightly different from that of the first edition. Chapter 6 in the first edition has been deleted, and the sections dealing with the fundamental forces in nature and the gravitational field are now incorporated in Chapter 5. The material on oscillatory motion and waves (Chapters 21, 22, and 23 in the first edition) is now covered earlier (Chapters 12, 13, and 14 in the second edition), following mechanics. The two chapters in the first edition covering interference and diffraction of light waves (Chapters 27 and 28) have been combined and condensed into one chapter entitled Wave Optics (Chapter 27). The section on polarization of light waves is now treated earlier in Chapter 24 entitled Electromagnetic Waves. The section on fission and fusion in Chapter 10 of the first edition has been moved to Chapter 30 entitled Nuclear Physics. Finally, by popular demand, Chapter 15 entitled Fluid Mechanics is new to this edition.

Content   In this second edition, a concerted effort was made to place more emphasis on critical thinking and teaching physical concepts. This was accomplished by the addition of approximately 150 conceptual examples, called **Thinking Physics,** and over 200 **Conceptual Problems.** Most of the **Thinking Physics** examples, which include reasoning statements, provide students with a means of reviewing the concepts presented in that section. Some examples demonstrate the connection between the content of that chapter and other scientific disciplines. These examples also serve as models for students to respond to the **Conceptual Problems** posed in the text and end-of-chapter conceptual questions. The answers to all **Conceptual Problems** are contained at the end of each chapter.

Problems and Conceptual Questions   A substantial revision of the end-of-chapter problems and conceptual questions was made in this second edition. Most of the new problems that have been added are intermediate in level, and many of them are problems that require students to make order-of-magnitude calculations or estimates. All problems have been carefully edited and reworded where necessary. Solutions to approximately 25 percent of the end-of-chapter problems and selected conceptual questions are included in the Study Guide and Student Solutions Manual. These problems and questions are identified by boxes around their numbers.

Significant Figures   Significant figures in both worked examples and end-of-chapter problems have been handled with care. Most numerical examples and problems are worked out to three significant figures.

TEXTBOOK FEATURES

The textbook includes many features which are intended to enhance its usefulness to both the student and instructor:

Style   I have attempted to write the book in a style that is clear, logical, and succinct. At the same time, my relaxed writing style is meant to make the reading more enjoyable and less intimidating.

Previews   Most chapters begin with a chapter preview, which includes a brief discussion of chapter objectives and content. This feature enables the student to better understand how the topics covered in that chapter fit into the overall structure and objectives of the course.

Mathematical Level   Calculus is introduced gradually, where it is needed, keeping in mind that many students often take a course in calculus concurrently. Most steps are shown when basic equations are developed, and reference is made to mathematical appendices provided at the end of the text. Vector products are covered where they are needed in physical applications.

Conceptual Examples and Problems   As mentioned earlier, one of the objectives of this edition is to place more emphasis on critical thinking and conceptual understanding. To meet this objective, I have included many **Thinking Physics** examples and **Conceptual Problems** throughout each chapter. The reasoning or responses to all examples immediately follow the example statement, while the answers to all conceptual problems are included at the end of each chapter. Ideally, the student will use these features to better understand physical concepts before being presented with quantitative examples and working homework problems.

Worked Examples   Every chapter includes several worked examples of varying difficulty which are intended to reinforce conceptual understanding and to serve as models for solving end-of-chapter problems.

Exercises   Many examples are followed immediately by exercises with answers. Those exercises contained in an example box represent extensions of the worked examples. Other related exercises with answers are included at the end of many sections. All exercises are intended to encourage students to practice problem solving for immediate reinforcement, and to test the student's understanding of concepts and problem-solving skills. Students who work through these exercises on a regular basis should find the end-of-chapter problems less intimidating.

Important Statements and Equations   Most important statements and definitions are set in **boldface** type for added emphasis and ease of review. Important equations are highlighted with a tan screen for review or reference.

Russell Schlepman

Illustrations   The text material, worked examples, and end-of-chapter questions and problems are accompanied by numerous figures, photographs, and tables. Full color is used to add clarity to the figures and to make the visual presentations as realistic as possible. Three-dimensional effects are produced with the use of color airbrushed areas, where appropriate. Vectors are color-coded, and curves in *xy*-plots are drawn in color. Color photographs have been carefully selected, and their accompanying captions have been written to serve as an added instructional tool.

Margin Notes   Margin notes are used to help students locate and review important statements, concepts, and equations.

Units   The international system of units (SI) is used throughout the text. The British engineering system of units is used only to a limited extent in chapters on mechanics, heat, and thermodynamics.

Problem-Solving Strategies and Hints   I have included general strategies and hints for solving the types of problems featured in both the worked examples and in the end-of-chapter problems. This feature will help students identify important steps and understand the logic for solving specific classes of problems.

Summaries   Each chapter contains a summary which reviews the important concepts and equations discussed in that chapter.

Conceptual Questions   A list of conceptual questions is provided at the end of each chapter. These questions, which require verbal or written responses, provide the student with a means of self-testing the concepts presented in the chapter. Others could serve as a basis for initiating classroom discussions. Answers to selected questions, indicated by boxes around their numbers, are included in the Study Guide and Student Solutions Manual.

End-of-Chapter Problems   An extensive set of problems is included at the end of each chapter. Answers to odd-numbered problems are given at the end of the book; these pages have colored edges for ease of location. For the convenience of both the student and the instructor, about two thirds of the problems are keyed to specific sections of the chapter. The remaining problems, labeled "Additional Problems," are not keyed to specific sections. There are three levels of problems according to their level of difficulty. Straightforward problems are numbered in black, intermediate level problems are numbered in blue, and the most challenging problems are numbered in magenta.

Spreadsheet Problems   Most chapters include several spreadsheet problems following the end-of-chapter problem sets. Spreadsheet modeling of physical phenomena enables the student to obtain graphical representations of physical phenomena and perform numerical analyses without the burden of having to learn a high-level computer language. Spreadsheets are particularly useful in exploratory investigations; "what if" questions can be addressed easily and depicted graphically.

The level of difficulty in the spreadsheet problems, as with all end-of-chapter problems, is indicated by the color of the problem number. For the most straightforward problems (black), a disk with spreadsheet problems is provided. The student must enter the pertinent data, vary the parameters, and interpret the results. Intermediate level problems (blue) usually require students to modify an existing template to perform the required analysis. The more challenging problems (magenta) require students to develop their own spreadsheet templates. Brief instructions on using the templates are provided in Appendix E.

Appendices and Endpapers   Several appendices are provided at the end of the textbook, including the new appendix with instructions for problem-solving with spreadsheets. Most of the appendix material represents a review of mathematical techniques used in the text, including scientific notation, algebra, geometry, trigonometry, differential calculus, and integral calculus. Reference to these appendices is made throughout the text. Most mathematical sections include worked examples and exercises with answers. In addition to the mathematical reviews, the appendices contain tables of physical data, conversion factors, atomic masses, the SI units of physical quantities, and the periodic chart. The endpapers contain other useful information including the color code used in the text, lists of fundamental constants and physical data, conversions, and the Greek alphabet.

PHYSICS: THE CORE CD-ROM

The *Physics: The Core* CD-ROM applies the power of multimedia to the calculus-based course, offering unparalleled full-motion animation and video, engaging interactive graphics, clear and concise text, and guiding narration. *Physics: The Core* focuses on those concepts students typically find most difficult in the course, drawing from mechanics, thermodynamics, electromagnetism, and optics. The CD-ROM also contains the highly acclaimed ProSolv™ problem-solving software, which presents step-by-step explorations of essential mathematics, problem-solving strategies, and animations of problems to promote conceptual understanding and sharpen problem-solving skills. Problems on this CD-ROM will be taken directly from Serway's *Principles of Physics,* 2/e. Available for Macintosh and Windows.

STUDENT ANCILLARIES

Study Guide and Student Solutions Manual by John R. Gordon, Ralph McGrew, Steve Van Wyk, and Ray Serway. The manual features detailed solutions to 25 percent of the end-of-chapter problems and selected conceptual questions from the text. These are indicated in the text with boxed numbers. The manual also features a skills section, important notes from key sections of the text, and a list of important equations and concepts.

Courtesy of NASA

Spreadsheet Templates   The Spreadsheet Template Disk contains spreadsheet files designed to be used with the end-of-chapter problems entitled Spreadsheet Problems. The files have been developed in Microsoft Excel 5.0 for the Macintosh, and in several formats for Windows and DOS users (see Appendix E). These can be used with most spreadsheet programs including all the recent versions of Lotus 1-2-3, Excel for Windows, Quattro Pro, and f(g) Scholar. Over 30 templates are provided on the disk.

Spreadsheet Investigations in Physics by Lawrence B. Golden and James R. Klein. This workbook with the accompanying disk illustrates how spreadsheets can be used for solving many physics problems. The workbook is divided into two parts. The first part consists of spreadsheet tutorials, while the second part is a short introduction to numerical methods. The tutorials include basic spreadsheet techniques and emphasize navigating the spreadsheet, entering data, constructing formulas, and graphing. The numerical methods include differentiation, integration, interpolation, and the solution of differential equations. Many examples and exercises are provided. Step-by-step instructions are given for constructing numerical models of selected physics problems. The exercises and examples used to illustrate the numerical methods are chosen from introductory physics and mathematics.

Pocket Guide by V. Gordon Lind. This 5″ by 7″ notebook is a section-by-section capsule of the textbook that provides a handy guide for looking up important concepts, formulas, and problem-solving hints.

Mathematical Methods for Introductory Physics with Calculus by Ronald C. Davidson. This brief book is designed for students who find themselves unable to keep pace in their physics class because of a lack of familiarity with the necessary mathematical tools. *Mathematical Methods* provides an overview of those mathematical topics that are needed in an introductory-level physics course through the use of many worked examples and exercises.

So You Want to Take Physics: A Preparation for Scientists and Engineers by Rodney Cole. This text is useful to those students who need additional preparation before or during a course in physics. The book includes typical problems with worked solutions, and a review of techniques in mathematics and physics. The friendly, straightforward style makes it easier to understand how mathematics is used in the context of physics.

Life Science Applications for Physics compiled by Jerry Faughn. This supplement provides examples, readings, and problems from the biological sciences as they relate to physics. Topics include "Friction in Human Joints," "Physics of the Human Circulatory System," "Physics of the Nervous System," and "Ultrasound and Its Applications." This supplement is useful in those courses having a significant number of pre-med students.

Physics Laboratory Manual by David Loyd supplements the learning of basic physical principles while introducing laboratory procedures and equipment. Each chapter of the laboratory manual includes a pre-laboratory assignment, a list of objectives, an equipment list, the theory behind the experiment, the experimental procedure to be followed, calculations, graphs, and questions. In addition, laboratory report templates are provided for each experiment so the student can record data, calculations, and experimental results.

INSTRUCTOR ANCILLARIES

Instructor's Solutions Manual by Steve Van Wyk, Ralph McGrew, and Ray Serway. This manual contains solutions to all of the problems in the text and provides answers to even-numbered problems. All solutions have been carefully reviewed for accuracy.

Printed Test Bank by Myron Schneiderwent. This test bank offers over 1500 multiple choice and conceptually oriented critical thinking questions.

Computerized Test Bank   Available for the IBM PC (DOS and Windows) and Macintosh computers, this computerized version of the printed test bank also contains over 1500 questions. The test bank enables the instructor to create unique tests and permits the editing of questions as well as addition of new questions. The program gives answers to all problems and prints each answer on a separate grading key.

Guy Sauvage, Photo Researchers, Inc.

Overhead Transparency Acetates   A set of 175 transparencies with approximately 300 full-color figures from the text.

Instructor's Manual to Accompany Physics Laboratory Manual by David Loyd. Each chapter contains a discussion of the experiment, teaching hints, answers to selected questions, and a post-laboratory quiz with short answer and essay questions. Also included is a list of suppliers of scientific equipment and a summary of all the equipment needed for the laboratory experiments in the manual.

Physics Demonstrations Videotape by J. C. Sprott. This video features two hours of demonstrations divided into 12 primary topics. Each topic contains between four and nine demonstrations for a total of 70 physics demonstrations.

Homework Service   The World Wide Web (WWW) is the platform for an interactive homework system developed out of the University of Texas at Austin. This system, developed to coordinate with *Principles of Physics,* uses the Web for the submission and grading of homework. The system has been class-tested at the University of Texas with over 2,000 students participating each semester since 1991. Over 35,000 questions are answered electronically per week. Instructors at any university using Serway's *Principles of Physics* may establish access to this system. Over 5,000 algorithm-based problems are accessible, as well as numerical problems directly from the ends of chapters. Problem parameters vary from student to student, so that each student must do original work. All grading is done by computer, with results automatically posted on the WWW. Students receive immediate right/wrong feedback, with multiple tries for incorrect answers. Information (and a demo) is available at the URL **http://hw.ph.utexas.edu/**.

Saunders College Publishing may provide complimentary instructional aids and supplements or supplement packages to those adopters qualified under our adoption policy. Please contact your sales representative for more information. If as an adopter or potential user you receive supplements you do not need, please return them to your sales representative or send them to

Attn: Returns Department
Troy Warehouse
465 South Lincoln Drive
Troy, MO 63379

TEACHING OPTIONS

Although many topics found in traditional textbooks have been omitted from this textbook, instructors may find that the current text still contains more material than can be covered in a two-semester sequence. For this reason, I would like to offer the following suggestions. If you wish to place more emphasis on contemporary topics in physics, you should consider omitting parts or all of Chapters 15, 16, 17, 18, 24, 25, and 26. On the other hand, if you wish to follow a more traditional approach which places more emphasis on classical physics, you could omit Chapters 9, 11, 28, 29, 30, and 31. Either approach can be used without any loss in continuity. Other teaching options would fall somewhere in between these two extremes by omitting the sections labeled "Optional."

ACKNOWLEDGMENTS

As any author well knows, a project of this magnitude would be an impossible task without the assistance of many talented and dedicated individuals whose comments, criticisms, and suggestions are so important in developing a finished product. In preparing the second edition of this textbook, I have been guided by the expertise of many people who reviewed part or all of the manuscript. Special thanks go to John Jewett, Jr., Ron Bieniek, and Ralph McGrew who provided in-depth reviews of parts or all of the manuscript at various stages and offered many suggestions for its improvement. I am most grateful to the following reviewers of the manuscript for their valuable feedback:

Ralph V. Chamberlin, *Arizona State University*

Gary G. DeLeo, *Lehigh University*

Alan J. DeWeerd, *Creighton University*

Philip Fraundorf, *University of Missouri—St. Louis*

Todd Hann, *United States Military Academy*

Richard W. Henry, *Bucknell University*

Laurent Hodges, *Iowa State University*

John W. McClory, *United States Military Academy*

L. C. McIntyre, Jr., *University of Arizona*

Alan S. Meltzer, *Rensselaer Polytechnic Institute*

Perry Rice, *Miami University*

Janet E. Seger, *Creighton University*

James Whitmore, *Pennsylvania State University*

I also thank the following individuals for their suggestions during the development of the first edition of this textbook: Edward Adelson, Ohio State University; Subash Antani, Edgewood College; Harry Bingham, University of California, Berkeley; Anthony Buffa, California Polytechnic State University, San Luis Obispo; Gordon Emslie, University of Alabama at Huntsville; Donald Erbsloe, United States Air Force Academy; Gerald Hart, Moorhead State University; Joey Huston, Michigan State University; David Judd, Broward Community College; V. Gordon Lind, Utah State University; David Markowitz, University of Connecticut; Roy Middleton, University of Pennsylvania; Clement J. Moses, Utica College of Syracuse University; Anthony Novaco, Lafayette College; Desmond Penny, Southern Utah University; Prabha Ramakrishnan, North Carolina State University; Rogers Redding, University of North Texas; J. Clinton Sprott, University of Wisconsin at Madison; Cecil Thompson, University of Texas at Arlington.

Some end-of-chapter problems in Chapters 3, 6, and 8 of this second edition came from *University Physics,* 2nd edition by Alvin Hudson and Rex Nelson (Saunders College Publishing, 1990) or from *University Physics,* 2nd edition by George Arfken, David Griffing, Donald Kelly and Joseph Priest (Harcourt Brace Jovanovich, 1989). I thank the authors of these problems.

I am indebted to Colonel James Head and Lt. Col. Rolf Enger of the U.S. Air Force Academy for inspiring me to begin work on this project while I was at the Academy during the summer of 1990, and for their warm hospitality during my visit. They and their colleagues designed the original IUPP model upon which this textbook is based. Special thanks go to Ralph McGrew for assembling the problem sets, writing new problems, and checking solutions to all problems. I appreciate the tedious work of Ralph McGrew and Steve Van Wyk in writing the Instructor's Solutions Manual, and I thank Gloria Langer for her fine work in preparing this manual. I am grateful to John R. Gordon, Ralph McGrew, and Steve Van Wyk who co-authored the Study Guide and Student Solutions Manual, and to Linda Miller and Michael Rudmin for their assistance in its preparation. Dena Digilio-Betz was very helpful in locating many of the photographs used in this text. During the development of this text, I benefited from many useful discussions with my colleagues and other physics instructors including Robert Bauman, Don Chodrow, Jerry Faughn, John R. Gordon, Kevin Giovanetti, Dick Jacobs, Clem Moses, Dorn Peterson, Joseph Rudmin, and Gerald Taylor. I owe a debt of gratitude to Irene Nunes for clarifying much of the material in the first edition, and for eliminating many inconsistencies, redundancies, and other nightmares that authors must face. Special thanks and recognition go to the professional staff at Saunders College Publishing for their careful attention to detail and for their dedication to producing first quality textbooks, especially Susan Dust Pashos, Beth Ahrens, Alexandra Buczek, Sally Kusch, Carol Bleistine, and Angus McDonald. I sincerely appreciate

Michel Gouverneur, Photo News, Gamma Sport

the wisdom and enthusiasm of my publisher and good friend John Vondeling, who continues to publish high quality instructional products for science education.

Finally, I thank my entire family for their endless love, understanding, and continued encouragement, especially my beautiful and devoted wife Elizabeth, who is always at my side to share my joys and sorrows.

Ray Serway
Harrisonburg, VA

JANUARY 1997

To the Student

Ifeel it is appropriate to offer some words of advice which should be of benefit to you, the student. Before doing so, I will assume that you have read the Preface, which describes the various features of the text that will help you through the course.

HOW TO STUDY

Very often instructors are asked "How should I study physics and prepare for examinations?" There is no simple answer to this question, but I would like to offer some suggestions based on my own experiences in learning and teaching over the years.

First and foremost, maintain a positive attitude towards the subject matter, keeping in mind that physics is the most fundamental of all natural sciences. Other science courses that follow will use the same physical principles, so it is important that you understand and be able to apply the various concepts and theories discussed in the text.

Robin Smith/ Tony Stone Images

CONCEPTS AND PRINCIPLES

It is essential that you understand the basic concepts and principles *before* attempting to solve assigned problems. This is best accomplished through a careful reading of the textbook before attending your lecture on that material. In the process, it is useful to jot down certain points which are not clear to you. Take careful notes in class, and then ask questions pertaining to those ideas that require clarification. Keep in mind that few people are able to absorb the full meaning of scientific material after one reading. Several readings of the text and notes may be necessary. Your lectures and laboratory work should supplement the text and clarify some of the more difficult material. You should reduce memorization of material to a minimum. Memorizing passages from a text, equations, and derivations does not necessarily mean you understand the material. Your understanding of the material will be enhanced through a combination of efficient study habits, discussions with other students and instructors, and your ability to solve the problems in the text. Ask questions whenever you feel it is necessary.

STUDY SCHEDULE

It is important to set up a regular study schedule, preferably on a daily basis. Make sure to read the syllabus for the course and adhere to the schedule set by your instructor. The lectures will be much more meaningful if you read the corresponding textual material before attending the lecture. As a general rule, you should devote about two hours of study time for every hour in class. If you are having trouble with the course, seek the advice of the instructor or students who have taken the course. You may find it necessary to seek further instruction from experienced students. Very often, instructors will offer review sessions in addition to regular class periods. It is important that you avoid the practice of delaying study until a day or two before an exam. More often than not, this will lead to disastrous results. Rather than an all-night study session, it is better to briefly review the basic concepts and equations, followed by a good night's rest. If you feel in need of additional help in understanding the concepts, preparing for exams, or in problem-solving, we suggest that you acquire a copy of the Study Guide/Student Solutions Manual which accompanies the text and should be available at your college bookstore.

USE THE FEATURES

You should make full use of the various features of the text discussed in the Preface. For example, marginal notes are useful for locating and describing important equations and concepts, while important statements and definitions are highlighted in **boldface** type or with a blue box. Many useful tables are contained in appendices, but most are incorporated in the text where they are used most often. Appendix B is a convenient review of mathematical techniques. Answers to odd-numbered problems are provided at the end of the book. Exercises (with answers), which follow some worked examples, represent extensions of those examples, and in most cases you are expected to perform a simple calculation. Their purpose is to test your problem-solving skills as you read through the text. Problem-Solving Strategies and Hints are included in selected chapters throughout the text to give you additional information to help you solve problems. An overview of the entire text is given in the table of contents, while the index will enable you to locate specific material quickly. Footnotes are sometimes used to supplement the discussion or to cite other references on the subject. Many chapters include problems that require the use of programmable calculators, computers, or simulation software. Consult your instructor; it may be possible to obtain some of the relevant software from the Saunders physics Web site. Spreadsheet problems can be solved either analytically or with the use of the *Spreadsheet Investigations* supplement. These are intended for those courses that place some emphasis on numerical methods. You may want to develop appropriate programs for some of these problems even if they are not assigned by your instructor.

Courtesy of NASA

After reading a chapter, you should be able to define any new quantities introduced in that chapter, and discuss the principles and assumptions that were used to arrive at certain key relations. The chapter summaries and the review sections of the Study Guide/Student Solutions Manual should help you in this regard. In some cases, it will be necessary to refer to the index of the text to locate certain topics. You should be able to correctly associate with each physical quantity a symbol used to represent that quantity and the unit in which the quantity is specified. Furthermore, you should be able to express each important relation in a concise and accurate prose statement.

THE IMPORTANCE OF PROBLEM SOLVING

R.P. Feynman, Nobel laureate in physics, once said, "You do not know anything until you have practiced." In keeping with this statement, I strongly advise you to develop the skills necessary to solve a wide range of problems. Your ability to solve problems will be one of the main tests of your knowledge of physics; therefore, you should try to solve as many problems as possible. It is essential that you understand basic concepts and principles before attempting to solve problems. It is good practice to try to find alternate solutions to the same problem. For example, problems in mechanics can be solved using Newton's laws, but very often an alternative method using energy considerations is more direct. You should not deceive yourself into thinking you understand the problem after seeing its solution in class. You must be able to solve the problem and similar problems on your own.

The method of solving problems should be carefully planned. A systematic plan is especially important when a problem involves several concepts. First, read the problem several times until you are confident you understand what is being asked. Look for any key words that will help you interpret the problem, and perhaps allow you to make certain assumptions. Your ability to interpret the question properly is an integral part of problem solving. You should acquire the habit of writing down the information given in a problem, and decide what quantities need to be found. You might want to construct a table listing quantities given and quantities to be found. This procedure is sometimes used in the worked examples of the text. After you have decided on the method you feel is appropriate for the situation, proceed with your solution. General problem-solving strategies of this type are included in the text and are highlighted by a tan screen.

I often find that students fail to recognize the limitations of certain formulas or physical laws in a particular situation. It is very important that you understand and remember the assumptions which underlie a particular theory or formalism. For example, certain equations in kinematics apply only to a particle moving with constant acceleration. These equations are not valid for situations in which the acceleration is not constant, such as the motion of an object connected to a spring, or the motion of an object through a fluid.

GENERAL PROBLEM-SOLVING STRATEGY

Most courses in general physics require the student to learn the skills of problem solving, and examinations are largely composed of problems that test such skills. This brief section describes some useful ideas which will enable you to increase your accuracy in solving problems, enhance your understanding of physical concepts, eliminate initial panic or lack of direction in aproaching a problem, and organize your work. One way to help accomplish these goals is to adopt a problem-solving strategy. Many chapters in this text will include a section labeled "Problem-Solving Strategies and Hints" which should help you through the "rough spots."

In developing problem-solving strategies, five basic steps are commonly used, and are exemplified in the sample Example below.

1. Draw a suitable diagram with appropriate labels and coordinate axes if needed.
2. As you examine what is being asked in the problem, identify the basic physical principle (or principles) that are involved, listing the knowns and unknowns.
3. Select a basic relationship or derive an equation that can be used to find the unknown, and solve the equation for the unknown symbolically.
4. Substitute the given values along with the appropriate units into the equation.
5. Obtain a numerical value for the unknown. The problem is verified and receives a check mark if the following questions can be properly answered: Do the units match? Is the answer reasonable? Is the plus or minus sign proper or meaningful?

One of the purposes of this strategy is to promote accuracy. Properly drawn diagrams can eliminate many sign errors. Diagrams also help to isolate the physical principles of the problem. Symbolic solutions and carefully labeled knowns and unknowns will help eliminate other careless errors. The use of symbolic solutions should help you think in terms of the physics of the problem. A check of units at the end of the problem can indicate a possible algebraic error. The physical layout and organization of your problem will make the final product more understandable and easier to follow. Once you have developed an organized system for examining problems and extracting relevant information, you will become a more confident problem solver.

Example

A person driving in a car at a speed of 20 m/s applies the brakes and stops in a distance of 100 m. What was the acceleration of the car?

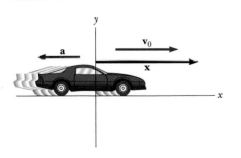

Given:

$x_0 = 0$ m

$x = 100$ m

$v_0 = 20$ m/s

$v = 0$ m/s

$a = ?$

$$v^2 = v_0{}^2 + 2a(x - x_0)$$

$$a = \frac{v_2 - v_0{}^2}{2(x - x_0)}$$

$$a = \frac{(0 \text{ m/s})^2 - (20 \text{ m/s})^2}{2(100 \text{ m})} = -2.0 \text{ m/s}^2$$

Check units for consistency:

$$\frac{\text{m}^2/\text{s}^2}{\text{m}} = \frac{\text{m}}{\text{s}^2}$$

EXPERIMENTS

Physics is a science based upon experimental observations. In view of this fact, I recommend that you try to supplement the text through various type of "hands-on" experiments, either at home or in the laboratory. These can be used to test ideas and models discussed in class or in the text. For example, the common "Slinky" toy is excellent for studying traveling waves; a ball swinging on the end of a long string can be used to investigate pendulum motion; various masses attached to the end of a vertical spring or rubber band can be used to determine their elastic nature; an old pair of Polaroid sunglasses and some discarded lenses and magnifying glass are the components of various experiments in optics; you can get an approximate measure of the acceleration of gravity simply by dropping a ball from a known height and measuring the time of its fall with a stopwatch. The list is endless. When physical models are not available, be imaginative and try to develop models of your own.

THINKING PHYSICS AND CONCEPTUAL PROBLEMS

These new features are intended to develop your conceptual understanding of physics. You should be able to answer all **Conceptual Problems** based on previous content in the chapter. (Answers are provided at the chapter's end, but please try to answer them on your own first.) **Thinking Physics** examples provide the model for the form your answers should take. The answers ("Reasoning") for **Thinking Physics** are provided immediately for you because these examples often extend and further develop the chapter content. They may, for instance, illuminate the connection between physics and other scientific disciplines.

Tom Raymond/ Tony Stone Images

AN INVITATION TO PHYSICS

It is my sincere hope that you too will find physics an exciting and enjoyable experience, and that you will profit from this experience, regardless of your chosen profession. Welcome to the exciting world of physics.

> *The scientist does not study nature because it is useful; he studies it because he delights in it, and he delights in it because it is beautiful. If nature were not beautiful, it would not be worth knowing, and if nature were not worth knowing, life would not be worth living.*
>
> HENRI POINCARÉ

Contents Overview

NASA

Contents

David Madison / Tony Stone Images

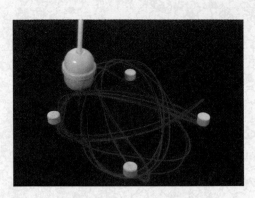

© *Yoav Levy/Phototake*

Martin Dohrn/SPL/Photo Researchers

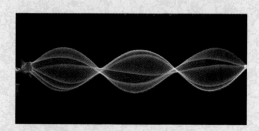

Richard Megna, Fundamental Photographs, NYC

© Ben Rose 1992/ The IMAGE Bank

*Kim Vandiver and Harold Edgerton,
Palm Press, Inc.*

12   Oscillatory Motion   332

13   Wave Motion   360

14   Superposition and Standing Waves   389

Earl Young/FPG

Principles of Physics

SECOND EDITION

Volume 1

An Invitation to Physics

Physics, the most fundamental physical science, is concerned with the basic principles of the Universe. It is the foundation on which the other physical sciences—astronomy, chemistry, and geology—are based. The beauty of physics lies in the simplicity of its fundamental theories and in the manner in which a small number of basic concepts, equations, and assumptions can alter and expand our view of the world.

All physical phenomena in our world can be described in terms of one or more of the following theories:

- Classical mechanics, which is concerned with the motion of objects moving at speeds that are low compared to the speed of light;
- Relativity, which is a theory describing particles moving at any speed, even those whose speeds approach the speed of light;
- Thermodynamics, which deals with heat, temperature, and the behavior of large numbers of particles;
- Electromagnetism, which involves the theory of electricity, magnetism, and electromagnetic fields;
- Quantum mechanics, a theory dealing with the behavior of submicroscopic particles as well as the macroscopic world.

When a discrepancy between theory and experiment arises in any of the theories listed above, the theories must be modified and experiments must be performed to test the predictions of the modified theories. Many times a theory is satisfactory under limited conditions; a more general theory might be satisfactory without such limi-

◄ **White light directed through a triangular glass prism creates "Pheaenomena of Colours" on the manuscript of Isaac Newton's Opticks.** *(© Erich Lessing, Magnum Photos)*

tations. A classic example is Newtonian mechanics, which accurately describes the motions of objects at low speeds but does not apply to objects moving at speeds comparable to the speed of light. The special theory of relativity developed by Albert Einstein (1879–1955) successfully predicts the motions of objects at speeds approaching the speed of light and hence is a more general theory of motion.

Classical physics, developed prior to 1900, includes the theories, concepts, laws, and experiments in classical mechanics, thermodynamics, and electromagnetism. For example, Galileo Galilei (1564–1642) made significant contributions to classical mechanics through his work on the laws of motion with constant acceleration. In the same era, Johannes Kepler (1571–1630) used astronomical observations to develop empirical laws for the motions of planetary bodies.

The most important contributions to classical mechanics were provided by Isaac Newton (1642–1727), who developed classical mechanics as a systematic theory and was one of the originators of the calculus as a mathematical tool. Although major developments in classical physics continued in the 18th century, thermodynamics and electromagnetism were not developed until the latter part of the 19th century, principally because the apparatus for controlled experiments was either too crude or unavailable. Although many electric and magnetic phenomena had been studied earlier, it was the work of James Clerk Maxwell (1831–1879) that provided a unified theory of electromagnetism. In this book we shall discuss the various disciplines of classical physics in separate sections; however, we will see that the disciplines of mechanics and electromagnetism are basic to all the branches of classical and modern physics.

A major revolution in physics, usually referred to as *modern physics*, began near the end of the 19th century. Modern physics developed in response to the discovery that many physical phenomena could not be explained by classical physics. The two most important developments in this modern era were the theories of relativity and quantum mechanics. Einstein's theory of relativity completely revolutionized the traditional concepts of space, time, and energy. Einstein's theory correctly described the motion of objects moving at speeds comparable to the speed of light. Newton's laws fail to make correct predictions for objects moving at such high speeds. The theory of relativity also assumes that the speed of light is the upper limit of the speed of an object or signal and shows that mass and energy are related. Quantum mechanics was formulated by a number of distinguished scientists to provide descriptions of physical phenomena at the atomic level.

Scientists continually work at improving our understanding of fundamental laws, and new discoveries are made every day. In many research areas there is a great deal of overlap among physics, chemistry, and biology. The many technological advances in recent times are the result of the efforts of many scientists, engineers, and technicians. Some of the most notable recent developments are (1) unmanned space missions and manned moon landings, (2) microcircuitry and high-speed computers, and (3) sophisticated imaging techniques used in scientific research and medicine. The impacts of such developments and discoveries on society have indeed been great, and it is very likely that future discoveries and developments will be exciting, challenging, and of great benefit to humanity.

1

Introduction and Vectors

The goal of physics is to provide a quantitative understanding of certain basic phenomena that occur in our Universe. Physics is a science based on experimental observations and mathematical analyses. The main objective behind such experiments and analyses is to develop theories that explain the phenomenon being studied and to relate those theories to other established theories. Fortunately, it is possible to explain the behavior of various physical systems using relatively few fundamental laws. Analytical procedures require the expression of those laws in the language of mathematics, the tool that provides a bridge between theory and experiment. In this chapter we shall discuss a few mathematical concepts and techniques that will be used throughout the book.

Because later chapters will be concerned with the laws of physics, it is necessary to provide clear definitions of the basic quantities involved in these laws. For example, such physical quantities as force, velocity, volume, and acceleration can be described in terms of more fundamental quantities. In the next several chapters we will encounter three of the latter: **length** (L), **time** (T), and **mass** (M). Later in the book we shall add units for two other standard quantities to our list, one for temperature (the kelvin), the other for electric current (the ampere). In our study of mechanics, however, we shall be concerned only with the units of mass, length, and time.

Some physical quantities require the measurement of more than one attribute. For example, physical quantities that have both numerical and direc-

◀ This photograph was taken in Kinsley, Kansas, located midway between New York City and San Francisco. *(Mike & Carol Werner/Comstock)*

tional properties—such as force, velocity, and acceleration—are represented by vectors. A large part of this chapter is concerned with vector algebra and the general properties of vectors. The addition and subtraction of vectors will be discussed, together with some common applications to physical situations. Because vectors will be used throughout this book, it is imperative that you understand both their graphical and their algebraic properties.

1.1 • STANDARDS OF LENGTH, MASS, AND TIME

If we make a measurement of a certain quantity and wish to describe it to someone, a unit for the quantity must be defined. For example, it would be meaningless for a visitor from another planet to talk to us about a length of 8 "glitches" if we did not know the meaning of the unit "glitch." However, if someone familiar with our system of measurement and weights reports that a wall is 2.0 meters high and our unit of length is defined to be 1.0 meter, we then know that the height of the wall is twice our fundamental unit of length. Likewise, if we are told that a person has a mass of 75 kilograms and our unit of mass is defined as 1.0 kilogram, then that person has a mass 75 times larger than our fundamental unit of mass. In 1960 an international committee agreed on a system of definitions and standards to describe fundamental physical quantities. It is called the **SI system** (Système International) of units. Its units of length, mass, and time are the meter, kilogram, and second, respectively.

Length

In A.D. 1120 the king of England decreed that the standard of length in his country would be the yard and that the yard would be precisely equal to the distance from the tip of his nose to the end of his outstretched arm. In a similar way, the original standard for the foot adopted by the French was the length of the royal foot of King Louis XIV. This standard prevailed until 1799, when the legal standard of length in France became the meter, defined as one ten-millionth of the distance from the Equator to the North Pole.

Many other systems have been developed in addition to those just discussed, but the French system has prevailed in most countries and in scientific circles everywhere. As recently as 1960, the length of the meter was still defined as the distance between two lines on a specific bar of platinum-iridium alloy stored under controlled conditions. This standard was abandoned for several reasons, a principal one being that the limited accuracy with which the separation between the lines can be determined does not meet the current requirements of science and technology. Until recently, the meter was defined as 1 650 763.73 wavelengths of orange-red light emitted from a krypton-86 lamp. However, in October 1983 the **meter was redefined to be the distance traveled by light in a vacuum during a time of 1/299 792 458 s.** In effect, this latest definition establishes that the speed of light in a vacuum is 299 792 458 m/s.

Definition of the meter •

Mass

Definition of the kilogram •

The SI unit of mass, **the kilogram, is defined as the mass of a specific platinum-iridium alloy cylinder kept at the International Bureau of Weights and Measures at Sèvres, France.** At this point, we should add a word of caution. Most

TABLE 1.1   Approximate Values of Some Measured Lengths

	Length (m)
Distance from Earth to most remote quasar known	1.4×10^{26}
Distance from Earth to most remote normal galaxies known	4×10^{25}
Distance from Earth to nearest large galaxy (M 31 in Andromeda)	2×10^{22}
Distance from Sun to nearest star (Proxima Centauri)	4×10^{16}
One lightyear	9.46×10^{15}
Mean orbit radius of Earth	1.5×10^{11}
Mean distance from Earth to Moon	3.8×10^{8}
Distance from Equator to North Pole	1×10^{7}
Mean radius of Earth	6.4×10^{6}
Typical altitude of orbiting Earth satellite	2×10^{5}
Length of a football field	9.1×10^{1}
Length of a housefly	5×10^{-3}
Size of smallest dust particles	1×10^{-4}
Size of cells of most living organisms	1×10^{-5}
Diameter of a hydrogen atom	1×10^{-10}
Diameter of a uranium nucleus	1.4×10^{-14}
Diameter of a proton	1×10^{-15}

beginning students of physics tend to confuse the physical quantities called *weight* and *mass*. For the present we shall not discuss the distinction between them; they will be clearly defined in later chapters. For now you should note that they are distinctly different quantities.

Time

Before 1960 the standard of time was defined in terms of the average length of a solar day in the year 1900. (A solar day is the time interval between successive appearances of the Sun at the highest point it reaches in the sky each day.) The basic unit of time, the second, was defined to be $(1/60)(1/60)(1/24) = 1/86\ 400$ of the average solar day. In 1967 the second was redefined to take advantage of the great precision obtainable with a device known as an atomic clock, which uses the characteristic frequency of the cesium-133 atom as the "reference clock." **The second is now defined as 9 192 631 770 times the period of oscillation of radiation from the cesium atom.**

Approximate Values for Length, Mass, and Time

Approximate values of various lengths, masses, and time intervals are presented in Tables 1.1, 1.2, and 1.3, respectively. Note the wide range of values for these quantities.[1] You should study the tables and get a feel for what is meant by a mass of 100 kilograms, for example, or by a time interval of 3.2×10^{7} seconds. Systems of units commonly used are the SI system in which the units of length, mass, and

TABLE 1.2   Masses of Various Bodies (approximate values)

	Mass (kg)
Universe	10^{52}
Milky Way galaxy	10^{42}
Sun	2×10^{30}
Earth	6×10^{24}
Moon	7×10^{22}
Shark	3×10^{2}
Human	7×10^{1}
Frog	1×10^{-1}
Mosquito	1×10^{-5}
Bacterium	1×10^{-15}
Hydrogen atom	1.67×10^{-27}
Electron	9.11×10^{-31}

[1]If you are unfamiliar with the use of powers of ten (scientific notation), you should review Appendix B.1.

TABLE 1.3   Approximate Values of Some Time Intervals

	Interval (s)
Age of the Universe	5×10^{17}
Age of the Earth	1.3×10^{17}
Time since the fall of the Roman Empire	5×10^{12}
Average age of a college student	6.3×10^{8}
One year	3.2×10^{7}
One day (time for one revolution of Earth around its axis)	8.6×10^{4}
Time between normal heartbeats	8×10^{-1}
Period[a] of audible sound waves	1×10^{-3}
Period of typical radio waves	1×10^{-6}
Period of vibration of an atom in a solid	1×10^{-13}
Period of visible light waves	2×10^{-15}
Duration of a nuclear collision	1×10^{-22}
Time for light to cross a proton	3.3×10^{-24}

[a] A period is defined as the time interval required for one complete vibration.

TABLE 1.4   Some Prefixes for Powers of Ten

Power	Prefix	Abbreviation
10^{-18}	atto	a
10^{-15}	femto	f
10^{-12}	pico	p
10^{-9}	nano	n
10^{-6}	micro	μ
10^{-3}	milli	m
10^{-2}	centi	c
10^{-1}	deci	d
10^{3}	kilo	k
10^{6}	mega	M
10^{9}	giga	G
10^{12}	tera	T
10^{15}	peta	P
10^{18}	exa	E

time are the meter (m), kilogram (kg), and second (s), respectively; the cgs or Gaussian system, in which the units of length, mass, and time are the centimeter (cm), gram (g), and second (s), respectively; and the British engineering system (sometimes called the *conventional system*), in which the units of length, mass, and time are the foot (ft), slug, and second, respectively. Throughout most of this book we shall use SI units, because they are almost universally accepted in science and industry. We will make limited use of conventional units in the study of classical mechanics.

Some of the most frequently used prefixes for the powers of ten and their abbreviations are listed in Table 1.4. For example, 10^{-3} m is equivalent to 1 millimeter (mm), and 10^{3} m is 1 kilometer (km). Likewise, 1 kg is 10^{3} g, and 1 megavolt (MV) is 10^{6} volts.

1.2 • DENSITY AND ATOMIC MASS

The **density** ρ (Greek letter rho) of any substance is defined as its *mass per unit volume* (a table of the letters in the Greek alphabet is provided at the back of the book):

Density •

$$\rho \equiv \frac{m}{V} \qquad\qquad [1.1]$$

For example, aluminum has a density of $2.70 \times 10^{3} \ \text{kg/m}^3$ and lead has a density of $11.3 \times 10^{3} \ \text{kg/m}^3$. A list of densities for various substances is given in Table 1.5.

Atomic mass •

The difference in density between aluminum and lead is due in part to their different *atomic masses;* the atomic mass of lead is 207 and that of aluminum is 27. However, the ratio of atomic masses, $207/27 = 7.67$, does not correspond to the ratio of densities, $11.3/2.70 = 4.19$. The discrepancy is due to the difference in atomic spacings and atomic arrangements in the crystal structures of the two types of elements.

All ordinary matter consists of atoms, and each atom is made up of electrons and a nucleus. Practically all of the mass of an atom is contained in the nucleus, which consists of protons and neutrons. Thus, we can understand why the atomic masses of the elements differ. The mass of a nucleus is measured relative to the mass of an atom of carbon-12 (this isotope of carbon has six protons and six neutrons).

The mass of ^{12}C is defined to be exactly 12 atomic mass units (u), where 1 u = $1.6605402 \times 10^{-27}$ kg. In these units, the proton and neutron have masses of about 1 u. To be more precise,

$$\text{Mass of proton} = 1.007\ 276\ \text{u}$$

$$\text{Mass of neutron} = 1.008\ 664\ \text{u}$$

The mass of the nucleus of ^{27}Al is approximately 27 u. In fact, a more precise measurement shows that the nuclear mass is always slightly *less* than the combined mass of the protons and neutrons making up the nucleus. The processes of nuclear fission and nuclear fusion are based on this mass difference.

One **mole** (mol) of any element (or compound) consists of Avogadro's number, N_A, of molecules of the substance. Avogadro's number, $N_A = 6.02 \times 10^{23}$, was defined so that 1 mole of carbon-12 atoms would have a mass of exactly 12 g. A mole of one element differs in mass from a mole of another. For example, 1 mole of aluminum has a mass of 27 g, and 1 mole of lead has a mass of 207 g. But 1 mole of aluminum contains the same number of atoms as 1 mole of lead, because there are 6.02×10^{23} atoms in 1 mole of *any* element. The mass per atom, which is *not* equal to the atomic mass, is then given by

$$m = \frac{\text{atomic mass}}{N_A}$$

[1.2]

For example, the mass of an aluminum atom is

$$m = \frac{27\ \text{g/mol}}{6.02 \times 10^{23}\ \text{atoms/mol}} = 4.5 \times 10^{-23}\ \text{g/atom}$$

Note that 1 u is equal to N_A^{-1} g.

TABLE 1.5 Densities of Various Substances

Substance	Density $\rho(\text{kg/m}^3)$
Platinum	21.45×10^3
Gold	19.3×10^3
Uranium	18.7×10^3
Lead	11.3×10^3
Copper	8.93×10^3
Iron	7.86×10^3
Aluminum	2.70×10^3
Magnesium	1.75×10^3
Water	1.00×10^3
Air at atmospheric pressure	0.0012×10^3

Example 1.1 How Many Atoms in the Cube?

A solid cube of aluminum (density 2.7 g/cm^3) has a volume of 0.20 cm^3. How many aluminum atoms are contained in the cube?

Solution Because density equals mass per unit volume, the mass of the cube is

$$M = \rho V = (2.7\ \text{g/cm}^3)(0.20\ \text{cm}^3) = 0.54\ \text{g}$$

To find the number of atoms, N, we can set up a proportion using the fact that 1 mole of aluminum (27 g) contains 6.02×10^{23} atoms:

$$\frac{6.02 \times 10^{23}\ \text{atoms}}{27\ \text{g}} = \frac{N}{0.54\ \text{g}}$$

$$N = \frac{(0.54\ \text{g})(6.02 \times 10^{23}\ \text{atoms})}{27\ \text{g}} = 1.2 \times 10^{22}\ \text{atoms}$$

1.3 · DIMENSIONAL ANALYSIS

The word *dimension* has a special meaning in physics. It usually denotes the physical nature of a quantity. Whether a distance is measured in units of feet or meters or furlongs, it is a distance. We say its dimension is *length.*

The symbols that will be used in this section to specify length, mass, and time are L, M, and T, respectively. We will often use square brackets [ ] to denote the dimensions of a physical quantity. For example, in this notation the dimensions of velocity, v, are written $[v] = L/T$, and the dimensions of area, A, are $[A] = L^2$. The dimensions of area, volume, velocity, and acceleration are listed in Table 1.6, along with their units in the three common systems of measurement. The dimensions of other quantities, such as force and energy, will be described as they are introduced in the book.

In many situations, you may be faced with having to derive or check a specific formula. Although you may have forgotten the details of the derivation, there is a useful and powerful procedure called *dimensional analysis* that can be used as a consistency check, to assist in the derivation, or to check your final expression. This procedure should always be used and should help minimize the rote memorization of equations. Dimensional analysis makes use of the fact that **dimensions can be treated as algebraic quantities.** That is, quantities can be added or subtracted only if they have the same dimensions. Furthermore, the terms on both sides of an equation must have the same dimensions. By following these simple rules, you can use dimensional analysis to help determine whether or not an expression has the correct form, because the relationship can be correct only if the dimensions on the two sides of the equation are the same.

To illustrate this procedure, suppose you wish to derive a formula for the distance x traveled by a car in a time t if the car starts from rest and moves with constant acceleration a. In Chapter 2 we shall find that the correct expression for this special case is $x = \frac{1}{2}at^2$. Let us check the validity of this expression from a dimensional analysis approach.

The quantity x on the left side has the dimension of length. In order for the equation to be dimensionally correct, the quantity on the right side must also have the dimension of length. We can perform a dimensional check by substituting the basic dimensions for acceleration, L/T^2, and time, T, into the equation. That is, the dimensional form of the equation $x = \frac{1}{2}at^2$ can be written as

$$L = \frac{L}{T^2} \cdot T^2 = L$$

The units of time cancel as shown, leaving the unit of length.

TABLE 1.6   Dimensions of Area, Volume, Velocity, and Acceleration

System	Area (L^2)	Volume (L^3)	Velocity (L/T)	Acceleration (L/T^2)
SI	m^2	m^3	m/s	m/s^2
cgs	cm^2	cm^3	cm/s	cm/s^2
British engineering	ft^2	ft^3	ft/s	ft/s^2

Example 1.2 Analysis of an Equation

Show that the expression $v = v_0 + at$ is dimensionally correct, where v and v_0 represent velocities, a is acceleration, and t is a time interval.

Solution Because

$$[v] = [v_0] = \frac{L}{T}$$

and the dimensions of acceleration are L/T^2, the dimensions of at are

$$[at] = \frac{L}{T^2} \cdot T = \frac{L}{T}$$

and the expression is dimensionally correct. However, if the expression were given as $v = v_0 + at^2$, it would be dimensionally *incorrect*. Try it and see!

1.4 • CONVERSION OF UNITS

Sometimes it is necessary to convert units from one system to another. Conversion factors between SI and conventional units of length are as follows:

$$1 \text{ mile} = 1609 \text{ m} = 1.609 \text{ km} \qquad 1 \text{ ft} = 0.3048 \text{ m} = 30.48 \text{ cm}$$

$$1 \text{ m} = 39.37 \text{ in.} = 3.281 \text{ ft} \qquad 1 \text{ in.} = 0.0254 \text{ m} = 2.54 \text{ cm}$$

A more complete list of conversion factors can be found in Appendix A.

Units can be treated as algebraic quantities that can cancel each other. For example, suppose we wish to convert 15.0 in. to centimeters. Because 1 in. = 2.54 cm (exactly), we find that

$$15.0 \text{ in.} = (15.0 \text{ in.}) \left(2.54 \frac{\text{cm}}{\text{in.}} \right) = 38.1 \text{ cm}$$

(*Left*) Conversion of miles to kilometers. (*Right*) This modern car speedometer gives speed readings in both miles per hour and kilometers per hour. You should confirm the conversion between the two units for a few readings on the dial. (*Paul Silverman, Fundamental Photographs*)

Example 1.3   The Density of a Cube

The mass of a solid cube is 856 g, and each edge has a length of 5.35 cm. Determine the density ρ of the cube in SI units.

Solution   Because $1\ g = 10^{-3}\ kg$ and $1\ cm = 10^{-2}\ m$, the mass, m, and volume, V, in SI units are given by

$$m = 856\ \cancel{g} \times 10^{-3}\ kg/\cancel{g} = 0.856\ kg$$

$$V = L^3 = (5.35\ \cancel{cm} \times 10^{-2}\ m/\cancel{cm})^3$$
$$= (5.35)^3 \times 10^{-6}\ m^3 = 1.53 \times 10^{-4}\ m^3$$

Therefore

$$\rho = \frac{m}{V} = \frac{0.856\ kg}{1.53 \times 10^{-4}\ m^3} = 5.59 \times 10^3\ kg/m^3$$

1.5 • ORDER-OF-MAGNITUDE CALCULATIONS

It is often useful to compute an approximate answer to a given physical problem even in instances in which little information is available. This answer can then be used to determine whether or not a more precise calculation is necessary. Such an approximation is usually based on certain assumptions, which must be modified if greater precision is needed. Thus, we will sometimes refer to an *order of magnitude* of a certain quantity as the power of ten of the number that describes that quantity. Usually, when an order-of-magnitude calculation is made, the results are reliable to within a factor of 10. If a quantity increases in value by three orders of magnitude, this means that its value increases by a factor of $10^3 = 1000$. We use the symbol $\sim$ for *is on the order of*. Thus,

$$0.0086 \sim 10^{-2} \qquad 0.0031 \sim 10^{-3} \qquad 700 \sim 10^3$$

The spirit of attempting order-of-magnitude calculations, sometimes referred to as '*guesstimates*' or '*ball-park figures*' is captured by the following quotation: "Make an estimate before every calculation, try a simple physical argument . . . before every derivation, guess the answer to every puzzle. Courage: No one else needs to know what the guess is."[2]

Example 1.4   The Number of Atoms in a Solid

Estimate the number of atoms in 1 cm^3 of a solid.

Solution   From Table 1.1 we note that the diameter of an atom is about 10^{-10} m. Thus, if in our model we assume that the atoms in the solid are solid spheres of this diameter, then the volume of each sphere is about 10^{-30} m^3 (more precisely, volume $= 4\pi r^3/3 = \pi d^3/6$, where $r = d/2$). Therefore, be-cause $1\ cm^3 = 10^{-6}\ m^3$, the number of atoms in the solid is on the order of $10^{-6}/10^{-30} = 10^{24}$ atoms.

A more precise calculation would require knowledge of the density of the solid and the mass of each atom. However, our estimate agrees with the more precise calculation to within a factor of ten.

---

[2]E. Taylor and J. A. Wheeler, *Spacetime Physics*, San Francisco, W. H. Freeman, 1966, p. 60.

Example 1.5 **How Much Gas Do We Use?**

Estimate the number of gallons of gasoline used by all cars in the United States each year.

Solution Because there are about 240 million people in the United States, an estimate of the number of cars in the country is 60 million (assuming one car and four people per family). We also estimate that the average distance traveled per year is 10 000 miles. If we assume gasoline consumption of 0.05 gal/mi, each car uses about 500 gal/year. Multiplying this by the total number of cars in the United States gives an estimated total consumption of 3×10^{10} gal, which corresponds to a yearly consumer expenditure of more than $30 billion! This is probably a low estimate because we have not accounted for commercial consumption.

1.6 • SIGNIFICANT FIGURES

When certain quantities are measured, the measured values are known only to within the limits of the experimental uncertainty. The value of the uncertainty can depend on various factors, such as the quality of the apparatus, the skill of the experimenter, and the number of measurements performed.

Suppose that in a laboratory experiment we are asked to measure the area of a rectangular plate using a meter stick as a measuring instrument. Let us assume that the accuracy to which we can measure a particular dimension of the plate is ± 0.1 cm. If the length of the plate is measured to be 16.3 cm, we can claim only that its length lies somewhere between 16.2 cm and 16.4 cm. In this case, we say that the measured value has three significant figures. Likewise, if its width is measured to be 4.5 cm, the actual value lies between 4.4 cm and 4.6 cm. This measured value has only two significant figures. Note that the significant figures include the first reliable digit. Thus, we could write the measured values as 16.3 ± 0.1 cm and 4.5 ± 0.1 cm.

Suppose now that we would like to find the area of the plate by multiplying the two measured values. If we were to claim that the area is (16.3 cm)(4.5 cm) = 73.35 cm², our answer would be unjustifiable because it contains four significant figures, which is greater than the number of significant figures in either of the measured lengths. A good rule of thumb in determining the number of significant figures that can be claimed is as follows:

When multiplying several quantities, the number of significant figures in the final answer is the same as the number of significant figures in the *least* accurate of the quantities being multiplied, where *least accurate* means *having the lowest number of significant figures*. The same rule applies to division.

Applying this rule to the multiplication example given previously, we see that the answer for the area can have only two significant figures because the length of 4.5 cm has only two significant figures. Thus, all we can claim is that the area is 73 cm², realizing that the value can range between (16.2 cm)(4.4 cm) = 71 cm² and (16.4 cm)(4.6 cm) = 75 cm².

Zeros may or may not be significant figures. Those used to position the decimal point in such numbers as 0.03 and 0.0075 are not significant. Thus there are one and two significant figures, respectively, in these two values. When the positioning of zeros comes after other digits, however, there is the possibility of misinterpretation. For example, suppose the mass of an object is given as 1500 g. This value is ambiguous because we do not know whether the last two zeros are being used to locate the decimal point or whether they represent significant figures in the measurement. In order to remove this ambiguity, it is common to use scientific notation to indicate the number of significant figures. In this case, we would express the mass as 1.5×10^3 g if there are two significant figures in the measured value, 1.50×10^3 g if there are three significant figures, and 1.500×10^3 g if there are four. Likewise, 0.000 15 should be expressed in scientific notation as 1.5×10^{-4} if it has two significant figures or as 1.50×10^{-4} if it has three significant figures. The three zeros between the decimal point and the digit 1 in the number 0.000 15 are not counted as significant figures, because they are present only to locate the decimal point. In general, a **significant figure** is a reliably known digit (other than a zero used to locate the decimal point).

For addition and subtraction, the number of decimal places must be considered when you are determining how many significant figures to report.

> When numbers are **added** or **subtracted,** the number of decimal places in the result should equal the smallest number of decimal places of any term in the sum.

For example, if we wish to compute $123 + 5.35$, the answer is 128 and not 128.35. If we compute the sum $1.0001 + 0.0003 = 1.0004$, the result has the correct number of decimal places; consequently, it has five significant figures even though one of the terms in the sum, 0.0003, has only one significant figure. Likewise, if we perform the subtraction $1.002 - 0.998 = 0.004$, the result has only one significant figure even though one term has four significant figures and the other has three. In this book, **most of the numerical examples and end-of-chapter problems will yield answers having either two or three significant figures.**

Example 1.6   Installing a Carpet

A carpet is to be installed in a room, the length of which is measured to be 12.71 m (four significant figures) and the width of which is measured to be 3.46 m (three significant figures). Find the area of the room.

Solution   If you multiply 12.71 m by 3.46 m on your calculator, you will get an answer of 43.9766 m². How many of these numbers should you claim? Our rule of thumb for multiplication tells us that you can claim only the number of significant figures in the least accurate of the quantities being measured. In this example, we have only three significant figures in our least accurate measurement, so we should express our final answer as 44.0 m². Note that in the answer given, we used a general rule for rounding off numbers, which states that the last digit retained is to be increased by 1 if the first digit dropped was equal to 5 or greater. If the last digit is 5, the result should be rounded to the nearest even number. (This helps avoid accumulation of errors.)

1.7 • COORDINATE SYSTEMS AND FRAMES OF REFERENCE

Many aspects of physics deal in some way or another with locations in space. For example, the mathematical description of the motion of an object requires a method for describing the position of the object. Thus, it is fitting that we first discuss how to describe the position of a point in space. It is done by means of coordinates. A point on a line can be located with one coordinate; a point in a plane is located with two coordinates; and three coordinates are required to locate a point in space.

A coordinate system used to specify locations in space consists of:

- A fixed reference point O, called the *origin,*
- A set of specified axes or directions with an appropriate scale and labels on the axes,
- Instructions that tell us how to label a point in space relative to the origin and axes.

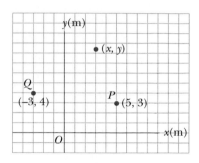

Figure 1.1 Designation of points in a cartesian coordinate system. Every point is labeled with coordinates (x, y).

One convenient coordinate system that we will use frequently is the *cartesian coordinate system,* sometimes called the *rectangular coordinate system.* Such a system in two dimensions is illustrated in Figure 1.1. An arbitrary point in this system is labeled with the coordinates (x, y). Positive x is taken to the right of the origin, and positive y is upward from the origin. Negative x is to the left of the origin, and negative y is downward from the origin. For example, the point P, which has coordinates $(5, 3)$, may be reached by going first 5 meters to the right of the origin and then 3 meters above the origin. In the same way, the point Q has coordinates $(-3, 4)$, which correspond to going 3 meters to the left of the origin and 4 meters above the origin.

Sometimes it is more convenient to represent a point in a plane by its *plane polar coordinates* (r, θ), as in Figure 1.2a. In this coordinate system, r is the length of the line from the origin to the point, and θ is the angle between that line and a fixed axis, usually the positive x axis, with θ measured counterclockwise. From the right triangle in Figure 1.2b, we find $\sin \theta = y/r$ and $\cos \theta = x/r$. (A review of trigonometric functions is given in Appendix B.4.) Therefore, starting with plane polar coordinates, one can obtain the cartesian coordinates through the equations

$$x = r \cos \theta \qquad \textbf{[1.3]}$$

$$y = r \sin \theta \qquad \textbf{[1.4]}$$

Furthermore, it follows that

$$\tan \theta = \frac{y}{x} \qquad \textbf{[1.5]}$$

and

$$r = \sqrt{x^2 + y^2} \qquad \textbf{[1.6]}$$

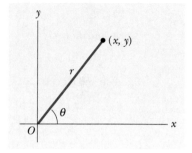

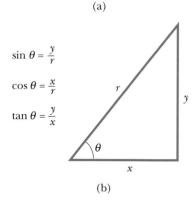

Figure 1.2 (a) The plane polar coordinates of a point are represented by the distance r and the angle θ. (b) The right triangle used to relate (x, y) to (r, θ).

You should note that these expressions relating the coordinates (x, y) to the coordinates (r, θ) apply only when θ is defined as in Figure 1.2a, in situations in which positive θ is an angle measured *counterclockwise* from the positive x axis. Other choices are made in navigation and astronomy. If the reference axis for the polar angle θ is chosen to be other than the positive x axis, or the sense of increasing θ is chosen differently, then the corresponding expressions relating the two sets of coordinates will change.

1.8 · PROBLEM-SOLVING STRATEGY

Most courses in general physics require the student to learn the skills of problem solving, and examinations usually include problems that test such skills. This brief section describes some useful suggestions that will enable you to increase your accuracy in solving problems, enhance your understanding of physical concepts, eliminate initial panic or lack of direction in approaching a problem, and organize your work. One way to help accomplish these goals is to adopt a problem-solving strategy. Many chapters in this book will include a section labeled "Problem-Solving Strategy," which should help you through the rough spots.

The following steps are commonly used to develop a problem-solving strategy:

A MENU FOR
PROBLEM SOLVING

Read Problem

⬇

Draw Diagram

⬇

Identify Data

⬇

Choose Equation(s)

⬇

Solve Equation(s)

⬇

Evaluate and Check
Answer

1. Read the problem carefully at least twice. Be sure you understand the nature of the problem before proceeding further.
2. Draw a suitable diagram with appropriate labels and coordinate axes if needed.
3. Imagine a movie, running in your mind, of what happens in the problem.
4. As you examine what is being asked in the problem, identify the basic physical principle or principles that are involved, listing the knowns and unknowns.
5. Select a basic relationship or derive an equation that can be used to find the unknown and symbolically solve the equation for the unknown.
6. Substitute the given values, along with the appropriate units, into the equation.
7. Obtain a numerical value with units for the unknown. You can have confidence in your result if the following questions can be properly answered: Do the units match? Is the answer reasonable? Is the plus or minus sign proper or meaningful?

One of the purposes of this strategy is to promote accuracy. Properly drawn diagrams can eliminate many sign errors. Diagrams also help to isolate the physical principles of the problem. Symbolic solutions and carefully labeled knowns and unknowns will help eliminate other careless errors. The use of symbolic solutions should help you think in terms of the physics of the problem. A check of units at the end of the problem can indicate a possible algebraic error. The physical layout and organization of your problem will make the final product more understandable and easier to follow. Once you have developed an organized system for examining problems and extracting relevant information, you will become a more confident problem solver.

1.9 • VECTORS AND SCALARS

Each of the physical quantities that we shall encounter in this book can be placed in one of two categories: It is either a scalar or a vector. A scalar is a quantity that is completely specified by a positive or negative number with appropriate units:

> A **scalar** quantity has a positive or negative value and no direction. However, a **vector** is a physical quantity that must be specified by both magnitude and direction.

The number of apples in a basket is an example of a scalar quantity. If you are told there are 38 apples in the basket, this completes the required information; no specification of direction is required. Other examples of scalars are temperature,

(*Left*) The number of apples in the basket is one example of a scalar quantity. Can you think of other examples? (*Superstock*)   (*Right*) Jennifer pointing in the right direction tells us to travel 5 blocks to the north to reach the courthouse. A vector is a physical quantity that must be specified by both magnitude and direction. (*Photo by Raymond A. Serway*)

volume, mass, and time intervals. The rules of ordinary arithmetic are used to manipulate scalar quantities.

Force is an example of a vector quantity. To completely describe the force on an object, we must specify both the direction of the applied force, the magnitude of the force, and the location of the force. When the motion (velocity) of an object is described, we must specify both the speed and the direction of its motion.

Another simple example of a vector quantity is the **displacement** of a particle, defined as its *change in position*. Suppose the particle moves from some point O to a point P along a straight path, as in Figure 1.3. We represent this displacement by drawing an arrow from O to P, where the tip of the arrow represents the direction of the displacement and the length of the arrow represents the magnitude of the displacement. If the particle travels along some other path from O to P, such as the broken line in Figure 1.3, its displacement is still OP. The vector displacement along any indirect path from O to P is defined as being equivalent to the displacement represented by the direct path from O to P. The magnitude of the displacement is the shortest distance between the end points. Thus, **the displacement of a particle is completely known if its initial and final coordinates are known.** The path need not be specified. In other words, **the displacement is independent of the path** if the end points of the path are fixed.

It is important to note that the distance traveled by a particle is distinctly different from its displacement. The **distance** traveled (a scalar quantity) is the length of the path, which in general can be much greater than the magnitude of the displacement (see Fig. 1.3).

If the particle moves along the x axis from position x_i to position x_f, as in Figure 1.4, its displacement is given by $x_f - x_i$. (The indices i and f refer to the initial and final values.) We use the Greek letter delta (Δ) to denote the *change* in a quantity.

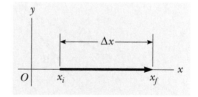

Figure 1.3 As a particle moves from O to P along the broken line, its displacement vector is the arrow drawn from O to P.

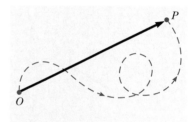

Figure 1.4 A particle moving along the x axis from x_i to x_f undergoes a displacement $\Delta x = x_f - x_i$.

Therefore, we define the change in the position of the particle (the displacement) as

Definition of displacement • along a line

$$\Delta x \equiv x_f - x_i \qquad [1.7]$$

From this definition we see that Δx is positive if x_f is greater than x_i and negative if x_f is less than x_i. For example, if a particle changes its position from $x_i = -3$ units to $x_f = 5$ units, its displacement is 8 units.

Many physical quantities in addition to displacement are vectors. They include velocity, acceleration, force, and momentum, all of which will be defined in later chapters. In this book we will use boldface letters, such as **A**, to represent arbitrary vectors. Another common method of notating vectors with which you should be familiar is to use an arrow over the letter: $\vec{A}$.

The magnitude of the vector **A** is written A or, alternatively, |**A**|. The magnitude of a vector is always positive and has physical units, such as meters for displacement or meters per second for velocity, as discussed earlier. Vectors combine according to special rules, which will be discussed in Sections 1.10 and 1.11.

Thinking Physics 1

Consider your commute to work or school in the morning. Which is larger, the distance you traveled or the magnitude of the displacement vector?

Reasoning The distance traveled, unless you have a very unusual commute, *must* be larger than the magnitude of the displacement vector. The distance includes all of the twists and turns that you made in following the roads from home to work or school. However, the magnitude of the displacement vector is the length of a straight line from your home to work or school. This is often described informally as "the distance as the crow flies." The only way that the distance could be the same as the magnitude of the displacement vector is if your commute is a perfect straight line, which is highly unlikely! The distance could *never* be less than the magnitude of the displacement vector, because the shortest distance between two points is a straight line.

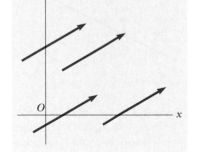

Figure 1.5 Four representations of the same vector.

1.10 • SOME PROPERTIES OF VECTORS

Equality of Two Vectors   Two vectors **A** and **B** are defined to be equal if they have the same magnitude and the same direction. That is, **A** = **B** only if $A = B$ *and* **A** and **B** act in parallel. For example, all the vectors in Figure 1.5 are equal even though they have different starting points. This property allows us to translate a vector parallel to itself in a diagram without affecting the vector.

Addition   When two or more vectors are added together, they must *all* have the same units. For example, it would be meaningless to add a velocity vector to a displacement vector, because they are different physical quantities. Scalars obey the same rule. For example, it would be meaningless to add time intervals and temperatures.

The rules for vector sums are conveniently described by geometric methods. To add vector **B** to vector **A**, first draw vector **A**, with its magnitude represented by a convenient scale, on graph paper and then draw vector **B** to the same scale with its tail starting from the tip of **A**, as in Figure 1.6. The *resultant vector* **R** = **A** + **B** is the vector drawn from the tail of **A** to the tip of **B**. This is known as the *triangle*

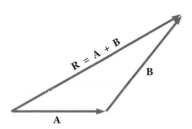

Figure 1.6 When vector **A** is added to vector **B**, the resultant **R** runs from the tail of **A** to the tip of **B**.

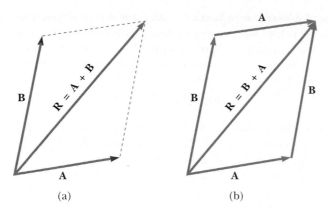

(a)   (b)

Figure 1.7   (a) In this construction, the resultant **R** is the diagonal of a parallelogram with sides **A** and **B**. (b) This construction shows that **A** + **B** = **B** + **A**.

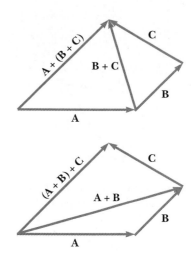

Figure 1.8   Geometric constructions for verifying the associative law of addition.

method of addition. An alternative graphical procedure for adding two vectors, known as the **parallelogram rule of addition,** is shown in Figure 1.7a. In this construction, the tails of the two vectors **A** and **B** are together, and the resultant vector **R** is the diagonal of a parallelogram formed with **A** and **B** as its sides.

When two vectors are added, **the sum is independent of the order of the addition.** This can be seen from the geometric construction in Figure 1.7b and is known as the **commutative law of addition:**

$$\mathbf{A} + \mathbf{B} = \mathbf{B} + \mathbf{A} \qquad [1.8]$$

- *Commutative law*

If three or more vectors are added, **their sum is independent of the way in which they are grouped.** A geometric proof of this for three vectors is given in Figure 1.8. This is called the **associative law of addition:**

$$\mathbf{A} + (\mathbf{B} + \mathbf{C}) = (\mathbf{A} + \mathbf{B}) + \mathbf{C} \qquad [1.9]$$

- *Associative law*

Geometric constructions can also be used to add more than three vectors. This is shown in Figure 1.9 for the case of four vectors. The resultant vector sum **R** = **A** + **B** + **C** + **D** is **the vector that completes the polygon.** In other words, **R is the vector drawn from the tail of the first vector to the tip of the last vector.** Again, the order of the summation is unimportant.

Thus we conclude that **a vector is a quantity that has both magnitude and direction and also obeys the laws of vector addition,** as described in Figures 1.6 through 1.9.

Negative of a Vector   The negative of the vector **A** is defined as the vector that, when added to **A**, gives zero for the vector sum. That is, **A** + (−**A**) = 0. The vectors **A** and −**A** have the same magnitude but opposite directions.

Subtraction of Vectors   The operation of vector subtraction makes use of the definition of the negative of a vector. We define the operation **A** − **B** as vector −**B** added to vector **A**:

$$\mathbf{A} - \mathbf{B} = \mathbf{A} + (-\mathbf{B}) \qquad [1.10]$$

The geometric construction for subtracting two vectors is shown in Figure 1.10.

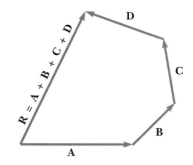

Figure 1.9   Geometric construction for summing four vectors. The resultant vector **R** completes the polygon.

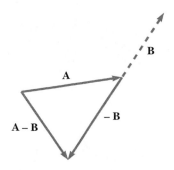

Figure 1.10 This construction shows how to subtract vector **B** from vector **A**. The vector $-$**B** is equal in magnitude and opposite to the vector **B**.

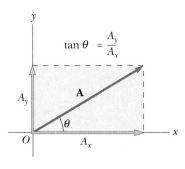

Figure 1.11 Any vector **A** lying in the *xy* plane can be represented by its components A_x and A_y.

Multiplication of a Vector by a Scalar   If a vector **A** is multiplied by a positive scalar quantity *m*, the product *m***A** is a vector that has the same direction as **A** and magnitude *mA*. If *m* is a negative scalar quantity, the vector *m***A** is directed opposite to **A**. For example, the vector 5**A** is five times as great in magnitude as **A** and has the same direction as **A**. By contrast, the vector $-\frac{1}{3}$**A** has one third the magnitude of **A** and points in the direction opposite **A** (because of the negative sign).

Multiplication of Two Vectors   Two vectors **A** and **B** can be multiplied to produce either a scalar or a vector quantity. The **scalar product** (or dot product) **A** $\cdot$ **B** is a scalar quantity equal to $AB \cos \theta$, where θ is the angle between **A** and **B**. The **vector product** (or cross product) **A** $\times$ **B** is a vector quantity, the magnitude of which is equal to $AB \sin \theta$. We shall discuss these products more fully in Chapters 6 and 10, where they are first used.

Each chapter of this textbook contains a number of exercises. The purpose of such exercises is to test your understanding of the material just discussed by asking you to do a calculation or answer some question related to a worked example. Here is your first exercise.

EXERCISE 1   The magnitudes of two vectors **A** and **B** are $A = 8$ units and $B = 3$ units. Find the largest and smallest possible values for the magnitude of the resultant vector **R** = **A** + **B**.   Answer   The largest value of *R* is 11 units, corresponding to the case when **A** and **B** point in the same direction. The smallest value of *R* is 5 units, corresponding to the case when **A** is directed opposite to **B**.

1.11 • COMPONENTS OF A VECTOR AND UNIT VECTORS

The geometric method of adding vectors is not the recommended procedure in situations in which great precision is required or in three-dimensional problems. In this section we describe a method of adding vectors that makes use of the *projections* of a vector along the axes of a rectangular coordinate system.

Consider a vector **A** lying in the *xy* plane and making an arbitrary angle θ with the positive *x* axis, as in Figure 1.11. The vector **A** can be represented by its **rectangular components,** A_x and A_y. The component A_x represents the projection of **A** along the *x* axis, whereas A_y represents the projection of **A** along the *y* axis. The components of a vector, which are scalar quantities, can be positive or negative. For example, in Figure 1.11, A_x and A_y are both positive.

From Figure 1.11 and the definition of the sine and cosine of an angle, we see that $\cos \theta = A_x/A$ and $\sin \theta = A_y/A$. Hence, the components of **A** are given by

Components of the vector **A**   •

$$A_x = A \cos \theta \qquad \text{and} \qquad A_y = A \sin \theta \qquad \textbf{[1.11]}$$

It is important to note that when using these component equations, θ must be measured counterclockwise from the positive *x* axis. These components form two sides of a right triangle, the hypotenuse of which has a magnitude *A*. Thus, it follows that the magnitude of **A** and its direction are related to its components through the expressions

Magnitude of **A**   •

$$A = \sqrt{A_x^{\,2} + A_y^{\,2}} \qquad \textbf{[1.12]}$$

Direction of **A**   •

$$\tan \theta = \frac{A_y}{A_x} \qquad \textbf{[1.13]}$$

To solve for θ, we can write $\theta = \tan^{-1}(A_y/A_x)$, which is read "$\theta$ equals the angle whose tangent is the ratio A_y/A_x." **Note that the signs of the components A_x and A_y depend on the angle θ.** For example, if $\theta = 120°$, A_x is negative and A_y is positive. By contrast, if $\theta = 225°$, both A_x and A_y are negative. If both components are negative, the angle measured from the positive x axis will be the angle you determine on your calculator plus 180°. Figure 1.12 summarizes the signs of the components when **A** lies in the various quadrants.

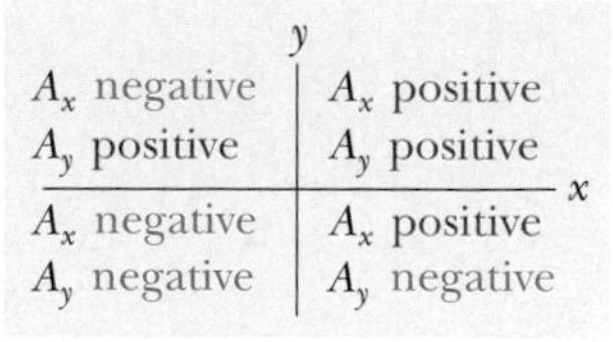

If you choose reference axes or an angle other than those shown in Figure 1.11, the components of the vector must be modified accordingly. In many applications it is more convenient to express the components of a vector in a coordinate system having axes that are not horizontal and vertical but that are still perpendicular to each other. Suppose a vector **B** makes an angle θ' with the x' axis defined in Figure 1.13. The components of **B** along these axes are given by $B_{x'} = B\cos\theta'$ and $B_{y'} = B\sin\theta'$, as in Equation 1.9. The magnitude and direction of **B** are obtained from expressions equivalent to Equations 1.10 and 1.11. Thus, we can express the components of a vector in *any* coordinate system that is convenient for a particular situation.

The components of a vector differ when viewed from different coordinate systems. Furthermore, the components of a vector can change with respect to a fixed coordinate system if the vector changes in magnitude, orientation, or both.

Vector quantities are often expressed in terms of unit vectors. **A unit vector is a dimensionless vector 1 unit in length used to specify a given direction.** Unit vectors have no other physical significance. They are used simply as a convenience in describing a direction in space. We will use the symbols **i**, **j**, and **k** to represent unit vectors pointing in the x, y, and z directions, respectively. Thus, the unit vectors **i**, **j**, and **k** form a set of mutually perpendicular vectors as shown in Figure 1.14a, where the magnitude of the unit vectors equals unity. That is, $|\mathbf{i}| = |\mathbf{j}| = |\mathbf{k}| = 1$.

Consider a vector **A** lying in the xy plane, as in Figure 1.14b. The product of the component A_x and the unit vector **i** is the vector $A_x\mathbf{i}$ parallel to the x axis with magnitude A_x. Likewise, $A_y\mathbf{j}$ is a vector of magnitude A_y parallel to the y axis. When

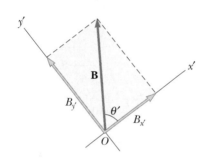

Figure 1.12 The signs of the components of a vector **A** depend on the quadrant in which the vector is located.

• *Unit vectors*

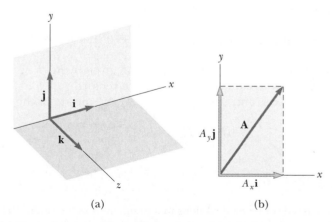

Figure 1.13 The components of vector **B** in a coordinate system that is tilted.

Figure 1.14 (a) The unit vectors **i**, **j**, and **k** are directed along the x, y, and z axes, respectively. (b) A vector **A** lying in the xy plane has component vectors $A_x\mathbf{i}$ and $A_y\mathbf{j}$ where A_x and A_y are the components of **A**.

using the unit form of a vector, we are simply multiplying a vector (the unit vector) by a scalar. Thus, the unit-vector notation for the vector **A** is written

$$\mathbf{A} = A_x\mathbf{i} + A_y\mathbf{j} \qquad [1.14]$$

The vectors $A_x\mathbf{i}$ and $A_y\mathbf{j}$ are the component vectors of **A**. These should not be confused with A_x and A_y, which we shall always refer to as the components of **A**.

Now suppose we wish to add vector **B** to vector **A**, where **B** has components B_x and B_y. The procedure for performing this sum is simply to add the x and y components separately. The resultant vector $\mathbf{R} = \mathbf{A} + \mathbf{B}$ is therefore

$$\mathbf{R} = (A_x + B_x)\mathbf{i} + (A_y + B_y)\mathbf{j} \qquad [1.15]$$

Thus, the components of the resultant vector are given by

$$R_x = A_x + B_x$$
$$R_y = A_y + B_y \qquad [1.16]$$

The magnitude of **R** and the angle it makes with the x axis can then be obtained from its components using the relationships

$$R = \sqrt{R_x^2 + R_y^2} = \sqrt{(A_x + B_x)^2 + (A_y + B_y)^2} \qquad [1.17]$$

$$\tan\theta = \frac{R_y}{R_x} = \frac{A_y + B_y}{A_x + B_x} \qquad [1.18]$$

The procedure just described for adding two vectors **A** and **B** using the component method can be checked using a geometric construction, as in Figure 1.15. Again you must take note of the *signs* of the components when using either the algebraic or the geometric method.

The extension of these methods to three-dimensional vectors is straightforward. If **A** and **B** both have x, y, and z components, we express them in the form

$$\mathbf{A} = A_x\mathbf{i} + A_y\mathbf{j} + A_z\mathbf{k} \qquad \text{and} \qquad \mathbf{B} = B_x\mathbf{i} + B_y\mathbf{j} + B_z\mathbf{k}$$

The sum of **A** and **B** is

$$\mathbf{R} = \mathbf{A} + \mathbf{B} = (A_x + B_x)\mathbf{i} + (A_y + B_y)\mathbf{j} + (A_z + B_z)\mathbf{k} \qquad [1.19]$$

Thus, the resultant vector also has a z component, given by $R_z = A_z + B_z$. The same procedure can be used to sum up three or more vectors.

If a vector **R** has x, y, and z components, the magnitude of the vector is

$$R = \sqrt{R_x^2 + R_y^2 + R_z^2}$$

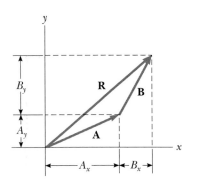

Figure 1.15 A geometric construction showing the relation between the components of the resultant **R** of two vectors and the individual component vectors.

Thinking Physics 2

You may have asked someone directions to a destination in a city and have been told something like, "Walk 3 blocks east and then 5 blocks south." If so, are you experienced with vector components?

Reasoning Yes, you are! Although you may not have thought of vector component language when you heard these directions, this is exactly what the directions represent. The perpendicular streets of the city reflect an *x*–*y* coordinate system—we can assign the *x* axis to the east–west streets and the *y* axis to the north–south streets. Thus, the comment of the person giving you directions can be translated as, "Undergo a displacement vector that has an *x* component of $+3$ blocks and a *y* component of -5 blocks." You would arrive at the same destination by undergoing the *y* component first, followed by the *x* component, demonstrating the commutative law of addition.

Many sections of this book will include conceptual problems that are nonnumerical like the Thinking Physics example above. The answers to all conceptual problems are found at the end of each chapter. Here is your first set of conceptual problems.

CONCEPTUAL PROBLEM 1

If one component of a vector is not zero, can its magnitude be zero? Explain.

CONCEPTUAL PROBLEM 2

If $\mathbf{A} + \mathbf{B} = 0$, what can you say about the components of the two vectors?

CONCEPTUAL PROBLEM 3

Figure CP1.3 shows two vectors lying in the *xy*-plane. Determine the signs of (a) the *x* components of **A** and **B**, (b) the *y* components of **A** and **B**, and (c) the *x* and *y* components of **A** + **B**.

Figure CP1.3

CONCEPTUAL PROBLEM 4

Can the component of a vector ever be equal to the magnitude of the vector?

PROBLEM-SOLVING STRATEGY • Adding Vectors, Using Components

When two or more vectors are to be added, the following steps are recommended:

1. Select a coordinate system.
2. Draw a sketch of the vectors to be added (or subtracted) and label each vector.
3. Find the *x* and *y* components of all vectors.
4. Find the resultant components (the algebraic sum of the components) in both the *x* and *y* directions.
5. Use the Pythagorean theorem to find the magnitude of the resultant vector.
6. Use a suitable trigonometric function to find the angle the resultant vector makes with the *x* axis.

Example 1.7   The Sum of Two Vectors

Find the sum of two vectors **A** and **B** lying in the *xy* plane and given by

$$\mathbf{A} = 2.0\mathbf{i} + 2.0\mathbf{j} \quad \text{and} \quad \mathbf{B} = 2.0\mathbf{i} - 4.0\mathbf{j}$$

Solution   Note that $A_x = 2.0$, $A_y = 2.0$, $B_x = 2.0$, and $B_y = -4.0$. Therefore, the resultant vector **R** is

$$\mathbf{R} = \mathbf{A} + \mathbf{B} = (2.0 + 2.0)\mathbf{i} + (2.0 - 4.0)\mathbf{j} = 4.0\mathbf{i} - 2.0\mathbf{j}$$

or

$$R_x = 4.0, \qquad R_y = -2.0$$

The magnitude of **R** is

$$R = \sqrt{R_x^2 + R_y^2} = \sqrt{(4.0)^2 + (-2.0)^2} = \sqrt{20} = 4.5$$

EXERCISE 2   Find the angle θ that the resultant vector **R** makes with the positive *x* axis.   Answer   330°

Example 1.8   The Resultant Displacement

A particle undergoes three consecutive displacements: $\mathbf{d}_1 = (1.5\mathbf{i} + 3.0\mathbf{j} - 1.2\mathbf{k})$ cm, $\mathbf{d}_2 = (2.3\mathbf{i} - 1.4\mathbf{j} - 3.6\mathbf{k})$ cm, and $\mathbf{d}_3 = (-1.3\mathbf{i} + 1.5\mathbf{j})$ cm. Find the components of the resultant displacement and its magnitude.

Solution

$$\begin{aligned} \mathbf{R} &= \mathbf{d}_1 + \mathbf{d}_2 + \mathbf{d}_3 \\ &= (1.5 + 2.3 - 1.3)\mathbf{i} + (3.0 - 1.4 + 1.5)\mathbf{j} \\ &\quad + (-1.2 - 3.6 + 0)\mathbf{k} \\ &= (2.5\mathbf{i} + 3.1\mathbf{j} - 4.8\mathbf{k})\,\text{cm} \end{aligned}$$

That is, the resultant displacement has components $R_x = 2.5$ cm, $R_y = 3.1$ cm, and $R_z = -4.8$ cm. Its magnitude is

$$\begin{aligned} R &= \sqrt{R_x^2 + R_y^2 + R_z^2} \\ &= \sqrt{(2.5\ \text{cm})^2 + (3.1\ \text{cm})^2 + (-4.8\ \text{cm})^2} = 6.2\ \text{cm} \end{aligned}$$

Example 1.9   Taking a Hike

A hiker begins a trip by first walking 25.0 km due southeast from her base camp. On the second day she walks 40.0 km in a direction 60.0° north of east, at which point she discovers a forest ranger's tower.

(a) Determine the components of the hiker's displacements in the first and second days.

Solution   If we denote the displacement vectors on the first and second days by **A** and **B**, respectively, and use the camp as the origin of coordinates, we get the vectors shown in Figure 1.16. Displacement **A** has a magnitude of 25.0 km and is 45.0° southeast. Its components are

$$A_x = A \cos(-45.0°) = (25.0\ \text{km})(0.707) = 17.7\ \text{km}$$

$$A_y = A \sin(-45.0°) = -(25.0\ \text{km})(0.707) = -17.7\ \text{km}$$

The negative value of A_y indicates that the *y* coordinate decreased in this displacement. The signs of A_x and A_y are also evident from Figure 1.16.

The second displacement, **B**, has a magnitude of 40.0 km and is 60.0° north of east. Its components are

$$B_x = B \cos 60.0° = (40.0\ \text{km})(0.500) = 20.0\ \text{km}$$

$$B_y = B \sin 60.0° = (40.0\ \text{km})(0.866) = 34.6\ \text{km}$$

(b) Determine the components of the hiker's total displacement for the trip.

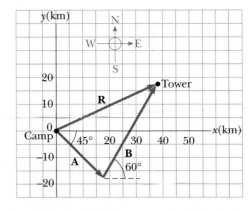

Figure 1.16   The total displacement of the hiker is the vector $\mathbf{R} = \mathbf{A} + \mathbf{B}$.

Solution The resultant displacement vector for the trip, **R** = **A** + **B**, has components

$$R_x = A_x + B_x = 17.7 \text{ km} + 20.0 \text{ km} = \boxed{37.7 \text{ km}}$$

$$R_y = A_y + B_y = 17.7 \text{ km} + 34.6 \text{ km} = \boxed{16.9 \text{ km}}$$

In unit-vector form, we can write the total displacement as **R** = (37.7**i** + 16.9**j**) km.

EXERCISE 3 Determine the magnitude and direction of the total displacement. Answer 41.3 km, 24.1° north of east from the base camp

SUMMARY

Mechanical quantities can be expressed in terms of three fundamental quantities, *length, mass,* and *time,* which in the SI system have the units *meters* (m), *kilograms* (kg), and *seconds* (s), respectively. It is often useful to use the *method of dimensional analysis* to check equations and to assist in deriving expressions.

The density of a substance is defined as its mass per unit volume. Different substances have different densities mainly because of differences in their atomic masses and atomic arrangements.

The number of molecules in 1 mole of any element or compound, called **Avogadro's number** (N_A), is 6.02×10^{23}.

Vectors are quantities that have both magnitude and direction and obey the vector law of addition. **Scalars** are quantities that have only magnitude.

Two vectors **A** and **B** can be added using either the triangle method or the parallelogram rule. In the triangle method (Fig. 1.6), the vector **C** = **A** + **B** runs from the tail of **A** to the tip of **B**. In the parallelogram method (Fig. 1.7a), **C** is the diagonal of a parallelogram having **A** and **B** as its sides.

The *x* component of the vector **A**, A_x, is equal to its projection along the *x* axis of a coordinate system as in Figure 1.11, where $A_x = A \cos \theta$ and θ is the angle **A** makes with the *x* axis. Likewise, the *y* component of **A**, A_y, is its projection along the *y* axis, where $A_y = A \sin \theta$. The resultant of two or more vectors can be found by resolving all vectors into their *x* and *y* components, adding their resultant *x* and *y* components, and then using the Pythagorean theorem to find the magnitude of the resultant vector. The angle that the resultant vector makes with the *x* axis can be found by the use of a suitable trigonometric function.

If a vector **A** has an *x* component equal to A_x and a *y* component equal to A_y, the vector can be expressed in unit-vector form as **A** = A_x**i** + A_y**j**. In this notation, **i** is a unit vector in the positive *x* direction and **j** is a unit vector in the positive *y* direction. Because **i** and **j** are unit vectors, |**i**| = |**j**| = 1.

CONCEPTUAL QUESTIONS

1. Suppose that the three fundamental standards of the metric system were length, *density,* and time rather than length, *mass,* and time. The standard of density in this system is to be defined as that of water. What considerations about water would need to be addressed to make sure that the standard unit of density were to be as accurate as possible?

2. What types of natural phenomena could serve as alternative time standards?

3. The height of a horse is sometimes given in units of "hands." Why is this a poor standard of length?

4. Express the following quantities using the prefixes given in Table 1.4: (a) 3×10^{-4} m, (b) 5×10^{-5} s, (c) 72×10^2 g.

5. As one moves upward in atomic number in the periodic table, the atomic masses of the elements increases. Because the atoms are becoming more and more massive, why doesn't the *density* of elemental materials increase in the same way?

6. Suppose that two quantities *A* and *B* have different dimensions. Determine which of the following arithmetic opera-

Boxed numbers indicate questions that have answers available in the Student Solutions Manual and Study Guide.

tions *could* be physically meaningful: (a) $A + B$, (b) A/B, (c) $B - A$, (d) AB.

7. What accuracy is implied in an order-of-magnitude calculation?

8. Apply an order-of-magnitude calculation to an everyday situation you might encounter. For example, how far do you walk or drive each day?

9. Which of the following are vectors and which are not: force, temperature, the volume of water in a can, the ratings of a TV show, the height of a building, the velocity of a sports car, the age of the Universe?

10. A vector **A** lies in the *xy* plane. For what orientations of **A** will both of its rectangular components be negative? For what orientations will its components have opposite signs?

11. A book is moved once around the perimeter of a tabletop with the dimensions 1.0 m × 2.0 m. If the book ends up at its initial position, what is its displacement? What is the distance traveled?

12. While traveling along a straight interstate highway you notice that the mile marker reads 260. You travel until you reach the 150-mile marker and then retrace your path to the 175-mile marker. What is the magnitude of your resultant displacement from the 260-mile marker?

13. If the component of vector **A** along the direction of vector **B** is zero, what can you conclude about the two vectors?

14. If **A** = **B**, what can you conclude about the components of **A** and **B**?

15. Is it possible to add a vector quantity to a scalar quantity? Explain.

16. A roller coaster travels 135 ft at an angle of 40.0° above the horizontal. How far does it move horizontally and vertically?

17. The resolution of vectors into components is equivalent to replacing the original vector with the sum of two vectors, whose sum is the same as the original vector. There are an infinite number of pairs of vectors that will satisfy this condition; we choose that pair with one vector parallel to the *x* axis and the second parallel to the *y* axis. What difficulties would be introduced by defining components relative to axes that are not perpendicular—for example, the *x* axis and a *y* axis oriented at 45° to the *x* axis?

PROBLEMS

Section 1.2   Density and Atomic Mass

1. Calculate the mass of an atom of (a) helium, (b) iron, (c) lead. Give your answers in atomic mass units and in grams. The atomic masses of the atoms given are 4.00, 55.9, and 207, respectively.

2. The standard kilogram is a platinum-iridium cylinder 39.0 mm in height and 39.0 mm in diameter. What is the density of the material?

3. One cubic meter (1.00 m^3) of aluminum has a mass of 2.70×10^3 kg, and one cubic meter of iron has a mass of 7.86×10^3 kg. Find the radius of a solid aluminum sphere that will balance a solid iron sphere of radius 2.00 cm on an equal-arm balance.

4. Let ρ_{Al} represent the density of aluminum and ρ_{Fe} that of iron. Find the radius of a solid aluminum sphere that balances a solid iron sphere of radius r_{Fe} on an equal-arm balance.

5. The mass of the planet Saturn is 5.64×10^{26} kg, and its radius is 6.00×10^7 m. (a) Calculate its density. (b) If Saturn were placed in a large enough ocean of water, would it float? Explain.

Section 1.3   Dimensional Analysis

6. The radius *r* of a circle inscribed in any triangle whose sides are *a*, *b*, and *c* is given by $r = [(s - a)(s - b)(s - c)/s]^{1/2}$,

where *s* is an abbreviation for $(a + b + c)/2$. Check this formula for dimensional consistency.

7. The consumption of natural gas by a company satisfies the empirical equation $V = 1.50t + 0.00800t^2$, where *V* is the volume in millions of cubic feet and *t* the time in months. Express this equation in units of cubic feet and seconds. Put the proper units on the coefficients. Assume that a month is 30.0 days.

8. Newton's law of universal gravitation is given by

$$F = G\frac{Mm}{r^2}$$

Here *F* is the force of gravity, *M* and *m* are masses, and *r* is a length. Force has the units $\text{kg} \cdot \text{m/s}^2$. What are the SI units of the proportionality constant *G*?

9. The period *T* of a simple pendulum (the time for one complete oscillation) is measured in time units and is given by

$$T = 2\pi\sqrt{\frac{\ell}{g}}$$

where ℓ is the length of the pendulum and *g* is the acceleration due to gravity in units of length divided by the square of time. Show that this equation is dimensionally consistent.

Boxed numbers indicate problems that have full solutions available in the Student Solutions Manual and Study Guide. Cyan numbers indicate intermediate level problems and magenta numbers indicate more challenging problems.

Section 1.4   Conversion of Units

10. An auditorium measures 40.0 m × 20.0 m × 12.0 m. The density of air is 1.20 kg/m³. What are (a) the volume of the room in cubic feet and (b) the weight of air in the room in pounds?

11. An astronomical unit (AU) is defined as the average distance between the Earth and Sun. (a) How many astronomical units are there in one lightyear? (b) Determine the distance from Earth to the Andromeda galaxy in astronomical units.

12. Assume that it takes 7.00 minutes to fill a 30.0-gal gasoline tank. (a) Calculate the rate at which the tank is filled in gallons per second. (b) Calculate the rate at which the tank is filled in cubic meters per second. (c) Determine the time, in hours, required to fill a 1-cubic-meter volume at the same rate. (1 U.S. gal = 231 in.³)

13. One gallon of paint (volume = 3.78 × 10⁻³ m³) covers an area of 25.0 m². What is the thickness of the paint on the wall?

14. You can obtain a rough estimate of the size of a molecule by the following simple experiment. Let a droplet of oil spread out on a smooth water surface. The resulting oil slick will be approximately 1 molecule thick. If an oil droplet of mass 9.00 × 10⁻⁷ kg and density 918 kg/m³ spreads out into a circle of radius 41.8 cm on the water surface, calculate the diameter of an oil molecule.

15. (a) Find a conversion factor to convert from miles per hour to kilometers per hour. (b) Until recently, federal law mandated that highway speeds would be 55 mi/h. Use the conversion factor of part (a) to find the speed in kilometers per hour. (c) The maximum highway speed has been raised to 65 mi/h in some places. In kilometers per hour, how much increase is this over the 55-mi/h limit?

16. (a) How many seconds are there in a year? (b) If one micrometeorite (a sphere with a diameter of 10⁻⁶ m) struck each square meter of the Moon each second, how many years would it take to cover the Moon to a depth of 1.00 m? (*Hint:* Consider a cubic box on the Moon 1.00 m on a side and find how long it would take to fill the box.)

17. The mass of the Sun is about 1.99 × 10³⁰ kg, and the mass of a hydrogen atom, of which the Sun is mostly composed, is 1.67 × 10⁻²⁷ kg. How many atoms are there in the Sun?

Section 1.5   Order-of-Magnitude Calculations

18. Compute the order of magnitude of the mass of (a) a bathtub filled with water, and (b) a bathtub filled with pennies. In your solution list the quantities you estimate and the value you estimate for each.

19. Estimate the number of piano tuners living in New York City. This problem was posed by the physicist Enrico Fermi, who was well known for making order-of-magnitude calculations.

20. Soft drinks are commonly sold in aluminum containers. Estimate the number of such containers thrown away or recycled each year by U.S. consumers. Approximately how many tons of aluminum does this represent?

Section 1.6   Significant Figures

21. Carry out the following arithmetic operations: (a) the sum of the numbers 756, 37.2, 0.83, and 2.5; (b) the product 3.2 × 3.563; (c) the product 5.6 × π.

22. How many significant figures are there in (a) 78.9 ± 0.2, (b) 3.788 × 10⁹, (c) 2.46 × 10⁻⁶, (d) 0.0053?

23. The radius of a circle is measured to be 10.5 ± 0.2 m. Calculate the (a) area and (b) circumference of the circle and give the uncertainty in each value.

24. The *radius* of a solid sphere is measured by a student to be (6.50 ± 0.20) cm, and its mass is measured to be (1.85 ± 0.02) kg. Determine the density of the sphere, in kilograms per cubic meter, and the uncertainty in the density.

Section 1.7   Coordinate Systems and Frames of Reference

25. The polar coordinates of a point are $r = 5.50$ m and $\theta = 240.0°$. What are the cartesian coordinates of this point?

26. A point in the xy plane has cartesian coordinates (−3.00, 5.00) m. What are the polar coordinates of this point?

27. Two points in a plane have polar coordinates (2.50 m, 30.0°) and (3.80 m, 120.0°). Determine (a) the cartesian coordinates of these points and (b) the distance between them.

28. If the polar coordinates of the point (x, y) are (r, θ), determine the polar coordinates for the points: (a) $(−x, y)$, (b) $(−2x, −2y)$, and (c) $(3x, −3y)$.

Section 1.9   Vectors and Scalars
Section 1.10   Some Properties of Vectors

29. A person walks along a circular path of radius 5.00 m, around one half of the circle. (a) Find the magnitude of the displacement vector. (b) How far did the person walk? (c) What is the magnitude of the displacement if the circle is completed?

30. A jogger runs 100 m due west, then changes direction for the second leg of the run. At the end of the run, she is 175 m away from the starting point at an angle of 15.0° north of west. What were the direction and magnitude of her second displacement? Use a graph to solve this problem.

31. Each of the displacement vectors **A** and **B** shown in Figure P1.31 on page 26 has a magnitude of 3.0 m. Graphically find (a) **A** + **B**, (b) **A** − **B**, (c) **B** − **A**, (d) **A** − 2**B**.

32. Indiana Jones is trapped in a maze. To find his way out, he walks 10.0 m, makes a 90.0° right turn, walks 5.00 m, makes

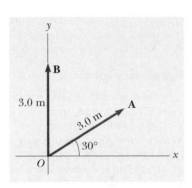

Figure P1.31

another 90.0° right turn, and walks 7.00 m. What is his displacement from his initial position?

Section 1.11   Components of a Vector and Unit Vectors

33. A vector has an *x* component of −25.0 units and a *y* component of 40.0 units. Find the magnitude and direction of this vector.

34. Find the horizontal and vertical components of the 100-m displacement of a superhero who flies from the top of a tall building along the path shown in Figure P1.34.

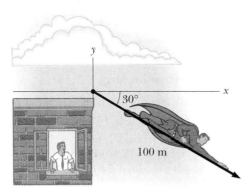

Figure P1.34

35. Instructions for finding a buried treasure include the following: Go 75 paces at 240°, turn to 135° and walk 125 paces, then travel 100 paces at 160°. Determine the resultant displacement from the starting point.

36. Given the vectors $\mathbf{A} = 2.00\mathbf{i} + 6.00\mathbf{j}$ and $\mathbf{B} = 3.00\mathbf{i} - 2.00\mathbf{j}$, (a) sketch the vector sum $\mathbf{C} = \mathbf{A} + \mathbf{B}$ and the vector

subtraction $\mathbf{D} = \mathbf{A} - \mathbf{B}$. (b) Find analytical solutions for $\mathbf{C}$ and $\mathbf{D}$ first in terms of unit vectors and then in terms of polar coordinates, with angles measured with respect to the positive *x* axis.

37. Three vectors are oriented as shown in Figure P1.37, where $|\mathbf{A}| = 20.0$, $|\mathbf{B}| = 40.0$, and $|\mathbf{C}| = 30.0$ units. Find (a) the *x* and *y* components of the resultant vector and (b) the magnitude and direction of the resultant vector.

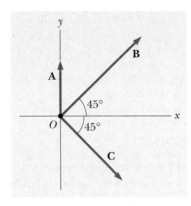

Figure P1.37

38. A novice golfer on the green takes three strokes to sink the ball. The successive displacements are 4.00 m due north, 2.00 m northeast, and 1.00 m 30.0° west of south. Starting at the same initial point, an expert golfer might make the hole in what single vector displacement?

39. A person going for a walk follows the path shown in Figure P1.39. The total trip consists of four straight-line paths. At

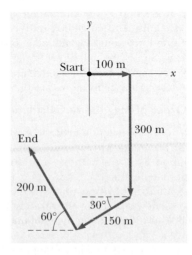

Figure P1.39

the end of the walk, what is the person's resultant displacement, measured from the starting point?

40. A vector is given by $\mathbf{R} = 2.00\mathbf{i} + 1.00\mathbf{j} + 3.00\mathbf{k}$. Find (a) the magnitudes of the x, y, and z components, (b) the magnitude of $\mathbf{R}$, and (c) the angles between $\mathbf{R}$ and the x, y, and z axes.

41. A particle undergoes two displacements. The first has a magnitude of 150 cm and makes an angle of 120.0° with the positive x axis. The *resultant* displacement has a magnitude of 140 cm and is directed at an angle of 35.0° to the positive x axis. Find the magnitude and direction of the second displacement.

42. A jet airliner moving initially at 300 mph due east enters a region where the wind is blowing at 100 mph in a direction 30.0° north of east. Determine the new speed and direction of the aircraft.

Additional Problems

43. A useful fact is that there are about $\pi \times 10^7$ s in one year. Use a calculator to find the percentage error in this approximation. *Note:*

$$\text{Percentage error} = \frac{|\text{assumed value} - \text{true value}|}{\text{true value}} \times 100\%$$

44. The eye of a hurricane passes over Grand Bahama Island. It is moving in a direction 60.0° north of west with a speed of 41.0 km/h. Three hours later the course of the hurricane suddenly shifts due north, and its speed slows to 25.0 km/h. How far is the hurricane from Grand Bahama 4.50 hours after it passes over the island?

45. Assume that there are 50 million passenger cars in the United States and that the average fuel consumption is 20 mi/gal of gasoline. If the average distance traveled by each car is 10 000 miles/year, how much gasoline would be saved per year if average fuel consumption could be increased to 25 mi/gal?

46. The basic function of the carburetor of an automobile is to atomize the gasoline and mix it with air to promote rapid combustion. As an example, assume that 30.0 cm³ of gasoline is atomized into N spherical droplets, each with a radius of 2.00×10^{-5} m. What is the total surface area of these N spherical droplets?

47. Assume that 70 percent of the Earth's surface is covered with water at an average depth of 1 mile, and **estimate** the mass of the water on Earth in kilograms.

48. The data in the following table represent measurements of the masses and dimensions of solid cylinders of aluminum, copper, brass, tin, and iron. Use these data to calculate the densities of the substances. Compare your results for aluminum, copper, and iron with those given in Table 1.5.

Substance	Mass (g)	Diameter (cm)	Length (cm)
Aluminum	51.5	2.52	3.75
Copper	56.3	1.23	5.06
Brass	94.4	1.54	5.69
Tin	69.1	1.75	3.74
Iron	216.1	1.89	9.77

49. An air-traffic controller notices two aircraft on his radar screen. The first is at altitude 800 m, horizontal distance 19.2 km, and 25.0° south of west. The second aircraft is at altitude 1100 m, horizontal distance 17.6 km, and 20.0° south of west. What is the distance between the two aircraft? (Place the x axis west, the y axis south, and the z axis vertical.)

50. Two people pull on a stubborn mule, as shown by the helicopter view in Figure P1.50. Find (a) the single force that is equivalent to the two forces shown, and (b) the force that a third person would have to exert on the mule to make the net force equal to zero.

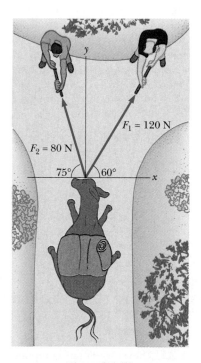

Figure P1.50

51. The distance from the Sun to the nearest star is on the order of 10^{17} m. The Milky Way galaxy is roughly a disk of radius 10^{21} m and thickness 10^{19} m. **Estimate** the number of stars

in the Milky Way. Assume the distance between the Sun and the nearest neighbor is typical.

52. A rectangular parallelepiped has dimensions a, b, and c, as in Figure P1.52. (a) Obtain a vector expression for the face diagonal vector $\mathbf{R}_1$. What is the magnitude of this vector? (b) Obtain a vector expression for the body diagonal vector $\mathbf{R}_2$. Note that $\mathbf{R}_1$, $c\mathbf{k}$, and $\mathbf{R}_2$ form a right triangle and prove that the magnitude of $\mathbf{R}_2$ is $\sqrt{a^2 + b^2 + c^2}$.

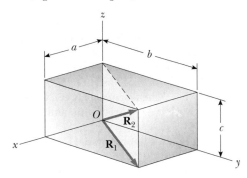

Figure P1.52

☐ Spreadsheet Problems

S1. Spreadsheet programs are useful in examining data graphically. Most spreadsheet programs can fit the best straight line (a regression line) to a set of data. The following table gives experimental results for the measurement of the period T of a pendulum of length L. These data are consistent with an equation of the form $T = CL^n$, where C and n are constants and n is not necessarily an integer.

Length L(m)	Period T(s)
0.25	1.00
0.50	1.40
0.75	1.75
1.00	2.00
1.50	2.50
2.00	2.80

(a) Use the least-squares or the regression-line procedure of your spreadsheet program to find the best fit to the data. Because $T = CL^n$, we see that

$$\log T = \log C + n \log L$$

First calculate a column of log T values and another of log L as the independent variable and log T as the dependent variable. The least-squares fit finds the slope n and the intercept log C. (b) Which data points deviate most from a straight line plot of T versus L^n? (c) Is the experimental value of n found in part (a) consistent with a dimensional analysis of $T = CL^n$?

S2. The period T and orbit radius R for the motions of four moons of Jupiter are

	Io	Europa	Ganymede	Callisto
T (days)	1.77	3.55	7.16	16.69
R (km)	422 000	671 000	1 070 000	1 880 000

(a) These data can be fitted by the formula $T = CR^n$. Follow the procedures in Problem S1a to find C and n. (b) A fifth satellite, Amalthea, has a period of 0.50 days. Use $T = CR^n$ to find the radius for its orbit.

S3. Use Spreadsheet 1.1 to find the total displacement corresponding to the sum of the three displacement vectors: $\mathbf{A} = 3.0\mathbf{i} + 4.0\mathbf{j}$, $\mathbf{B} = -2.3\mathbf{i} - 7.8\mathbf{j}$, and $\mathbf{C} = 5.0\mathbf{i} - 2.0\mathbf{j}$, where the units are in meters. (a) Give your answer in component form. (b) Give your answer in polar form. (c) Find a fourth displacement that returns you to the origin.

ANSWERS TO CONCEPTUAL PROBLEMS

1. No. The magnitude of a vector $\mathbf{A}$ is equal to $\sqrt{A_x^2 + A_y^2 + A_z^2}$. Therefore, if any component is nonzero, $\mathbf{A}$ cannot be zero.

2. $\mathbf{A} = -\mathbf{B}$, therefore the components of the two vectors must have opposite signs and equal magnitudes.

3. (a) A_x is negative, B_x is positive; (b) A_y is positive, B_y is negative; (c) both components are negative.

4. It is possible to have a situation in which one component of a vector is equal to the magnitude of the vector, as long as the other component is *zero*. For example, a velocity vector lying directly along the east direction has an eastward component that is equal to the magnitude of the velocity vector.

2

Motion in One Dimension

Dynamics is the study of the motions of objects and the relationships of those motions to such physical concepts as force and mass. Before beginning our study of dynamics, however, it is convenient to describe motion using the concepts of space and time, without regard to the causes of the motion; this is a portion of mechanics called *kinematics*. In this chapter we shall consider motion along a straight line—that is, one-dimensional motion. Starting with the concept of displacement, discussed in Chapter 1, we shall define velocity and acceleration. Using these concepts, we shall proceed to study the motion of objects undergoing constant acceleration. In Chapter 3 we shall extend our discussion to two-dimensional motion.

From everyday experience we recognize that motion represents continuous change in the position of an object. The movement of an object through space may be accompanied by the rotation or vibration of the object. Such motions can be quite complex. However, it is sometimes possible to simplify matters by temporarily neglecting the internal motions of the moving object. In many situations, an object can be

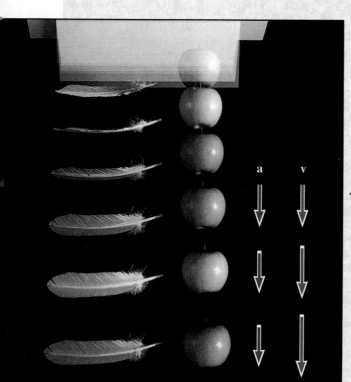

◀ An apple and a feather, released from rest in a 4-ft vacuum chamber, fall at the same rate, regardless of their masses. Neglecting air resistance, all objects fall to Earth with the same acceleration, $a = g$, of magnitude 9.80 m/s^2 as indicated by the violet arrows in this multiflash photograph. The velocity, v, of the two objects increases linearly with time, as indicated by the series of red arrows.

treated as a *particle* if the only motion being considered is translation through space.

Although an idealized particle is a mathematical point with no size, we can sometimes perform useful calculations by representing macroscopic objects as particles. For example, if we wish to describe the motion of the Earth around the Sun, we can approximate the Earth by treating it as a particle and thereby attain reasonable accuracy in predicting the Earth's orbit. This approximation is justified because the radius of the Earth's orbit is large compared with the dimensions of the Earth and Sun. By contrast, we could not use a particle description to explain the internal structure of the Earth or such phenomena as tides, earthquakes, and volcanic activity. On a much smaller scale, it is possible to explain the pressure exerted by a gas on the walls of a container by treating the gas molecules as particles. However, the particle description of the gas molecules is generally inadequate for understanding those properties of the gas that depend on the internal motions (vibrations) and rotations of the gas molecules.

2.1 • AVERAGE VELOCITY

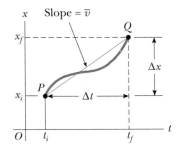

Figure 2.1 Position–time graph for a particle moving along the x axis. The average velocity $\bar{v}_x$ in the interval $\Delta t = t_f - t_i$ is obtained from the slope of the straight line connecting the points P and Q.

The motion of a particle is completely known if the position of the particle in space is known at all times. Consider a particle moving along the x axis from point P to point Q. Let its position at point P be x_i at some time t_i, and let its position at point Q be x_f at time t_f. At times other than t_i and t_f, the position of the particle between these two points may vary, as in Figure 2.1. Such a plot is often called a *position–time graph*. In the time interval $\Delta t = t_f - t_i$, the displacement of the particle is $\Delta x = x_f - x_i$. (Recall that displacement is defined as the change in the position of the particle, which is equal to its final position value minus its initial position value.) Using unit-vector notation, the **displacement vector** can be expressed as $\Delta\mathbf{x} \equiv (x_f - x_i)\mathbf{i}$.

> The x component of the **average velocity** of the particle, $\bar{\mathbf{v}}_x$, is defined as the ratio of its displacement vector, $\Delta\mathbf{x}$, and the time interval, Δt:

Definition of average velocity •

$$\bar{\mathbf{v}}_x \equiv \frac{\Delta\mathbf{x}}{\Delta t} = \frac{(x_f - x_i)\mathbf{i}}{t_f - t_i} \qquad [2.1]$$

From this definition we see that the average velocity has the dimensions of length divided by time (L/T)—m/s in SI units and ft/s in conventional units. The average velocity is *independent* of the path taken between the points P and Q. This is true because the average velocity is proportional to the displacement, $\Delta\mathbf{x}$, which in turn depends only on the initial and final coordinates of the particle. It therefore follows that if a particle starts at some point and returns to the same point via any path, its average velocity for this trip is zero, because its displacement along such a path is zero. Displacement should not be confused with the distance traveled, because the distance traveled for any motion is clearly nonzero. Thus, average velocity gives us no details of the motion between points P and Q. (How we evaluate the velocity at some instant in time is discussed in the next section.) Finally, note that the average velocity in one dimension can be positive or negative, depending on the sign of the displacement vector. (The time interval, Δt, is always positive.) If the x coordinate of the particle increases in time (that is, if $x_f > x_i$), then $\Delta \mathbf{x}$ is positive and $\bar{\mathbf{v}}_x$ is positive. This corresponds to an average velocity in the positive x direction. By contrast, if the coordinate decreases in time ($x_f < x_i$), $\Delta \mathbf{x}$ is negative; hence $\bar{\mathbf{v}}_x$ is negative. This corresponds to an average velocity in the negative x direction.

The average velocity can also be interpreted geometrically. A straight line drawn between the points P and Q in Figure 2.1 forms the hypotenuse of a right triangle of height Δx and base Δt. The slope of this line is the ratio $\Delta x/\Delta t$. Therefore, we see that **the average velocity of the particle during the time interval t_i to t_f is equal to the slope of the straight line joining the initial and final points on the position–time graph**. (The word *slope* will often be used in reference to the graphs of physical data. Regardless of what data are plotted, the word *slope* will represent the ratio of the change in the quantity represented on the vertical axis to the change in the quantity represented on the horizontal axis.)

Example 2.1   Calculate the Average Velocity

A particle moving along the x axis is located at $x_i = 12$ m at $t_i = 1$ s and at $x_f = 4$ m at $t_f = 3$ s. Find its displacement and average velocity during this time interval.

Solution   The displacement is

$$\Delta \mathbf{x} = (x_f - x_i)\mathbf{i} = (4 \text{ m} - 12 \text{ m})\mathbf{i} = \; -8\mathbf{i} \text{ m}$$

The average velocity is

$$\bar{\mathbf{v}}_x = \frac{\Delta \mathbf{x}}{\Delta t} = \frac{(x_f - x_i)\mathbf{i}}{t_f - t_i} = \frac{(4 \text{ m} - 12 \text{ m})\mathbf{i}}{3 \text{ s} - 1 \text{ s}} = \; -4\mathbf{i} \text{ m/s}$$

Because the displacement is negative for this time interval, we conclude that the particle has moved to the left, toward decreasing values of x.

EXERCISE 1   A jogger runs in a straight line, with an average speed of 5.00 m/s for 4.00 min, and then with an average speed of 4.00 m/s for 3.00 min. (a) What is her total displacement? (b) What is her average speed during this time?
Answer   (a) 1920 m   (b) $\frac{32}{7}$ m/s

2.2 · INSTANTANEOUS VELOCITY

We would like to be able to define the velocity of a particle at a particular instant of time, rather than just during a finite interval of time. The velocity of a particle at any instant of time, or at some point on a position–time graph, is called the **instantaneous velocity.** This concept is especially important when the average velocity is *not constant* through different time intervals.

Consider the motion of a particle between the two points P and Q on the position–time graph shown in Figure 2.2. As the point Q is brought closer and closer to the point P, the time intervals ($\Delta t_1, \Delta t_2, \Delta t_3, \ldots$) get progressively smaller. The average velocity for each time interval is given by the slope of the appropriate dotted line in Figure 2.2. As the point Q approaches P, the time interval approaches zero, but at the same time the slope of the dotted line approaches that of the line tangent to the curve at the point P. **The slope of the line tangent to the curve at P gives the instantaneous velocity at the time t_i.** In other words,

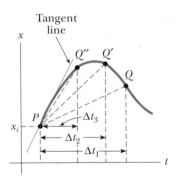

Figure 2.2   Position–time graph for a particle moving along the *x* axis. As the time intervals, starting at t_i, get smaller and smaller, the average velocity for one interval approaches the slope of the line tangent at *P*. The instantaneous velocity at *P* is obtained from the slope of the blue tangent line at the time t_i.

> the instantaneous velocity, $\mathbf{v}_x$, equals the limiting value of the ratio $\Delta\mathbf{x}/\Delta t$ as Δt approaches zero:[1]

The instantaneous velocity vector is the derivative of the displacement vector with respect to time.

$$\mathbf{v}_x \equiv \lim_{\Delta t \to 0} \frac{\Delta\mathbf{x}}{\Delta t} \qquad [2.2]$$

In the calculus notation, this limit is called the *derivative* of *x* with respect to *t*, written $d\mathbf{x}/dt$:

$$\mathbf{v}_x \equiv \lim_{\Delta t \to 0} \frac{\Delta\mathbf{x}}{\Delta t} = \frac{d\mathbf{x}}{dt} \qquad [2.3]$$

The instantaneous velocity can be positive, negative, or zero. When the slope of the position–time graph is positive, such as at the point *P* in Figure 2.3, *v* is positive. At point *R*, *v* is negative, because the slope is negative. Finally, the instantaneous velocity is zero at the peak *Q* (the turning point), where the slope is zero. **From here on, we shall usually use the word *velocity* to designate instantaneous velocity.**

The **instantaneous speed** of a particle is defined as the magnitude of the instantaneous velocity vector. Hence, by definition, *speed* can never be negative.

Those of you familiar with the calculus should recognize that there are specific rules for taking the derivatives of functions. These rules, which are listed in Appendix B.6, enable us to evaluate derivatives quickly.

Suppose *x* is proportional to some power of *t*, such as

$$x = At^n$$

where *A* and *n* are constants. (This is a very common functional form.) The derivative of *x* with respect to *t* is

$$\frac{dx}{dt} = nAt^{n-1}$$

For example, if $x = 5t^3$, we see that $dx/dt = 3(5)t^{3-1} = 15t^2$.

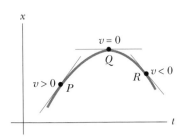

Figure 2.3   In the position–time graph shown, the velocity is positive at *P*, where the slope of the tangent line is positive; the velocity is zero at *Q*, where the slope of the tangent line is zero; and the velocity is negative at *R*, where the slope of the tangent line is negative.

---

[1]*Note:* The magnitude of the displacement, Δx, also approaches zero as Δt approaches zero. However, as Δx and Δt become smaller and smaller, the ratio $\Delta x/\Delta t$ approaches a value equal to the *true* slope of the line tangent to the *x* versus *t* curve.

Thinking Physics 1

Consider the following motions of an object in one dimension: (a) A ball is thrown directly upward, rises to a highest point, and falls back into the thrower's hand. (b) A race car starts from rest and speeds up to 100 m/s. (c) A Voyager spacecraft drifts through space at constant velocity. Are there any points in the motion of these particles at which the average velocity (over the entire interval) and the instantaneous velocity (at an instant of time within the interval) are the same? If so, identify the point(s).

Reasoning (a) The average velocity for the thrown ball is zero—the ball returns to the starting point at the end of the time interval. There is one point at which the instantaneous velocity is zero—at the top of the motion. (b) The average velocity for the motion of the race car cannot be evaluated unambiguously with the information given, but it must be some value between 0 and 100 m/s. Because the car will have every instantaneous velocity between 0 and 100 m/s at some time during the interval, there must be some instant at which the instantaneous velocity is equal to the average velocity. (c) Because the instantaneous velocity of the spacecraft is constant, its instantaneous velocity over *any* time interval and its average velocity at *any* time are the same.

Thinking Physics 2

Does an automobile speedometer measure average or instantaneous speed?

Reasoning Based on the fact that changes in the speed of the automobile are reflected in changes in the speedometer reading, we might be tempted to say that a speedometer measures instantaneous speed. We would be wrong, however. The reading on a speedometer is related to the rotation of the wheels. Often, a rotating magnet, whose rotation is related to that of the wheels, is used in a speedometer. The reading is based on the *time interval* between rotations of the magnet. Thus, the reading is an average velocity over this time interval. This time interval is generally very short, so that the average velocity is close to the instantaneous velocity. As long as the velocity changes are gradual, this approximation is very good. If the automobile brakes to a quick stop, however, the speedometer may not be able to keep up with the rapid changes in speed, and the measured average speed over the time interval may be very different than the instantaneous speed at the end of the interval.

CONCEPTUAL PROBLEM 1

Under what conditions is the magnitude of the average velocity of a particle moving in one dimension smaller than the average speed?

Example 2.2 Average and Instantaneous Velocity

A particle moves along the x axis. Its x coordinate varies with time according to the expression $x = -4t + 2t^2$, where x is in meters and t is in seconds. The position–time graph for this motion is shown in Figure 2.4. Note that the particle moves in the negative x direction for the first second of motion, stops instantaneously at $t = 1$ s, and then heads back in the positive x direction for $t > 1$ s.

(a) Determine the displacement of the particle in the time intervals $t = 0$ to $t = 1$ s and $t = 1$ s to $t = 3$ s.

Solution In the first time interval we set $t_i = 0$ and $t_f = 1$ s. Because $x = -4t + 2t^2$, we get for the first displacement

$$\Delta \mathbf{x}_{01} = (x_f - x_i)\mathbf{i} = [-4(1) + 2(1)^2]\mathbf{i} - [-4(0) + 2(0)^2]\mathbf{i}$$
$$= -2\mathbf{i} \text{ m}$$

Likewise, in the second time interval we can set $t_i = 1$ s and $t_f = 3$ s. Therefore, the displacement in this interval is

$$\Delta \mathbf{x}_{13} = (x_f - x_i)\mathbf{i} = [-4(3) + 2(3)^2]\mathbf{i} - [-4(1) + 2(1)^2]\mathbf{i}$$
$$= 8\mathbf{i} \text{ m}$$

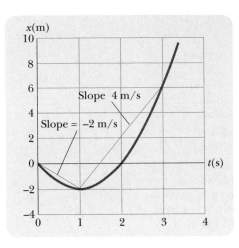

Figure 2.4 (Example 2.2) Position–time graph for a particle having an x coordinate that varies in time according to $x = -4t + 2t^2$.

Note that these displacements can also be read directly from the position–time graph (Fig. 2.4).

(b) Calculate the average velocity in the time intervals $t = 0$ to $t = 1$ s and $t = 1$ s to $t = 3$ s.

Solution In the first time interval, $\Delta t = t_f - t_i = 1$ s. Therefore, using Equation 2.1 and the results from (a) gives

$$\overline{\mathbf{v}}_x = \frac{\Delta \mathbf{x}_{01}}{\Delta t} = \frac{-2\mathbf{i} \text{ m}}{1 \text{ s}} = -2\mathbf{i} \text{ m/s}$$

Likewise, in the second time interval, $\Delta t = 2$ s; therefore,

$$\overline{\mathbf{v}}_x = \frac{\Delta \mathbf{x}_{13}}{\Delta t} = \frac{8\mathbf{i} \text{ m}}{2 \text{ s}} = 4\mathbf{i} \text{ m/s}$$

These values (the coefficients of $\mathbf{i}$) agree with the slopes of the lines joining these points in Figure 2.4.

(c) Find the instantaneous velocity of the particle at $t = 2.5$ s.

Solution We can find the instantaneous velocity at any time t by taking the first derivative of x with respect to t:

$$v = \frac{dx}{dt} = \frac{d}{dt}(-4t + 2t^2) = -4 + 4t$$

Thus, at $t = 2.5$ s, we find that $v = -4 + 4(2.5) = 6$ m/s.

We can also obtain this result by measuring the slope of the position–time graph at $t = 2.5$ s. (You should show that the velocity is $-4\mathbf{i}$ m/s at $t = 0$ and zero at $t = 1$ s.) Do you see any symmetry in the motion? For example, does the speed ever repeat itself?

Example 2.3   The Limiting Process

The position of a particle moving along the x axis varies in time according to the expression $x = (3 \text{ m/s}^2)\, t^2$, where x is in meters and t is in seconds. Find the velocity in terms of t at any time.

Solution The position–time graph for this motion is shown in Figure 2.5. We can compute the velocity at any time t by using the definition of the instantaneous velocity. If the initial coordinate of the particle at time t is $x_i = 3t^2$, then the coordinate at a later time, $t + \Delta t$, is

$$x_f = 3(t + \Delta t)^2 = 3[t^2 + 2t\,\Delta t + (\Delta t)^2]$$
$$= 3t^2 + 6t\,\Delta t + 3(\Delta t)^2$$

Therefore, the displacement in the time interval Δt is

$$\Delta \mathbf{x} = (x_f - x_i)\mathbf{i} = [3t^2 + 6t\,\Delta t + 3(\Delta t)^2 - 3t^2]\mathbf{i}$$
$$= (6t\,\Delta t + 3(\Delta t)^2)\mathbf{i}$$

The average velocity in this time interval is

$$\overline{\mathbf{v}}_x = \frac{\Delta \mathbf{x}}{\Delta t} = (6t + 3\,\Delta t)\mathbf{i}$$

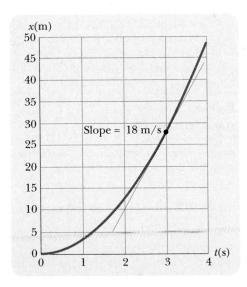

Figure 2.5 (Example 2.3) Position–time graph for a particle having an x coordinate that varies in time according to $x = 3t^2$. Note that the instantaneous velocity at $t = 3.0$ s is obtained from the slope of the blue line tangent to the curve at this point.

To find the instantaneous velocity, we take the limit of this expression as Δt approaches zero. In doing so, we see that the term $3\,\Delta t$ goes to zero; therefore,

$$\mathbf{v}_x = \lim_{\Delta t \to 0} \frac{\Delta \mathbf{x}}{\Delta t} = 6t\mathbf{i}\ \text{m/s}$$

Notice that this expression gives us the velocity at *any* general time t. It tells us that $\mathbf{v}_x$ is increasing linearly in time. It is then a straightforward matter to find the velocity at some specific time from the expression $\mathbf{v}_x = 6t\mathbf{i}$. For example, at $t = 3.0$ s, the velocity is $\mathbf{v}_x = 6(3)\mathbf{i} = 18\mathbf{i}$ m/s. Again, this can be checked from the slope of the graph at $t = 3.0$ s.

We can also find v by taking the first derivative of x with respect to time. In this example, $x = 3t^2$, and we see that $v_x = dx/dt = 6t$, in agreement with our result of taking the limit explicitly.

2.3 • ACCELERATION

When the velocity of a particle changes with time, the particle is said to be *accelerating*. For example, the speed of a car increases when you step on the gas. The car slows down when you apply the brakes, and it changes direction when you turn the wheel. However, we need a more precise definition of acceleration.

Suppose a particle moving along the x axis has a velocity $\mathbf{v}_{xi}$ at time t_i and a velocity $\mathbf{v}_{xf}$ at time t_f.

The **average acceleration** of a particle in the time interval $\Delta t = t_f - t_i$ is defined as the ratio $\Delta \mathbf{v}_x / \Delta t$, where $\Delta \mathbf{v}_x = \mathbf{v}_{xf} - \mathbf{v}_{xi}$ is the *change* in velocity in this time interval:

$$\bar{\mathbf{a}}_x \equiv \frac{\mathbf{v}_{xf} - \mathbf{v}_{xi}}{t_f - t_i} = \frac{\Delta \mathbf{v}_x}{\Delta t} \qquad\qquad [2.4]$$

• *Definition of average acceleration*

Acceleration is a vector quantity having dimensions of length divided by (time)^2, or L/T^2. Some of the common units of acceleration are meters per second per second (m/s^2) and feet per second per second (ft/s^2).

In some situations, the value of the average acceleration may be different for different time intervals. It is therefore useful to define the **instantaneous acceleration** as the limit of the average acceleration as Δt approaches zero. This concept is analogous to the definition of instantaneous velocity discussed in the previous section. If we take the limit of the ratio $\Delta \mathbf{v}_x / \Delta t$ as Δt approaches zero, we get the **instantaneous acceleration:**

$$\mathbf{a}_x \equiv \lim_{\Delta t \to 0} \frac{\Delta \mathbf{v}_x}{\Delta t} = \frac{d\mathbf{v}_x}{dt} \qquad\qquad [2.5]$$

• *The instantaneous acceleration is the derivative of the velocity with respect to time.*

That is, **the instantaneous acceleration equals the derivative of the velocity with respect to time, which by definition is the slope of the velocity–time graph.** One can interpret the derivative of the velocity with respect to time as the *time rate of change of velocity*. Note that if $\mathbf{a}_x$ is positive, the acceleration is in the positive x direction, whereas negative $\mathbf{a}_x$ implies acceleration in the negative x direction. This does not mean that the particle is *moving* in the x direction. From now on we shall use the term *acceleration* to mean instantaneous acceleration.

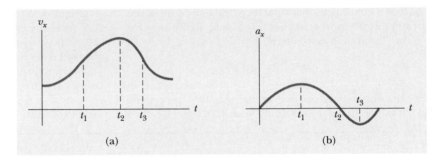

Figure 2.6   The instantaneous acceleration can be obtained from the velocity–time graph (a). At each instant, the acceleration in the a_x versus t graph (b) equals the slope of the line tangent to the v versus t curve.

Because $\mathbf{v}_x = d\mathbf{x}/dt$, the acceleration can also be written as

$$\mathbf{a}_x = \frac{d\mathbf{v}_x}{dt} = \frac{d}{dt}\left(\frac{d\mathbf{x}}{dt}\right) = \frac{d^2\mathbf{x}}{dt^2} \qquad \textbf{[2.6]}$$

That is, the acceleration equals the *second derivative* of the displacement with respect to time.

Figure 2.6 shows how the acceleration–time curve can be derived from the velocity–time curve. In these sketches, the acceleration at any time is simply the slope of the velocity–time graph at that time. Positive values of the acceleration correspond to those points at which the velocity is increasing in the positive x direction. The acceleration reaches a maximum at time t_1, when the slope of the velocity–time graph is a maximum. The acceleration then goes to zero at time t_2, when the velocity is a maximum (that is, when the velocity is momentarily not changing and the slope of the v versus t graph is zero). Finally, the acceleration is negative when the velocity in the positive x direction is decreasing in time.

Acceleration can mean •
speeding up, slowing down,
or changing direction.

As an example of the computation of acceleration, consider the car pictured in Figure 2.7. In this case, the velocity of the car has changed from an initial value of $30\mathbf{i}$ m/s to a final value of $15\mathbf{i}$ m/s in a time interval of 2.0 s. The average acceleration during this time interval is

$$\bar{\mathbf{a}}_x = \frac{15\mathbf{i}\ \text{m/s} - 30\mathbf{i}\ \text{m/s}}{2.0\ \text{s}} = -7.5\mathbf{i}\ \text{m/s}^2$$

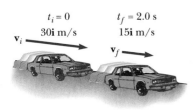

Figure 2.7   The velocity of the car decreases from $30\mathbf{i}$ m/s to $15\mathbf{i}$ m/s in a time interval of 2.0 s.

The minus sign in this example indicates that the acceleration vector is in the negative x direction (to the left). For the case of motion in a straight line, the direction of the velocity of an object and the direction of its acceleration are related as follows. **When the object's velocity and acceleration are in the same direction, the object is speeding up in that direction.** However, **when the object's velocity and acceleration are in opposite directions, the speed of the object decreases in time.**

CONCEPTUAL PROBLEM 2

If a car is traveling eastward, can its acceleration be westward? Explain.

CONCEPTUAL PROBLEM 3

Can a particle with constant acceleration ever stop and stay stopped?

Example 2.4   **Average and Instantaneous Acceleration**

The velocity of a particle moving along the x axis varies in time according to the expression $\mathbf{v}_x = (40 - 5t^2)\mathbf{i}$ m/s, where t is in seconds.

(a) Find the average acceleration in the time interval $t = 0$ to $t = 2.0$ s.

Solution   The velocity–time graph for this function is given in Figure 2.8. The velocities at $t_i = 0$ and $t_f = 2.0$ s are found by substituting these values of t into the expression given for the velocity:

$$\mathbf{v}_{xi} = (40 - 5t_i^2)\mathbf{i} \text{ m/s} = [40 - 5(0)^2]\mathbf{i} \text{ m/s} = 40\mathbf{i} \text{ m/s}$$

$$\mathbf{v}_{xf} = (40 - 5t_f^2)\mathbf{i} \text{ m/s} = [40 - 5(2)^2]\mathbf{i} \text{ m/s} = 20\mathbf{i} \text{ m/s}$$

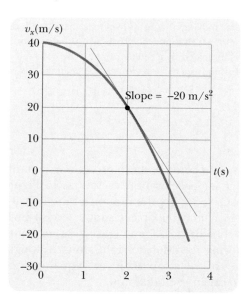

Figure 2.8   (Example 2.4) The velocity–time graph for a particle moving along the x axis according to the relation $v_x = (40 - 5t^2)$ m/s. Note that the acceleration at $t = 2.0$ s is obtained from the slope of the blue tangent line at that time.

Therefore, the average acceleration in the specified time interval, $\Delta t = t_f - t_i = 2.0$ s, is

$$\bar{\mathbf{a}}_x = \frac{\mathbf{v}_{xf} - \mathbf{v}_{xi}}{t_f - t_i} = \frac{(20 - 40)\mathbf{i} \text{ m/s}}{(2 - 0)\text{s}} = -10\mathbf{i} \text{ m/s}^2$$

The negative sign is consistent with the fact that the slope of the line joining the initial and final points on the velocity–time graph is negative.

(b) Determine the acceleration at $t = 2.0$ s.

Solution   The velocity at time t is $\mathbf{v}_{xi} = (40 - 5t^2)\mathbf{i}$ m/s, and the velocity at time $t + \Delta t$ is

$$\mathbf{v}_{xf} = 40\mathbf{i} - 5(t + \Delta t)^2\mathbf{i} = [40 - 5t^2 - 10t\,\Delta t - 5(\Delta t)^2]\mathbf{i}$$

Therefore, the change in velocity over the time interval Δt is

$$\Delta\mathbf{v}_x = \mathbf{v}_{xf} - \mathbf{v}_{xi} = [-10t\,\Delta t - 5(\Delta t)^2]\mathbf{i} \text{ m/s}$$

Dividing this expression by Δt and taking the limit of the result as Δt approaches zero, we get the acceleration at *any* time t:

$$\mathbf{a}_x = \lim_{\Delta t \to 0} \frac{\Delta\mathbf{v}_x}{\Delta t} = \lim_{\Delta t \to 0}(-10t - 5\,\Delta t)\mathbf{i} = -10t\mathbf{i} \text{ m/s}^2$$

Therefore, at $t = 2.0$ s we find that

$$\mathbf{a}_x = (-10)(2)\mathbf{i} \text{ m/s}^2 = -20\mathbf{i} \text{ m/s}^2$$

This result can also be obtained by measuring the slope of the velocity–time graph at $t = 2.0$ s. Note that the acceleration is not constant in this example. Situations involving constant acceleration will be treated in Section 2.5.

EXERCISE 2   When struck by a club, a golf ball initially at rest acquires a speed of 31.0 m/s. If the ball is in contact with the club for 1.17 ms, what is the magnitude of the average acceleration of the ball?   Answer   26 500 m/s²

2.4 · MOTION DIAGRAMS

The concepts of velocity and acceleration are often confused with each other, but in fact they are quite different quantities. It is instructive to make use of motion diagrams to describe the velocity and acceleration vectors as time progresses while

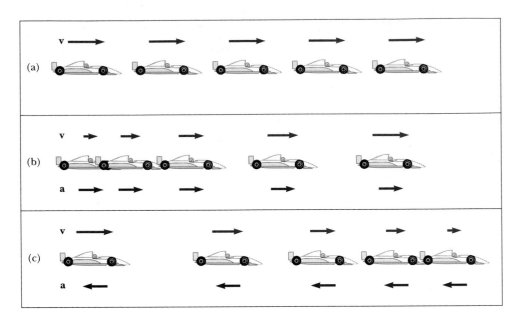

Figure 2.9   (a) Motion diagram for a car moving at constant velocity (zero acceleration). (b) Motion diagram for a car, the constant acceleration of which is in the direction of its velocity. The velocity vector at each instant is indicated by a red arrow, and the constant acceleration vector by a violet arrow. (c) Motion diagram for a car, the constant acceleration of which is in the direction *opposite* the velocity at each instant.

an object is in motion. In order not to confuse these two vector quantities, we use red for velocity vectors and violet for acceleration vectors as in Figure 2.9. The vectors are sketched at several instants during the motion of the object.

A stroboscopic photograph of a moving object shows several images of the object, each image taken as the strobe light flashes. Figure 2.9 represents three sets of strobe photographs of cars moving from left to right along a straight roadway. The time intervals between flashes of the stroboscope are equal in each diagram. Let us describe the motion of the car in each diagram.

In Figure 2.9a, the images of the car are equally spaced—the car moves the same distance in each time interval. Thus, the car moves with *constant positive velocity* and has zero acceleration.

In Figure 2.9b, the images of the car become farther apart as time progresses. In this case, the velocity vector increases in time because the car's displacement between adjacent positions increases as time progresses. Thus, the car is moving with a *positive velocity* and a *positive acceleration*.

In Figure 2.9c, the car slows as it moves to the right because its displacement between adjacent positions decreases as time progresses. In this case, the car moves initially to the right with a constant negative acceleration. The velocity vector decreases in time and eventually reaches zero. (This type of motion is exhibited by a car that skids to a stop after applying its brakes.) From this diagram we see that the acceleration and velocity vectors are *not* in the same direction.

2.5 • ONE-DIMENSIONAL MOTION WITH CONSTANT ACCELERATION

If the acceleration of a particle varies in time, the motion can be complex and difficult to analyze. A common and simple type of one-dimensional motion occurs when the acceleration is constant or uniform. In this case the average acceleration over any time interval equals the instantaneous acceleration at any instant of time within the interval. As a consequence, the velocity increases or decreases at the same rate throughout the motion.

If we replace $\overline{\mathbf{a}}_x$ with $\mathbf{a}_x$ in Equation 2.4, we find that

$$\mathbf{a}_x = \frac{\mathbf{v}_{xf} - \mathbf{v}_{xi}}{t_f - t_i}$$

Because we are dealing with motion in one dimension, we shall drop the boldface notation for vectors and use a minus sign when the vector is in the negative x direction. For convenience, let $t_i = 0$ and t_f be any arbitrary time t. Also, let $v_i = v_{x0}$ (the initial velocity at $t = 0$) and $v_f = v_x$ (the velocity at any arbitrary time t). With this notation, we can express the acceleration as

$$a_x = \frac{v_x - v_{x0}}{t}$$

or

$$v_x = v_{x0} + a_x t \qquad \text{(for constant } a_x) \qquad \text{[2.7]}$$

• *Velocity as a function of time*

This expression enables us to predict the velocity at *any* time t if the initial velocity, acceleration, and elapsed time are known. A graph of position versus time for this motion is shown in Figure 2.10a. The velocity–time graph shown in Figure 2.10b is a straight line, the slope of which is the acceleration, a_x. This is consistent with the fact that $a_x = dv_x/dt$ is a constant. From this graph and from Equation 2.7, we see that the velocity at any time t is the sum of the initial velocity, v_{x0}, and the change in velocity, $a_x t$. The graph of acceleration versus time (Fig. 2.10c) is a straight line with a slope of zero, because the acceleration is constant. Note that if the acceleration were negative, the slope of Figure 2.10b would be negative.

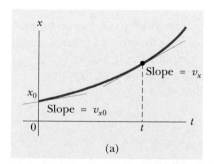
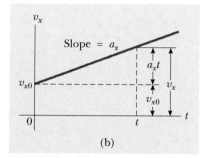
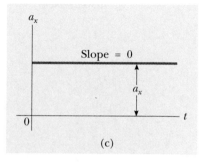

Figure 2.10 A particle moving along the x axis with constant acceleration a_x: (a) the position–time graph, (b) the velocity–time graph, and (c) the acceleration–time graph.

Because the velocity varies linearly with time according to Equation 2.7, we can express the average velocity in any time interval as the arithmetic mean of the initial velocity, v_{x0}, and the final velocity, v_x.

$$\bar{v}_x = \frac{v_{x0} + v_x}{2} \qquad \text{(for constant } a_x)\qquad [2.8]$$

Note that this expression is useful only when the acceleration is constant—that is, when the velocity varies linearly with time.

We can now use Equations 2.1 and 2.8 to obtain the displacement as a function of time. Again we choose $t_i = 0$, at which time the initial position is $x_i = x_0$. This gives

$$\Delta x = \bar{v}_x \Delta t = \left(\frac{v_{x0} + v_x}{2}\right) t$$

$$x - x_0 = \frac{1}{2}(v_{x0} + v_x)t \qquad \text{(for constant } a_x)\qquad [2.9]$$

We can obtain another useful expression for the displacement by substituting Equation 2.7 into Equation 2.9:

$$x - x_0 = \tfrac{1}{2}(v_{x0} + v_{x0} + a_x t)t$$

**Displacement as a function of •
time**

$$x - x_0 = v_{x0}t + \tfrac{1}{2}a_x t^2 \qquad \text{(for constant } a_x)\qquad [2.10]$$

The validity of this expression can be checked by differentiating it with respect to time, to give

$$v_x = \frac{dx}{dt} = \frac{d}{dt}\left(x_0 + v_{x0}t + \tfrac{1}{2}a_x t^2\right) = v_{x0} + a_x t$$

Finally, we can obtain an expression that does not contain the time by substituting the value of t from Equation 2.7 into Equation 2.9. This gives

$$x - x_0 = \tfrac{1}{2}(v_{x0} + v_x)\left(\frac{v_x - v_{x0}}{a_x}\right) = \frac{v_x^2 - v_{x0}^2}{2a_x}$$

**Velocity as a function of •
displacement**

$$v_x^2 = v_{x0}^2 + 2a_x(x - x_0) \qquad \text{(for constant } a_x)\qquad [2.11]$$

A position–time graph for motion under constant acceleration, assuming positive a_x, is shown in Figure 2.10a. The curve representing Equation 2.10 is a parabola. The slope of the tangent to this curve at $t = 0$ equals the initial velocity, v_{x0}, and the slope of the tangent line at any time t equals the velocity at that time.

If motion occurs in which the acceleration is *zero*, then we see that

$$\left.\begin{array}{l} v_x = v_{x0} \\ x - x_0 = v_x t \end{array}\right\} \quad \text{when } a_x = 0$$

That is, when the acceleration is zero, the velocity remains constant and the displacement changes linearly with time.

Equations 2.7 through 2.11 **may be used to solve any problem in one-dimensional motion with constant acceleration.** Keep in mind that these rela-

TABLE 2.1   Kinematic Equations for Motion in a Straight Line Under Constant Acceleration

Equation	Information Given by Equation
$v_x = v_{x0} + a_x t$	Velocity as a function of time
$x - x_0 = \frac{1}{2}(v_x + v_{x0})t$	Displacement as a function of velocity and time
$x - x_0 = v_{x0}t + \frac{1}{2}a_x t^2$	Displacement as a function of time
$v_x^2 = v_{x0}^2 + 2a_x(x - x_0)$	Velocity as a function of displacement

Note: Motion is along the x axis. At $t = 0$, the position of the particle is x_0 and its velocity is v_{x0}.

tionships were derived from the definitions of velocity and acceleration, together with some simple algebraic manipulations and the requirement that the acceleration be constant. It is often convenient to choose the initial position of the particle as the origin of the motion, so that $x_0 = 0$ at $t = 0$. In such a case, the displacement is simply x.

The four kinematic equations that are used most often are listed in Table 2.1. The choice of which kinematic equation or equations you should use in a given situation depends on what is known beforehand. Sometimes it is necessary to use two of these equations to solve for two unknowns, such as the displacement and velocity at some instant. For example, suppose the initial velocity, v_{x0}, and acceleration, a_x, are given. You can then find (1) the velocity after a time t has elapsed using $v_x = v_{x0} + a_x t$, and (2) the displacement after a time t has elapsed using $x - x_0 = v_{x0}t + \frac{1}{2}a_x t^2$. You should recognize that the quantities that vary during the motion are velocity, displacement, and time.

You will get a great deal of practice in the use of these equations by solving exercises and problems. You will discover again and again that there is more than one method for obtaining a solution.

PROBLEM-SOLVING STRATEGY • **Accelerated Motion**

The following procedure is recommended for solving problems that involve accelerated motion:

1. Make sure all the units in the problem are consistent. That is, if distances are measured in meters, be sure that velocities have units of meters per second and accelerations have units of meters per second per second.
2. Choose a coordinate system.
3. Choose an instant to call the *initial point* and another the *final point*. Make a list of all the quantities given in the problem and a separate list of those to be determined.
4. Think about what is going on physically in the problem and then select from the list of kinematic equations the one or ones that will enable you to determine the unknowns.
5. Construct an appropriate motion diagram and check to see if your answers are consistent with the diagram.

Example 2.5   Watch Out for the Speed Limit

A car traveling at a constant speed of 30.0 m/s ($\approx$67 mi/h) passes a trooper hidden behind a billboard. One second after the speeding car passes the billboard, the trooper sets off in chase with a constant acceleration of 3.00 m/s². How long does it take the trooper to overtake the speeding car?

Reasoning   To solve this problem algebraically, we write an expression for the position of each vehicle as a function of time. It is convenient to choose the origin at the position of the billboard and take $t = 0$ as the time the trooper begins moving. At that instant, the speeding car has already traveled a distance of 30.0 m, because it travels at a constant speed of 30.0 m/s. Thus, the initial position of the speeding car is $x_0 = 30.0$ m.

Solution   Because the car moves with constant speed, its acceleration is zero, and applying Equation 2.10 gives

$$x_C = 30.0 \text{ m} + (30.0 \text{ m/s})t$$

Note that at $t = 0$, this expression does give the car's correct initial position, $x_C = x_0 = 30.0$ m.

For the trooper, who starts from the origin at $t = 0$, we have $x_0 = 0$, $v_{x0} = 0$, and $a_x = 3.00$ m/s². Hence, the position of the trooper as a function of time is

$$x_T = \tfrac{1}{2}a_x t^2 = \tfrac{1}{2}(3.00 \text{ m/s}^2)t^2$$

The trooper overtakes the car at the instant that $x_T = x_C$, or

$$\tfrac{1}{2}(3.00 \text{ m/s}^2)t^2 = 30.0 \text{ m} + (30.0 \text{ m/s})t$$

This gives the quadratic equation

$$1.50t^2 - 30.0t - 30.0 = 0$$

the positive solution of which is $t = 21.0$ s. Note that in this time interval, the trooper has traveled a distance of about 660 m.

EXERCISE 3   This problem can also be easily solved graphically. On the *same* graph, plot the position versus time for each vehicle, and from the intersection of the two curves determine the time at which the trooper overtakes the speeding car.

EXERCISE 4   A hockey puck sliding on a frozen lake comes to rest after traveling 200 m. Its initial speed was 3.00 m/s. (a) What was its acceleration if it is assumed to have been constant? (b) How long was it in motion? (c) What was its speed after traveling 150 m?
Answer   (a) -2.25×10^{-2} m/s²   (b) 133 s   (c) 1.50 m/s

2.6 • FREELY FALLING OBJECTS

It is well known that all objects, when dropped, fall toward the Earth with nearly constant acceleration in the absence of air resistance. There is a legend that Galileo Galilei first discovered this fact by observing that two different weights dropped simultaneously from the Tower of Pisa hit the ground at approximately the same time. Although there is some doubt that this particular experiment was carried out, it is well established that Galileo did perform many systematic experiments on objects moving on inclined planes. Through careful measurements of distances and time intervals, he was able to show that the displacement of an object starting from rest is proportional to the square of the time the object is in motion. This observation is consistent with one of the kinematic equations we derived for motion under constant acceleration (Eq. 2.10). Galileo's achievements in the science of mechanics paved the way for Newton in his development of the laws of motion.

You might want to try the following experiment. Drop a coin and a crumpled-up piece of paper simultaneously from the same height. In the absence of air resistance, the two experience the same motion and hit the floor at the same time. In a real (nonideal) experiment, air resistance cannot be neglected. In the idealized case, however, where air resistance *is* neglected, such motion is referred to as *free*

Galileo Galilei (1564–1642).

fall. If this same experiment could be conducted in a good vacuum, where air friction is truly negligible, the paper and coin would fall with the same acceleration, regardless of the shape or weight of the paper. This point is illustrated convincingly on page 29, in the photograph of the apple and feather falling in a vacuum. On August 2, 1971, such an experiment was conducted on the Moon by astronaut David Scott. He simultaneously released a geologist's hammer and a falcon's feather, and in unison they fell to the lunar surface. This demonstration surely would have pleased Galileo!

We shall denote the free-fall acceleration with the symbol **g**. The magnitude of **g** decreases with increasing altitude. Furthermore, there are slight variations in the magnitude of **g** with latitude. However, at the surface of the Earth the magnitude of **g** is approximately 9.80 m/s^2, or 980 cm/s^2, or 32 ft/s^2. Unless stated otherwise, we shall use the value 9.80 m/s^2 when doing calculations. Furthermore, we shall assume that the vector **g** is directed downward toward the center of the Earth.

When we use the expression *freely falling object*, we do not necessarily mean an object dropped from rest.

> A freely falling object is an object moving freely under the influence of gravity only, regardless of its initial motion. Objects thrown upward or downward and those released from rest are all falling freely once they are released.

It is important to emphasize that any freely falling object experiences an acceleration directed downward, as shown in the multiflash photograph of a falling billiard

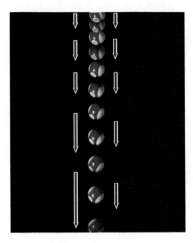

Figure 2.11 A multiflash photograph of a falling billiard ball. As the ball falls, the spacing between successive images increases, indicating that the ball accelerates downward. The motion diagram shows that the ball's velocity (red arrows) increases with time, whereas its acceleration (violet arrows) remains constant. (© *Richard Megna 1990, Fundamental Photographs*)

ball (Fig. 2.11). This is true regardless of the initial motion of the object. An object thrown upward (or downward) experiences the same acceleration as an object released from rest. **Once they are in free fall, all objects have a downward acceleration equal to the free-fall acceleration.**

If we neglect air resistance and assume that the gravitational acceleration does not vary with altitude, then a freely falling object moves in one dimension with constant acceleration. Therefore, the equations developed in Section 2.5 for objects moving with constant acceleration can be applied. The only modification that we need to make in these equations for freely falling objects is to note that the motion is in the vertical direction (the y direction) rather than the horizontal direction (the x direction) and that the acceleration is downward and has a magnitude of 9.80 m/s^2. Thus, for a freely falling object we commonly take $a_y = -g = -9.80 \text{ m/s}^2$, where the minus sign means that the acceleration of the object is downward.

Thinking Physics 3

A skydiver jumps out of a helicopter. A few seconds later, another skydiver jumps out, so that they both fall along the same vertical line. Ignore air resistance, so that both skydivers fall with the same acceleration. Does the vertical distance between them stay the same? Does the difference in their velocities stay the same? If they were connected by a long bungee cord, would the tension in the cord become greater, less, or stay the same?

Reasoning At any given instant of time, the velocities of the jumpers are definitely different, because one had a head start. In a time interval after this instant, however, each jumper increases his or her velocity by the same amount, because they have the same acceleration. Thus, the difference in velocities remains the same. The first jumper will always be moving with a higher velocity than the second. Thus, in a given time interval, the first jumper will cover more distance than the second. Thus, the separation distance between them increases. As a result, once the distance between the divers reaches the value at which the bungee cord becomes straight, the tension in the bungee cord will increase as the jumpers move apart. Of course, the bungee cord will pull downward on the second jumper and upward on the first, resulting in a force in addition to gravity, which will affect their motion. If the bungee cord is robust enough, this will result in the separation distance of the divers becoming smaller as the cord pulls them together.

CONCEPTUAL PROBLEM 4

A child throws a ball into the air with some initial velocity. Another child drops a toy at the same instant. Compare the accelerations of the two objects while they are in the air.

CONCEPTUAL PROBLEM 5

A ball is thrown upward. While the ball is in the air, (a) what happens to its velocity? (b) Does its acceleration increase, decrease, or remain constant?

Example 2.6   **Try to Catch the Dollar**

Emily challenges David to catch a dollar bill as follows. She holds the bill vertically, as in Figure 2.12, with the center of the bill between David's index finger and thumb. David must catch the bill after Emily releases it without moving his hand downward. Who would you bet on?

Reasoning   Place your bets on Emily. There is a time delay between the instant Emily releases the bill and the time David reacts and closes his fingers. The reaction time of most people is at best about 0.2 s. Because the bill is in free fall and undergoes a downward acceleration of 9.80 m/s², in 0.2 s it falls a distance of $\frac{1}{2}gt^2 \cong 0.2$ m $= 20$ cm. This distance is about twice the distance between the center of the bill and its top edge ($\cong 8$ cm). Thus, David will be unsuccessful. You might want to try this on one of your friends.

Figure 2.12   (Example 2.6)

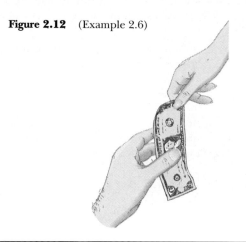

Example 2.7   **Not a Bad Throw for a Rookie!**

A stone is thrown from the top of a building with an initial velocity of 20.0 m/s straight upward. The building is 50.0 m high, and the stone just misses the edge of the roof on its way down, as in Figure 2.13. Determine (a) the time needed for the stone to reach its maximum height, (b) the maximum height, (c) the time needed for the stone to return to the level of the thrower, (d) the velocity of the stone at this instant, and (e) the velocity and position of the stone at $t = 5.00$ s.

Solution   (a) To find the time necessary to reach the maximum height, use Equation 2.7, $v_y = v_{y0} + a_y t$, noting that $v_y = 0$ at maximum height:

$$20.0 \text{ m/s} + (-9.80 \text{ m/s}^2)t_1 = 0$$

$$t_1 = \frac{20.0 \text{ m/s}}{9.80 \text{ m/s}^2} = \boxed{2.04 \text{ s}}$$

(b) This value of time can be substituted into Equation 2.10, $y = v_{y0}t + \frac{1}{2}a_y t^2$, to give the maximum height measured from the position of the thrower:

$$y_{max} = (20.0 \text{ m/s})(2.04 \text{ s}) + \frac{1}{2}(-9.80 \text{ m/s}^2)(2.04 \text{ s})^2$$

$$= \boxed{20.4 \text{ m}}$$

(c) When the stone is back at the height of the thrower, the y coordinate is zero. From the expression $y = v_{y0}t + \frac{1}{2}a_y t^2$ (Eq. 2.10), with $y = 0$, we obtain the expression

$$20.0t - 4.90t^2 = 0$$

This is a quadratic equation and has two solutions for t. The equation can be factored to give

$$t(20.0 - 4.90t) = 0$$

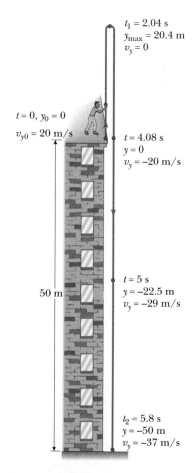

$t_1 = 2.04$ s
$y_{max} = 20.4$ m
$v_y = 0$

$t = 0, y_0 = 0$
$v_{y0} = 20$ m/s

$t = 4.08$ s
$y = 0$
$v_y = -20$ m/s

50 m

$t = 5$ s
$y = -22.5$ m
$v_y = -29$ m/s

$t_2 = 5.8$ s
$y = -50$ m
$v_y = -37$ m/s

Figure 2.13   (Example 2.7) Position and velocity versus time for a freely falling particle initially thrown upward with a velocity $v_{y0} = 20$ m/s. (Not drawn to scale.)

One solution is $t = 0$, corresponding to the time the stone starts its motion. The other solution—the one we are after—is $t = 4.08$ s.

(d) The value for t found in (c) can be inserted into $v_y = v_{y0} + a_y t$ (Eq. 2.7) to give

$$v_y = 20.0 \text{ m/s} + (-9.80 \text{ m/s}^2)(4.08 \text{ s}) = \boxed{-20.0 \text{ m/s}}$$

Note that the velocity of the stone when it arrives back at its original height is equal in magnitude to its initial velocity but opposite in direction. This indicates that the motion is symmetric.

(e) From $v_y = v_{y0} + a_y t$ (Eq. 2.7), the velocity after 5.00 s is

$$v_y = 20 \text{ m/s} + (-9.80 \text{ m/s}^2)(5.00 \text{ s}) = \boxed{-29.0 \text{ m/s}}$$

We can use $y = v_{y0} t + \frac{1}{2} a_y t^2$ (Eq. 2.10) to find the position of the particle at $t = 5.00$ s:

$$y = (20.0 \text{ m/s})(5.00 \text{ s}) + \frac{1}{2}(-9.80 \text{ m/s}^2)(5.00 \text{ s})^2$$

$$= \boxed{-22.5 \text{ m}}$$

EXERCISE 5   Find (a) the velocity of the stone just before it hits the ground and (b) the total time the stone is in the air.
Answer   (a) -37.1 m/s   (b) 5.83 s

SUMMARY

The **average velocity** of a particle during some time interval is equal to the ratio of the displacement vector, $\Delta \mathbf{x}$, and the time interval, Δt:

$$\bar{\mathbf{v}}_x \equiv \frac{\Delta \mathbf{x}}{\Delta t} \tag{2.1}$$

The **instantaneous velocity** of a particle is defined as the limit of the ratio $\Delta \mathbf{x}/\Delta t$ as Δt approaches zero:

$$\mathbf{v}_x \equiv \lim_{\Delta t \to 0} \frac{\Delta \mathbf{x}}{\Delta t} = \frac{d\mathbf{x}}{dt} \tag{2.3}$$

The **speed** of a particle is defined as the magnitude of the instantaneous velocity vector.

The **average acceleration** of a particle during some time interval is defined as the ratio of the change in its velocity, $\Delta \mathbf{v}_x$, and the time interval, Δt:

$$\bar{\mathbf{a}}_x \equiv \frac{\Delta \mathbf{v}_x}{\Delta t} \tag{2.4}$$

The **instantaneous acceleration** is equal to the limit of the ratio $\Delta \mathbf{v}_x/\Delta t$ as $\Delta t \to 0$. By definition, this equals the derivative of $\mathbf{v}_x$ with respect to t or the time rate of change of the velocity:

$$\mathbf{a}_x \equiv \lim_{\Delta t \to 0} \frac{\Delta \mathbf{v}_x}{\Delta t} = \frac{d\mathbf{v}_x}{dt} \tag{2.5}$$

The slope of the tangent to the x versus t curve at any instant gives the instantaneous velocity of the particle. The slope of the tangent to the v versus t curve gives the instantaneous acceleration of the particle.

The **equations of kinematics** for a particle moving along the x axis with uniform acceleration a (constant in magnitude and direction) are

$$v_x = v_{x0} + a_x t \tag{2.7}$$

$$x - x_0 = \frac{1}{2}(v_{x0} + v_x)t \tag{2.9}$$

$$x - x_0 = v_{x0} t + \frac{1}{2} a_x t^2 \tag{2.10}$$

$$v_x^2 = v_{x0}^2 + 2a_x(x - x_0) \tag{2.11}$$

(constant a_x only)

A body falling freely experiences an acceleration directed toward the center of the Earth. If air friction is neglected, and if the altitude of the motion is small compared with the Earth's radius, then one can assume that the free-fall acceleration, **g**, is constant over the range of motion, where g is equal to 9.80 m/s², or 32 ft/s². Assuming y to be positive upward, the acceleration is given by $-g$, and the equations of kinematics for a body in free fall are the same as those already given, with the substitutions $x \rightarrow y$ and $a_y \rightarrow -g$.

CONCEPTUAL QUESTIONS

1. Are there any conditions under which the average velocity of a particle moving in one dimension is larger than the average speed?
2. The average velocity for a particle moving in one dimension has a positive value. Is it possible for the instantaneous velocity of the particle to have been negative at any time in the interval? Suppose the particle started at the origin, $x = 0$. If the average velocity is positive, could the particle ever have been in the $-x$ region?
3. If the average velocity of an object is zero in some time interval, what can you say about the displacement of the object for that interval?
4. Can the instantaneous velocity of an object ever be greater in magnitude than the average velocity? Can it ever be less?
5. Consider the following combinations of signs and values for velocity and acceleration of a particle with respect to a one-dimensional x axis:

Velocity	Acceleration
a. Positive	Positive
b. Positive	Negative
c. Positive	Zero
d. Negative	Positive
e. Negative	Negative
f. Negative	Zero
g. Zero	Positive
h. Zero	Negative

Describe what a particle is doing in each case, and give a real-life example for an automobile on an east–west one-dimensional axis, with east considered the positive direction.
6. Can the equations of kinematics (Eqs. 2.7–2.11) be used in a situation in which the acceleration varies in time? Can they be used when the acceleration is zero?
7. A student at the top of a building of height h throws one ball upward with an initial speed v_0 and then throws a second ball downward with the same initial speed. How do the final velocities of the balls compare when they reach the ground?
8. Two cars are moving in the same direction in parallel lanes along a highway. At some instant, the velocity of car A exceeds the velocity of car B. Does this mean that the acceleration of A is greater than that of B? Explain.
9. Car A traveling south from New York to Miami has a speed of 25 m/s. Car B traveling west from New York to Chicago also has a speed of 25 m/s. Are their velocities equal? Explain.
10. You drop a ball from a window on an upper floor of a building. It strikes the ground with velocity v. You now repeat the drop, but you have a friend down on the street who throws another ball upward at velocity v. Your friend throws the ball upward at exactly the same time that you drop yours from the window. At some location, the balls pass each other. Is this location *at* the halfway point between window and ground, *above* this point, or *below* this point?
11. Galileo experimented with balls rolling down inclined planes in order to reduce the acceleration along the plane and thus reduce the rate of descent of the balls. Suppose the angle that the incline makes with the horizontal is θ. How would you expect the acceleration along the plane to decrease as θ decreases? What specific trigonometric dependence on θ would you expect for the acceleration?
12. A ball rolls in a straight line along the horizontal direction. Using motion diagrams (or multiflash photographs) as in Figure 2.9, describe the velocity and acceleration of the ball for each of the following situations: (a) The ball moves to the right at a constant speed. (b) The ball moves from right to left and continually slows down. (c) The ball moves from right to left and continually speeds up. (d) The ball moves to the right, first speeding up at a constant rate and then slowing down at a constant rate.

Boxed numbers indicate questions that have answers available in the Student Solutions Manual and Study Guide.

Figure Q2.14   See Question 14.

13. A rapidly growing plant doubles in height each week. At the end of the 25th day, the plant reaches the height of the building. At what time was the plant one fourth the height of the building?

14. A pebble is dropped into a water well, and the splash is heard 16 s later, as illustrated in the cartoon strip. What is the *approximate* distance from the rim of the well to the water's surface?

15. The motion of the Earth's crustal plates is described by a model referred to as *plate tectonic motion*. Measurements indicate that coastal portions of southern California have northward plate tectonic motions of 2.5 cm per year. Estimate the time it would take for this motion to carry southern California to Alaska.

16. A heavy object falls from a height h under the influence of gravity. It is released at $t = 0$ and strikes the ground at time t. Ignore air resistance. (a) When the object is at height $0.5\,h$, is the time earlier than $0.5\,t$, later than $0.5\,t$, or equal to $0.5\,t$? (b) When the time is $0.5\,t$, is the height of the object greater than $0.5\,h$, less than $0.5\,h$, or equal to $0.5\,h$?

PROBLEMS

Section 2.1   Average Velocity

1. The position of a pinewood derby car was observed at various times; the results are summarized in the table. Find the average velocity of the car for (a) the first second, (b) the last 3 seconds, and (c) the entire period of observation.

x (m)	0	2.3	9.2	20.7	36.8	57.5
t (s)	0	1.0	2.0	3.0	4.0	5.0

2. A motorist drives north for 35.0 minutes at 85.0 km/h and then stops for 15.0 minutes. He then continues north, traveling 130 km in 2.00 h. (a) What is his total displacement? (b) What is his average velocity?

3. The displacement versus time for a certain particle moving along the x axis is shown in Figure P2.3. Find the average velocity in the time intervals (a) 0 to 2 s, (b) 0 to 4 s, (c) 2 s to 4 s, (d) 4 s to 7 s, (e) 0 to 8 s.

4. A person walks first at a constant speed of 5.00 m/s along a straight line from point A to point B and then back along the line from B to A at a constant speed of 3.00 m/s. (a) What is her average speed over the entire trip? (b) Her average velocity over the entire trip?

5. A person first walks at a constant speed v_1 along a straight line from A to B and then back along the line from B to A at a constant speed v_2. (a) What is her average speed over the entire trip? (b) Her average velocity over the entire trip?

Boxed numbers indicate problems that have full solutions available in the Student Solutions Manual and Study Guide. Cyan numbers indicate intermediate level problems and magenta numbers indicate more challenging problems.

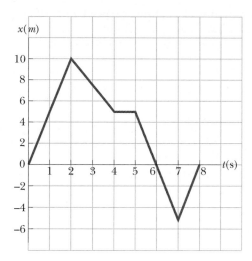

Figure P2.3

Section 2.2 Instantaneous Velocity

6. At $t = 1.00$ s, a particle moving with constant velocity is located at $x = -3.00$ m, and at $t = 6.00$ s the particle is located at $x = 5.00$ m. (a) From this information, plot the position as a function of time. (b) Determine the velocity of the particle from the slope of this graph.

7. A position–time graph for a particle moving along the x axis is shown in Figure P2.7. (a) Find the average velocity in the time interval $t = 1.5$ s to $t = 4.0$ s. (b) Determine the instantaneous velocity at $t = 2.0$ s by measuring the slope of the tangent line shown in the graph. (c) At what value of t is the velocity zero?

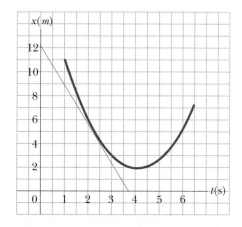

Figure P2.7

8. (a) Use the data in Problem 1 to construct a smooth graph of position versus time. (b) By constructing tangents to the $x(t)$ curve, find the instantaneous velocity of the car at several instants. (c) Plot the instantaneous velocity versus time and, from this, determine the average acceleration of the car. (d) What was the initial velocity of the car?

9. Find the instantaneous velocity of the particle described in Figure P2.3 at the following times: (a) $t = 1.0$ s, (b) $t = 3.0$ s, (c) $t = 4.5$ s, and (d) $t = 7.5$ s.

Section 2.3 Acceleration

10. A particle is moving with a velocity $\mathbf{v}_0 = 60.0\mathbf{i}$ m/s at $t = 0$. Between $t = 0$ and $t = 15.0$ s the velocity decreases uniformly to zero. What was the acceleration during this 15.0-s interval? What is the significance of the sign of your answer?

11. A 50-g superball traveling at 25.0 m/s bounces off a brick wall and rebounds at 22.0 m/s. A high-speed camera records this event. If the ball is in contact with the wall for 3.50 ms, what is the magnitude of the average acceleration of the ball during this time interval? (*Note:* 1 ms = 10^{-3} s.)

12. A velocity–time graph for an object moving along the x axis is shown in Figure P2.12. (a) Plot a graph of the acceleration versus time. (b) Determine the average acceleration of the object in the time intervals $t = 5$ s to $t = 15$ s and $t = 0$ to $t = 20$ s.

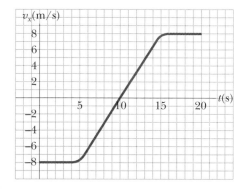

Figure P2.12

13. A particle moves along the x axis according to the equation $x = 2.0 + 3.0t - t^2$, where x is in meters and t is in seconds. At $t = 3.00$ s, find (a) the position of the particle, (b) its velocity, and (c) its acceleration.

14. When struck by a club, a golf ball initially at rest acquires a speed of 31.0 m/s. If the ball is in contact with the club for 1.17 ms, what is the magnitude of the average acceleration of the ball?

15. Figure P2.15 shows a graph of v versus t for the motion of a motorcyclist as he starts from rest and moves along the road in a straight line. (a) Find the average acceleration for the time interval $t_0 = 0$ to $t_1 = 6.0$ s. (b) Estimate the time at which the acceleration has its greatest positive value and the value of the acceleration at that instant. (c) When is the acceleration zero? (d) Estimate the maximum negative value of the acceleration and the time at which it occurs.

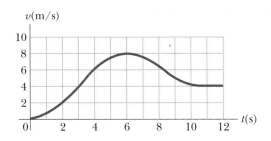

Figure P2.15

Section 2.4   Motion Diagrams

Section 2.5   One-Dimensional Motion with Constant Acceleration

Section 2.6   Freely Falling Objects

16. A truck travels 40.0 m in 8.50 s while uniformly slowing down to a final speed of 2.80 m/s. Find (a) its original speed and (b) its acceleration.

17. A body moving with uniform acceleration has a velocity of 12.0 cm/s in the positive x direction when its x coordinate is 3.00 cm. If its x coordinate 2.00 s later is −5.00 cm, what is the magnitude of its acceleration?

18. The new BMW M3 can accelerate from 0 to 60.0 mi/h in 5.00 s. (a) What is this acceleration in m/s²? (b) How long does it take this car to go from 60.0 mi/h to 130 mi/h?

19. The minimum distance required to stop a car moving at 35.0 mi/h is 40.0 ft. What is the minimum stopping distance for the same car moving at 70.0 mi/h, assuming the same rate of acceleration?

20. A drag racer starts her car from rest and accelerates at 10.0 m/s² for the entire distance of 400 m ($\frac{1}{4}$ mile). (a) How long did it take the race car to travel this distance? (b) What is the speed of the race car at the end of the run?

21. A jet plane lands with a speed of 100 m/s and can accelerate at a maximum rate of −5.00 m/s² as it comes to rest. (a) From the instant the plane touches the runway, what is the minimum time needed before it can come to rest? (b) Can this plane land on a small tropical island airport where the runway is 0.800 km long?

22. A locomotive slows from a speed of 26.0 m/s to 0 in 18.0 s. What distance does it travel?

23. The driver of a car slams on the brakes when he sees a tree blocking the road. The car slows uniformly with acceleration −5.60 m/s² for 4.20 s, making straight skid marks 62.4 m long. With what speed does the car then strike the tree?

24. *Help! One of our equations is missing!* We describe motion under constant acceleration with the parameters v_x, v_{x_0}, a_x, t, and $x - x_0$. Of the equations listed in Table 2.1, the first does not involve $x - x_0$; the second does not contain a_x; the third omits v_x; and the last does not contain t. To complete the set, there should be an equation not involving v_{x_0}. Derive this equation from the others. Use it to solve Problem 23 in one step.

25. Until recently, the world's land-speed record was held by Col. John P. Stapp, USAF. On March 19, 1954, he rode a rocket-propelled sled that moved down a track at 632 mi/h. He and the sled were safely brought to rest in 1.40 s. Determine (a) the negative acceleration he experienced and (b) the distance he traveled during this negative acceleration.

26. An electron in a cathode-ray-tube (CRT) accelerates from 2.00×10^4 m/s to 6.00×10^6 m/s over 1.50 cm. (a) How long does the electron take to travel this 1.50 cm? (b) What is its acceleration?

27. A student throws a set of keys vertically upward to her sorority sister, who is in a window 4.00 m above. The keys are caught 1.50 s later by the sister's outstretched hand. (a) With what initial velocity were the keys thrown? (b) What was the velocity of the keys just before they were caught?

28. A ball is thrown directly downward, with an initial speed of 8.00 m/s, from a height of 30.0 m. After what interval does the ball strike the ground?

29. A baseball is hit so that it travels straight upward after being struck by the bat. A fan observes that it takes 3.00 s for the ball to reach its maximum height. Find (a) its initial velocity and (b) the height reached by the ball. Ignore the effects of air resistance.

30. A hot-air balloon is traveling vertically upward at a constant speed of 5.00 m/s. When it is 21.0 m above the ground, a package is released from the balloon. (a) For how long after being released is the package in the air? (b) What is the velocity of the package just before impact with the ground? (c) Repeat (a) and (b) for the case of the balloon descending at 5.00 m/s.

31. A daring stunt woman sitting on a tree limb wishes to drop vertically onto a horse galloping under the tree. The speed of the horse is 10.0 m/s, and the distance from the limb to the saddle is 3.00 m. (a) What must be the horizontal distance between the saddle and limb when the woman makes her move? (b) How long is she in the air?

Figure P2.30 Hot air balloon over Carefree, Arizona. *(Russell Schlepman)*

32. A ball is thrown vertically upward from the ground with an initial speed of 15.0 m/s. (a) How long does it take the ball to reach its maximum altitude? (b) Find its maximum altitude. (c) Determine the velocity and acceleration of the ball at $t = 2.00$ s.

33. A ball accelerates at 0.500 m/s² while moving down an inclined plane 9.00 m long. When it reaches the bottom, the ball rolls up another plane, where, after moving 15.0 m, it comes to rest. (a) What is the speed of the ball at the bottom of the first plane? (b) How long does it take to roll down the first plane? (c) What is the acceleration along the second plane? (d) What is the ball's speed 8.00 m along the second plane?

34. The height of a helicopter above the ground is given by $h = 3.00t^3$, where h is in meters and t is in seconds. After 2.00 s, the helicopter releases a small mailbag. How long after its release does the mailbag reach the ground?

35. Speedy Sue, driving at 30.0 m/s enters a one-lane tunnel. She then observes a slow-moving van 155 m ahead traveling at 5.00 m/s. Sue applies her brakes but can decelerate only at 2.00 m/s² because the road is wet. Will there be a collision? If yes, determine how far into the tunnel and at what time the collision occurs. If no, determine the distance of closest approach between Sue's car and the van.

Additional Problems

36. A motorist is traveling at 18.0 m/s when he sees a deer in the road 38.0 m ahead. (a) If the maximum negative acceleration of the vehicle is -4.50 m/s², what is the maximum reaction time, Δt, of the motorist that will allow him to avoid hitting the deer? (b) If his reaction time is 0.300 s, how fast will he be traveling when he hits the deer?

37. Another scheme to catch the roadrunner has failed and a safe falls from rest from the top of a 25.0-m high cliff toward Wile E. Coyote, who is standing at the base. Wile first notices the safe after it has fallen 15.0 m. How long does he have to get out of the way?

38. A rocket is fired vertically upward with an initial velocity of 80.0 m/s. It accelerates upward at 4.00 m/s² until it reaches an altitude of 1000 m. At that point, its engines fail and the rocket goes into free flight, with an acceleration of -9.80 m/s². (a) How long is the rocket in motion? (b) What is its maximum altitude? (c) What is its velocity just before it collides with the Earth? (*Hint:* Consider the motion while the engine is operating separate from the free-fall motion.)

39. In a 100-m race, Maggie and Judy cross the finish line in a dead heat, both taking 10.2 s. They both accelerated uniformly, and Maggie took 2.00 s and Judy 3.00 s to attain maximum speed, which they maintained for the rest of the race. (a) What was the acceleration of each sprinter? (b) What were their respective maximum speeds? (c) Which sprinter was ahead at the 6.00-s mark and by how much?

40. The position of a softball tossed vertically upward is described by the equation $y = 7.00t - 4.90t^2$, where y is in meters and t in seconds. Find (a) the initial speed v_0 at $t_0 = 0$, (b) the velocity at $t = 1.26$ s, and (c) the acceleration of the ball.

41. A hard rubber ball, released at chest height, falls to the pavement and bounces back to nearly the same height. When it is in contact with the pavement, the lower side of the ball is temporarily flattened. Before this dent in the ball pops out, suppose that its maximum depth is about 1 centimeter. Make an order-of-magnitude estimate of the maximum acceleration of the ball. State your assumptions, the quantities you estimate, and the values you estimate for them.

42. A teenager has a car that accelerates at 3.00 m/s² and decelerates at -4.50 m/s². On a trip to the store, he accelerates from rest to 12.0 m/s, drives at a constant speed for 5.00 s, and then comes to a momentary stop at the corner. He then accelerates to 18.0 m/s, drives at a constant speed for 20.0 s, decelerates for 2.67 s; he continues for 4.00 s at this speed and then comes to a stop. (a) How long does the trip take? (b) How far has he traveled? (c) What is his average speed for the trip? (d) How long would it take to walk to the store and back if he walks at 1.50 m/s?

43. An inquisitive physics student and mountain climber climbs a 50.0-m cliff that overhangs a calm pool of water. He throws two stones vertically downward, 1.00 s apart, and observes that they cause a single splash. The first stone has an initial speed of 2.00 m/s. (a) How long after release of the first stone do the two stones hit the water? (b) What initial velocity must the second stone have if they are to hit simultaneously? (c) What is the speed of each stone at the instant the two hit the water?

44. In the first hour of travel, a train moves with a speed v, in the next half hour it has a speed $3v$, in the next 90.0 min it travels with a speed $v/2$, and in the final 2.00 h it travels with a speed $v/3$. (a) Plot the speed–time graph for this trip. (b) How far does the train travel in this trip? (c) What is the average speed of the train over the entire trip?

45. A rock is dropped from rest into a well. (a) If the sound of the splash is heard 2.40 s later, how far below the top of the well is the surface of the water? The speed of sound in air (for the ambient temperature) is 336 m/s. (b) If the travel time for the sound is neglected, what percentage error is introduced when the depth of the well is calculated?

46. A car and train move together along parallel paths at 25.0 m/s. The car then undergoes a uniform acceleration of -2.50 m/s^2 because of a red light and comes to rest. It remains at rest for 45.0 s, then accelerates back to a speed of 25.0 m/s at a rate of 2.50 m/s^2. How far behind the train is the car when it reaches the speed of 25.0 m/s, assuming that the speed of the train has remained 25.0 m/s?

47. Kathy Kool buys a sports car that can accelerate at the rate of 4.90 m/s^2. She decides to test the car by dragging with another speedster, Stan Speedy. Both start from rest, but experienced Stan leaves the starting line 1.00 s before Kathy. If Stan moves with a constant acceleration of 3.50 m/s^2 and Kathy maintains an acceleration of 4.90 m/s^2, find (a) the time it takes Kathy to overtake Stan, (b) the distance she travels before she catches him, and (c) the speeds of both cars at the instant she overtakes him.

48. In a 100-m linear accelerator, an electron is accelerated to 1.00% the speed of light in 40.0 m before it coasts 60.0 m to a target. (a) What is the electron's acceleration during the first 40.0 m? (b) How long does the total flight take?

49. In order to protect his food from hungry bears, a Boy Scout raises his food pack, of mass m, with a rope that is thrown over a tree limb at height h above his hands. He walks away from the vertical rope with constant velocity v_0, holding the free end of the rope in his hands (Fig. P2.49). (a) Show that the velocity v of the food pack is $x(x^2 + h^2)^{-1/2}v_0$ where x is the distance he has walked away from the vertical rope. (b) Show that the acceleration a of the food pack is $h^2(x^2 + h^2)^{-3/2}v_0^2$. (c) What values do the acceleration and velocity v have shortly after the Boy Scout leaves the vertical rope? (d) What values do the velocity and acceleration approach as the distance x continues to increase?

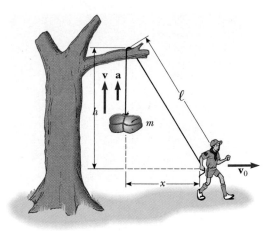

Figure P2.49

50. In Problem 49, let the height, h, equal 6.00 m and the speed, v_0, equal 2.00 m/s. Assume that the food pack starts from rest. (a) Tabulate and graph the speed–time graph. (b) Tabulate and graph the acceleration–time graph. (Let the range of time be from 0 to 5.00 s and the time intervals be 0.50 s.)

51. Two objects, A and B, are connected by a rigid rod that has a length L. The objects slide along perpendicular guide rails, as shown in Figure P2.51. If A slides to the left with a constant speed v, find the velocity of B when $\alpha = 60.0°$.

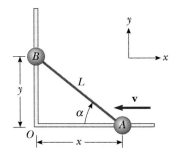

Figure P2.51

▣ Spreadsheet Problems

S1. Use Spreadsheet 2.1 to plot position and velocity as functions of time for one object traveling at constant velocity and another traveling with constant acceleration. Choose a variety of velocity and acceleration values, and view the graphs. Be sure to investigate zero and negative values. *Note:* There are two graphs associated with Spreadsheet 2.1, one for position versus time and one for velocity versus time.

S2. Spreadsheet 2.2 models the sport of drag racing. Two dragsters can have different accelerations *A*1 and *A*2 as well as different maximum speeds *V*1 and *V*2. Both cars start from rest at the same starting position. However, a time delay *t'* in starting times may be introduced if the cars are quite different. Enter the following data:

	Acceleration (ft/s^2)	Maximum Speed (ft/s)	Delay Time *t'* (s)
Car 1	5	300	—
Car 2	6	250	0.5

(a) Which car wins a 1/4-mile race? (b) If the acceleration of Car 1 is increased to 5.2 ft/s^2, which car now wins?

S3. Modify Spreadsheet 2.1 to solve this problem. The police have set up a "speed trap" on the highway. From a police car hidden behind a billboard, an officer with a radar gun measures a motorist's speed to be 35.0 m/s. Three seconds later, she alerts her partner, who is in another police car 100 m down the road. The second police car starts from rest accelerating at 2.00 m/s^2 in pursuit of the speeder 2.00 s after receiving the alert. (a) How much time elapses before the speeder is overtaken? (b) What is the police car's speed when it overtakes the speeder? (c) How far does the second police car travel before overtaking the speeder?

ANSWERS TO CONCEPTUAL PROBLEMS

1. If the particle moves along a line without changing direction, the displacement and distance over any time interval will be the same. As a result, the magnitude of the average velocity and the average speed will be the same. If the particle reverses direction, however, then the displacement will be less than the distance. In turn, the magnitude of the average velocity will be smaller than the average speed.

2. Yes. This occurs when a car is slowing down, so that the direction of its acceleration is opposite to its direction of motion.

3. For an accelerating particle to stop at all, the velocity and acceleration must have opposite signs, so that the speed is decreasing. Given that this is the case, the particle will eventually come to rest. If the acceleration remains constant, however, the particle must begin to move again, opposite to the direction of its original motion. If the particle comes to rest and then stays at rest, the acceleration has disappeared at the moment the motion stops. This is the case for a braking car—the acceleration is negative and goes to zero as the car comes to rest.

4. Once the objects leave the hand, both are freely falling objects, and both experience the same downward acceleration equal to the free-fall acceleration of magnitude 9.80 m/s^2.

5. (a) As it travels upward, its speed decreases by 9.80 m/s during each second of its motion. When it reaches the peak of its motion, its speed becomes zero. As the ball moves downward, its speed increases by 9.80 m/s each second. (b) The acceleration of the ball remains constant while it is in the air. Its magnitude of its acceleration is the free-fall acceleration, $g = 9.80$ m/s^2.

3

Motion in Two Dimensions

CHAPTER OUTLINE

3.1 The Displacement, Velocity, and Acceleration Vectors

3.2 Two-Dimensional Motion with Constant Acceleration

3.3 Projectile Motion

3.4 Uniform Circular Motion

3.5 Tangential and Radial Acceleration

In this chapter we deal with the kinematics of a particle moving in a plane, which is two-dimensional motion. Some common examples of motion in a plane are the motion of projectiles and satellites and the motion of charged particles in uniform electric fields. We begin by showing that displacement, velocity, and acceleration are vector quantities. As in the case of one-dimensional motion, we derive the kinematic equations for two-dimensional motion from the fundamental definitions of displacement,

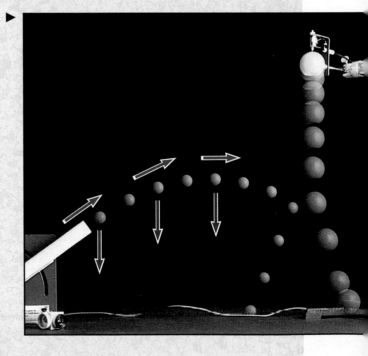

A multiflash photograph of a popular lecture demonstration in which a projectile is fired at a target that is being held by a magnet in the device at the top right of the photograph. The conditions of the experiment are that the gun is aimed at the target and the projectile leaves the gun at the instant the target is released from rest. Under these conditions the projectile will hit the target, independent of the initial speed of the projectile. The reason is that they both experience the same downward acceleration, and hence the velocities of the projectile and target change by the same amount in the same time interval. Note that the velocity of the projectile (red arrows) changes in direction and magnitude, whereas its downward acceleration (violet arrows) remains constant. (Central Scientific Company)

velocity, and acceleration. As special cases of motion in two dimensions, we then treat constant-acceleration motion in a plane and uniform circular motion.

3.1 • THE DISPLACEMENT, VELOCITY, AND ACCELERATION VECTORS

In Chapter 2 we found that the motion of a particle moving along a straight line is completely known if its coordinate is known as a function of time. Now let us extend this idea to motion in the xy plane. We begin by describing the position of a particle with a *position vector* **r**, drawn from the origin of some reference frame to the particle located in the xy plane, as in Figure 3.1. At time t_i, the particle is at the point P, and at some later time, t_f, the particle is at Q, where the indices i and f refer to initial and final values. As the particle moves from P to Q in the time interval $\Delta t = t_f - t_i$, the position vector changes from $\mathbf{r}_i$ to $\mathbf{r}_f$. As we learned in Chapter 2, the displacement of a particle is the difference between its final position and initial position. Therefore, the **displacement vector** for the particle of Figure 3.1 equals the difference between its final position vector and its initial position vector:

$$\Delta \mathbf{r} \equiv \mathbf{r}_f - \mathbf{r}_i \qquad [3.1]$$

The direction of $\Delta\mathbf{r}$ is indicated in Figure 3.1. As we see from the figure, the magnitude of the displacement vector is *less* than the distance traveled along the curved path.

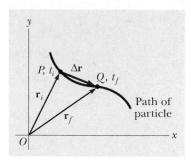

Figure 3.1 A particle moving in the xy plane is located with the position vector **r** drawn from the origin to the particle. The displacement of the particle as it moves from P to Q in the time interval $\Delta t = t_f - t_i$ is equal to the vector $\Delta\mathbf{r} = \mathbf{r}_f - \mathbf{r}_i$.

• *Definition of the displacement vector*

We define the **average velocity** of the particle during the time interval Δt as the ratio of the displacement to that time interval:

$$\bar{\mathbf{v}} \equiv \frac{\Delta \mathbf{r}}{\Delta t} \qquad [3.2]$$

• *Definition of average velocity*

Because displacement is a vector quantity and the time interval is a scalar quantity, we conclude that the average velocity is a *vector* quantity directed along $\Delta\mathbf{r}$. Note that the average velocity between points P and Q is *independent of the path* between the two points. This is because the average velocity is proportional to the displacement, which in turn depends only on the initial and final position vectors and not on the path taken between those two points. As we did with one-dimensional motion, we conclude that if a particle starts its motion at some point and returns to this point via any path, its average velocity is zero for this trip, because its displacement is zero. Consider again the motion of a particle between two points in the xy plane, as in Figure 3.2. As the time intervals over which we observe the motion become smaller and smaller, the direction of the displacement approaches that of the line tangent to the path at the point P.

The **instantaneous velocity, v**, is defined as the limit of the average velocity, $\Delta\mathbf{r}/\Delta t$, as Δt approaches zero:

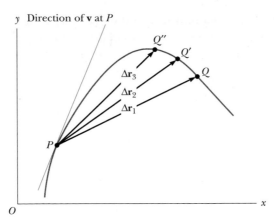

Figure 3.2 As a particle moves between two points, its average velocity is in the direction of the displacement vector $\Delta\mathbf{r}$. As the end point of the path is moved from Q to Q' to Q'', the respective displacements and corresponding time intervals become smaller and smaller. In the limit that the end point approaches P, Δt approaches zero and the direction of $\Delta\mathbf{r}$ approaches that of the line tangent to the curve at P. By definition, the instantaneous velocity at P is in the direction of this tangent line.

Definition of instantaneous •
velocity

$$\mathbf{v} \equiv \lim_{\Delta t \to 0} \frac{\Delta\mathbf{r}}{\Delta t} = \frac{d\mathbf{r}}{dt} \qquad [3.3]$$

That is, the instantaneous velocity equals the derivative of the position vector with respect to time. The direction of the instantaneous velocity vector at any point in a particle's path is along a line that is tangent to the path at that point and in the direction of motion; this is illustrated in Figure 3.3. The magnitude of the instantaneous velocity vector is called the *speed*.

As a particle moves from one point to another along some path, as in Figure 3.3, its instantaneous velocity vector changes from $\mathbf{v}_i$ at time t_i to $\mathbf{v}_f$ at time t_f.

> The **average acceleration** of a particle as it moves from P to Q is defined as the ratio of the change in the instantaneous velocity vector, $\Delta\mathbf{v}$, to the elapsed time, Δt:

Definition of average •
acceleration

$$\bar{\mathbf{a}} \equiv \frac{\mathbf{v}_f - \mathbf{v}_i}{t_f - t_i} = \frac{\Delta\mathbf{v}}{\Delta t} \qquad [3.4]$$

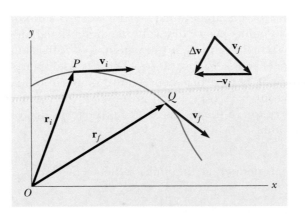

Figure 3.3 The average acceleration vector $\bar{\mathbf{a}}$ for a particle moving from P to Q is in the direction of the change in velocity, $\Delta\mathbf{v} = \mathbf{v}_f - \mathbf{v}_i$.

Because the average acceleration is the ratio of a vector quantity, $\Delta\mathbf{v}$, and a scalar quantity, Δt, we conclude that $\overline{\mathbf{a}}$ is a vector quantity directed along $\Delta\mathbf{v}$. As is indicated in Figure 3.3, the direction of $\Delta\mathbf{v}$ is found by adding the vector $-\mathbf{v}_i$ (the negative of $\mathbf{v}_i$) to the vector $\mathbf{v}_f$, because by definition $\Delta\mathbf{v} = \mathbf{v}_f - \mathbf{v}_i$.

> The **instantaneous acceleration, a,** is defined as the limiting value of the ratio $\Delta\mathbf{v}/\Delta t$ as Δt approaches zero:

$$\mathbf{a} \equiv \lim_{\Delta t \to 0} \frac{\Delta\mathbf{v}}{\Delta t} = \frac{d\mathbf{v}}{dt} \qquad [3.5]$$

• *Definition of instantaneous acceleration*

In other words, the instantaneous acceleration equals the derivative of the velocity vector with respect to time.

It is important to recognize that various changes can cause a particle to accelerate. First, the magnitude of the velocity vector (the speed) may change with time as in straight-line (one-dimensional) motion. Second, only the direction of the velocity vector may change with time as its magnitude (speed) remains constant, as in curved-path (two-dimensional) motion. Third, both the magnitude and the direction of the velocity vector may change.

Thinking Physics 1

The gas pedal in an automobile is called the *accelerator*. Are there any other automobile controls that could also be considered "accelerators"?

Reasoning The gas pedal is called the accelerator because the cultural usage of the word *acceleration* refers to an *increase in speed*. The scientific definition, however, is that acceleration occurs *whenever the velocity changes* in any way. Thus, the *brake pedal* can also be considered an accelerator, because it causes the car to slow down. The *steering wheel* is also an accelerator, because it changes the direction of the velocity vector.

CONCEPTUAL PROBLEM 1

(a) Can an object accelerate if its speed is constant? (b) Can an object accelerate if its velocity is constant?

CONCEPTUAL PROBLEM 2

Can a particle have a constant velocity and varying speed? Can a particle have a constant speed and a varying velocity? Give examples of a positive answer and explain a negative answer.

3.2 • TWO-DIMENSIONAL MOTION WITH CONSTANT ACCELERATION

Let us consider two-dimensional motion during which the acceleration remains constant. That is, we assume that the magnitude and direction of the acceleration remain unchanged during the motion.

The motion of a particle can be determined by its position vector **r**. The position vector for a particle moving in the *xy* plane can be written

$$\mathbf{r} = x\mathbf{i} + y\mathbf{j} \qquad [3.6]$$

where *x*, *y*, and **r** change with time as the particle moves. If the position vector is known, the velocity of the particle can be obtained from Equations 3.3 and 3.6, which give

$$\mathbf{v} = v_x\mathbf{i} + v_y\mathbf{j} \qquad [3.7]$$

Because **a** is assumed constant, its components a_x and a_y are also constants. Therefore, we can apply the equations of kinematics to the *x* and *y* components of the velocity vector. Substituting $v_x = v_{x0} + a_x t$ and $v_y = v_{y0} + a_y t$ into Equation 3.7 gives

$$\mathbf{v} = (v_{x0} + a_x t)\mathbf{i} + (v_{y0} + a_y t)\mathbf{j}$$
$$= (v_{x0}\mathbf{i} + v_{y0}\mathbf{j}) + (a_x\mathbf{i} + a_y\mathbf{j})t$$

Velocity vector as a function •
of time

$$\mathbf{v} = \mathbf{v}_0 + \mathbf{a}t \qquad [3.8]$$

This result states that the velocity of a particle at some time *t* equals the vector sum of its initial velocity, $\mathbf{v}_0$, and the additional velocity $\mathbf{a}t$ acquired in the time *t* as a result of its constant acceleration.

In a similar way, from Equation 2.9 we know that the *x* and *y* coordinates of a particle moving with constant acceleration are

$$x = x_0 + v_{x0}t + \tfrac{1}{2}a_x t^2 \qquad \text{and} \qquad y = y_0 + v_{y0}t + \tfrac{1}{2}a_y t^2$$

Substituting these expressions into Equation 3.6 gives

$$\mathbf{r} = (x_0 + v_{x0}t + \tfrac{1}{2}a_x t^2)\mathbf{i} + (y_0 + v_{y0}t + \tfrac{1}{2}a_y t^2)\mathbf{j}$$
$$= (x_0\mathbf{i} + y_0\mathbf{j}) + (v_{x0}\mathbf{i} + v_{y0}\mathbf{j})t + \tfrac{1}{2}(a_x\mathbf{i} + a_y\mathbf{j})t^2$$

Position vector as a function •
of time

$$\mathbf{r} = \mathbf{r}_0 + \mathbf{v}_0 t + \tfrac{1}{2}\mathbf{a}t^2 \qquad [3.9]$$

This equation implies that the displacement vector $\mathbf{r} - \mathbf{r}_0$ is the vector sum of a displacement $\mathbf{v}_0 t$, arising from the initial velocity of the particle, and a displacement $\tfrac{1}{2}\mathbf{a}t^2$, resulting from the uniform acceleration of the particle. Graphical representations of Equations 3.8 and 3.9 are shown in Figure 3.4a and 3.4b. For simplicity in drawing the figure, we have taken $\mathbf{r}_0 = 0$ in Figure 3.4b. That is, we assume that the particle is at the origin at $t = 0$. Note from Figure 3.4b that **r** is generally not along the direction of $\mathbf{v}_0$ or **a**, because the relationship between these quantities is a vector expression. For the same reason, from Figure 3.4a we see that **v** is generally not along the direction of $\mathbf{v}_0$ or **a**. Finally, if we compare the two figures, we see that **v** and **r** are not in the same direction.

Because Equations 3.8 and 3.9 are *vector* expressions, we may write their *x* and *y* component forms with $\mathbf{r}_0 = 0$:

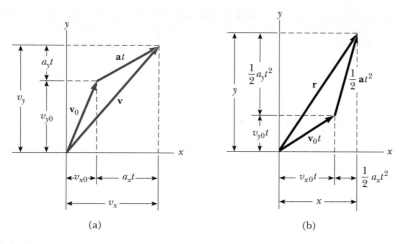

(a)

(b)

Figure 3.4 Vector representations and rectangular components of (a) the velocity and (b) the displacement of a particle moving with a uniform acceleration **a**.

$$\mathbf{v} = \mathbf{v}_0 + \mathbf{a}t \qquad \begin{cases} v_x = v_{x0} + a_x t \\ v_y = v_{y0} + a_y t \end{cases}$$

$$\mathbf{r} = \mathbf{v}_0 t + \tfrac{1}{2}\mathbf{a}t^2 \qquad \begin{cases} x = v_{x0}t + \tfrac{1}{2}a_x t^2 \\ y = v_{y0}t + \tfrac{1}{2}a_y t^2 \end{cases}$$

These components are illustrated in Figure 3.4. In other words, two-dimensional motion having constant acceleration is equivalent to two independent motions in the x and y directions having constant accelerations a_x and a_y.

Example 3.1 Motion in a Plane

A particle starts from the origin at $t = 0$ with an initial velocity having an x component of 20 m/s and a y component of -15 m/s. The particle moves in the xy plane with an x component of acceleration only, given by $a_x = 4.0$ m/s^2. (a) Determine the components of velocity as a function of time and the total velocity vector at any time.

Solution With $v_{x0} = 20$ m/s and $a_x = 4.0$ m/s^2, the equations of kinematics give

$$v_x = v_{x0} + a_x t = (20 + 4.0t) \text{ m/s}$$

Also, with $v_{y0} = -15$ m/s and $a_y = 0$,

$$v_y = v_{y0} = -15 \text{ m/s}$$

Therefore, using these results and noting that the velocity vector **v** has two components, we get

$$\mathbf{v} = v_x\mathbf{i} + v_y\mathbf{j} = [(20 + 4.0t)\mathbf{i} - 15\mathbf{j}] \text{ m/s}$$

We could also obtain this result using Equation 3.8 directly, noting that $\mathbf{a} = 4.0\mathbf{i}$ m/s^2 and $\mathbf{v}_0 = (20\mathbf{i} - 15\mathbf{j})$ m/s. Try it!

(b) Calculate the velocity and speed of the particle at $t = 5.0$ s.

Solution With $t = 5.0$ s, the result from (a) gives

$$\mathbf{v} = \{[20 + 4(5.0)]\mathbf{i} - 15\mathbf{j}\} \text{ m/s} = (40\mathbf{i} - 15\mathbf{j}) \text{ m/s}$$

That is, at $t = 5.0$ s, $v_x = 40$ m/s, and $v_y = -15$ m/s. Knowing these two components for this two-dimensional motion, we know the numerical value of the velocity vector. To determine the angle θ that **v** makes with the x axis, we use the fact that $\tan \theta = v_y/v_x$, or

$$\theta = \tan^{-1}\left(\frac{v_y}{v_x}\right) = \tan^{-1}\left(\frac{-15 \text{ m/s}}{40 \text{ m/s}}\right) = -21°$$

The speed is the magnitude of **v**:

$$v = |\mathbf{v}| = \sqrt{v_x^2 + v_y^2} = \sqrt{(40)^2 + (-15)^2} \text{ m/s} = 43 \text{ m/s}$$

(*Note:* If you calculate v_0 from the x and y components of $\mathbf{v}_0$, you will find that $v > v_0$. Why?)

EXERCISE 1   Determine the x and y coordinates of the particle at any time t and the displacement vector at this time.   Answer $x = (20t + 2.0t^2)$ m; $y = (-15t)$ m; $\mathbf{r} = [(20t + 2.0t^2)\mathbf{i} - 15t\mathbf{j}]$ m

EXERCISE 2   A particle starts from rest at $t = 0$ at the origin and moves in the xy plane with a constant acceleration of $\mathbf{a} = (2.0\mathbf{i} + 4.0\mathbf{j})$ m/s². After a time t has elapsed, determine (a) the x and y components of velocity, (b) the coordinates of the particle, and (c) the speed of the particle.

Answer   (a) $v_x = 2t$ m/s, $v_y = 4t$ m/s (b) $x = t^2$ m, $y = 2t^2$ m   (c) $4.47t$ m/s

3.3 • PROJECTILE MOTION

Anyone who has observed a baseball in motion (or any object thrown into the air) has observed projectile motion. The ball moves in a curved path when thrown at some angle with respect to the Earth's surface. This very common form of motion is surprisingly simple to analyze if the following two assumptions are made:

Assumptions of projectile motion •

(1) the free-fall acceleration, g, is constant over the range of motion and is directed downward,[1] and (2) the effect of air resistance is negligible.[2] With these assumptions, we find that the path of a projectile, which we call its *trajectory*, is *always* a parabola. **We use these assumptions throughout this chapter.**

If we choose our reference frame such that the y direction is vertical and positive upward, then $a_y = -g$ (as in one-dimensional free fall) and $a_x = 0$ (because air resistance is neglected). Furthermore, let us assume that at $t = 0$, the projectile leaves the origin ($x_0 = y_0 = 0$) with speed v_0, as in Figure 3.5. If the vector $\mathbf{v}_0$ makes an angle θ_0 with the horizontal, where θ_0 is the angle at which the projectile leaves the origin as in Figure 3.5, then from the definitions of the cosine and sine functions we have

$$\cos \theta_0 = \frac{v_{x0}}{v_0} \quad \text{and} \quad \sin \theta_0 = \frac{v_{y0}}{v_0}$$

Therefore, the initial x and y components of velocity are

$$v_{x0} = v_0 \cos \theta_0 \quad \text{and} \quad v_{y0} = v_0 \sin \theta_0$$

Substituting these expressions into Equations 3.8 and 3.9 with $a_x = 0$ and $a_y = -g$ gives the velocity components and coordinates for the projectile at any time t:

Horizontal velocity component •

$$v_x = v_{x0} = v_0 \cos \theta_0 = \text{constant} \tag{3.10}$$

Vertical velocity component •

$$v_y = v_{y0} - gt = v_0 \sin \theta_0 - gt \tag{3.11}$$

Horizontal position component •

$$x = v_{x0}t = (v_0 \cos \theta_0)\, t \tag{3.12}$$

Vertical position component •

$$y = v_{y0}t - \tfrac{1}{2}gt^2 = (v_0 \sin \theta_0)\, t - \tfrac{1}{2}gt^2 \tag{3.13}$$

[1]This approximation is reasonable as long as the range of motion is small compared with the radius of the Earth (6.4×10^6 m). In effect, this approximation is equivalent to assuming that the earth is flat within the range of motion considered.

[2]This approximation is generally *not* justified, especially at high velocities. In addition, the spin of a projectile, such as a baseball, can give rise to some very interesting effects associated with aerodynamic forces (for example, a curveball thrown by a pitcher).

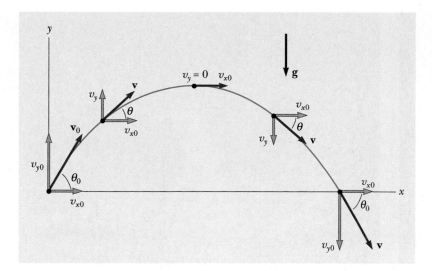

Figure 3.5 The parabolic path of a projectile that leaves the origin with a velocity $\mathbf{v}_0$. The velocity vector $\mathbf{v}$ changes with time in both magnitude and direction. The change in the velocity vector is the result of acceleration in the negative y direction. The x component of velocity remains constant in time because there is no acceleration along the horizontal direction. Also, the y component of velocity is zero at the peak of the path.

From Equation 3.10 we see that v_x remains constant in time and is equal to v_{x0}; there is no horizontal component of acceleration. For the y motion note that v_y and y are similar to Equations 2.12 and 2.14 for freely falling bodies. In fact, *all* of the equations of kinematics developed in Chapter 2 are applicable to projectile motion.

If we solve for t in Equation 3.12 and substitute this expression for t into Equation 3.13, we find that

$$y = (\tan\,\theta_0)x - \left(\frac{g}{2v_0{}^2 \cos^2\theta_0}\right)x^2 \qquad \textbf{[3.14]}$$

which is valid for the angles in the range $0 < \theta_0 < \pi/2$. This expression is of the form $y = ax - bx^2$, which is the equation of a parabola that passes through the origin. Thus, we have proved that the trajectory of a projectile is a parabola. Note that the trajectory is *completely* specified if v_0 and θ_0 are known.

One can obtain the speed, v, of the projectile as a function of time by noting that Equations 3.10 and 3.11 give the x and y components of velocity at any instant. Therefore, by definition, because v is equal to the magnitude of $\mathbf{v}$,

$$v = \sqrt{v_x{}^2 + v_y{}^2} \qquad \textbf{[3.15]} \qquad \bullet \quad \textit{Speed}$$

Also, because the velocity vector is tangent to the path at any instant, as shown in Figure 3.5, the angle θ that v makes with the horizontal can be obtained from v_x and v_y through the expression

$$\tan\theta = \frac{v_y}{v_x} \qquad \textbf{[3.16]}$$

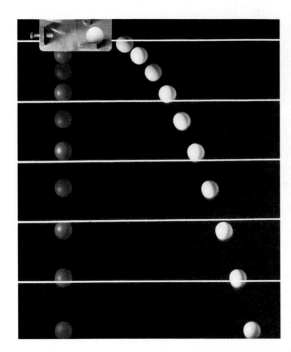

(Left) This multiflash photograph of two balls released simultaneously illustrates both free fall (red ball) and projectile motion (yellow ball). The yellow ball was projected horizontally, whereas the red ball was released from rest. Can you explain why both balls reach the floor simultaneously? *(© Richard Megna 1990, Fundamental Photographs)*   *(Right)* Multiflash exposure of a tennis player executing a backhand swing. Note that the ball follows a parabolic path characteristic of a projectile. Such photographs can be used to study the quality of sports equipment and the performance of an athlete. *(© Zimmerman, FPG International)*

The vector expression for the position vector of the projectile as a function of time follows directly from Equation 3.9, with $\mathbf{a} = \mathbf{g}$:

$$\mathbf{r} = \mathbf{v}_0 t + \tfrac{1}{2}\mathbf{g} t^2$$

This expression gives the same information as the combination of Equations 3.12 and 3.13 and is plotted in Figure 3.6. Note that this expression for $\mathbf{r}$ is consistent

Figure 3.6   The displacement vector $\mathbf{r}$ of a projectile, the initial velocity of which is $\mathbf{v}_0$ at the origin. The vector $\mathbf{v}_0 t$ would be the displacement of the projectile if gravity were absent, and the vector $\tfrac{1}{2}\mathbf{g} t^2$ is its vertical displacement due to its downward gravitational acceleration.

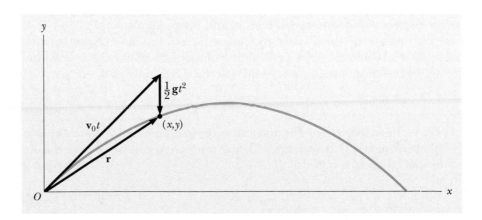

with Equation 3.13, because the expression for **r** is a vector equation and **a** = **g** = $-g\mathbf{j}$ when the upward direction is taken to be positive.

It is interesting to note that the motion of a particle can be considered the superposition of the term $\mathbf{v}_0 t$, which would be the displacement if no acceleration were present, and the term $\frac{1}{2}\mathbf{g}t^2$, which arises from the acceleration due to gravity. In other words, if there were no gravitational acceleration, the particle would continue to move along a straight path in the direction of $\mathbf{v}_0$. Therefore, the vertical distance $\frac{1}{2}\mathbf{g}t^2$ through which the particle "falls" off the straight-path line is the same distance a freely falling body would fall during the same time interval. **We conclude that projectile motion is the superposition of two motions: (1) constant velocity motion in the initial direction and (2) the motion of a particle freely falling in the vertical direction under constant acceleration.**

Horizontal Range and Maximum Height of a Projectile Let us assume that a projectile is fired from the origin at $t = 0$ with a positive v_y component, as in Figure 3.7. There are two special points that are interesting to analyze: the peak that has Cartesian coordinates $(R/2, h)$ and the point having coordinates $(R, 0)$. The distance R is called the *horizontal range* of the projectile and h is its *maximum height*. Let us find h and R in terms of v_0, θ_0, and g.

We can determine h by noting that at the peak $v_y = 0$. Therefore, Equation 3.11 can be used to determine the time t_1 it takes to reach the peak:

$$t_1 = \frac{v_0 \sin \theta_0}{g}$$

Substituting this expression for t_1 into Equation 3.13 and replacing y with h gives h in terms of v_0 and θ_0:

$$h = (v_0 \sin \theta_0) \frac{v_0 \sin \theta_0}{g} - \frac{1}{2}g \left(\frac{v_0 \sin \theta_0}{g} \right)^2$$

$$h = \frac{v_0^2 \sin^2 \theta_0}{2g} \qquad \text{[3.17]}$$

The range, R, is the horizontal distance traveled in twice the time it takes to reach the peak—that is, in a time $2t_1$. (This can be seen by setting $y = 0$ in Equation 3.13 and solving the quadratic for t. One solution of this quadratic is $t = 0$, and the other is $t = 2t_1$.) Using Equation 3.12 and noting that $x = R$ at $t = 2t_1$, we find that

$$R = (v_0 \cos \theta_0)2t_1 = (v_0 \cos \theta_0) \frac{2v_0 \sin \theta_0}{g} = \frac{2v_0^2 \sin \theta_0 \cos \theta_0}{g}$$

Because $\sin 2\theta = 2 \sin \theta \cos \theta$, R can be written in the most compact form

$$R = \frac{v_0^2 \sin 2\theta_0}{g} \qquad \text{[3.18]}$$

Keep in mind that Equations 3.17 and 3.18 are useful for calculating h and R only if v_0 and θ_0 are known and only for a symmetric path, as shown in Figure 3.7 (which means that only $\mathbf{v}_0$ has to be specified). The general expressions given by Equations 3.10 through 3.13 are the *more important* results, because they give the coordinates and velocity components of the projectile at *any* time t.

You should note that the maximum value of R from Equation 3.18 is given by $R_{\text{max}} = v_0^2/g$. This result follows from the fact that the maximum value of $\sin 2\theta_0$

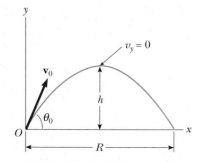

Figure 3.7 A projectile fired from the origin at $t = 0$ with an initial velocity $\mathbf{v}_0$. The maximum height of the projectile is h, and its horizontal range is R. At the peak of the trajectory, the y component of its velocity is zero and the projectile has coordinates $(\frac{R}{2}, h)$.

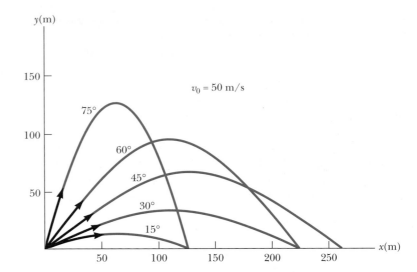

Figure 3.8 A projectile fired from the origin with an initial speed of 50 m/s at various angles of projection. Note that complementary values of θ_0 will result in the same value of x (range of the projectile).

is unity, which occurs when $2\theta_0 = 90°$. Therefore, we see that R is a maximum when $\theta_0 = 45°$, as you would expect if air resistance is neglected.

Figure 3.8 illustrates various trajectories for a projectile of a given initial speed. As you can see, the range is a maximum for $\theta_0 = 45°$. In addition, for any θ_0 other than 45°, a point with coordinates $(R, 0)$ can be reached by using either one of two complementary values of θ_0, such as 75° and 15°. Of course, the maximum height and the time of flight will be different for these two values of θ_0.

Before we look at some numerical examples dealing with projectile motion, let us pause to summarize what we have learned so far about this kind of motion:

- Provided air resistance is negligible, the horizontal component of velocity, v_x, remains constant because there is no horizontal component of acceleration.
- The vertical component of acceleration is equal to the free-fall acceleration, **g**.
- The vertical component of velocity, v_y, and the displacement in the y direction are identical to those of a freely falling body.
- Projectile motion can be described as a superposition of the two motions in the x and y directions.

PROBLEM-SOLVING STRATEGY • Projectile Motion

We suggest that you use the following approach to solving projectile motion problems:

1. Select a coordinate system.
2. Resolve the initial velocity vector into x and y components.
3. Treat the horizontal motion and the vertical motion independently.

4. Follow the techniques for solving problems with constant velocity to analyze the horizontal motion of the projectile.
5. Follow the techniques for solving problems with constant acceleration to analyze the vertical motion of the projectile.

Thinking Physics 2

A home run is hit in a baseball game. The ball is hit from home plate into the center-field stands along a long, parabolic path. What is the acceleration of the ball (a) while it is rising, (b) at the highest point of the trajectory, and (c) while it is descending after reaching the highest point? Ignore air resistance.

Reasoning The answers to all three parts are the same—the acceleration is that due to gravity, 9.8 m/s^2, because the force of gravity is pulling downward on the ball during the entire motion. During the rising part of the trajectory, the downward acceleration results in the decreasing positive values of the vertical component of the velocity of the ball. During the falling part of the trajectory, the downward acceleration results in the increasing negative values of the vertical component of the velocity. Many people have trouble with the topmost point, mistakenly thinking that the acceleration of the ball at the highest point is zero. This interpretation arises from a confusion between zero vertical velocity and zero acceleration. Even though the ball has momentarily come to rest at the highest point (vertically; it is still moving horizontally), it is still accelerating, because the force of gravity is still acting. If the ball were to come to rest and experience zero acceleration, then the velocity would not change—the ball would remain suspended at the highest point! We don't see that happening, because the acceleration is *not* equal to zero.

CONCEPTUAL PROBLEM 3

Suppose that as you are running at constant speed, you wish to throw a ball such that you will catch it as it comes back down. How should you throw the ball?

CONCEPTUAL PROBLEM 4

As a projectile moves in its parabolic path, is there any point along its path at which the velocity and acceleration vectors are (a) perpendicular to each other? (b) parallel to each other?

Example 3.2 That's Quite an Arm

A stone is thrown from the top of a building upward at an angle of 30.0° to the horizontal and with an initial speed of 20.0 m/s, as in Figure 3.9. If the height of the building is 45.0 m, (a) how long is the stone "in flight"?

Solution The initial x and y components of the velocity are

$$v_{x0} = v_0 \cos \theta_0 = (20.0 \text{ m/s})(\cos 30.0°) = 17.3 \text{ m/s}$$

$$v_{y0} = v_0 \sin \theta_0 = (20.0 \text{ m/s})(\sin 30.0°) = 10.0 \text{ m/s}$$

To find t, we can use $y = v_{y0}t - \frac{1}{2}gt^2$ (Eq. 3.13) with $y = -45.0$ m and $v_{y0} = 10.0$ m/s (we have chosen the top of the building as the origin, as in Fig. 3.9):

$$-45.0 \text{ m} = (10.0 \text{ m/s})t - \tfrac{1}{2}(9.80 \text{ m/s}^2)t^2$$

Solving the quadratic equation for t gives, for the positive root, $t = 4.22$ s. Does the negative root have any physical

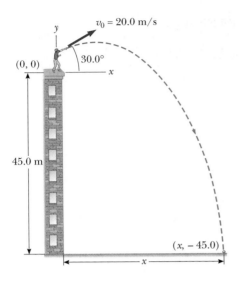

Figure 3.9   (Example 3.2)

meaning? (Can you think of another way of finding t from the information given?)

(b) What is the speed of the stone just before it strikes the ground?

Solution   The y component of the velocity just before the stone strikes the ground can be obtained using the equation $v_y = v_{y0} - gt$ (Eq. 3.11) with $t = 4.22$ s:

$$v_y = 10.0 \text{ m/s} - (9.80 \text{ m/s}^2)(4.22 \text{ s}) = -31.4 \text{ m/s}$$

Because $v_x = v_{x0} = 17.3$ m/s, the required speed is

$$v = \sqrt{v_x^2 + v_y^2} = \sqrt{(17.3)^2 + (-31.4)^2} \text{ m/s} = \boxed{35.9 \text{ m/s}}$$

EXERCISE 3   Where does the stone strike the ground?

Answer   73.0 m from the base of the building.

Example 3.3   The Stranded Explorers

An Alaskan rescue plane drops a package of emergency rations to a stranded party of explorers, as shown in Figure 3.10. If the plane is traveling horizontally at 40.0 m/s at a height of 100 m above the ground, where does the package strike the ground relative to the point at which it is released?

Reasoning and Solution   The coordinate system for this problem is selected as shown in Figure 3.10, with the positive x direction to the right and the positive y direction upward.

Consider first the horizontal motion of the package. The only equation available to us is $x = v_{x0}t$ (Eq. 3.12).

The initial x component of the package velocity is the same as that of the plane when the package is released, 40.0 m/s. Thus, we have

$$x = (40.0 \text{ m/s})t$$

If we know t, the length of time the package is in the air, we can determine x, the distance traveled by the package in the horizontal direction. To find t, we move to the equations for the vertical motion of the package. We know that at the instant the package hits the ground, its y coordinate is -100 m. We also know that the initial component of velocity of the package in the vertical direction, v_{y0}, is zero because the package was released with only a horizontal component of velocity. From Equation 3.13, we have

$$y = -\tfrac{1}{2}gt^2$$
$$-100 \text{ m} = -\tfrac{1}{2}(9.80 \text{ m/s}^2)t^2$$
$$t^2 = 20.4 \text{ s}^2$$
$$t = 4.51 \text{ s}$$

The value for the time of flight substituted into the equation for the x coordinate gives

$$x = (40.0 \text{ m/s})(4.51 \text{ s}) = \boxed{180 \text{ m}}$$

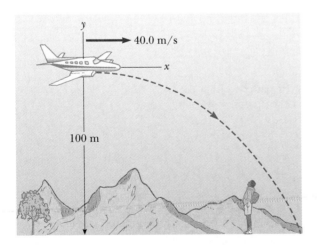

Figure 3.10   (Example 3.3) According to a ground observer, a package released from the rescue plane travels along the path shown. How does the path followed by the package appear to an observer on the plane (assumed to be moving at constant speed)?

The package hits the ground not directly under the drop point but 180 m to the right of that point.

EXERCISE 4   What are the horizontal and vertical components of the velocity of the package just before it hits the ground?   Answer $v_x = 40.0$ m/s; $v_y = -44.1$ m/s

EXERCISE 5   It has been said that in his youth George Washington threw a silver dollar across a river. Assuming that the river was 75 m wide, (a) what *minimum initial* speed was necessary to get the coin across the river, and   (b) how long was the coin in flight?   Answer   (a) 27.1 m/s   (b) 3.91 s

3.4 · UNIFORM CIRCULAR MOTION

Figure 3.11a shows an object moving in a circular path with *constant linear speed v.* Such motion is called **uniform circular motion.** It is often surprising to students to find that **even though the object moves at a constant speed, it still has an acceleration.** To see why, consider the defining equation for average acceleration, **a̅** $= \Delta \mathbf{v}/\Delta t$ (Eq. 3.4).

Note that the acceleration depends on *the change in the velocity vector.* Because velocity is a vector quantity, there are two ways in which an acceleration can be produced, as mentioned in Section 3.1: by a change in the *magnitude* of the velocity and by a change in the *direction* of the velocity. It is the latter situation that is occurring for an object moving with constant speed in a circular path. The velocity vector is always tangent to the path of the object and perpendicular to the radius r of the circular path. We now show that the acceleration vector in uniform circular motion is always perpendicular to the path and always points toward the center of the circle. An acceleration of this nature is called a *centripetal acceleration* (center seeking) and its magnitude is

$$a_r = \frac{v^2}{r}$$

[3.19]   • *Magnitude of centripetal acceleration*

where r is the radius of the circle.

To derive Equation 3.19, consider Figure 3.11b. Here an object is seen first at point P and then at point Q. The particle is at P at time t_i, and its velocity at that time is $\mathbf{v}_i$; it is at Q at some later time t_f, and its velocity at that time is $\mathbf{v}_f$. Let us

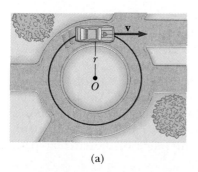

(a)

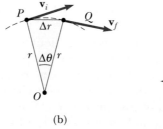

(b)

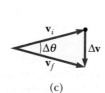

(c)

Figure 3.11   (a) An object moving along a circular path at constant speed experiences uniform circular motion. (b) As the particle moves from P to Q, the direction of its velocity vector changes from $\mathbf{v}_i$ to $\mathbf{v}_f$. (c) The construction for determining the direction of the change in velocity $\Delta\mathbf{v}$ that is toward the center of the circle.

also assume here that $\mathbf{v}_i$ and $\mathbf{v}_f$ differ only in direction; their magnitudes are the same (that is, $v_i = v_f = v$). In order to calculate the acceleration of the particle, let us begin with the defining equation for average acceleration (Eq. 3.4):

$$\bar{\mathbf{a}} = \frac{\mathbf{v}_f - \mathbf{v}_i}{t_f - t_i} = \frac{\Delta \mathbf{v}}{\Delta t}$$

This equation indicates that we must vectorially subtract $\mathbf{v}_i$ from $\mathbf{v}_f$, where $\Delta \mathbf{v} = \mathbf{v}_f - \mathbf{v}_i$ is the change in the velocity. Because $\mathbf{v}_i + \Delta \mathbf{v} = \mathbf{v}_f$, the vector $\Delta \mathbf{v}$ can be found using the vector triangle in Figure 3.11c. When Δt is very small, Δr and $\Delta \theta$ are also very small. In this case, $\mathbf{v}_f$ is almost parallel to $\mathbf{v}_i$ and the vector $\Delta \mathbf{v}$ is approximately perpendicular to them, pointing toward the center of the circle.

Now consider the triangle in Figure 3.11b, which has sides Δr and r. This triangle and the one in Figure 3.11c that has sides Δv and v are similar. (Two triangles are *similar* if the angle between any two sides is the same for both triangles and if the ratio of lengths of these sides is the same.) This enables us to write a relationship between the lengths of the sides:

$$\frac{\Delta v}{v} = \frac{\Delta r}{r}$$

This equation can be solved for Δv and the expression obtained from it can be substituted into $\bar{a} = \Delta v / \Delta t$ (Eq. 3.4) to give

$$\bar{a} = \frac{v\,\Delta r}{r\Delta t}$$

Now imagine that points P and Q in Figure 3.11b come extremely close together. In this case $\Delta \mathbf{v}$ would point toward the center of the circular path, and because the acceleration is in the direction of $\Delta \mathbf{v}$, it too is toward the center. Furthermore, as P and Q approach each other, Δt approaches zero, and the ratio $\Delta r / \Delta t$ approaches the speed v. Hence, in the limit $\Delta t \to 0$, the magnitude of the acceleration is

$$a_r = \frac{v^2}{r}$$

Thus, we conclude that in uniform circular motion the acceleration is directed inward toward the center of the circle and has magnitude v^2/r. You should show that the dimensions of a_r are L/T^2—as required because this is a true acceleration.

In many situations it is convenient to describe the motion of a particle moving with constant speed in a circle of radius r in terms of the **period,** T, which is defined as the time required for one complete revolution. In the time T the particle moves a distance of $2\pi r$, which is equal to the circumference of the particle's circular path. Therefore, because its speed is equal to the circumference of the circular path divided by the period, or $v = 2\pi r / T$, it follows that

$$T \equiv \frac{2\pi r}{v} \qquad\qquad \textbf{[3.20]}$$

Thinking Physics 3

An airplane travels from Los Angeles to Sydney, Australia. For 15 hours after cruising altitude is reached, the instruments on the plane indicate that the ground speed holds rock-steady at 700 km/hr^{-1} and that the heading of the airplane does not change. Is the velocity of the airplane constant?

Reasoning The velocity is not constant, due to the curvature of the Earth. Even though the speed does not change and the heading is always toward Sydney (is this actually true?), the airplane is traveling around a significant portion of the circumference of the Earth. Thus, the direction of the velocity vector does change. We could extend this by imagining that the airplane passes over Sydney and continues (assuming it has enough fuel!) around the Earth until it arrives at Los Angeles again. It is impossible for an airplane to have a constant velocity (relative to the Universe, not to the Earth's surface) and return to its starting point.

EXERCISE 6   Find the acceleration of a particle moving with a constant speed of 8.0 m/s in a circle 2.0 m in radius.   Answer   32 m/s^2 toward the center of the circle

3.5 • TANGENTIAL AND RADIAL ACCELERATION

Let us consider the motion of a particle along a curved path where the velocity changes both in direction and in magnitude, as described in Figure 3.12. In this situation, the velocity vector is always tangent to the path; however, the acceleration vector **a** is at some angle to the path.

As the particle moves along the curved path in Figure 3.12, we see that the direction of the total acceleration vector, **a**, changes from point to point. This vector can be resolved into two components: a radial component, a_r, and a tangential component, a_t. The *total* acceleration vector, **a**, can be written as the vector sum of the component vectors:

$$\mathbf{a} = \mathbf{a}_r + \mathbf{a}_t \qquad [3.21]$$

• *Total acceleration*

The tangential acceleration arises from the change in the speed of the particle, and the projection of the tangential acceleration along the direction of the velocity is

$$a_t = \frac{d\,|\mathbf{v}|}{dt} \qquad [3.22]$$

• *Tangential acceleration*

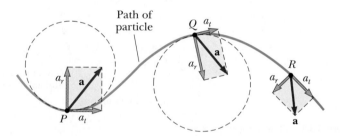

Path of particle

Figure 3.12   The motion of a particle along an arbitrary curved path lying in the *xy* plane. If the velocity vector **v** (always tangent to the path) changes in direction and magnitude, the acceleration **a** has a tangential component a_t and a radial component a_r.

The radial acceleration is due to the change in direction of the velocity vector and has an absolute magnitude given by

Radial or centripetal • acceleration

$$a_r = \frac{v^2}{r}$$

[3.23]

where r is the radius of curvature of the path at the point in question. Because $\mathbf{a}_r$ and $\mathbf{a}_t$ are perpendicular component vectors of $\mathbf{a}$, it follows that $a = \sqrt{a_r^2 + a_t^2}$. As in the case of uniform circular motion, $\mathbf{a}_r$ always points toward the center of curvature, as shown in Figure 3.12. Also, at a given speed, a_r is large when the radius of curvature is small (as at points P and Q in Fig. 3.12) and small when r is large (such as at point R). The direction of $\mathbf{a}_t$ is either in the same direction as $\mathbf{v}$ (if v is increasing) or opposite $\mathbf{v}$ (if v is decreasing).

Note that in the case of uniform circular motion, where v is constant, $a_t = 0$ and the acceleration is always radial, as we described in Section 3.4. (*Note:* Eq. 3.23 is identical to Eq. 3.19.) In other words, uniform circular motion is a special case of motion along a curved path. Furthermore, if the direction of $\mathbf{v}$ does not change, then there is no radial acceleration and the motion is one-dimensional ($a_r = 0$, but a_t may not be zero).

Thinking Physics 4

In uniform circular motion, the acceleration vector is always perpendicular to the velocity vector.   (a) Suppose a particle is moving at uniform speed around an *elliptical* path. What is the relationship between the acceleration and velocity vectors in this case?   (b) Now, let the elliptical path be that of Halley's Comet in orbit around the Sun. This is *not* a case of uniform speed—the comet moves slowly when it is far from the Sun and rapidly when it is close. Are the acceleration and velocity vectors perpendicular in this case?

Reasoning (a) We consider first the case of the particle moving at constant speed around the elliptical path (Fig. 3.13a). Because the speed is constant, there must be no component of the acceleration tangent to the path. Thus, the acceleration must be

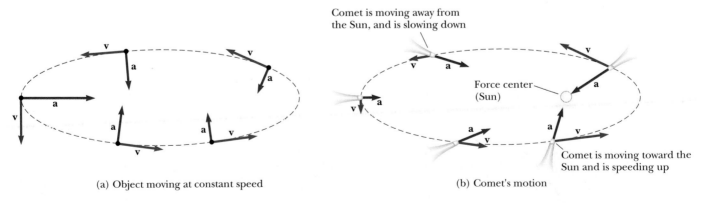

(a) Object moving at constant speed

(b) Comet's motion

Figure 3.13   (Thinking Physics 4)

perpendicular to the path and, therefore, always perpendicular to the velocity vector. The magnitude of the acceleration vector is not constant—it will be largest at the ends of the ellipse, because the velocity is changing direction the most rapidly there. Figure 3.13a shows the velocity and acceleration vectors at several locations around the path.

(b) In the case of Halley's Comet, or any other object in an elliptical orbit around a gravitational-force center, the direction of the acceleration is the same as the direction of the gravitational force on the object, which is toward the *focus* of the ellipse, where the gravitational force center lies (Fig. 3.13b). Thus, the acceleration and velocity vectors are not perpendicular, except for the points at the ends of the ellipse. As the comet moves away from the Sun, the acceleration has a component opposite to the velocity. As it moves toward the Sun, the acceleration has a component parallel to the velocity. This results in the slowing down of the comet as it moves away from the Sun and its speeding up as it moves toward the Sun. Notice in Figure 3.13b that the velocity vectors have large magnitudes when the comet is near the Sun and small magnitudes when it is far away.

Example 3.4   The Swinging Ball

A ball tied to the end of a string 0.50 m in length swings in a vertical circle under the influence of gravity, as in Figure 3.14. When the string makes an angle of $\theta = 20°$ with the vertical, the ball has a speed of 1.5 m/s. (a) Find the magnitude of the radial component of acceleration at this instant.

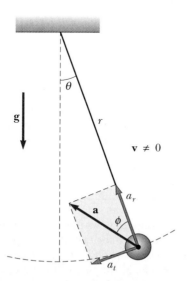

Figure 3.14   (Example 3.4) Motion of a ball suspended by a string of length *r*. The ball swings with nonuniform circular motion in a vertical plane, and its acceleration **a** has a radial component a_r and a tangential component a_t.

Solution   Because $v = 1.5$ m/s and $r = 0.50$ m, we find that

$$a_r = \frac{v^2}{r} = \frac{(1.5 \text{ m/s})^2}{0.50 \text{ m}} = 4.5 \text{ m/s}^2$$

(b) When the ball is at an angle θ to the vertical, it has a tangential acceleration of magnitude $g \sin \theta$ (the component of **g** tangent to the circle). Therefore, at $\theta = 20°$, $a_t = g \sin 20° = 3.4 \text{ m/s}^2$. Find the magnitude and direction of the *total* acceleration at $\theta = 20°$.

Solution   Because $\mathbf{a} = \mathbf{a}_r + \mathbf{a}_t$, the magnitude of **a** at $\theta = 20°$ is

$$a = \sqrt{a_r^2 + a_t^2} = \sqrt{(4.5)^2 + (3.4)^2} \text{ m/s}^2 = 5.6 \text{ m/s}^2$$

If ϕ is the angle between **a** and the string, then

$$\phi = \tan^{-1} \frac{a_t}{a_r} = \tan^{-1} \left(\frac{3.4 \text{ m/s}^2}{4.5 \text{ m/s}^2} \right) = 36°$$

Note that all of the vectors—**a**, $\mathbf{a}_t$, and $\mathbf{a}_r$—change in direction *and* magnitude as the ball swings through the circle. When the ball is at its lowest elevation ($\theta = 0$), $a_t = 0$, because there is no tangential component of **g** at this angle, and a_r is a *maximum* because v is a maximum. If the ball has enough speed to reach its highest position ($\theta = 180°$), a_t is again zero, but a_r is a minimum because v is now a minimum. Finally, in the two horizontal positions ($\theta = 90°$ and $270°$), $|a_t| = g$ and a_r has a value between its minimum and maximum values.

SUMMARY

If a particle moves with *constant* acceleration $\mathbf{a}$ and has velocity $\mathbf{v}_0$ and position $\mathbf{r}_0$ at $t = 0$, its velocity and position vectors at some later time t are

$$\mathbf{v} = \mathbf{v}_0 + \mathbf{a}t \qquad [3.8]$$

$$\mathbf{r} = \mathbf{r}_0 + \mathbf{v}_0 t + \tfrac{1}{2}\mathbf{a}t^2 \qquad [3.9]$$

For two-dimensional motion in the xy plane under constant acceleration, these vector expressions are equivalent to two component expressions, one for the motion along x and one for the motion along y.

Projectile motion is two-dimensional motion under constant acceleration, where $a_x = 0$ and $a_y = -g$. In this case, if $x_0 = y_0 = 0$, the components of Equations 3.8 and 3.9 reduce to

$$v_x = v_{x0} = \text{constant} \qquad [3.10]$$

$$v_y = v_{y0} - gt \qquad [3.11]$$

$$x = v_{x0}t \qquad [3.12]$$

$$y = v_{y0}t - \tfrac{1}{2}gt^2 \qquad [3.13]$$

where $v_{x0} = v_0 \cos \theta_0$, $v_{y0} = v_0 \sin \theta_0$, v_0 is the initial speed of the projectile, and θ_0 is the angle $\mathbf{v}_0$ makes with the positive x axis. Note that these expressions give the velocity components (and hence the velocity vector) and the coordinates (and hence the position vector) at *any* time t that the projectile is in motion.

It is useful to think of projectile motion as the superposition of two motions: (1) uniform motion in the x direction and (2) motion in the vertical direction subject to a constant downward acceleration of magnitude $g = 9.80$ m/s^2.

A particle moving in a circle of radius r with constant speed v undergoes a centripetal (or radial) acceleration, $\mathbf{a}_r$, because the direction of $\mathbf{v}$ changes in time. The magnitude of $\mathbf{a}_r$ is

$$a_r = \frac{v^2}{r} \qquad [3.19]$$

and its direction is always toward the center of the circle.

If a particle moves along a curved path in such a way that the magnitude and direction of $\mathbf{v}$ change in time, the particle has an acceleration vector that can be described by two components: (1) a radial component, a_r, arising from the change in direction of $\mathbf{v}$, and (2) a tangential component, a_t, arising from the change in magnitude of $\mathbf{v}$.

CONCEPTUAL QUESTIONS

1. If you know the position vectors of a particle at two points along its path and also know the time it took to get from one point to the other, can you determine the particle's instantaneous velocity? Its average velocity? Explain.

2. Explain whether or not the following particles have an acceleration: (a) a particle moving in a straight line with constant speed and (b) a particle moving around a curve with constant speed.

3. Correct the following statement: "The racing car rounds the turn at a constant velocity of 90 miles per hour."

4. A spacecraft drifts through space at a constant velocity. Suddenly a gas leak in the side of the spacecraft causes a constant acceleration of the spacecraft in a direction perpendicular to the initial velocity. The orientation of the spacecraft does not change, so that the acceleration remains perpendicular to the original direction of the velocity. What

Boxed numbers indicate questions that have answers available in the Student Solutions Manual and Study Guide.

is the shape of the path followed by the spacecraft in this situation?

5. A ball is projected horizontally from the top of a building. One second later another ball is projected horizontally from the same point with the same velocity. At what point in the motion will the balls be closest to each other? Will the first ball always be traveling faster than the second ball? What will be the time difference between when the balls hit the ground? Can the horizontal projection velocity of the second ball be changed so that the balls arrive at the ground at the same time?

6. At the end of a pendulum's arc, its velocity is zero. Is its acceleration also zero at that point?

7. If a rock is dropped from the top of a sailboat's mast, will it hit the deck at the same point whether the boat is at rest or in motion at constant velocity?

8. Two projectiles are thrown with the same magnitude of initial velocity, one at an angle θ with respect to the level ground and the other at angle $90° - \theta$. Both projectiles will strike the ground at the same distance from the projection point. Will both projectiles be in the air for the same time interval?

9. A projectile is fired at some angle to the horizontal with some initial speed v_0, and air resistance is neglected. Is the projectile a freely falling body? What is its acceleration in the vertical direction? What is its acceleration in the horizontal direction?

10. State which of the following quantities, if any, remain constant as a projectile moves through its parabolic trajectory: (a) speed, (b) acceleration, (c) horizontal component of velocity, (d) vertical component of velocity.

11. The maximum range of a projectile occurs when it is launched at an angle of 45.0° with the horizontal, if air resistance is neglected. If air resistance is not neglected, will the optimum angle be greater or less than 45.0°? Explain.

12. An object moves in a circular path with constant speed v. (a) Is the velocity of the object constant? (b) Is its acceleration constant? Explain.

13. A projectile is fired on the Earth with some initial velocity. Another projectile is fired on the Moon with the *same* initial velocity. Neglecting air resistance, which projectile has the greater range? Which reaches the greater altitude? (Note that the free-fall acceleration on the Moon is about 1.6 m/s².)

14. A coin on a table is given an initial horizontal velocity such that it ultimately leaves the end of the table and hits the floor. At the instant the coin leaves the end of the table, a ball is released from the same height and falls to the floor. Explain why the two objects hit the floor simultaneously, even though the coin has an initial velocity.

15. Describe how a driver can steer a car traveling at constant speed so that (a) the acceleration is zero or (b) the magnitude of the acceleration remains constant.

16. An ice skater is executing a figure eight, consisting of two equal, tangent circular paths. Throughout the first loop she increases her speed uniformly, and during the second loop she moves at a constant speed. Make a sketch of her acceleration vector at several points along the path of motion.

17. Construct motion diagrams showing the velocity and acceleration of a projectile at several points along its path if (a) the projectile is fired horizontally and (a) the projectile is fired at an angle θ with the horizontal.

18. A baseball is thrown such that its initial x and y components of velocity are known. Neglecting air resistance, describe how you would calculate, at the instant the ball reaches the top of its trajectory, (a) its coordinates, (b) its velocity, and (c) its acceleration. How would these results change if air resistance were taken into account?

PROBLEMS

Section 3.1   The Displacement, Velocity, and Acceleration Vectors

1. A motorist drives south at 20.0 m/s for 3.00 min, then turns west and travels at 25.0 m/s for 2.00 min, and finally travels northwest at 30.0 m/s for 1.00 min. For this 6.00-min trip, find (a) the net vector displacement of the motorist, (b) the motorist's average speed, and (c) the average velocity of the motorist.

2. Suppose that the position-vector function for a particle is given as $\mathbf{r}(t) = x(t)\mathbf{i} + y(t)\mathbf{j}$, with $x(t) = at + b$ and $y(t) = ct^2 + d$, where $a = 1.00$ m/s, $b = 1.00$ m, $c =$ 0.125 m/s², and $d = 1.00$ m. (a) Calculate the average velocity during the time interval $t = 2.00$ s to $t = 4.00$ s. (b) Determine the velocity and speed at $t = 2.00$ s.

3. A golf ball is hit off a tee at the edge of a cliff. Its x and y coordinates versus time are given by the following expressions:

$$x = (18.0 \text{ m/s})t \qquad y = (4.00 \text{ m/s})t - (4.90 \text{ m/s}^2)t^2$$

(a) Write a vector expression for the position $\mathbf{r}$ versus time t, using the unit vectors $\mathbf{i}$ and $\mathbf{j}$. By using derivatives, repeat

---

Boxed numbers indicate problems that have full solutions available in the Student Solutions Manual and Study Guide. Cyan numbers indicate intermediate level problems and magenta numbers indicate more challenging problems.

for (b) the velocity vector **v** versus time and (c) the acceleration vector **a** versus time. (d) Find the x and y coordinates of the golf ball at $t = 3.00$ s. Using the unit vectors **i** and **j**, write expressions for (e) the velocity **v** and (f) the acceleration **a** at the instant $t = 3.00$ s.

4. The coordinates of an object moving in the xy plane vary with time according to the equations $x = -(5.00 \text{ m}) \sin(t)$ and $y = (4.00 \text{ m}) - (5.00 \text{ m}) \cos(t)$, where t is in seconds. (a) Determine the components of velocity and components of acceleration at $t = 0$ s. (b) Write expressions for the position vector, the velocity vector, and the acceleration vector at any time $t > 0$. (c) Describe the path of the object in an xy plot.

Section 3.2 Two-Dimensional Motion with Constant Acceleration

5. At $t = 0$, a particle moving in the xy plane with constant acceleration has a velocity of $\mathbf{v}_0 = (3.00\mathbf{i} - 2.00\mathbf{j})$ m/s at the origin. At $t = 3.00$ s, the particle's velocity is $\mathbf{v} = (9.00\mathbf{i} + 7.00\mathbf{j})$ m/s. Find (a) the acceleration of the particle and (b) its coordinates at any time t.

6. The vector position of a particle varies in time according to the expression $\mathbf{r} = (3.00\mathbf{i} - 6.00t^2\mathbf{j})$ m. (a) Find expressions for the velocity and acceleration as functions of time. (b) Determine the particle's position and velocity at $t = 1.00$ s.

7. A fish swimming in a horizontal plane has velocity $\mathbf{v}_0 = (4.00\mathbf{i} + 1.00\mathbf{j})$ m/s when its displacement from a certain rock is $\mathbf{r}_0 = (10.0\mathbf{i} - 4.00\mathbf{j})$ m. After the fish swims with constant acceleration for 20.0 s, its velocity is $\mathbf{v} = (20.0\mathbf{i} - 5.00\mathbf{j})$ m/s. (a) What are the components of the acceleration? (b) What is the direction of the acceleration with respect to unit vector **i**? (c) Where is the fish at $t = 25.0$ s and in what direction is it moving?

8. A particle initially located at the origin has an acceleration of $\mathbf{a} = 3.00\mathbf{j}$ m/s^2 and an initial velocity of $\mathbf{v}_0 = 5.00\mathbf{i}$ m/s. Find (a) the vector position and velocity at any time t and (b) the coordinates and speed of the particle at $t = 2.00$ s.

Section 3.3 Projectile Motion

(Neglect air resistance in all problems. Use $g = 9.80$ m/s^2.)

9. In a local bar, a customer slides an empty beer mug on the counter for a refill. The bartender is momentarily distracted and does not see the mug, which slides off the counter and strikes the floor 1.40 m from the base of the counter. If the height of the counter is 0.860 m, (a) with what velocity did the mug leave the counter, and (b) what was the direction of the mug's velocity just before it hit the floor?

10. A student decides to measure the muzzle velocity of a pellet from his gun. He points the gun horizontally. He places a target on a vertical wall a distance x away from the gun. The pellet hits the target a vertical distance y below the gun. (a) Show that the position of the pellet when traveling through the air is given by $y = Ax^2$, where A is a constant. (b) Express the constant A in terms of the initial velocity and the free-fall acceleration. (c) If $x = 3.00$ m and $y = 0.210$ m, what is the initial speed of the pellet?

11. A golfer wants to drive a golf ball a distance of 310 yards (283 m). If the four-wood launches the ball at 15.0° above the horizontal, what must be the initial speed of the ball to achieve the required distance?

12. One strategy in a snowball fight is to throw a snowball at a high angle over level ground. While your opponent is watching the first one, you throw a second snowball at a low angle timed to arrive before or at the same time as the first one. Assume both snowballs are thrown with a speed of 25.0 m/s. The first one is thrown at an angle of 70.0° with respect to the horizontal. (a) At what angle should the second snowball be thrown to arrive at the same point as the first? (b) How many seconds later should the second snowball be thrown after the first to arrive at the same time?

13. A tennis player standing 12.6 m from the net hits the ball at 3.00° above the horizontal. To clear the net, the ball must rise at least 0.330 m. If the ball just clears the net at the apex of its trajectory, how fast was the ball moving when it left the racket?

14. An artillery shell is fired with an initial velocity of 300 m/s at 55.0° above the horizontal. It explodes on a mountainside 42.0 s after firing. What are the x and y coordinates of the shell where it explodes relative to its firing point?

15. An astronaut on a strange planet finds that she can jump a *maximum* horizontal distance of 15.0 m if her initial speed is 3.00 m/s. What is the free-fall acceleration on the planet?

16. A ball is tossed from an upper-story window. The ball is given an initial velocity of 8.00 m/s at an angle of 20.0° below the horizontal. It strikes the ground 3.00 s later. (a) How far horizontally from the base of the building does the ball strike the ground? (b) Find the height from which the ball was thrown. (c) How long does it take the ball to reach a point 10.0 m below the level of launching?

17. A ball is thrown horizontally from the top of a building 35.0 m high. The ball strikes the ground at a point 80.0 m from the base of the building. Find (a) the time the ball is in flight, (b) its initial velocity, and (c) the x and y components of velocity just before the ball strikes the ground.

18. A cannon with a muzzle speed of 1000 m/s is used to destroy a target on a mountain top. The target is 2000 m from the cannon horizontally and 800 m above the ground. At what angle, relative to the ground, should the cannon be fired?

19. A placekicker must kick a football from a point 36.0 m (about 40 yards) from the goal and the ball must clear the crossbar, which is 3.05 m high. The ball leaves the ground when kicked with a speed of 20.0 m/s at an angle of 53.0°

to the horizontal. (a) By how much does the ball clear or fall short of clearing the crossbar? (b) Does the ball approach the crossbar while still rising or while falling?

20. A fireman, 50.0 m away from a burning building, directs a stream of water from a fire hose at an angle of 30.0° above the horizontal. If the speed of the stream is 40.0 m/s, at what height will the stream of water strike the building?

21. The speed of a projectile when it reaches its maximum height is one half of its speed when it is at half its maximum height. What is the initial projection angle of the projectile?

22. A soccer player kicks a rock horizontally off a 40.0-m–high cliff into a pool of water. If the player hears the sound of the splash 3.00 s later, what was the initial speed given to the rock? Assume the speed of sound in air to be 343 m/s.

23. During the 1968 Olympics in Mexico City, Bob Beamon executed a record long jump. The horizontal distance he achieved was 8.90 m. His center of gravity started at an elevation of 1.00 m, reached a maximum height of 1.90 m, and finished at 0.150 m. From these data determine (a) his time of flight, (b) his horizontal and vertical velocity components at the takeoff time, and (c) his takeoff angle.

Section 3.4 Uniform Circular Motion

24. The orbit of the Moon around the Earth is approximately circular, with a mean radius of 3.84×10^8 m. It takes 27.3 days for the Moon to complete one revolution around the Earth. Find (a) the mean orbital speed of the Moon and (b) its centripetal acceleration.

25. An athlete rotates a 1.00-kg discus along a circular path of radius 1.06 m (Figure P3.25). The maximum speed of the discus is 20.0 m/s. Determine the magnitude of the maximum radial acceleration of the discus.

26. A tire 0.500 m in radius rotates at a constant rate of 200 rev/min. Find the speed and magnitude of acceleration of a small stone lodged in the outer edge of the tread of the tire.

27. Young David who slew Goliath experimented with slings before tackling the giant. He found that he could revolve a sling of length 0.600 m at the rate of 8.00 rev/s. If he increased the length to 0.900 m, he could revolve the sling only 6.00 times per second. (a) Which rate of rotation gives the greater linear speed? (b) What is the centripetal acceleration at 8.00 rev/s? (c) What is the centripetal acceleration at 6.00 rev/s?

28. An astronaut standing on the Moon fires a gun so that the bullet leaves the barrel initially moving in a horizontal direction. (a) What must be the muzzle speed of the bullet so that it travels completely around the Moon and returns to its original location? (b) How long is the bullet in flight? Assume that the free-fall acceleration on the Moon is one sixth that on the Earth.

29. If the rotation of the Earth increased so that the centripetal acceleration was equal to the gravitational acceleration at

Figure P3.25

the Equator, (a) what would be the tangential speed of a person standing at the Equator, and (b) how long would a day be?

Section 3.5 Tangential and Radial Acceleration

30. An automobile, the speed of which is increasing at a rate of 0.600 m/s², travels along a circular road of radius 20.0 m. When the instantaneous speed of the automobile is 4.00 m/s, find (a) the tangential acceleration component, (b) the centripetal acceleration component, and (c) the magnitude and direction of the total acceleration.

31. A train slows down as it rounds a sharp horizontal turn, slowing from 90.0 km/h to 50.0 km/h in the 15.0 s that it takes to round the bend. The radius of the curve is 150 m. Compute the acceleration at the moment the train speed reaches 50.0 km/h, assuming it continues to slow down at this time.

32. A point on a rotating turntable 20.0 cm from the center accelerates from rest to a final speed of 0.700 m/s in 1.75 s. At $t = 1.25$ s, find the magnitude and direction of

(a) the centripetal acceleration, (b) the tangential acceleration, and (c) the total acceleration of the point.

33. Figure P3.33 represents the total acceleration of a particle moving clockwise in a circle of radius 2.50 m at a given instant of time. At this instant, find (a) the centripetal acceleration, (b) the speed of the particle, and (c) its tangential acceleration.

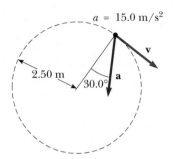

$a = 15.0 \text{ m/s}^2$

2.50 m

30.0°

Figure P3.33

34. A ball swings in a vertical circle at the end of a 1.50-m long rope. When it is 36.9° past the lowest point on its way up, the ball's total acceleration is $(-22.5\mathbf{i} + 20.2\mathbf{j})$ m/s². At that instant, (a) sketch a vector diagram showing the components of its acceleration, (b) determine the magnitude of its centripetal acceleration, and (c) determine the speed and velocity of the ball.

Additional Problems

35. At $t = 0$ a particle leaves the origin with a velocity of 6.00 m/s in the positive y direction. Its acceleration is given by $\mathbf{a} = (2.00\mathbf{i} - 3.00\mathbf{j})$ m/s². When the particle reaches its *maximum* y coordinate, its y component of velocity is zero. At this instant, find (a) the velocity of the particle and (b) its x and y coordinates.

36. A ball on the end of a string is whirled around in a horizontal circle of radius 0.300 m. The plane of the circle is 1.20 m above the ground. The string breaks and the ball lands 2.00 m (horizontally) away from the point on the ground directly beneath the ball's location when the string breaks. Find the centripetal acceleration of the ball during its circular motion.

37. A home run is hit in such a way that the baseball just clears a wall 21.0 m high, located 130 m from home plate. The ball is hit at an angle of 35.0° to the horizontal, and air resistance is negligible. Find (a) the initial speed of the ball, (b) the time it takes the ball to reach the wall, and (c) the velocity components and the speed of the ball when it reaches the wall. (Assume the ball is hit at a height of 1.00 m above the ground.)

38. The astronaut orbiting the Earth in the photograph is preparing to dock with a spinning Westar VII satellite. The satellite is in a circular orbit 600 km above the Earth's surface, where the free-fall acceleration is 8.21 m/s². The radius of the Earth is 6400 km. Determine the speed of the satellite and the time required to complete one orbit around the Earth.

Figure P3.38   *(Courtesy of NASA)*

39. A particle has velocity components

$$v_x = +4.00 \text{ m/s} \qquad v_y = -(6.00 \text{ m/s}^2)t + 4.00 \text{ m/s}$$

Calculate the speed of the particle and the direction $\theta = \tan^{-1}(v_y/v_x)$ of the velocity vector at $t = 2.00$ s.

40. When baseball outfielders throw the ball in, they usually allow it to take one bounce on the theory that the ball arrives sooner this way. Suppose that after the bounce the ball rebounds at the same angle θ as before, as in Figure P3.40 but loses half its speed. (a) Assuming the ball is always

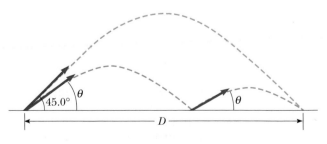

45.0° θ θ

D

Figure P3.40

thrown with the same initial speed, at what angle θ should the ball be thrown in order to go the same distance D with one bounce (blue path) as one thrown upward at 45.0° with no bounce (green path)? (b) Determine the ratio of the times for the one-bounce and no-bounce throws.

41. A girl can throw a ball a maximum horizontal distance of 40.0 m on a level field. How far can she throw the same ball vertically upward? Assume that her muscles give the ball the same speed in each case.

42. A girl can throw a ball a maximum horizontal distance of R on a level field. How far can she throw the same ball vertically upward? Assume that her muscles give the ball the same speed in each case.

43. A bomber is flown horizontally, with a ground speed of 275 m/s, at an altitude of 3000 m over level terrain. Neglect the effects of air resistance. (a) How far from the point that is vertically under the point of release will a bomb hit the ground? (b) If the plane maintains its original course and speed, where will it be when the bomb hits the ground? (c) For the preceding conditions, at what angle from the vertical at the point of release must the telescopic bomb sight be set so that the bomb will hit the target seen in the sight at the time of release?

44. A football is thrown toward a receiver with an initial speed of 20.0 m/s, at an angle of 30.0° above the horizontal. At that instant, the receiver is 20.0 m from the quarterback. In what direction and with what constant speed should the receiver run in order to catch the football at the level at which it was thrown?

45. A hawk is flying horizontally at 10.0 m/s in a straight line, 200 m above the ground. A mouse it has been carrying is released from its grasp. The hawk continues on its path at the same speed for 2.00 s before attempting to retrieve its prey. To accomplish the retrieval, it dives in a straight line at constant speed and recaptures the mouse 3.00 m above the ground. (a) Assuming no air resistance, find the diving speed of the hawk. (b) What angle did the hawk make with the horizontal during its descent? (c) For how long did the mouse "enjoy" free fall?

46. A rocket is launched at an angle of 53.0° to the horizontal with an initial speed of 100 m/s. For 3.00 s it moves along its initial line of motion with an acceleration of 30.0 m/s². Then its engines fail and the rocket proceeds to move in free fall. Find (a) the maximum altitude reached by the rocket, (b) its total time of flight, and (c) its horizontal range.

47. A car is parked on a steep incline overlooking the ocean, where the incline makes an angle of 37.0° with the horizontal. The negligent driver leaves the car in neutral, and the parking brakes are defective. The car rolls from rest down the incline with a constant acceleration of 4.00 m/s², traveling 50.0 m to the edge of a cliff. The cliff is 30.0 m above the ocean. Find (a) the speed of the car when it reaches the edge of the cliff and the time it takes to get there,

(b) the velocity of the car when it lands in the ocean, (c) the total time the car is in motion, and (d) the position of the car when it lands in the ocean, relative to the base of the cliff.

48. The determined coyote is out once more to try to capture the elusive roadrunner. The coyote wears a pair of Acme jet-powered roller skates, which provide a constant horizontal acceleration of 15.0 m/s², as pictured in Figure P3.48. The coyote starts off at rest 70.0 m from the edge of a cliff at the instant the roadrunner zips past him in the direction of the cliff. (a) If the roadrunner moves with constant speed, determine the minimum speed he must have in order to reach the cliff before the coyote. (b) If the cliff is 100 m above the base of a canyon, determine where the coyote lands in the canyon (assume his skates are still in operation when he is in flight). (c) Determine the coyote's velocity components just before he lands in the canyon. (As usual, the roadrunner saves himself by making a sudden turn at the cliff.)

Coyoté   Chicken
Stupidus   Delightus

BEEP
BEEP

Figure P3.48

49. A skier leaves the ramp of a ski jump with a velocity of 10.0 m/s, 15.0° above the horizontal, as in Figure P3.49. The slope is inclined at 50.0°, and air resistance is negligi-

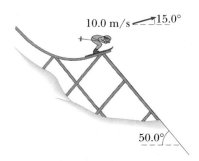

10.0 m/s — 15.0°

50.0°

Figure P3.49

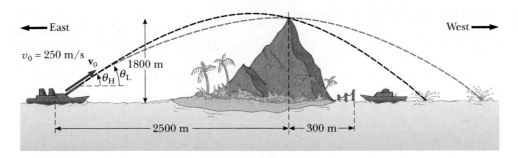

Figure P3.51

ble. Find (a) the distance from the ramp to where the jumper lands and (b) the velocity components just before the landing. (How do you think the results might be affected if air resistance were included? Note that jumpers lean forward in the shape of an airfoil, with their hands at their sides, to increase their distance. Why does this work?)

50. Describe what you might do to give your hand a large acceleration without hurting yourself or striking your hand against anything. Compute an order-of-magnitude estimate of this acceleration, stating the quantities you measure or estimate and their values.

51. An enemy ship is on the east side of a mountain island, as shown in Figure P3.51. The enemy ship can maneuver to within 2500 m of the 1800-m–high mountain peak and can shoot projectiles with an initial speed of 250 m/s. If the western shoreline is horizontally 300 m from the peak, what are the distances from the western shore at which a ship can be safe from the bombardment of the enemy ship?

Spreadsheet Problems

S1. The punter for the St. Louis Rams needs to kick the football 40 yd to 45 yd (36.6 m to 41.1 m) to keep the opposing team inside the 5-yd line. Use Spreadsheet 3.1 to estimate the angle at which he should kick the ball. *Note:* He must keep the ball in the air as long as possible in order to help prevent a runback (in other words, he wants a long "hang time"). The initial speed of the ball should be in the range of 20 m/s to 30 m/s.

S2. Modify Spreadsheet 3.1 by adding two columns and then calculate the maximum height reached by the football in Problem S1 and the total time it is in the air.

S3. The x and y coordinates of a projectile are given by Equations 3.12 and 3.13. Using these equations, set up a spreadsheet to calculate the x and y coordinates of a projectile and the components of its velocity v_x and v_y as functions of time. The initial speed and initial angle of the projectile should be input parameters. Use the graph function of the spreadsheet to plot the x and y coordinates as functions of time. Also graph v_x and v_y as functions of time. Use this spreadsheet to solve the following problem.

The New York Giants are tied with the Chicago Bears with only a few seconds left in the game. The Giants have the football and call the placekicker into the game. He must kick the ball 52 yd (47.5 m) for a field goal. If the crossbar of the goal is 10 ft (3.05 m) high, at what angle and speed should he kick the ball to score the winning points? Is there only one solution to this problem? Solutions can be found by varying the parameters and studying the graph of y versus x.

ANSWERS TO CONCEPTUAL PROBLEMS

1. (a) Yes. Although its speed may be constant, the direction of its motion (that is, the direction of **v**) may change, causing an acceleration. For example, an object moving in a circle with constant speed has an acceleration directed toward the center of the circle. (b) No. An object that moves with constant velocity has zero acceleration. Constant velocity means that both the direction and magnitude of **v** remain constant.

2. A constant velocity is interpreted as constant in both magnitude and direction. Thus, a constant velocity clearly implies a nonvarying speed. On the other hand, a constant speed indicates that only the magnitude of the velocity vector is constant—the direction can change. Thus, as long as a particle covers the same distance in equal time intervals, the speed is constant, and its direction can change continuously, giving a varying velocity. A familiar example is a particle moving in a circular path at constant speed, such that the velocity is constantly changing direction.

3. You should simply throw it straight up in the air. Because the ball was moving along with you, it would follow a parabolic trajectory with a horizontal component of velocity that is the same as yours.

4. (a) At the top of its flight, **v** is horizontal and **a** is vertical. This is the only point at which the velocity and acceleration vectors are perpendicular. (b) If the object is thrown straight up or down, then **v** and **a** will be parallel throughout the downward motion. Otherwise, the velocity and acceleration vectors are never parallel.

The Wizard of Id   by Parker and Hart

By permission of John Hart and Field Enterprises, Inc.

4

The Laws of Motion

In the preceding two chapters on kinematics, we described the motions of particles based on the definitions of displacement, velocity, and acceleration. However, we would like to be able to answer specific questions related to the causes of motion, such as, "What mechanism causes motion?" and, "Why do some objects accelerate at higher rates than others?" In this chapter we shall describe the change in motion of particles using the concepts of force and mass. We then discuss the three fundamental laws of motion, which are based on experimental observations and were formulated about three centuries ago by Sir Isaac Newton.

Classical mechanics describes the relationship between the motion of a body and the forces acting on it. Classical mechanics deals only with objects that are large compared with the dimensions of atoms ($\approx 10^{-10}$ m) and that move at speeds much slower than the speed of light (3.00×10^8 m/s).

We learn in this chapter how to describe the acceleration of an object in terms of the resultant force acting on the object and its mass. This force represents the interaction of the object with its environment. Mass is a measure of the object's tendency to resist an acceleration when a force acts on it.

We also discuss *force laws*, which describe how to

Calf-roping, as shown in this photograph taken in Steamboat, Colorado, is a standard rodeo event. The external forces exerted on the horse are the force of friction between the horse and ground, the downward force of gravity, the force exerted by the rope attached to the calf, the downward force exerted by the cowboy on the horse, and the upward force of the ground. Can you identify the forces acting on the calf? *(Fourbyfive, Inc.)*

calculate the force on an object if its environment is known. We shall see that, although the force laws are simple in form, they successfully explain a wide variety of phenomena and experimental observations. They, together with the laws of motion, are the foundations of classical mechanics.

4.1 • THE CONCEPT OF FORCE

Everyone has a basic understanding of the concept of force as a result of everyday experiences. When you push or pull an object, you exert a force on it. You exert a force when you throw or kick a ball. In these examples, the word *force* is associated with the result of muscular activity and with some change in the state of motion of an object. Forces do not always cause an object to move, however. For example, as you sit reading this book, the force of gravity acts on your body, and yet you remain stationary. Likewise, you can push on a block of stone and yet fail to move it.

What force (if any) causes a distant star to drift freely through space? Newton answered such questions by stating that the change in velocity of an object is caused by unbalanced forces. Therefore, if an object moves with uniform motion (constant velocity), no force is required to maintain the motion. Because only a force can cause a change in velocity, we can think of force as that which causes a body to accelerate.

- *A body accelerates due to an external force.*

Now consider a situation in which several forces act simultaneously on an object. In this case, the object accelerates only if the **net force** acting on it is not equal to zero. We shall often refer to the net force as the *resultant force* or as the *unbalanced force.* **If the net force is zero, the acceleration is zero and the velocity of the object remains constant.** That is, if the net force acting on the object is zero, either the object will be at rest or it will move with constant velocity. **When a body has constant velocity or is at rest, it is said to be in equilibrium.**

- *Definition of equilibrium*

This chapter is concerned with the relation between the force on an object and the acceleration of that object. If you pull on a spring, as in Figure 4.1a, the spring stretches. If the spring is calibrated, the distance it stretches can be used to measure the strength of the force. If a child pulls hard enough on a cart to overcome friction, as in Figure 4.1b, the cart moves. When a football is kicked, as in Figure 4.1c, it is both deformed and set in motion. These are all examples of a class of forces called *contact forces.* That is, they represent the result of physical contact between two objects.

Another class of forces, which do not involve physical contact between two objects but act through empty space, are known as *field forces.* The force of gravitational attraction between two objects is an example of this class of force, illustrated in Figure 4.1d. This gravitational force keeps objects bound to the Earth and gives rise to what we commonly call the *weight* of an object. The planets of our solar system are bound under the action of gravitational forces. Another common example of a field force is the electric force that one electric charge exerts on another electric charge, as in Figure 4.1e. These charges might be an electron and proton forming a hydrogen atom. A third example of a field force is the force that a bar magnet exerts on a piece of iron, as shown in Figure 4.1f. The forces holding an atomic nucleus together are also field forces but are usually very short-range. They are the dominating interaction for particle separations of the order of 10^{-15} m.

Early scientists, including Newton, were uneasy with the concept of a force acting between two disconnected objects. To overcome this conceptual problem,

A football is set in motion as a result of the contact force **F** on it due to the kicker's foot. The ball is distorted in the short time it is in contact with the foot.   *(Ralph Cowan, Tony Stone Worldwide)*

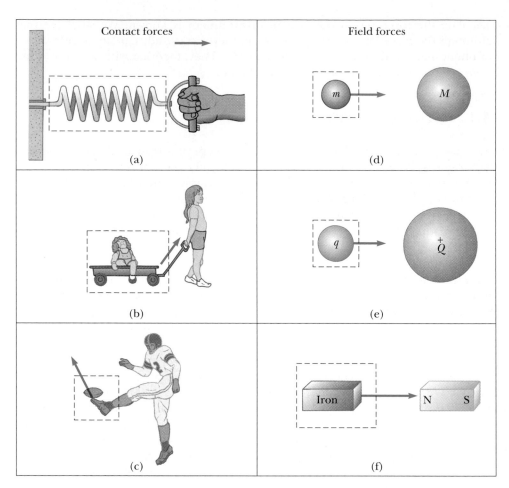

Figure 4.1   Some examples of forces applied to various objects. In each case a force is exerted on the particle or object within the boxed area. The environment external to the boxed area provides this force.

Michael Faraday (1791–1867) introduced the concept of a *field*. According to this approach, when a mass m_1 is placed at some point P near a mass m_2, one can say that m_1 interacts with m_2 by virtue of the gravitational field that exists at P. The field at P is created by mass m_2. Likewise, a field exists at the position of m_2 created by m_1. In fact, all objects create a gravitational field in the space around them.

The distinction between contact forces and field forces is not as sharp, however, as the previous section might lead you to believe. At the atomic level, all the forces we classify as contact forces turn out to be due to repulsive electrical (field) forces of the type illustrated in Figure 4.1e. Nevertheless, in developing models for macroscopic phenomena, it is convenient to use both classifications of forces. However, the only known *fundamental* forces in nature are all field forces: (1) gravitational attractions between objects, (2) electromagnetic forces between charges, (3) strong nuclear forces between subatomic particles, and (4) weak nuclear forces that arise in certain radioactive decay processes. In classical physics we are concerned only with gravitational and electromagnetic forces.

Fundamental forces in nature •

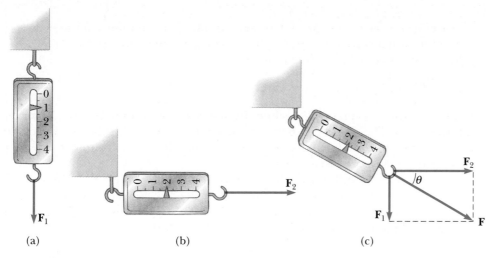

(a)   (b)   (c)

Figure 4.2   The vector nature of a force is tested with a spring scale. (a) The downward vertical force $\mathbf{F}_1$ elongates the spring 1 unit. (b) The horizontal force $\mathbf{F}_2$ elongates the spring 2 units. (c) The combination of $\mathbf{F}_1$ and $\mathbf{F}_2$ elongates the spring $\sqrt{1^2 + 2^2} = \sqrt{5}$ units.

It is sometimes convenient to use the deformation of a spring to measure force. Suppose a force is applied vertically to a spring that has a fixed upper end, as in Figure 4.2a. We can calibrate the spring by defining the unit force, $\mathbf{F}_1$, as the force that produces an elongation of 1.00 cm. If a force $\mathbf{F}_2$, applied horizontally as in Figure 4.2b, produces an elongation of 2.00 cm, the magnitude of $\mathbf{F}_2$ is 2 units. If the two forces $\mathbf{F}_1$ and $\mathbf{F}_2$ are applied simultaneously, as in Figure 4.2c, the elongation of the spring is $\sqrt{5} = 2.24$ cm. The single force $\mathbf{F}$ that would produce this same elongation is the vector sum of $\mathbf{F}_1$ and $\mathbf{F}_2$, as described in Figure 4.2c. That is, $|\mathbf{F}| = \sqrt{F_1^2 + F_2^2} = \sqrt{5}$ units, and its direction is $\theta = \arctan(-0.500) = -26.6°$. **Because forces are vectors, you must use the rules of vector addition to get the resultant force on a body.**

Springs that elongate in proportion to an applied force are said to obey **Hooke's law.** Such springs can be constructed and calibrated to measure unknown forces.

4.2 • NEWTON'S FIRST LAW AND INERTIAL FRAMES

Before about 1600, scientists believed that the natural state of matter was the state of rest. Galileo was the first to take a different approach to motion and the natural state of matter. He devised thought experiments, such as an object moving on a frictionless surface, and concluded that it is not the nature of an object to stop once it is set in motion; rather, it is an object's nature to resist changes in its motion. In his words, "Any velocity, once imparted to a moving body, will be rigidly maintained as long as the external causes of retardation are removed."

This new approach to motion was later formalized by Newton in a statement that has come to be known as **Newton's first law of motion:**

An object at rest remains at rest, and an object in motion continues in motion with a constant velocity (that is, constant speed in a straight line), unless it experiences a net external force.

A statement of Newton's first •
law

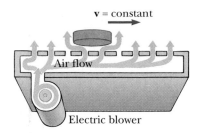

v = constant

Air flow

Electric blower

Figure 4.3 A disk moving on a layer of air is an example of uniform motion—that is, motion in which the acceleration is zero.

In simpler terms, we can say that **when the net force on a body is zero, its acceleration is zero.** That is, when $\Sigma\mathbf{F} = 0$, then $\mathbf{a} = 0$. From the first law, we conclude that an isolated body (a body that does not interact with its environment) is either at rest or moving with constant velocity.

One example of uniform motion on a nearly frictionless plane is the motion of a light disk on a layer of air, as in Figure 4.3. If the disk is given an initial velocity, it coasts a great distance before coming to rest. This idea is used in the game of air hockey, where the disk makes many collisions with the walls before coming to rest.

As a second example, consider a spaceship traveling in space, far removed from any planets or other matter. The spaceship requires some propulsion system to change its velocity. However, if the propulsion system is turned off when the spaceship reaches a velocity **v**, the spaceship "coasts" in space with that velocity, and the astronauts get a free ride (that is, no propulsion system is required to keep them moving at the velocity **v**).

Inertial Frames

Inertial frame •

Newton's first law, sometimes called the *law of inertia,* defines a special set of reference frames called inertial frames. **An inertial frame of reference is one in which Newton's first law is valid.** Any reference frame that moves with constant velocity with respect to an inertial frame is itself an inertial frame. A reference frame that moves with constant velocity relative to the distant stars is the best approximation of an inertial frame. The Earth is not an inertial frame because of its orbital motion around the Sun and rotational motion around its own axis. As the Earth travels in its nearly circular orbit around the Sun, it experiences a centripetal acceleration of about 4.4×10^{-3} m/s^2 toward the Sun. In addition, because the Earth rotates around its own axis once every 24 h, a point on the Equator experiences an additional centripetal acceleration of 3.37×10^{-2} m/s^2 toward the center of the Earth. However, these accelerations are small compared with g and can often be neglected. In most situations **we shall assume that a frame on or near the Earth's surface is an inertial frame.**

Thus, if an object is moving with constant velocity, an observer in one inertial frame (say, one at rest with respect to the object) will claim that the acceleration and the resultant force on the object are zero. An observer in *any other* inertial frame will also find that $\mathbf{a} = 0$ and $\mathbf{F} = 0$ for the object. According to the first law, a body at rest and one moving with constant velocity are equivalent. Unless stated otherwise, we shall usually write the laws of motion with respect to an observer "at rest" in an inertial frame.

Thinking Physics 1

When the Copernican Theory of the Solar System was proposed, a natural question arose: "What keeps the Earth and other planets moving in their paths around the Sun?" An interesting response to this question comes from Richard Feynman[1]: "In those days, one of the theories proposed was that the planets went around because behind them there were invisible angels, beating their wings and driving the planets forward. . . . It turns out that in order to keep the planets going around, the invisible angels must fly in a different direction. . . ." What did Feynman mean by this?

Reasoning First, the question asked by those at the time of Copernicus indicates that they did not have a proper understanding of inertia, as described by Newton's first law. At that time in history, before Galileo and Newton, the interpretation was that *motion* was caused by force. This is different from our current understanding that *changes in motion* are caused by force. Thus, it was natural for Copernicus' contemporaries to ask what force propelled a planet in its orbit. According to our current understanding, it is equally natural for us to realize that no force is necessary—the motion simply continues due to inertia. Thus, in Feynman's imagery, we do not need the angels pushing the planet *from behind.* We do need the angels to push *inward,* however, to provide the centripetal acceleration needed to change its direction into a circle. Of course, the angels are not real, from a scientific point of view—they are a metaphor for the *gravitational force.*

CONCEPTUAL PROBLEM 1

Is it possible to have motion in the absence of a force?

CONCEPTUAL PROBLEM 2

If a single force acts on an object, is there an acceleration? If an object experiences an acceleration, is a force acting on it? If an object experiences no acceleration, is there no force acting?

4.3 • INERTIAL MASS

If you attempt to change the velocity of an object, the object resists this change. **Inertia** is solely a property of an individual object; it is a measure of the response of an object to an external force. For instance, consider two large, solid cylinders of equal size, one balsa wood and the other steel. If you were to push the cylinders along a horizontal, rough surface, the force required to give the steel cylinder some acceleration would be larger than the force needed to give the balsa wood cylinder the same acceleration. Therefore, we say that the steel cylinder has more inertia than the balsa wood cylinder.

• *Inertia*

Mass is used to measure inertia, and the SI unit of mass is the kilogram. The greater the mass of a body, the less it accelerates (changes its state of motion) under the action of an applied force.

A quantitative measurement of mass can be made by comparing the accelera-

[1]R. P. Feynman, R. B. Leighton, and M. Sands, *The Feynman Lectures on Physics,* Vol. 1, Reading, Massachusetts, 1963, Addison-Wesley Publishing Co., p. 7-2.

tions that a given force produces on different bodies. Suppose a force acting on a body of mass m_1 produces an acceleration $\mathbf{a}_1$, and the *same force* acting on a body of mass m_2 produces an acceleration $\mathbf{a}_2$. The ratio of the two masses is defined as the *inverse* ratio of the magnitudes of the accelerations produced by the same force:

$$\frac{m_1}{m_2} \equiv \frac{a_2}{a_1} \qquad \textbf{[4.1]}$$

If one mass is standard and known—say, 1 kg—the mass of an unknown object can be obtained from acceleration measurements. For example, if the standard 1-kg mass undergoes an acceleration of 3 m/s² under the influence of some force, a 2-kg mass will undergo an acceleration of 1.5 m/s² under the action of the same force.

Mass is an inherent property of a body, independent of the body's surroundings and of the method used to measure it. It is an experimental fact that **mass is a scalar quantity.** Finally, **mass is a quantity that obeys the rules of ordinary arithmetic.** That is, several masses can be combined in simple numerical fashion. For example, if you combine a 3-kg mass with a 5-kg mass, their total mass is 8 kg. This can be verified experimentally by comparing the acceleration of each object produced by a known force with the acceleration of the combined system using the same force.

Mass should not be confused with weight. **Mass and weight are two different quantities.** The weight of a body is equal to the magnitude of the force exerted by the Earth on the body and varies with location. For example, a person who weighs 180 lb on Earth weighs only about 30 lb on the Moon. However, the mass of a body is the same everywhere. An object having a mass of 2 kg on Earth also has a mass of 2 kg on the Moon.

Isaac Newton (1642–1727)

An English physicist and mathematician, Newton was one of the most brilliant scientists in history. Before the age of 30, he formulated the basic concepts and laws of mechanics, discovered the law of universal gravitation, and invented the mathematical methods of calculus. As a consequence of his theories, Newton was able to explain the motions of the planets, the ebb and flow of the tides, and many special features of the motions of the Moon and the Earth. He also interpreted many fundamental observations concerning the nature of light. His contributions to physical theories dominated scientific thought for two centuries and remain important today.   *(Giraudon/Art Resource)*

4.4 • NEWTON'S SECOND LAW

Newton's first law explains what happens to an object when the net external force on it is zero: It either remains at rest or moves in a straight line with constant speed. Newton's second law answers the question of what happens to an object that has a nonzero net force acting on it.

Imagine you are pushing a block of ice across a frictionless horizontal surface. When you exert some horizontal force $\mathbf{F}$, the block moves with some acceleration $\mathbf{a}$. If you apply a force twice as large, the acceleration doubles. If you increase the applied force to $3\mathbf{F}$, the original acceleration is tripled, and so on. From such observations, we conclude that **the acceleration of an object is directly proportional to the net force acting on it.**

As stated in the preceding section, the acceleration of an object also depends on its mass. This can be understood by considering the following set of experiments. If you apply a force $\mathbf{F}$ to a block of ice on a frictionless surface, the block will undergo some acceleration $\mathbf{a}$. If the mass of the block is doubled, the same applied force will produce an acceleration $\mathbf{a}/2$. If the mass is tripled, the same applied force will produce an acceleration $\mathbf{a}/3$, and so on. We conclude that **the acceleration of an object is inversely proportional to its mass.**

These observations are summarized in **Newton's second law:**

> The acceleration of an object is directly proportional to the net force acting on it and inversely proportional to its mass.

Thus, we can relate mass and force through the following mathematical statement of Newton's second law:[2]

$$\Sigma \mathbf{F} = m\mathbf{a} \qquad [4.2]$$

• *Newton's second law*

You should note that Equation 4.2 is a *vector* expression and hence is equivalent to the following three component equations:

$$\Sigma F_x = ma_x \qquad \Sigma F_y = ma_y \qquad \Sigma F_z = ma_z \qquad [4.3]$$

• *Newton's second law — component form*

Units of Force and Mass

The SI unit of force is the **newton,** which is defined as the force that, when acting on a 1-kg mass, produces an acceleration of 1 m/s². From this definition and Newton's second law, we see that the newton can be expressed in terms of the fundamental units of mass, length, and time:

$$1 \text{ N} \equiv 1 \text{ kg} \cdot \text{m/s}^2 \qquad [4.4]$$

• *Definition of a newton*

The units of force and mass are summarized in Table 4.1. Most of the calculations we shall make in our study of mechanics will be in SI units. Conversion factors between the three systems are given in Appendix A.

Thinking Physics 2

You have most likely had the experience of standing in an elevator that accelerates upward as it leaves to move toward a higher floor. In this case, you *feel* heavier. If you are standing on a bathroom scale at the time, it will *measure* a force larger than your weight. Thus, you have tactile and measured evidence that lead you to believe that you are heavier in this situation. *Are* you heavier?

Reasoning No, you are not—your weight is unchanged. In order to provide the acceleration upward, the floor or the bathroom scale must apply an upward force larger than your weight. It is this larger force that you feel, which you interpret as feeling heavier. A bathroom scale reads this upward force, not your weight, so its reading also increases.

TABLE 4.1   Units of Force, Mass, and Acceleration[a]

Systems of Units	Mass	Acceleration	Force
SI	kg	m/s²	N = kg·m/s²
cgs	g	cm/s²	dyne = g·cm/s²
British engineering	slug	ft/s²	lb = slug·ft/s²

[a] $1 \text{ N} = 10^5 \text{ dyne} = 0.225 \text{ lb}$.

---

[2] Equation 4.2 is valid only when the speed of the object is much less than the speed of light. We will treat the relativistic situation in Chapter 9.

Thinking Physics 3

In a train, the cars are connected by *couplers*. The couplers between the cars are under tension as the train is pulled by the locomotive in the front. As you move from the locomotive to the caboose, does the tension force in the couplers *increase, decrease,* or *stay the same* as the train speeds up? What if the engineer applies the brakes, so that the couplers are under compression? How does the compression force vary from locomotive to caboose in this case?

Reasoning The tension force *decreases* from the front of the train to the back. The coupler between the locomotive and the first car must apply enough force to accelerate all of the remaining cars. As we move back along the train, each coupler is accelerating less mass behind it. The last coupler only has to accelerate the caboose, so it is under the least tension. If the brakes are applied, the couplers are under compression. The force of compression decreases from the front to the back of the train in this case also. The first coupler, at the back of the locomotive, must apply a large force to slow down all of the remaining cars. The final coupler must only apply a force large enough to slow down the mass of the caboose.

Example 4.1   An Accelerating Hockey Puck

A hockey puck with a mass of 0.30 kg slides on the horizontal frictionless surface of an ice rink. Two forces act on the puck as shown in Figure 4.4. The force $\mathbf{F}_1$ has a magnitude of 5.0 N, and $\mathbf{F}_2$ has a magnitude of 8.0 N. Determine the acceleration of the puck.

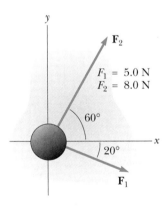

Figure 4.4 (Example 4.1) An object moving on a frictionless surface accelerates in the direction of the *resultant* force, $\mathbf{F}_1 + \mathbf{F}_2$.

Solution The resultant force in the *x* direction is

$$\sum F_x = F_{1x} + F_{2x} = F_1 \cos 20° + F_2 \cos 60°$$
$$= (5.0 \text{ N})(0.940) + (8.0 \text{ N})(0.500) = 8.7 \text{ N}$$

The resultant force in the *y* direction is

$$\sum F_y = F_{1y} + F_{2y} = -F_1 \sin 20° + F_2 \sin 60°$$
$$= -(5.0 \text{ N})(0.342) + (8.0 \text{ N})(0.866) = 5.2 \text{ N}$$

Now we can use Newton's second law in component form to find the *x* and *y* components of acceleration:

$$a_x = \frac{\Sigma F_x}{m} = \frac{8.7 \text{ N}}{0.30 \text{ kg}} = 29 \text{ m/s}^2$$

$$a_y = \frac{\Sigma F_y}{m} = \frac{5.2 \text{ N}}{0.30 \text{ kg}} = 17 \text{ m/s}^2$$

The acceleration has a magnitude of

$$a = \sqrt{(29)^2 + (17)^2} \text{ m/s}^2 = 34 \text{ m/s}^2$$

and its direction is

$$\theta = \tan^{-1}\left(\frac{a_y}{a_x}\right) = \tan^{-1}\left(\frac{17}{29}\right) = 31°$$

relative to the positive *x* axis.

EXERCISE 1 Determine the components of a third force that, when applied to the puck, will cause it to be in equilibrium.   Answer $F_x = -8.7$ N, $F_y = -5.2$ N

EXERCISE 2 A 6.0-kg object undergoes an acceleration of 2.0 m/s². (a) What is the magnitude of the resultant force acting on the object? (b) If this same force is applied to a 4.0-kg object, what acceleration does it produce?   Answer   (a) 12 N   (b) 3.0 m/s²

EXERCISE 3   An 1800-kg car is traveling in a straight line with a speed of 25.0 m/s. What is the magnitude of the constant horizontal force that is needed to bring the car to rest in a distance of 80.0 m?   Answer   7030 N

4.5 • THE GRAVITATIONAL FORCE AND WEIGHT

We are well aware of the fact that all objects are attracted to the Earth. The force exerted by the Earth on an object is the gravitational force $\mathbf{F}_g$ (Fig. 4.5). This force is directed toward the center of the Earth.[3] The magnitude of the gravitational force is called the **weight** of the object, w.

We have seen that a freely falling object experiences an acceleration $\mathbf{g}$ directed toward the center of the Earth. Applying Newton's second law to a freely falling object of mass m, we have $\mathbf{F} = m\mathbf{a}$. Because $\mathbf{F}_g = m\mathbf{g}$, it follows that $\mathbf{a} = \mathbf{g}$ and

$$w = mg \qquad [4.5]$$

Because it depends on g, weight varies with geographic location. (You should not confuse the italicized symbol g that we use for gravitational acceleration with the symbol g that is used for grams.) Bodies weigh less at higher altitudes than at sea level because g decreases with increasing distance from the center of the Earth. Hence, weight, unlike mass, is not an inherent property of a body. For example, if a body has a mass of 70 kg, then its weight in a location where $g = 9.80$ m/s^2 is $mg = 686$ N (about 154 lb). At the top of a mountain where $g = 9.76$ m/s^2, the

Astronaut Edwin E. Aldrin, Jr., walking on the Moon after the Apollo 11 lunar landing. The weight of the astronaut on the Moon is less than it is on Earth but his mass remains the same.   *(Courtesy of NASA)*

Figure 4.5   The only force acting on an object in free fall is the gravitational force, $\mathbf{F}_g$. The magnitude of this force is the weight of the object, mg.

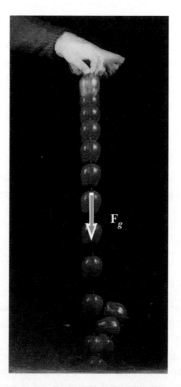

[3]This statement is a simplification in that it ignores the fact that the mass distribution of the Earth is not perfectly spherical.

body's weight would be 683 N. Thus, if you want to lose weight without going on a diet, climb a mountain or weigh yourself at 30 000 ft during an airplane flight.

Because $w = mg$, we can compare the masses of two bodies by measuring their weights with a spring scale or balance. At a given location the ratio of the weights of two bodies equals the ratio of their masses.

Thinking Physics 4

In the absence of air friction, it is claimed that all objects fall with the same acceleration. A heavier object is pulled to the Earth with more force than a light object. Why does the heavier object not fall faster?

Reasoning It is indeed true that the heavier object is pulled with a larger force. The *strength* of the force is determined by the gravitational mass of the object. But the *resistance* to the force and, therefore, to the change in motion of the object, is represented by the object's inertial mass. Inertial mass and gravitational mass are chosen to be equal in Newtonian mechanics. Thus, if an object has twice as much mass as another, it is pulled to the Earth with twice the force, but it also exhibits twice the resistance to having its motion changed. These factors cancel, so that the change in motion, the acceleration, is the same for all objects, regardless of mass.

CONCEPTUAL PROBLEM 3

Suppose you are talking by interplanetary telephone to your friend, who lives on the Moon. He tells you that he has just won a newton of gold in a contest. You tell him that you entered the Earth-version of the same contest and also won a newton of gold! Who is richer?

CONCEPTUAL PROBLEM 4

A baseball of mass m is thrown upward with some initial speed. If air resistance is neglected, what is the force on the ball (a) when it reaches half its maximum height and (b) when it reaches its peak?

4.6 • NEWTON'S THIRD LAW

A statement of Newton's third law •

> Newton's third law states that if two bodies interact, the force exerted on body 1 by body 2 is equal in magnitude but opposite in direction to the force exerted on body 2 by body 1.

$$\mathbf{F}_{12} = -\mathbf{F}_{21} \qquad [4.6]$$

This law, which is illustrated in Figure 4.6a, is equivalent to stating that **forces always occur in pairs,** or that **a single isolated force cannot exist.** The force that body 1 exerts on body 2 is sometimes called the *action force,* and the force of body 2 on body 1 is called the *reaction force.* In reality, either force can be labeled the action or reaction force. **The action force is equal in magnitude to the reaction force and opposite in direction. In all cases, the action and reaction forces act on different objects and must be of the same type.** For example, the force acting on a freely falling projectile is the force of the Earth on the projectile, $\mathbf{F}_g$, and the

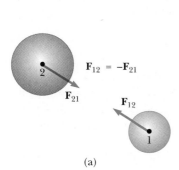

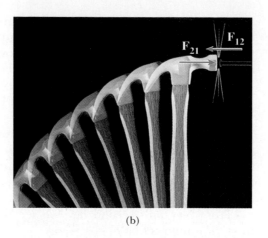

(a)

(b)

Figure 4.6 Newton's third law. (a) The force exerted by object 1 on object 2 is equal in magnitude and opposite the force exerted by object 2 on object 1. (b) The force exerted by the hammer on the nail is equal in magnitude and opposite the force exerted by the nail on the hammer. *(John Gillmoure, The Stock Market)*

magnitude of this force is *mg*. The reaction to this force is the force of the projectile on the Earth, $\mathbf{F}'_g = -\mathbf{F}_g$. The reaction force, $\mathbf{F}'_g$, must accelerate the Earth toward the projectile just as the action force, $\mathbf{F}_g$, accelerates the projectile toward the Earth. However, because the Earth has such a large mass, its acceleration due to this reaction force is negligibly small.

Another example of Newton's third law in action is shown in Figure 4.6b. The force exerted by the hammer on the nail (the action) is equal to and opposite the force exerted by the nail on the hammer (the reaction). You directly experience the law if you slam your fist against a wall or kick a football with your bare foot. You should be able to identify the action and reaction forces in these cases.

As we mentioned earlier, the Earth exerts a force $\mathbf{F}_g$ on any object. If the object is a TV at rest on a table, as in Figure 4.7a, the reaction force to $\mathbf{F}_g$ is the force the TV exerts on the Earth, $\mathbf{F}'_g$. The TV does not accelerate, because it is held up by the table. The table, therefore, exerts on the TV an upward action force, **n**, called

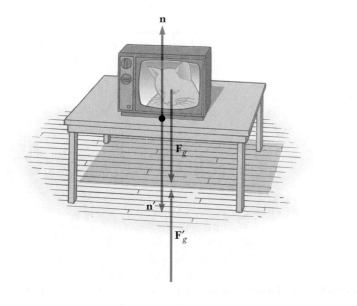

(a)

(b)

Figure 4.7 (a) When a TV set is sitting on a table, the forces acting on the set are the normal force, **n**, and the force of gravity, $\mathbf{F}_g$. The reaction to **n** is the force of the TV set on the table, **n**'. The reaction to $\mathbf{F}_g$ is the force of the TV set on the Earth, $\mathbf{F}'_g$. (b) The free-body diagram for the TV set.

Normal force •   the **normal force.**[4] This is the force that prevents the TV from falling through the table; it can have any value needed, up to the point of breaking the table. The normal force balances the weight and provides equilibrium. The reaction to **n** is the force of the TV on the table, **n**'. Therefore, we conclude that

$$\mathbf{F}_g = -\mathbf{F}'_g \qquad \text{and} \qquad \mathbf{n} = -\mathbf{n}'$$

The forces **n** and **n**' have the same magnitude, which is the same as $\mathbf{F}_g$ unless the table has broken. Note that the forces acting on the TV are $\mathbf{F}_g$ and **n**, as shown in Figure 4.7b. The two reaction forces, $\mathbf{F}'_g$ and **n**', are exerted on objects other than the TV. Remember, the two forces in an action–reaction pair always act on two different objects.

From Newton's second law, we see that, because the TV is in equilibrium (**a** = 0), it follows that $F_g = n = mg$.

Thinking Physics 5

A horse pulls a sled with a horizontal force, causing it to accelerate as in Figure 4.8a. Newton's third law says that the sled exerts an equal and opposite force on the horse. In view of this, how can the sled accelerate? Under what condition does the system (horse plus sled) move with constant velocity?

Reasoning When applying Newton's third law, it is important to remember that the forces involved act on different objects. When you are determining the motion of an object, you must add the forces on that object alone. The force that accelerates the system (horse and sled) is the force exerted by the Earth on the horse's feet. Newton's third law tells us that the horse exerts an equal and opposite force on the Earth. The horizontal forces exerted on the sled are the forward force exerted by the horse and the backward force of friction between sled and surface (Fig. 4.8b). When the forward force exerted by the horse on the sled exceeds the backward force, the sled accelerates to the right. The horizontal forces exerted on the horse are the forward force of the Earth and the backward force of the sled (Fig. 4.8c). The resultant of these two forces causes the horse to accelerate. When the forward force of the Earth on the horse balances the force of friction between sled and surface, the system moves with constant velocity.

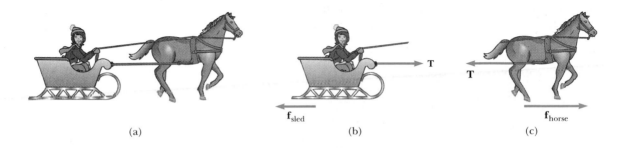

(a)   (b)   (c)

Figure 4.8   (Thinking Physics 5)

[4]The word *normal* is used because the direction of **n** is always *perpendicular* to the surface.

CONCEPTUAL PROBLEM 5

If a small sports car collides head-on with a massive truck, which vehicle experiences the greater impact force? Which vehicle experiences the greater acceleration?

CONCEPTUAL PROBLEM 6

Two forces, $\mathbf{F}_1$ and $\mathbf{F}_2$, are exerted on an object of mass m. $\mathbf{F}_1$ is directed toward the right, whereas $\mathbf{F}_2$ is directed toward the left. Under what condition does the object accelerate toward the right? Toward the left? Can its acceleration ever be zero?

CONCEPTUAL PROBLEM 7

What force causes an automobile to move? A propeller-driven airplane? A rocket? A rowboat?

CONCEPTUAL PROBLEM 8

A large man and a small boy stand facing each other on frictionless ice. They put their hands together and push toward each other, so that they move away from each other. Who exerts the larger force? Who experiences the larger acceleration? Who moves away with the higher velocity? Who moves over a longer distance while their hands are in contact?

4.7 · SOME APPLICATIONS OF NEWTON'S LAWS

In this section we present some simple applications of Newton's laws to bodies that are either in equilibrium ($\mathbf{a} = 0$) or moving linearly under the action of constant external forces. For our model, we shall assume that the bodies behave as particles so that we need not worry about rotational motions. In this section we also neglect the effects of friction for those problems involving motion. This is equivalent to stating that the surfaces are *frictionless*. Finally, we usually neglect the masses of any ropes involved. In this approximation, the magnitude of the force exerted at any point along the rope is the same at all points along the rope. In problem statements, the terms *light* and *of negligible mass* are used to indicate that a mass is to be ignored when you work the problem. These two terms are synonymous in this context.

When we apply Newton's laws to an object, we shall be interested only in those external forces that act *on the object*. For example, in Figure 4.7 the only external forces acting on the TV are $\mathbf{n}$ and $\mathbf{F}_g$. The reactions to these forces, $\mathbf{n}'$ and $\mathbf{F}_g'$, act on the table and on the Earth, respectively, and do not appear in Newton's second law as applied to the TV.

When an object such as a block is being pulled by a rope attached to it, the rope exerts a force on the object. In general, **tension** is a scalar and is defined as the magnitude of the force that the rope exerts on whatever is attached to it.

• *Tension*

Consider a crate being pulled to the right on the frictionless, horizontal surface, as in Figure 4.9a. Suppose you are asked to find the acceleration of the crate and the force the floor exerts on it. First, note that the horizontal force being applied to the crate acts through the rope. The force the rope exerts on the crate is denoted by the symbol $\mathbf{T}$. The magnitude of $\mathbf{T}$ is equal to the tension in the rope. In Figure 4.9a, a dotted circle is drawn around the block to remind you to isolate it from its surroundings.

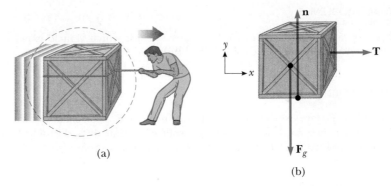

Figure 4.9   (a) A crate being pulled to the right on a frictionless surface. (b) The free-body diagram that represents the external forces on the crate.

Because we are interested only in the motion of the block, we must be able to identify all external forces acting on it. These are illustrated in Figure 4.9b. In addition to the force **T**, the force diagram for the block includes the force of gravity, $\mathbf{F}_g$, and the normal force, **n**, exerted by the floor on the crate. Such a force diagram is referred to as a **free-body diagram.** The construction of a correct free-body diagram is an essential step in applying Newton's laws. The *reactions* to the forces we have listed—namely, the force exerted by the rope on the hand, the force exerted by the crate on the Earth, and the force exerted by the crate on the floor— are not included in the free-body diagram, because they act on *other* bodies and not on the crate.

> *Free-body diagrams are important when applying Newton's laws.*

Now let us apply Newton's second law to the block. First we must choose an appropriate coordinate system. In this case it is convenient to use the coordinate system shown in Figure 4.9b, with the x axis horizontal and the y axis vertical. We can apply Newton's second law in the x direction, y direction, or both, depending on what we are asked to find in the problem. In addition, we may be able to use the equations of motion for constant acceleration that are found in Chapter 2. However, you should use these equations only when the acceleration is constant. For example, if the force **T** in Figure 4.9 is constant, then it follows that the acceleration in the x direction is also constant, because $\mathbf{a}_x = \mathbf{T}/m$. Hence, if we need to find the displacement or the velocity of the block at some instant of time, we can use the equations of motion with constant acceleration.

Objects in Equilibrium and Newton's First Law

Objects that are either at rest or moving with constant velocity are said to be in equilibrium, and Newton's first law is a statement of one condition that must be true for equilibrium conditions to prevail. In equation form, this condition of equilibrium can be expressed as

> *First condition of equilibrium*

$$\sum \mathbf{F} = 0 \qquad\qquad [4.7]$$

This statement signifies that the *vector* sum of all the forces (the net force) acting on an object in equilibrium is zero.[5]

---

[5]This is only one condition of equilibrium. A second condition of equilibrium is a statement of rotational equilibrium. This condition will be discussed in Chapter 10.

Usually, the problems we encounter in our study of equilibrium are easier to solve if we work with Equation 4.7 in terms of the components of the external forces acting on an object. By this we mean that, in a two-dimensional problem, the sum of all the external forces in the x and y directions must separately equal zero; that is,

$$\sum F_x = 0 \quad \text{and} \quad \sum F_y = 0 \qquad \text{[4.8]}$$

This set of equations is often referred to as the **first condition for equilibrium.** We shall not consider three-dimensional problems in this book, but the extension of Equations 4.8 to a three-dimensional situation can be made by adding a third equation, $\sum F_z = 0$.

PROBLEM-SOLVING STRATEGY • **Objects in Equilibrium**

The following procedure is recommended for problems involving objects in equilibrium:

1. Make a sketch of the object under consideration.
2. Draw a free-body diagram for the *isolated* object, and label all external forces acting on the object. Assume a direction for each force. If you select a direction that leads to a negative sign in your solution for a force, do not be alarmed; this merely means that the direction of the force is the opposite of what you assumed.
3. Resolve all forces into x and y components, choosing a convenient coordinate system.
4. Use the equations $\sum F_x = 0$ and $\sum F_y = 0$. Remember to keep track of the signs of the various force components.
5. Application of the last step leads to a set of equations with several unknowns. Solve the simultaneous equations for the unknowns in terms of the known quantities.

Example 4.2   A Traffic Light at Rest

A traffic light weighing 125 N hangs from a cable tied to two other cables fastened to a support, as in Figure 4.10a. The upper cables make angles of 37.0° and 53.0° with the horizontal. Find the tension in the three cables.

Reasoning   We must construct two free-body diagrams in order to work this problem. The first of these is for the traffic light, shown in Figure 4.10b; the second is for the knot that holds the three cables together, as in Figure 4.10c. This knot is a convenient point to choose because all the forces we are interested in act through this point. Because the acceleration of the system is zero, we can use the condition that the net force on the light is zero, and the net force on the knot is zero.

Solution   First we construct a free-body diagram for the traffic light, as in Figure 4.10b. The force exerted by the vertical cable, $\mathbf{T}_3$, supports the light, and so $T_3 = w = 125$ N. Next, we choose the coordinate axes as shown in Figure 4.10c and resolve the forces into their x and y components:

Force	x component	y component
$\mathbf{T}_1$	$-T_1 \cos 37.0°$	$T_1 \sin 37.0°$
$\mathbf{T}_2$	$T_2 \cos 53.0°$	$T_2 \sin 53.0°$
$\mathbf{T}_3$	0	-125 N

The first condition for equilibrium gives us the equations

(1) $\sum F_x = T_2 \cos 53.0° - T_1 \cos 37.0° = 0$

(2) $\sum F_y = T_1 \sin 37.0° + T_2 \sin 53.0° - 125 \text{ N} = 0$

From (1) we see that the horizontal components of $\mathbf{T}_1$ and $\mathbf{T}_2$ must be equal in magnitude, and from (2) we see that the sum of the vertical components of $\mathbf{T}_1$ and $\mathbf{T}_2$ must balance the weight of the light. We can solve (1) for T_2 in terms of T_1 to give

$$T_2 = T_1 \left(\frac{\cos 37.0°}{\cos 53.0°} \right) = 1.33 T_1$$

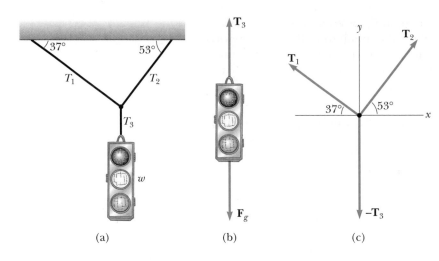

(a)   (b)   (c)

Figure 4.10 (Example 4.2) (a) A traffic light suspended by cables. (b) Free-body diagram for the traffic light. (c) Free-body diagram for the knot in the cable.

This value for T_2 can be substituted into (2) to give

$$T_1 \sin 37.0° + (1.33T_1)(\sin 53.0°) - 125 \text{ N} = 0$$

$$T_1 = 75.1 \text{ N}$$

$$T_2 = 1.33T_1 = 99.9 \text{ N}$$

EXERCISE 4 In what situation will $T_1 = T_2$?
Answer   The angles must be equal.

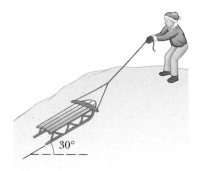

Figure 4.11 (Exercise 5) A child holding a sled on a frictionless hill.

EXERCISE 5 A child holds a sled weighing 77.0 N at rest on a frictionless snow-covered incline as in Figure 4.11. Find (a) the magnitude of the force the child must exert on the rope, and (b) the magnitude of the force the incline exerts on the sled. (c) What happens to the normal force as the angle of incline increases? (d) Under what conditions would the normal force equal the weight of the sled?
Answer (a) 38.5 N   (b) 66.7 N   (c) It decreases.   (d) If the sled were on a horizontal surface and the applied force were either zero or along the horizontal

Accelerating Objects and Newton's Second Law

In a situation in which a net force is acting on an object, the object is accelerated, and we use Newton's second law in order to determine the features of the motion. The representative problems and suggestions that follow should help you to solve problems of this kind.

PROBLEM-SOLVING STRATEGY • Newton's Second Law

The following procedure is recommended when dealing with problems involving the application of Newton's second law:

1. Draw a diagram of the system.
2. Isolate the object the motion of which is being analyzed. Draw a free-body diagram for this object, showing *all external forces acting on it.* For

systems containing more than one object, draw a *separate* diagram for each object.

3. Establish convenient coordinate axes for each object and find the components of the forces along those axes. Apply Newton's second law, $\Sigma \mathbf{F} = m\mathbf{a}$, in the x and y directions for each object.

4. Solve the component equations for the unknowns. Remember that in order to obtain a complete solution, you must have as many independent equations as you have unknowns.

5. If necessary, use the equations of kinematics (motion with constant acceleration) from Chapter 2 to find all the unknowns.

Example 4.3   A Crate on a Frictionless Incline

A crate of mass m is placed on a frictionless, inclined plane of angle θ, as in Figure 4.12a. (a) Determine the acceleration of the crate after it is released.

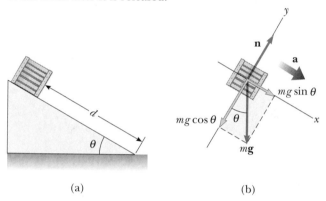

(a)   (b)

Figure 4.12 (Example 4.3) (a) A crate sliding down a frictionless incline. (b) Free-body diagram for the crate. Note that the magnitude of its acceleration along the incline is $g \sin \theta$.

Reasoning Because the forces acting on the crate are known, Newton's second law can be used to determine its acceleration. First, we construct the free-body diagram for the crate as in Figure 4.12b. The only forces on the crate are the normal force **n** acting perpendicular to the plane and the force of gravity $\mathbf{F}_g$ acting vertically downward. For problems of this type involving inclined planes, **it is convenient to choose the coordinate axes with x along the incline and y perpendicular to it.** Then, we replace $\mathbf{F}_g$ by a component of magnitude $mg \sin \theta$ along the *positive x* axis and one of magnitude $mg \cos \theta$ in the *negative y* direction.

Solution Applying Newton's second law to the crate in component form while noting that $a_y = 0$ gives

$$(1) \quad \sum F_x = mg \sin \theta = ma_x$$

$$(2) \quad \sum F_y = n - mg \cos \theta = 0$$

From (1) we see that the acceleration along the incline is provided by the component of the force of gravity down the incline:

$$(3) \quad a_x = g \sin \theta$$

From (2) we conclude that the component of the force of gravity perpendicular to the incline is *balanced* by the normal force; that is, $n = mg \cos \theta$. Note that the acceleration given by (3) is *independent* of the mass of the crate—it depends only on the angle of inclination and on g.

Special Cases When $\theta = 90°$, $a = g$ and $n = 0$. This case corresponds to the crate in free fall. When $\theta = 0$, $a_x = 0$ and $n = mg$ (its maximum value).

(b) Suppose the crate is released from rest at the top, and the distance from the front edge of the crate to the bottom is d. How long does it take the front edge of the crate to reach the bottom, and what is its speed just as it gets there?

Solution Because $a_x = $ constant, we can apply the Equation 2.10, $x - x_0 = v_{x0}t + \frac{1}{2}a_x t^2$ to the crate. Because the displacement $x - x_0 = d$ and $v_{x0} = 0$, we get $d = \frac{1}{2}a_x t^2$, or

$$(4) \quad t = \sqrt{\frac{2d}{a_x}} = \sqrt{\frac{2d}{g \sin \theta}}$$

Using Equation 2.11, $v_x^2 = v_{x0}^2 + 2a_x(x - x_0)$, with $v_{x0} = 0$, we find that $v_x^2 = 2a_x d$, or

$$(5) \quad v_x = \sqrt{2a_x d} = \sqrt{2gd \sin \theta}$$

Again, t and v_x are *independent* of the mass of the crate. This fact suggests a simple method of measuring g using an inclined air track or some other frictionless incline. Simply measure the angle of inclination, the distance traveled by the crate, and the time it takes to reach the bottom. The value of g can then be calculated from (4) and (5).

Example 4.4   Atwood's Machine

When two unequal masses are hung vertically over a light, frictionless pulley as in Figure 4.13a, the arrangement is called *Atwood's machine*. The device is sometimes used in the laboratory to measure the free-fall acceleration. Calculate the magnitude of the acceleration of the two masses and the tension in the string.

Reasoning The free-body diagrams for the two masses are shown in Figure 4.13b. Two forces act on each block: the upward force exerted by the string, **T**, and the downward force of gravity. Thus, the magnitude of the net force exerted on m_1 is $T - m_1g$, whereas the magnitude of the net force exerted on m_2 is $T - m_2g$. Because the blocks are connected by a string, their accelerations must be equal in magnitude. If we assume that $m_2 > m_1$, then m_1 must accelerate upward, whereas m_2 must accelerate downward.

Solution When Newton's second law is applied to m_1, with **a** upward for this mass (because $m_2 > m_1$), we find (taking upward to be the positive y direction)

$$(1) \quad \sum F_y = T - m_1g = m_1a$$

In a similar way, for m_2 we find

$$(2) \quad \sum F_y = T - m_2g = -m_2a$$

The negative sign on the right-hand side of (2) indicates that m_2 accelerates downward in the negative y direction.

When (2) is subtracted from (1), T drops out and we get

$$-m_1g + m_2g = m_1a + m_2a$$

$$(3) \quad a = \left(\frac{m_2 - m_1}{m_1 + m_2}\right)g$$

If (3) is substituted into (1), we get

$$(4) \quad T = \left(\frac{2m_1m_2}{m_1 + m_2}\right)g$$

Special Cases When $m_1 = m_2$, $a = 0$ and $T = m_1g = m_2g$, as we would expect for the balanced case. Also, if $m_2 \gg m_1$, $a \approx g$ (a freely falling body) and $T \approx 2m_1g$.

EXERCISE 6 Find the acceleration and tension of an Atwood's machine in which $m_1 = 2.00$ kg and $m_2 = 4.00$ kg.
Answer $a = 3.27$ m/s^2; $T = 26.1$ N

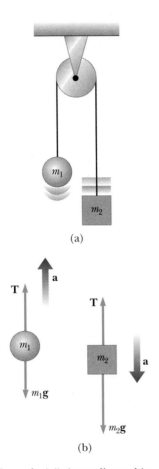

Figure 4.13 (Example 4.4) Atwood's machine. (a) Two masses connected by a light string over a frictionless pulley. (b) Free-body diagrams for m_1 and m_2.

Example 4.5   One Block Pushes Another

Two blocks of masses m_1 and m_2 are placed in contact with each other on a frictionless, horizontal surface, as in Figure 4.14a. A constant horizontal force **F** is applied to m_1 as shown. (a) Find the magnitude of the acceleration of the system.

Reasoning and Solution Both blocks must experience the *same* acceleration because they are in contact with each other. Because **F** is the only horizontal force exerted on the system (the two blocks), we have

$$\sum F_x(\text{system}) = F = (m_1 + m_2)a$$

$$(1) \quad a = \frac{F}{m_1 + m_2}$$

(b) Determine the magnitude of the contact force between the two blocks.

Reasoning and Solution To solve this part of the problem, first construct a free-body diagram for each block, as shown

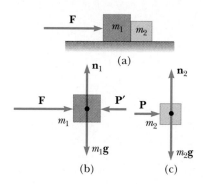

Figure 4.14   (Example 4.5)

in Figures 4.14b and 4.14c, where the contact force is denoted by **P**. From Figure 4.14c we see that the only horizontal force acting on m_2 is the contact force **P** (the force exerted by m_1 on m_2), which is directed to the right. Applying Newton's second law to m_2 gives

$$(2) \quad \sum F_x = P = m_2 a$$

Substituting the value of the acceleration a given by (1) into (2) gives

$$(3) \quad P = m_2 a = \left(\frac{m_2}{m_1 + m_2} \right) F$$

From this result, we see that the contact force **P** is *less* than the applied force **F**. This is consistent with the fact that the force required to accelerate m_2 alone must be less than the force required to produce the same acceleration for the system of two blocks.

It is instructive to check this expression for P by considering the forces acting on m_1, shown in Figure 4.14b. The horizontal forces acting on m_1 are the applied force **F** to the right and the contact force **P′** to the left (the force exerted by m_2 on m_1). From Newton's third law, **P′** is the reaction to **P**, so |**P′**| = |**P**|. Applying Newton's second law to m_2 gives

$$(4) \quad \sum F_x = F - P' = F - P = m_2 a$$

Substituting the value of a from (2) into (4) gives

$$P = F - m_2 a = F - \frac{m_1 F}{m_1 + m_2} = \left(\frac{m_2}{m_1 + m_2} \right) F$$

This agrees with (3), as it must.

EXERCISE 7   If $m_1 = 4.00$ kg, $m_2 = 3.00$ kg, and $F = 9.00$ N, find the magnitude of the acceleration of the system and that of the contact force.   Answer $a = 1.29$ m/s^2; $P = 3.86$ N

Example 4.6   Weighing a Fish in an Elevator

A person weighs a fish on a spring scale attached to the ceiling of an elevator, as shown in Figure 4.15. Show that if the elevator accelerates or decelerates, the spring scale reads a value different from the weight of the fish.

Reasoning   The external forces acting on the fish are the downward force of gravity $m\mathbf{g}$ and the upward force **T** exerted on it by the scale. By Newton's third law, the tension T is also the reading of the spring scale. If the elevator is either at rest or moving at constant velocity, then the fish is not accelerating and $T = mg$. However, if the elevator accelerates in either direction, the tension is no longer equal to the weight of the fish.

Solution   If the elevator accelerates with an acceleration **a** relative to an observer outside the elevator in an inertial frame, then the second law applied to the fish gives the total force on the fish:

$$\sum F = T - mg = ma_y$$

which leads to

$$(1) \quad T = mg + ma_y$$

Thus, we conclude from (1) that the scale reading, T, is greater than the weight, mg, if **a** is upward, as in Figure 4.15a. Furthermore, we see that T is less than mg if **a** is downward, as in Figure 4.15b.

For example, if the weight of the fish is 40.0 N, and a_y is 2.00 m/s^2 upward, then the scale reading is

$$T = mg + ma = mg \left(\frac{a}{g} + 1 \right)$$

$$= w \left(\frac{a}{g} + 1 \right) = (40.0 \text{ N}) \left(\frac{2.00 \text{ m/s}^2}{9.80 \text{ m/s}^2} + 1 \right)$$

$$= \boxed{48.2 \text{ N}}$$

If a_y is 2.00 m/s^2 downward, then

$$T = mg - ma = mg \left(1 - \frac{a}{g} \right) = \boxed{31.8 \text{ N}}$$

Hence, if you buy a fish in an elevator, make sure the fish is weighed while the elevator is at rest or accelerating down-

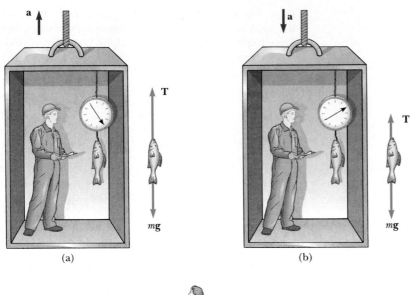

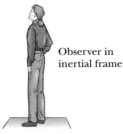

Observer in
inertial frame

Figure 4.15   (Example 4.6). (a) When the elevator accelerates *upward,* the spring scale reads a value *greater* than the weight. (b) When the elevator accelerates *downward,* the spring scale reads a value *less* than the weight.

ward! Furthermore, note that one cannot determine the *direction* of motion of the elevator from the information given here.

Special Cases   If the cable breaks, then the elevator falls freely and $a_y = -g$, and from (1) we see that the tension, T, is zero;

that is, the fish appears to be weightless. If the elevator accelerates *downward* with an acceleration *greater* than g, the fish (along with the person in the elevator) will eventually hit the ceiling, because the acceleration of the fish will still be that of a freely falling body relative to an outside observer.

SUMMARY

Newton's first law states that a body at rest will remain at rest, and a body in uniform motion in a straight line will maintain that motion, unless an external resultant force acts on the body.

Newton's second law states that the acceleration of an object is directly proportional to the resultant force acting on the object and inversely proportional to the object's mass. If the mass of the body is constant, the net force equals the product of the mass and its acceleration, or $\Sigma\mathbf{F} = m\mathbf{a}$.

Newton's first and second laws are valid in an inertial frame of reference. An **inertial frame** is one in which Newton's first law is valid.

The **weight** of a body is equal to the product of its mass (a scalar quantity) and the magnitude of the free-fall acceleration, or $w = mg$.

Newton's third law states that if two bodies interact, the force exerted on body 1 by body 2 is equal in magnitude but opposite in direction to the force exerted on body 2 by body 1. Thus, an isolated force cannot exist in nature.

CONCEPTUAL QUESTIONS

1. Draw a free-body diagram for each of the following objects: (a) a projectile in motion in the presence of air resistance, (b) a rocket leaving the launch pad with its engines operating, (c) an athlete running along a horizontal track.

2. In the motion picture *It Happened One Night* (Columbia Pictures, 1934), Clark Gable is standing inside a stationary bus in front of Claudette Colbert, who is seated. The bus suddenly starts moving forward and Clark falls into Claudette's lap. Why did this happen?

3. As you sit in a chair, the chair pushes up on you with a normal force. This force is equal to your weight and in the opposite direction. Is it the Newton's third law reaction force to your weight?

4. The observer in the elevator of Example 4.6 would claim that the "weight" of the fish is T, the scale reading. This is obviously wrong. Why does this observation differ from that of a person outside the elevator, at rest with respect to the Earth?

5. Identify the action–reaction pairs in the following situations: a man takes a step; a snowball hits a girl in the back; a baseball player catches a ball; a gust of wind strikes a window.

6. While a football is in flight, what forces act on it? What are the action–reaction pairs while the football is being kicked and while it is in flight?

7. A rubber ball is dropped onto the floor. What force causes the ball to bounce back into the air?

8. What is wrong with the statement "Because the car is at rest, there are no forces acting on it"? How would you correct this sentence?

9. A weightlifter stands on a bathroom scale. He pumps a barbell up and down. What happens to the reading on the bathroom scale as this is done? Suppose he is strong enough to actually *throw* the barbell upward. How does the reading on the scale vary now?

10. The mayor of a city decides to fire some city employees because they will not remove the sags from the cables that support the city traffic lights. If you were a lawyer, what defense would you give on behalf of the employees? Who do you think would win the case in court?

11. In a tug-of-war between two athletes, each athlete pulls on the rope with a force of 200 N. What is the tension in the rope? If the rope does not move what force does each athlete exert against the ground?

12. Suppose a truck loaded with sand accelerates at 0.5 m/s^2 on a highway. If the driving force on the truck remains constant, what happens to the truck's acceleration if its trailer leaks sand at a constant rate through a hole in its bottom?

PROBLEMS

Sections 4.1 Through 4.6

1. A force, **F**, applied to an object of mass m_1 produces an acceleration of 3.00 m/s^2. The same force applied to a second object of mass m_2 produces an acceleration of 1.00 m/s^2. (a) What is the value of the ratio m_1/m_2? (b) If m_1 and m_2 are combined, find their acceleration under the action of the force **F**.

2. A force of 10.0 N acts on a body of mass 2.00 kg. What are (a) the body's acceleration, (b) its weight in newtons, and (c) its acceleration if the force is doubled?

3. A 3.00-kg mass undergoes an acceleration given by $\mathbf{a} = (2.00\mathbf{i} + 5.00\mathbf{j})$ m/s^2. Find the resultant force, **F**, and its magnitude.

4. A heavy freight train has a mass of 15 000 metric tons. If the locomotives can exert a pulling force of 750 000 N, how long does it take to increase the speed from 0 to 80.0 km/h?

5. A pitcher throws a baseball of weight 1.40 N with velocity $(32.0\mathbf{i})$ m/s by uniformly accelerating her arm for 0.0900 s. If the ball starts from rest, (a) through what dis-

tance does the ball accelerate before its release? (b) What vector force does she exert on it?

6. A pitcher throws a baseball of weight w with velocity v_i by uniformly accelerating her arm for time t. If the ball starts from rest, (a) through what distance does the ball accelerate before its release? (b) What vector force does she exert on it?

7. A 4.00-kg object has a velocity of $3.00\mathbf{i}$ m/s at one instant. Eight seconds later, its velocity is $(8.00\mathbf{i} + 10.0\mathbf{j})$ m/s. Assuming the object was subject to a constant net force, find (a) the components of the force and (b) its magnitude.

8. The average speed of a nitrogen molecule in air is about 6.70×10^2 m/s, and its mass is about 4.68×10^{-26} kg. (a) If it takes 3.00×10^{-13} s for a nitrogen molecule to hit a wall and rebound with the same speed but in an opposite direction, what is the average acceleration of the molecule during this time interval? (b) What average force does the molecule exert on the wall?

9. An electron of mass 9.11×10^{-31} kg has an initial speed of 3.00×10^5 m/s. It travels in a straight line, and its speed

increases to 7.00×10^5 m/s in a distance of 5.00 cm. Assuming its acceleration is constant, (a) determine the force on the electron and (b) compare this force with the weight of the electron, which we neglected.

10. (a) A car with a mass of 850 kg is moving to the right with a constant speed of 1.44 m/s. What is the total force on the car? (b) What is the total force on the car if it is moving to the left?

11. A woman weighs 120 lb. Determine (a) her weight in newtons and (b) her mass in kilograms.

12. If a man weighs 900 N on Earth, what would he weigh on Jupiter, where the acceleration due to gravity is 25.9 m/s²?

13. Two forces, $\mathbf{F}_1$ and $\mathbf{F}_2$, act on a 5.00-kg mass. If $F_1 =$ 20.0 N and $F_2 = 15.0$ N, find the accelerations in (a) and (b) of Figure P4.13.

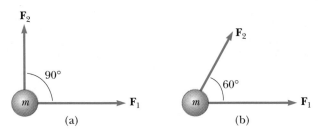

Figure P4.13

14. Besides its weight, a 2.80-kg object is subjected to one other constant force. The object starts from rest and in 1.20 s experiences a displacement of $(4.20 \text{ m})\mathbf{i} - (3.30 \text{ m})\mathbf{j}$, where the direction of $\mathbf{j}$ is the upward vertical direction. Determine the other force.

15. You stand on the seat of a chair and then hop off. (a) During the time you are in flight down to the floor, estimate the acceleration of the Earth as it lurches up toward you. (b) Estimate the upward distance through which the Earth moves as you fall to the floor. Visualize the Earth as a perfectly solid object.

16. Forces of 10.0 N north, 20.0 N east, and 15.0 N south are simultaneously applied to a 4.00-kg mass. Obtain the object's acceleration.

17. A 1000-kg boat moves through the water with two forces acting on it. One is a 2000-N forward push by the propeller; the other is an 1800-N resistive force due to the water. (a) What is the acceleration of the boat? (b) If it starts from rest, how far will it move in 10.0 s? (c) What will be its speed at the end of this time?

18. Three forces, $\mathbf{F}_1 = (-2.00\mathbf{i} + 2.00\mathbf{j})$ N, $\mathbf{F}_2 = (5.00\mathbf{i} - 3.00\mathbf{j})$ N, and $\mathbf{F}_3 = (-45.0\mathbf{i})$ N, act on an object to give it

an acceleration of magnitude 3.75 m/s². (a) What is the direction of the acceleration? (b) What is the mass of the object? (c) If the object is initially at rest, what is its speed after 10.0 s? (d) What are the velocity components of the object after 10.0 s?

19. A 15.0-lb block rests on the floor. (a) What force does the floor exert on the block? (b) If a rope is tied to the block and run vertically over a pulley and the other end is attached to a free-hanging 10.0-lb weight, what is the force of the floor on the 15.0-lb block? (c) If we replace the 10.0-lb weight in part (b) with a 20.0-lb weight, what is the force of the floor on the 15.0-lb block?

Section 4.7   Some Applications of Newton's Laws

20. A 3.00-kg mass is moving in a plane, with its x and y coordinates given by $x = 5t^2 - 1$ and $y = 3t^3 + 2$ (x and y are in meters and t is in seconds). Find the magnitude of the net force acting on this mass at $t = 2.00$ s.

21. Find the tension in each cord of the systems described in Figure P4.21. (Neglect the masses of the cords.)

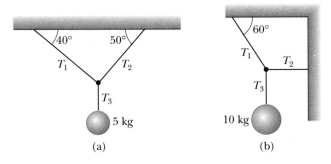

Figure P4.21

22. The distance between two telephone poles is 50.0 m. When a 1.00-kg bird lands on the telephone wire midway between the poles, the wire sags 0.200 m. Draw a free-body diagram of the bird. How much tension does the bird produce in the wire? Ignore the weight of the wire.

23. A bag of cement hangs from three wires as shown in Figure P4.23. Two of the wires make angles θ_1 and θ_2 with the horizontal. If the system is in equilibrium, (a) show that

$$T_1 = \frac{w \cos \theta_2}{\sin(\theta_1 + \theta_2)}$$

(b) Given that $w = 325$ N, $\theta_1 = 10.0°$ and $\theta_2 = 25.0°$, find the tensions T_1, T_2, and T_3 in the wires.

24. The systems shown in Figure P4.24 are in equilibrium. If

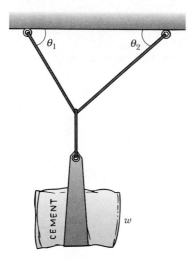

Figure P4.23

the spring scales are calibrated in newtons, what do they read? (Neglect the masses of the pulleys and strings, and assume the incline is frictionless.)

25. A fire helicopter carries a 620-kg bucket of water at the end of a 20.0-m long cable. Flying back from a fire at a constant speed of 40.0 m/s, the cable makes an angle of 40.0° with respect to the vertical. (a) Determine the force of air resistance on the bucket. (b) After filling the bucket with sea water, the helicopter returns to the fire at the same speed with the bucket now making an angle of 7.00° with the vertical. What is the mass of the water in the bucket?

26. A simple accelerometer is constructed by suspending a mass m from a string of length L that is tied to the top of a cart. As the cart is accelerated the string-mass system makes an angle of θ with the vertical. (a) Assuming that the string mass is negligible compared to m, derive an expression

for the cart's acceleration in terms of θ, and show that it is independent of the mass m and the length L. (b) Determine the acceleration of the cart when $\theta = 23.0°$.

27. A 1.00-kg mass is observed to accelerate at 10.0 m/s² in a direction 30.0° north of east (Fig. P4.27). The force $\mathbf{F}_2$ acting on the mass has a magnitude of 5.00 N and is directed north. Determine the magnitude and direction of the force $\mathbf{F}_1$ acting on the mass.

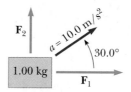

Figure P4.27

28. Draw a free-body diagram of a block that slides down a frictionless plane having an inclination of $\theta = 15.0°$ (Fig. P4.28). If the block starts from rest at the top and the length of the incline is 2.00 m, find (a) the acceleration of the block and (b) its speed when it reaches the bottom of the incline.

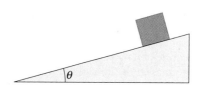

Figure P4.28

29. A block is given an initial velocity of 5.00 m/s up a frictionless 20.0° incline (Fig. P4.28). How far up the incline does the block slide before coming to rest?

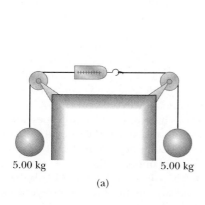

5.00 kg   5.00 kg

(a)

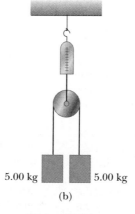

5.00 kg   5.00 kg

(b)

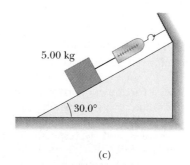

5.00 kg

30.0°

(c)

Figure P4.24

30. Two masses are connected by a light string that passes over a frictionless pulley, as in Figure P4.30. The incline is frictionless, $m_1 = 2.00$ kg, $m_2 = 6.00$ kg, and $\theta = 55.0°$. (a) Draw free-body diagrams of both masses. Find (b) the accelerations of the masses, (c) the tension in the string, and (d) the speed of each mass 2.00 s after being released from rest.

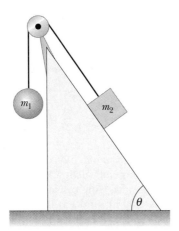

Figure P4.30

31. Two masses, m_1 and m_2, situated on a frictionless, horizontal surface, are connected by a light string. A force, **F**, is exerted on one of the masses to the right (Fig. P4.31). Determine the acceleration of the system and the tension, *T*, in the string.

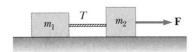

Figure P4.31

32. Two masses of 3.00 kg and 5.00 kg are connected by a light string that passes over a frictionless pulley, as in Figure 4.13a (page 98). Determine (a) the tension in the string, (b) the acceleration of each mass, and (c) the distance each mass will move in the first second of motion if they start from rest.

33. Mass m_1 on a frictionless horizontal table is connected to mass m_2 through a very light pulley, P_1, and a light fixed pulley, P_2, as shown in Figure P4.33. (a) If a_1 and a_2 are the accelerations of m_1 and m_2, respectively, what is the relation between these accelerations? Express (b) the tensions in the strings and (c) the accelerations a_1 and a_2 in terms of the masses m_1, and m_2, and *g*.

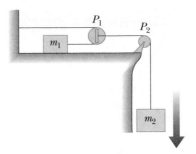

Figure P4.33

34. A 5.00-kg mass placed on a frictionless, horizontal table is connected to a cable that passes over a pulley and then is fastened to a hanging 10.0-kg mass, as in Figure P4.34. Draw free-body diagrams of both masses. Find the acceleration of the two objects and the tension in the string.

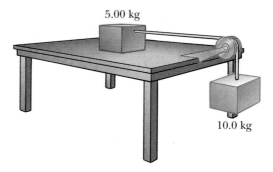

Figure P4.34

35. A 72.0-kg man stands on a spring scale in an elevator. Starting from rest, the elevator ascends, attaining its maximum speed of 1.20 m/s in 0.800 s. It travels with this constant speed for the next 5.00 s. The elevator then undergoes a uniform acceleration in the negative *y* direction for 1.50 s and comes to rest. What does the spring scale register (a) before the elevator starts to move? (b) during the first 0.800 s? (c) while the elevator is traveling at constant speed? (d) during the time it is slowing down?

36. Two people pull as hard as they can on ropes attached to a boat that has a mass of 200 kg. If they pull in the same direction, the boat has an acceleration of 1.52 m/s² to the right. If they pull in opposite directions, the boat has an acceleration of 0.518 m/s² to the left. What is the force exerted by each person on the boat? (Disregard any other forces on the boat.)

Additional Problems

37. A time-dependent force, $\mathbf{F} = (8.00\mathbf{i} - 4.00t\mathbf{j})$ N (where t is in seconds) is applied to a 2.00-kg object initially at rest. (a) At what time will the object be moving with a speed of 15.0 m/s? (b) Through what total displacement has the object traveled at this time? (c) How far is the object from its initial position when its speed is 15.0 m/s?

38. An elevator accelerates upward at 1.50 m/s². If the elevator has a mass of 200 kg, find the tension in the supporting cable.

39. The largest-caliber anti-aircraft gun operated by the German air force during World War II was the 12.8-cm Flak 40. This weapon fired a 25.8-kg shell with a muzzle speed of 880 m/s. What propulsive force was necessary to attain the muzzle speed within the 6.00-m barrel? (Assume constant acceleration and neglect the Earth's gravitational effect.)

40. One of the great dangers to mountain climbers is the *avalanche*, in which a mass of snow and ice breaks loose and goes on an essentially frictionless "ride" down the mountain on a cushion of compressed air. If you were on a 30.0° mountain slope and an avalanche started 400 m up the slope, how much time would you have to get out of the way?

41. An inventive child named Brian wants to reach an apple in a tree without climbing the tree. Sitting in a chair connected to a rope that passes over a frictionless pulley (Fig. P4.41), Brian pulls on the loose end of the rope with such a force that the spring scale reads 250 N. Brian's weight is 320 N, and the chair weighs 160 N. (a) Draw free-body diagrams for Brian and the chair considered as separate systems and another diagram for Brian and the chair considered as one system. (b) Show that the acceleration of the system is *upward* and find its magnitude. (c) Find the force Brian exerts on the chair.

42. Three blocks are in contact with each other on a frictionless, horizontal surface, as in Figure P4.42. A horizontal force $\mathbf{F}$ is applied to m_1. If $m_1 = 2.00$ kg, $m_2 = 3.00$ kg, $m_3 = 4.00$ kg, and $F = 18.0$ N, draw free-body diagrams of each block and find (a) the acceleration of the blocks, (b) the *resultant* force on each block, and (c) the magnitudes of the contact forces between the blocks.

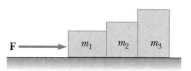

Figure P4.42

43. A high diver of mass 70.0 kg jumps off a board 10.0 m above the water. If her downward motion is stopped 2.00 s after she enters the water, what average upward force did the water exert on the diver?

44. Two forces, $\mathbf{F}_1 = (-6\mathbf{i} - 4\mathbf{j})$ N and $\mathbf{F}_2 = (-3\mathbf{i} + 7\mathbf{j})$ N, act on a particle of mass 2.00 kg that is initially at rest at coordinates $(-2.00$ m, $+4.00$ m). (a) What are the components of the particle's velocity at $t = 10.0$ s? (b) In what direction is the particle moving at $t = 10.0$ s? (c) What displacement does the particle undergo during the first 10.0 s? (d) What are the coordinates of the particle at $t = 10.0$ s?

45. A mass M is held in place by an applied force $\mathbf{F}$ and a pulley system as shown in Figure P4.45. The pulleys are massless and frictionless. Find (a) the tension in each section of rope, T_1, T_2, T_3, T_4, and T_5 and (b) the magnitude of $\mathbf{F}$.

46. A student is asked to measure the acceleration of a cart on a "frictionless" inclined plane, as in Figure 4.12, using an air track, a stopwatch, and a meter stick. The height of the incline is measured to be 1.774 cm, and the total length of the incline is measured to be $d = 127.1$ cm. Hence, the angle of inclination θ is determined from the relation $\sin \theta = 1.774/127.1$. The cart is released from rest at the top of the incline, and its displacement along the incline, x, is measured versus time, where $x = 0$ refers to the initial position of the cart. For x values of 10.0 cm, 20.0 cm, 35.0 cm, 50.0 cm, 75.0 cm, and 100 cm, the measured times to undergo these displacements (averaged over five runs) are 1.02 s, 1.53 s, 2.01 s, 2.64 s, 3.30 s, and 3.75 s, respec-

Figure P4.41

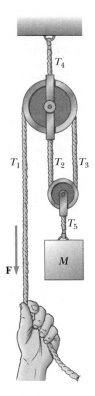

Figure P4.45

road exert on the car during this time, and how far did the car travel during the deceleration?

49. A van accelerates down a hill (Fig. P4.49), going from rest to 30.0 m/s in 6.00 s. During the acceleration, a toy ($m = 0.100$ kg) hangs by a string from the van's ceiling. The acceleration is such that the string remains perpendicular to the ceiling. Determine (a) the angle θ and (b) the tension in the string.

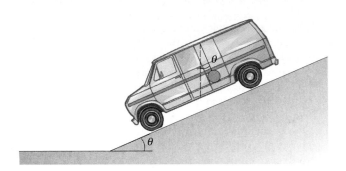

Figure P4.49

50. A car is at rest at the top of a driveway that has a slope of 20.0°. If the brake of the car is released with the car in neutral gear, find (a) the acceleration of the car down the drive and (b) the time it takes for the car to reach the street 10.0 m away.

tively. Construct a graph of x versus t^2, and perform a linear least-squares fit to the data. Determine the acceleration of the cart from the slope of this graph and compare it with the value you would get using $a' = g\sin\theta$, where $g = 9.80$ m/s².

47. What horizontal force must be applied to the cart shown in Figure P4.47 in order that the blocks remain stationary relative to the cart? Assume all surfaces, wheels, and pulley are frictionless. (*Hint:* Note that the force exerted by the string accelerates m_1 and that m_2 is in contact with the cart.)

Spreadsheet Problems

S1. An 8.4-kg mass slides down a fixed, frictionless inclined plane. Design and write a spreadsheet to determine the normal force exerted on the mass and its acceleration for a series of incline angles (measured from the horizontal) ranging from 0° to 90° in 5° increments. Use the graphing capability of your spreadsheet data to plot the normal force and the acceleration as functions of the incline angle. In the limiting cases of 0° and 90°, are your results consistent with the known behavior?

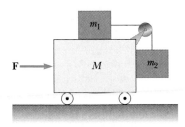

Figure P4.47

48. A 2000-kg car is slowed down uniformly from a speed of 20.0 m/s to 5.00 m/s in a time of 4.00 s. What force did the

ANSWERS TO CONCEPTUAL PROBLEMS

1. Motion requires no force. Newton's first law says an object in motion continues to move by itself in the absence of external forces.

2. If a single force acts on an object, it must accelerate. (From Newton's second law, $a = F/m$.) If an object accelerates, at least one force must act on it. However, if an object has no acceleration, you cannot conclude that no forces act on it. In this case, you can only say that the *net* force on the object is zero.

3. Because the gravitational field g_M is smaller on the Moon than on the Earth, more mass of gold would be required to represent 1 newton of weight. Thus, your friend on the Moon is richer, by about a factor of 6!

4. The only force acting on the ball during its motion is the downward force of gravity. The magnitude of this force is *mg*.

5. The car and truck experience forces that are equal in magnitude but in opposite directions. A calibrated spring scale placed between the colliding vehicles reads the same whichever way it faces. Because the car has the smaller mass, it experiences greater acceleration.

6. The *net* force on the object is $\mathbf{F}_1 + \mathbf{F}_2$, thus from Newton's second law, its acceleration is $(\mathbf{F}_1 + \mathbf{F}_2)/m$. The object accelerates toward the right when $|\mathbf{F}_1| > |\mathbf{F}_2|$, and toward the left when $|\mathbf{F}_2| > |\mathbf{F}_1|$. The acceleration of the object is zero when the two forces have equal magnitudes, that is, when $|\mathbf{F}_1| = |\mathbf{F}_2|$.

7. The force causing an automobile to move is the friction between the tires and the roadway as the automobile attempts to push the roadway backward. Notice that forces from the engine are not directly what accelerate the automobile. The force driving a propeller-driven airplane forward is the force of the air on the propeller as the rotating propeller pushes the air backward. The force propelling a rocket is the force of the exhausted gases from the back of the rocket on the rocket, as the rocket pushes the gases backward. In a rowboat, the rower pushes the water backward with the oars. As a result, the water pushes forward on the boat, and the force is applied at the oarlocks. Notice that all of these situations involve the propulsion of an object by Newton's third law.

8. According to Newton's third law, the force of the man on the boy and the force of the boy on the man are third law reaction forces, so they must be equal in magnitude. Both individuals experience the same force, so the boy, with the smaller mass, experiences the larger acceleration. Both individuals experience an acceleration over the same time interval. The larger acceleration of the boy will result in his moving away from the interaction with a larger velocity. Again, because of the larger acceleration, the boy moves through a larger distance during the interval while the hands are in contact.

B.C.   By John Hart

By permission of John Hart and Field Enterprises, Inc.

5

More Applications of Newton's Laws

In Chapter 4 we introduced Newton's laws of motion and applied them to some linear motion situations in which we were able to neglect frictional forces. In this chapter we shall expand our investigation to systems moving in the presence of friction forces. These systems include objects moving on rough surfaces and objects moving through viscous media such as liquids and air. In one section we discuss how numerical methods can be used to solve such real-world problems as motion in which the resistive force is velocity-dependent. We also apply Newton's laws to the dynamics of circular motion. Finally, we conclude this chapter with a discussion of the fundamental forces in nature.

5.1 • FORCES OF FRICTION

When a body is in motion either on a surface or through a viscous medium such as air or water, there is resistance to the motion because the body interacts with its surroundings. We call such resistance a **force of friction.** Forces of fric-

The passengers on this corkscrew rollercoaster experience the thrill of various forces as they travel along the curved track. The forces on one of the passenger cars include the force exerted by the track, the force of gravity, and the force of air resistance. *(Robin Smith/ Tony Stone Images)*

tion are very important in our everyday lives. They allow us to walk or run and are necessary for the motion of wheeled vehicles.

Consider a block on a horizontal table, as in Figure 5.1a. If we apply an external horizontal force **F** to the block, acting to the right, the block remains stationary if **F** is not too large. The force that counteracts **F** and keeps the block from moving acts to the left and is called the force of static friction, $\mathbf{f}_s$. As long as the block is not moving, $f_s = F$. Thus, if **F** is increased, $\mathbf{f}_s$ also increases. Likewise, if **F** decreases, $\mathbf{f}_s$ also decreases. Experiments show that the frictional force arises from the nature of the two surfaces: Because of their roughness, contact is made only at a few points, as shown in the magnified view of the surface in Figure 5.1a. Actually, the frictional force is much more complicated than presented here, because it ultimately involves the electrostatic force between atoms or molecules.

If we increase the magnitude of **F**, as in Figure 5.1b, the block eventually slips. When the block is on the verge of slipping, f_s is a maximum as shown in Figure 5.1c. When F exceeds $f_{s,\,max}$, the block moves and accelerates to the right. When

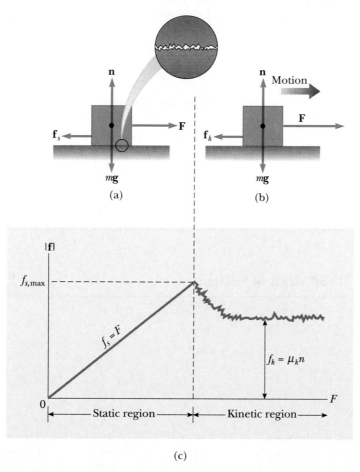

(c)

Figure 5.1   The force of friction, **f**, between a block and a rough surface is opposite the applied force, **F**. (a) The force of static friction equals the applied force. (b) When the applied force exceeds the force of kinetic friction, the block accelerates to the right. (c) A graph of the magnitude of the frictional force versus the applied force. Note that $f_{s,\,max} > f_k$.

the block is in motion, the frictional force becomes less than $f_{s, \, max}$ (Fig. 5.1c). We call the frictional force for an object in motion the **force of kinetic friction, f_k.** The unbalanced force in the x direction, $F - f_k$, produces an acceleration to the right. If $F = f_k$, the block moves to the right with constant speed. If the applied force is removed, then the frictional force acting to the left accelerates the block in the negative x direction and eventually brings it to rest.

Experimentally, one finds that, to a good approximation, both $f_{s, \, max}$ and f_k are proportional to the normal force acting on the block. The experimental observations can be summarized as follows:

- The magnitude of the force of static friction between any two surfaces in contact can have the values

Force of static friction •
$$f_s \leqslant \mu_s n \qquad \qquad [5.1]$$

where the dimensionless constant μ_s is called the **coefficient of static friction** and n is the magnitude of the normal force. The equality in Equation 5.1 holds when the block is on the verge of slipping—that is, when $f_s = f_{s, \, max} \equiv \mu_s n$. The inequality holds when the applied force is less than this value.

- The magnitude of the force of kinetic friction acting on an object is

Force of kinetic friction •
$$f_k = \mu_k n \qquad \qquad [5.2]$$

where μ_k is the **coefficient of kinetic friction.**
- The values of μ_k and μ_s depend on the nature of the surfaces, but μ_k is generally less than μ_s. Typical values of μ range from around 0.05 to 1.5. Table 5.1 lists some reported values.
- The coefficients of friction are nearly independent of the area of contact between the surfaces.

Finally, although the coefficient of kinetic friction varies with speed, we shall neglect any such variations. The approximate nature of the equations is easily dem-

TABLE 5.1   Coefficients of Friction[a]

	μ_s	μ_k
Steel on steel	0.74	0.57
Aluminum on steel	0.61	0.47
Copper on steel	0.53	0.36
Rubber on concrete	1.0	0.8
Wood on wood	0.25–0.5	0.2
Glass on glass	0.94	0.4
Waxed wood on wet snow	0.14	0.1
Waxed wood on dry snow	—	0.04
Metal on metal (lubricated)	0.15	0.06
Ice on ice	0.1	0.03
Teflon on Teflon	0.04	0.04
Synovial joints in humans	0.01	0.003

[a] All values are approximate.

onstrated by trying to get a block to slip down an incline at constant speed. Especially at low speeds, the motion is likely to be characterized by alternate *stick* and *slip* episodes.

Thinking Physics 1

In the motion picture *The Abyss* (Twentieth Century Fox, 1990), an underwater oil exploration rig sits at the bottom of very deep water. It is connected to a ship on the ocean surface by an "umbilical cord," as suggested in Figure 5.2a. On the ship, the umbilical cord is attached to a gantry. During a hurricane, the gantry structure breaks loose from the ship, falls into the water, and sinks to the bottom, passing over the edge of an extremely deep abyss. As a result, the rig is dragged by the umbilical cord along the ocean bottom, as described in Figure 5.2b. As the rig approaches the edge of the abyss, however, it is not pulled over the edge but stops just short of the edge, as shown in Figure 5.2c. Is this purely a cinematic edge-of-the-seat scenario, or is there a reason within the principles of physics for why the rig does not go over the edge?

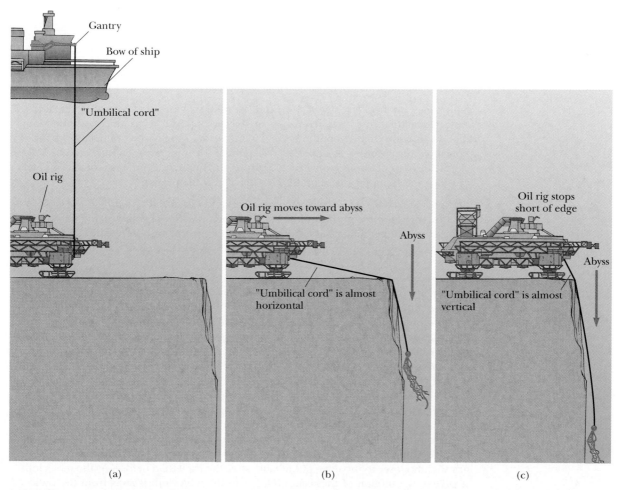

(a)   (b)   (c)

Figure 5.2   (Thinking Physics 1)

Reasoning There is a physics reason for this phenomenon. While the rig is being pulled across the ocean floor, it is pulled by the section of the umbilical cord that is almost horizontal—almost parallel to the ocean floor. Thus, the rig is subject to two horizontal forces—the tension in the umbilical cord pulling it forward and friction with the ocean floor pulling back. Let us assume that these forces are equal, so that the rig moves with constant velocity. As the rig nears the edge of the abyss, the angle that the umbilical cord makes with the horizontal increases. As a result, the component of force parallel to the ocean floor decreases and the downward vertical component increases. As a result of the increased vertical force, the rig is pulled downward more strongly into the ocean floor, increasing the normal force on it and, in turn, increasing the friction force between the rig and the ocean floor. Thus, with less force pulling it forward (from the umbilical cord) and more force pulling it back (due to friction), the rig slows down. By the time that the rig reaches the edge of the abyss, the force from the umbilical cord is almost *straight down*, resulting in little forward force. This downward force pulls the rig hard into the ocean floor, so that the friction force is large, and the rig stops.

This is not to say that this would happen in any situation. The characters on the rig had some luck in their favor. For example, if the umbilical cord had been attached at a lower point on the rig or if the coefficient of friction between the rig feet and the ocean floor were smaller, then the forward velocity of the rig could have been large enough that the friction force would not be sufficient to stop it, and it would have been propelled off the edge.

Thinking Physics 2

Sometimes, an incorrect statement about the frictional force between an object and a surface is made—"the friction force on an object is opposite to its motion or impending motion (motion that is about to occur)"—rather than the correct phrasing, "the friction force on an object is opposite to its motion or impending motion *relative to the surface.*" To emphasize the incorrect nature of the first statement, give some examples in which the friction force is in the *same direction* as the motion.

Reasoning Walking is a familiar example. To walk, you push backward on the ground, and the friction force pushes you forward, in the direction of your walking motion. In this case, the impending motion (because this is static friction) of the object, your foot, is backward relative to the ground, and the friction force exerted on your foot is forward.

Another example is a package sitting in the back of a pickup truck. As the truck accelerates forward, the natural inertia of the package tends to "leave it behind." Thus, the impending motion relative to the floor of the truck is toward the rear, and the friction force exerted on the package is forward, in the same direction as it moves. If the truck accelerates rapidly enough, the package will slide backward in the bed of the truck. In this case, we have real motion of the object rather than impending motion (relative to the surface), and the motion is backward. The friction force, then, is forward. Even though the package is sliding backward in the truck, an observer on the ground will see the package moving forward relative to the ground, so the friction force exerted on the package is again in the same direction as its motion.

As a final example, consider the popular dish-and-tablecloth trick, in which the demonstrator quickly pulls a tablecloth out from under some dishes. It may appear that the dishes do not move. Careful measurement, however, will show that the dishes are slightly closer to the edge of the table after the trick than before. As the tablecloth is pulled, the motion of the dishes relative to the tablecloth is away from the puller. Thus, the friction force on the dishes is toward the puller. This force causes the dishes to move toward the puller. Once the tablecloth is completely out from under the dishes, the tabletop becomes the new surface. The motion of the dishes relative to the tabletop is toward the puller. Thus, the friction force on the dishes is now away from the puller and opposite to the motion of the dishes, causing them to come to rest. The success of the trick depends on a relatively small friction force between the dishes and the tablecloth, so that the dishes do not achieve a very large velocity toward the puller, and a relatively large friction force between the dishes and the tabletop, so that they stop quickly.

CONCEPTUAL PROBLEM 1

You are playing with your daughter in the snow. She is sitting on a sled and asking you to slide her across a flat, horizontal field. You have a choice of pushing her from behind, by applying a force at 30° below the horizontal, or attaching a rope to the front of the sled and pulling with a force at 30° above the horizontal. Which would be easier for you and why?

CONCEPTUAL PROBLEM 2

A book is given a push at the bottom of a nonfrictionless ramp so that it slides up to a point and then slides back down to the starting point. Does it take the same time to go up as it does to come down?

CONCEPTUAL PROBLEM 3

The driver of an empty speeding truck slams on the brakes and skids to a stop through a distance d. (a) If the truck carried a heavy load such that its mass were doubled, what would be its skidding distance? (b) If the initial speed of the truck is halved, what would be its skidding distance?

Example 5.1 The Sliding Hockey Puck

A hockey puck on a frozen pond is hit and given an initial speed of 20.0 m/s, as in Figure 5.3. If the puck always remains on the ice and slides 115 m before coming to rest, determine the coefficient of kinetic friction between the puck and the ice.

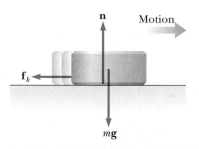

Figure 5.3 (Example 5.1) *After* the puck is given an initial velocity, the external forces acting on it are the weight, $m\mathbf{g}$, the normal force, $\mathbf{n}$, and the force of kinetic friction, $\mathbf{f}_k$.

Reasoning The forces acting on the puck after it is in motion are shown in Figure 5.3. If we assume that the force of friction, f_k, remains constant, then this force produces a uniform negative acceleration of the puck. First, we find the acceleration using Newton's second law. Knowing the acceleration of the puck and the distance it travels, we can then use kinematics to find the coefficient of kinetic friction.

Solution Applying Newton's second law in component form to the puck gives

$$(1) \quad \sum F_x = -f_k = ma$$

$$(2) \quad \sum F_y = n - mg = 0 \qquad (a_y = 0)$$

But $f_k = \mu_k n$, and from (2) we see that $n = mg$. Therefore, (1) becomes

$$-\mu_k n = -\mu_k mg = ma$$

$$a = -\mu_k g$$

The negative sign means that the acceleration is to the left, corresponding to a negative acceleration of the puck. The acceleration is independent of the mass of the puck and is constant, because we assume that μ_k remains constant.

Because the acceleration is constant, we can use Equation 2.12, $v^2 = v_0{}^2 + 2ax$, with the final speed $v = 0$:

$$v_0{}^2 + 2ax = v_0{}^2 - 2\mu_k gx = 0$$

$$\mu_k = \frac{v_0{}^2}{2gx}$$

$$\mu_k = \frac{(20.0 \text{ m/s})^2}{2(9.80 \text{ m/s}^2)(115 \text{ m})} = 0.177$$

Note that μ_k has no dimensions.

Example 5.2   Connected Objects

A ball and a cube are connected by a light string that passes over a frictionless pulley, as in Figure 5.4a. The coefficient of kinetic friction between the cube and the surface is 0.30. Find the acceleration of the two objects and the tension in the string.

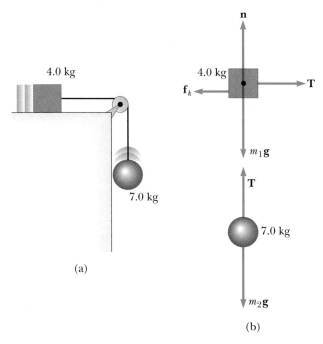

4.0 kg

7.0 kg

(a)

n

4.0 kg

$\mathbf{f}_k$ **T**

$m_1\mathbf{g}$

T

7.0 kg

$m_2\mathbf{g}$

(b)

Figure 5.4   (Example 5.2) (a) Two objects connected by a light string that passes over a frictionless pulley. (b) Free-body diagrams for the objects.

Reasoning   First, we draw the free-body diagrams for the two objects as in Figure 5.4b. Next, we apply Newton's second law in component form to each object and make use of the fact that the magnitude of the force of kinetic friction acting on the block is proportional to the normal force according to $f_k = \mu_k n$. Finally, we solve for the acceleration in terms of the parameters given.

Solution   Newton's second law applied to the cube in component form, with the positive x direction to the right, gives

$$\sum F_x = T - f_k = (4.0 \text{ kg})a$$

$$\sum F_y = n - (4.0 \text{ kg})g = 0$$

Because $n = mg = (4.0 \text{ kg})(9.80 \text{ m/s}^2) = 39.2 \text{ N}$ and $f_k = \mu_k n$, we have $f_k = \mu_k n = (0.30)(39.2 \text{ N}) = 11.8 \text{ N}$. Therefore,

$$(1) \quad T = f_k + (4.0 \text{ kg})a = 11.8 \text{ N} + (4.0 \text{ kg})a$$

Now we apply Newton's second law to the ball moving in the vertical direction, where the downward direction is selected as positive:

$$\sum F_y = (7.0 \text{ kg})g - T = (7.0 \text{ kg})a$$

or

$$(2) \quad T = 68.6 \text{ N} - (7.0 \text{ kg})a$$

Subtracting (1) from (2) eliminates T:

$$56.8 \text{ N} - (11 \text{ kg})a = 0$$

$$a = 5.2 \text{ m/s}^2$$

When this value for the acceleration is substituted into (1), we get

$$T = \boxed{33 \text{ N}}$$

EXERCISE 1   A block moves up a 45° incline with constant speed under the action of a force of 15 N applied *parallel* to the incline. If the coefficient of kinetic friction is 0.30, determine (a) the weight of the block and (b) the minimum force required to allow it to move *down* the incline at constant speed.   Answer   (a) 16.3 N   (b) 8.07 N up the incline

EXERCISE 2   A block is placed on a plane that is inclined at 60° with respect to the horizontal. If the block slides down the plane with an acceleration of $g/2$, determine the coefficient of kinetic friction between the block and the plane.   Answer   0.732

5.2 • NEWTON'S SECOND LAW APPLIED TO UNIFORM CIRCULAR MOTION

Solving problems involving friction is just one of many applications of Newton's second law. Let us now apply Newton's second law to another common situation: uniform circular motion. In Chapter 3, we found that a particle moving in a circular path of radius r with uniform speed v experiences an acceleration of magnitude

$$a_r = \frac{v^2}{r}$$

• *Centripetal acceleration*

Because the velocity vector **v** changes its direction continuously during the motion, the acceleration vector $\mathbf{a}_r$ is directed toward the center of the circle and hence is called **centripetal acceleration.** Furthermore, $\mathbf{a}_r$ is *always* perpendicular to **v**.

Consider a ball of mass m tied to a string of length r and being whirled in a horizontal circular path on a table top as in Figure 5.5. Let us assume that the ball moves with constant speed. The ball tends to maintain the motion in a straight-line path; however, the string prevents this motion along a straight line by exerting a force on the ball to make it follow its circular path. This force is directed along the length of the string toward the center of the circle, as shown in Figure 5.5. This force can be any one of our familiar forces playing the role of causing an object to follow a circular path.

If we apply Newton's second law along the radial direction, we find that the value of the force causing the centripetal acceleration can be evaluated:

$$F_r = ma_r = m\frac{v^2}{r} \qquad [5.3]$$

Figure 5.5   An overhead view of a ball moving in a circular path in a horizontal plane. A force $\mathbf{F}_r$ directed toward the center of the circle keeps the ball moving in the circle with constant speed.

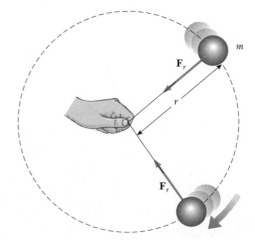

An athlete in the process of throwing the hammer. The force exerted by the chain on the hammer provides the centripetal acceleration. Only when the athlete releases the hammer will it move along a straight-line path tangent to the circle.   *(Focus on Sports)*

A force causing a centripetal acceleration acts toward the center of the circular path and causes a change in the direction of the velocity vector. In the case of a ball rotating at the end of a string, the force exerted by the string on the ball causes the centripetal acceleration of the ball. For a satellite in a circular orbit around the Earth, the force of gravity causes the centripetal acceleration of the satellite. The force acting on a car rounding a curve on a flat road to cause the centripetal acceleration of the car is the force of static friction between the tires and pavement, and so forth. In general, a body can move in a circular path under the influence of various types of forces or a combination of forces.

Regardless of the type of force causing the centripetal acceleration, if the force acting on an object should vanish, the object would no longer move in its circular path; instead it would move along a straight-line path tangent to the circle. This idea is illustrated in Figure 5.6 for the case of the ball whirling in a circle at the end of a string. If the string breaks at some instant, the ball will move along the straight-line path tangent to the circle at the point at which the string broke.

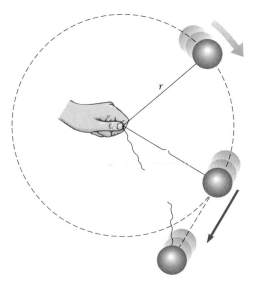

Figure 5.6   When the string breaks, the ball moves in the direction tangent to the circular path.

Thinking Physics 3

The force causing centripetal acceleration is often called a *centripetal force*. We are familiar with a variety of forces in nature—friction, gravity, normal forces, electrical forces, tension, and so on. Should we add *centripetal* force to this list?

Reasoning Centripetal force should *not* be added to this list. This is a pitfall for many students. Giving the force causing circular motion a name—centripetal force—leads many students to consider this as a new *kind* of force rather than a new *role* for force. A common mistake in force diagrams is to draw in all of the usual forces and then add another vector for the centripetal force. But it is not a separate force—it is simply one of our familiar forces *acting in the role of causing a circular motion.* Consider some examples. For the motion of the Earth around the Sun, the centripetal force is *gravity.* For an object sitting on a rotating turntable, the centripetal force is *friction.* For a rock whirled on the end of a string, the centripetal force is the *tension* in the string. For an amusement park patron pressed against the inner wall of a rapidly rotating circular room, the centripetal force is the *normal force* from the wall. In addition, the centripetal force could be a combination of two or more forces. For example, as a Ferris-wheel rider passes through the lowest point of the ride, the centripetal force on the rider is the difference between the normal force from the seat and the force of gravity.

Thinking Physics 4

High-speed, curved roadways are *banked*—that is, the roadway is tilted toward the inside of tight curves. Why is this?

Reasoning If an automobile is rounding a curve, there must be a force providing the centripetal acceleration. Normally, this is the static friction force between the tires and the roadway, parallel to the roadway surface and perpendicular to the centerline of the car. This is the only horizontal force—the weight and normal force are both vertical. If the curve is tight or the speed is high, the required friction force may be larger than the maximum possible friction force. In this case, Newton's first law takes over and the car skids along a straight line, which is off the side of the curved road. This situation is even more likely if the road is icy, which reduces the friction coefficient. Banking the roadway tilts the normal force away from the vertical, so that it has a horizontal component toward the center of the circular roadway. Thus, the normal force and friction work together to provide the needed centripetal acceleration. The larger the acceleration, the stronger the banking, in order to provide a larger component of the normal force in the horizontal direction.

As these cyclists in the Tour de France negotiate a curve on a flat racing track, the centripetal acceleration is provided by the force of static friction between the tires and the track surface. *(Michel Gouverneur, Photo News, Gamma Sport)*

CONCEPTUAL PROBLEM 4

An object executes circular motion with a constant speed whenever a net force of constant magnitude acts perpendicular to the velocity. What happens to the speed if the force is not perpendicular to the velocity?

Example 5.3   How Fast Can It Spin?

A ball of mass 0.500 kg is attached to the end of a cord, the length of which is 1.50 m. The ball is whirled in a horizontal circle as in Figure 5.6. If the cord can withstand a maximum tension of 50.0 N, what is the maximum speed the ball can have before the cord breaks?

Solution   Because the force that provides the centripetal acceleration in this case is the force **T** exerted by the cord on the ball, Equation 5.3 gives

$$T = m \frac{v^2}{r}$$

Solving for v, we have

$$v = \sqrt{\frac{Tr}{m}}$$

The maximum speed that the ball can have corresponds to the maximum value of the tension. Hence, we find

$$v_{max} = \sqrt{\frac{T_{max} r}{m}} = \sqrt{\frac{(50.0 \text{ N})(1.50 \text{ m})}{0.500 \text{ kg}}} = \boxed{12.2 \text{ m/s}}$$

EXERCISE 3   Calculate the tension in the cord if the speed of the ball is 5.00 m/s.   Answer   8.33 N

Example 5.4   The Conical Pendulum

A small body of mass m is suspended from a string of length L. The body revolves in a horizontal circle of radius r with constant speed v, as in Figure 5.7. (Because the string sweeps out the surface of a cone, the system is known as a *conical pendulum*.) Find the speed of the body and the period of revolution, T_P, defined as the time needed to complete one revolution.

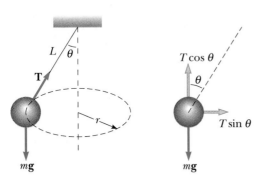

Figure 5.7   (Example 5.4) The conical pendulum and its free-body diagram.

Solution   The free-body diagram for the mass m is shown in Figure 5.7, where the force exerted by the string, **T**, has been resolved into a vertical component, $T \cos \theta$, and a component $T \sin \theta$ acting toward the center of rotation. Because the body does not accelerate in the vertical direction, the vertical component of **T** must balance the weight. Therefore,

$$(1) \quad T \cos \theta = mg$$

Because the force that provides the centripetal acceleration in this example is the component $T \sin \theta$, from Newton's second law we get

$$(2) \quad T \sin \theta = ma_r = \frac{mv^2}{r}$$

By dividing (2) by (1), we eliminate T and find that

$$\tan \theta = \frac{v^2}{rg}$$

But from the geometry we note that $r = L \sin \theta$; therefore,

$$v = \sqrt{rg \tan \theta} = \boxed{\sqrt{Lg \sin \theta \tan \theta}}$$

Because the ball travels a distance of $2\pi r$ (the circumference of the circular path) in a time equal to the period of revolution, T_P (not to be confused with the force **T**), we find

$$(3) \quad T_P = \frac{2\pi r}{v} = \frac{2\pi r}{\sqrt{rg \tan \theta}} = \boxed{2\pi \sqrt{\frac{L \cos \theta}{g}}}$$

The intermediate algebraic steps used in obtaining (3) are left to the reader. Note that T_P is independent of m. If we take $L = 1.00$ m and $\theta = 20.0°$, using (3) we find that

$$T_P = 2\pi \sqrt{\frac{(1.00 \text{ m})(\cos 20.0°)}{9.80 \text{ m/s}^2}} = 1.95 \text{ s}$$

Is it physically possible to have a conical pendulum with $\theta = 90°$?

Example 5.5   What Is the Maximum Speed of the Car?

A 1500-kg car moving on a flat, horizontal road negotiates a curve, the radius of which is 35.0 m, as in Figure 5.8. If the coefficient of static friction between the tires and the dry pavement is 0.500, find the maximum speed the car can have in order to make the turn successfully.

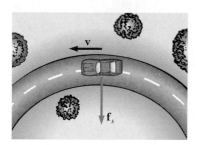

Figure 5.8   (Example 5.5) The force of static friction directed toward the center of the arc keeps the car moving in a circle.

Solution   In this case, the force that provides the centripetal acceleration and enables the car to remain in its circular path is the force of static friction. Hence, from Equation 5.3 we have

$$(1) \quad f_s = m\frac{v^2}{r}$$

The maximum speed that the car can have around the curve corresponds to the speed at which it is on the verge of skidding outward. At this point, the friction force has its maximum value

$$f_{s,\,max} = \mu_s n$$

Because the magnitude of the normal force equals the weight in this case, we find

$$f_{s,\,max} = \mu_s mg = (0.500)(1500\text{ kg})(9.80\text{ m/s}^2) = 7350\text{ N}$$

Substituting this value into (1), we find that the maximum speed is

$$v_{max} = \sqrt{\frac{f_{s,\,max}\, r}{m}} = \sqrt{\frac{(7350\text{ N})(35.0\text{ m})}{1500\text{ kg}}} = 13.1\text{ m/s}$$

EXERCISE 4   On a wet day, the car described in this example begins to skid on the curve when its speed reaches 8.00 m/s. What is the coefficient of static friction in this case?
Answer   0.187

Example 5.6   Let's Go Loop-the-loop

A pilot of mass m in a jet aircraft executes a loop-the-loop maneuver, as illustrated in Figure 5.9a. In this flying pattern, the aircraft moves in a vertical circle of radius 2.70 km at a *constant speed* of 225 m/s. Determine the force exerted by the seat on the pilot at (a) the bottom of the loop and (b) the top of the loop. Express the answers in terms of the weight of the pilot, *mg*.

Solution   (a) The free-body diagram for the pilot at the bottom of the loop is shown in Figure 5.9b. The only forces acting on the pilot are the downward force of gravity, *m***g**, and the upward force n_{bot} exerted by the seat on the pilot. Because the net upward force that provides the centripetal acceleration has a magnitude $n_{bot} - mg$, Newton's second law for the radial direction gives

$$n_{bot} - mg = m\frac{v^2}{r}$$

$$n_{bot} = mg + m\frac{v^2}{r} = mg\left[1 + \frac{v^2}{rg}\right]$$

Substituting the values given for the speed and radius gives

$$n_{bot} = mg\left[1 + \frac{(225\text{ m/s})^2}{(2.70 \times 10^3\text{ m})(9.80\text{ m/s}^2)}\right] = 2.91\,mg$$

Hence, the force exerted by the seat on the pilot is *greater* than his weight by a factor of 2.91. The pilot experiences an apparent weight that is greater than his weight by the factor 2.91. This is discussed further in Section 5.3.

(b) The free-body diagram for the pilot at the top of the loop is shown in Figure 5.9c. At this point, both the force of gravity and the force exerted by the seat on the pilot, $\mathbf{n}_{top}$, act *downward*, so the net force downward that provides the centripetal acceleration has a magnitude $n_{top} + mg$. Applying Newton's second law gives

$$n_{top} + mg = m\frac{v^2}{r}$$

$$n_{top} = m\frac{v^2}{r} - mg = mg\left[\frac{v^2}{rg} - 1\right]$$

$$n_{top} = mg\left[\frac{(225\text{ m/s})^2}{(2.70 \times 10^3\text{ m})(9.80\text{ m/s}^2)} - 1\right]$$

$$= 0.911\,mg$$

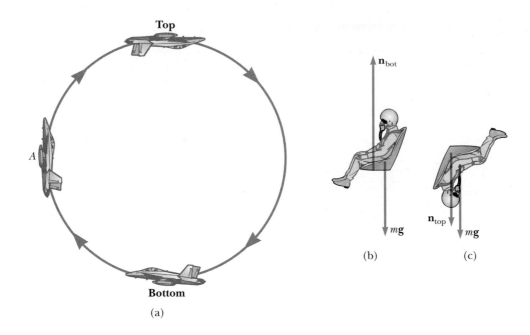

Figure 5.9   (Example 5.6)

In this case, the force exerted by the seat on the pilot is *less* than the weight by a factor of 0.911. Hence, the pilot will feel lighter at the top of the loop.

EXERCISE 5   Calculate the force on the pilot causing the centripetal acceleration if the aircraft is at point A in Figure 5.9a, midway up the loop.   Answer $1.911\,mg$ directed to the right

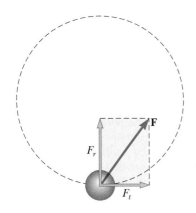

Figure 5.10   When the force acting on a particle moving in a circular path has a tangential component F_t, its speed changes. The total force on the particle also has a component F_r directed toward the center of the circular path. Thus, total force is $\mathbf{F} = \mathbf{F}_t + \mathbf{F}_r$.

EXERCISE 6   An automobile moves at constant speed over the crest of a hill. The driver moves in a vertical circle of radius 18.0 m. At the top of the hill, she notices that she barely remains in contact with the seat. Find the speed of the vehicle.   Answer   13.3 m/s

EXERCISE 7   A 1500-kg car rounds an unbanked curve with a radius of 52 m at a speed of 12 m/s. What minimum coefficient of friction must exist between the road and tires to prevent the car from slipping?   Answer   0.283

5.3 • NONUNIFORM CIRCULAR MOTION

In Chapter 3 we found that if a particle moves with varying speed in a circular path, there is, in addition to the centripetal component of acceleration, a tangential component of magnitude dv/dt. Therefore, the force acting on the particle must also have a tangential and a radial component, as shown in Figure 5.10. That is, because the total acceleration is $\mathbf{a} = \mathbf{a}_r + \mathbf{a}_t$, the total force exerted on the particle is $\mathbf{F} = \mathbf{F}_r + \mathbf{F}_t$. The component F_r is directed toward the center of the circle and is responsible for the centripetal acceleration. The component F_t tangent to the circle is responsible for the tangential acceleration, which causes the speed of the particle to change with time. The following examples demonstrate this type of motion.

Thinking Physics 5

Suppose you are the only rider on a Ferris wheel. The drive mechanism breaks after the ride starts, so that you continue to rotate freely, without friction. As you pass through the lowest point (let us call it the 6-o'clock point), you will feel the heaviest—that is, the seat will press upward on you with the largest force. As you pass over the highest point (the 12-o'clock point), you will feel the lightest. At the 3-o'clock and 9-o'clock points, will you feel as if you have the "correct" weight? If not, where *do* you feel as if your weight is correct?

Reasoning If the wheel were rotating at a constant speed, the answer to the question would be *yes*. If you are the only rider, however, the wheel is *unbalanced* and would not rotate with a constant speed. As you come down from the top point, your velocity increases. After you pass through the lowest point, your velocity decreases as you move back to the top. Thus, you will experience tangential acceleration as well as centripetal acceleration. At the 3-o'clock position, at which you are moving downward, you have a downward tangential acceleration. Thus, just as in a downward accelerating elevator, you will feel lighter. At the 9-o'clock position, when you are moving upward, you are accelerating downward tangentially. Again, you will feel lighter. In order to counteract the *downward* component of tangential acceleration near both positions, you need some *upward* component from the centripetal acceleration. This will be present if you are at positions *below* 3 o'clock or 9 o'clock, so that the centripetal acceleration has an upward component. The exact positions will depend on your mass and the details of the Ferris wheel.

Example 5.7 Follow the Rotating Ball

A small sphere of mass m is attached to the end of a cord of length R, which rotates in a *vertical* circle about a fixed point O, as in Figure 5.11a. Let us determine the tension in the cord at any instant when the speed of the sphere is v and the cord makes an angle θ with the vertical.

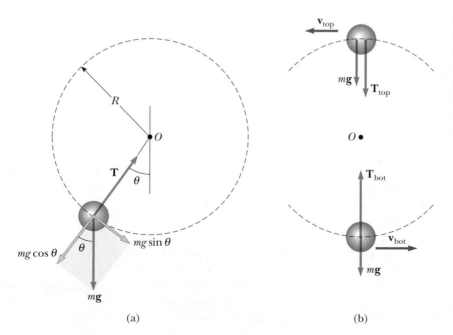

Figure 5.11 (Example 5.7) (a) Forces acting on a mass m connected to a string of length R and rotating in a vertical circle centered at O. (b) Forces acting on m when it is at the top and bottom of the circle. Note that the tension at the bottom is a maximum and the tension at the top is a minimum.

(a)   (b)

Solution   First we note that the speed is *not* uniform, because there is a tangential component of acceleration arising from the weight of the sphere. From the free-body diagram in Figure 5.11a, we see that the only forces acting on the sphere are the force of gravity, $m\mathbf{g}$, and the force exerted by the cord, **T**. Now we resolve $m\mathbf{g}$ into a tangential component, $mg\sin\theta$, and a radial component, $mg\cos\theta$. Applying Newton's second law to the forces in the tangential direction gives

$$\sum F_t = mg\sin\theta = ma_t$$
$$a_t = g\sin\theta$$

This component causes v to change in time, because $a_t = dv/dt$.

   Applying Newton's second law to the forces in the radial direction and noting that both **T** and $\mathbf{a}_r$ are directed toward O, we get

$$\sum F_r = T - mg\cos\theta = \frac{mv^2}{R}$$
$$T = m\left(\frac{v^2}{R} + g\cos\theta\right)$$

Limiting Cases   At the top of the path, where $\theta = 180°$, we have $\cos 180° = -1$, and the tension equation becomes

$$T_{\text{top}} = m\left(\frac{v_{\text{top}}^2}{R} - g\right)$$

This is the *minimum* value of T. Note that at this point $a_t = 0$ and therefore the acceleration is radial and directed downward, as in Figure 5.11b.

   At the bottom of the path, where $\theta = 0$, we see that because $\cos 0 = 1$,

$$T_{\text{bot}} = m\left(\frac{v_{\text{bot}}^2}{R} + g\right)$$

This is the *maximum* value of T. Again, at this point $a_t = 0$ and the acceleration is radial and directed upward.

EXERCISE 8   At what orientation of the system would the cord most likely break if the average speed increased?
Answer   At the bottom of the path, where T has its maximum value.

EXERCISE 9   A 0.40-kg object is swung in a vertical circular path on a string 0.50 m long. If a constant speed of 4.0 m/s is maintained, what is the tension in the string when the object is at the top of the circle?   Answer   8.88 N

EXERCISE 10   A car traveling on a straight road at 9.00 m/s goes over a hump in the road. The hump may be regarded as an arc of a circle of radius 11.0 m. What must be the speed of the car over the hump if a 600-N passenger is to experience weightlessness?
Answer   10.4 m/s

5.4 • MOTION IN THE PRESENCE OF VELOCITY-DEPENDENT RESISTIVE FORCES

Earlier we described the interaction between a moving object and the surface along which it moves. We completely ignored any interaction between the object and the medium through which it moves. Now let us consider the effect of a medium such as a liquid or gas. The medium exerts a **resistive force, R,** on the object moving through it. The magnitude of this force depends on the speed of the object, and the direction of **R** is always opposite the direction of motion of the object relative to the medium. In general, the magnitude of the resistive force increases with increasing speed. Some examples are the air resistance associated with moving vehicles (sometimes called air drag) and the viscous forces that act on objects moving through a liquid.

   The resistive force can have a complicated speed dependence. In the following discussions, we consider two situations. First, we assume that the resistive force is proportional to the speed; this is the case for objects that fall through a liquid with low speed and for very small objects, such as dust particles, that move through air. Second, we treat situations for which the resistive force is proportional to the square of the speed of the object; large objects, such as a skydiver moving through air in free fall, experience such a force.

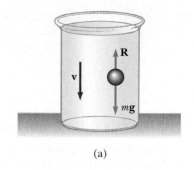

Resistive Force Proportional to Object Speed

If we assume that the resistive force acting on an object that is moving through a viscous medium is proportional to the object's velocity, then the resistive force can be expressed as

$$\mathbf{R} = -b\mathbf{v} \qquad\qquad [5.4]$$

where $\mathbf{v}$ is the velocity of the object and b is a constant that depends on the properties of the medium and on the shape and dimensions of the object. If the object is a sphere of radius r, then b is proportional to r.

Consider a sphere of mass m released from rest in a liquid, as in Figure 5.12a. Assuming the only forces acting on the sphere are the resistive force, $-b\mathbf{v}$, and the gravitational force, $m\mathbf{g}$, let us describe its motion.[1]

Applying Newton's second law to the vertical motion, choosing the downward direction to be positive, and noting that $\Sigma F_y = mg - bv$, we get

$$mg - bv = m\frac{dv}{dt}$$

where the acceleration is downward. Simplifying this expression gives

$$\frac{dv}{dt} = g - \frac{b}{m}v \qquad\qquad [5.5]$$

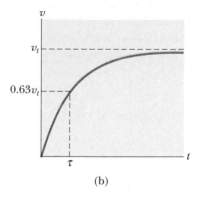

Figure 5.12   (a) A small sphere falling through a viscous fluid. (b) The speed–time graph for an object falling through a viscous medium. The object reaches a maximum, or terminal speed v_t, and τ is the time it takes to reach $0.63v_t$.

• *Terminal speed*

Equation 5.5 is called a *differential equation,* and the methods of solving such an equation may not be familiar to you as yet. However, note that initially, when $v = 0$, the resistive force is zero and the acceleration, dv/dt, is simply g. As t increases, the resistive force increases and the acceleration decreases. Eventually, the acceleration becomes zero when the resistive force balances the weight. At this point, the object reaches its **terminal speed,** v_t, and from then on it continues to move with zero acceleration. The terminal speed can be obtained from Equation 5.5 by setting $a = dv/dt = 0$. This gives

$$mg - bv_t = 0 \qquad \text{or} \qquad v_t = \frac{mg}{b}$$

The expression for v that satisfies Equation 5.5 with $v = 0$ at $t = 0$ is

$$v = \frac{mg}{b}\left(1 - e^{-bt/m}\right) = v_t(1 - e^{-t/\tau}) \qquad\qquad [5.6]$$

This function is plotted in Figure 5.12b. The time constant $\tau = m/b$ is the time it takes the object to reach 63.2% of its terminal speed. This can be seen by noting that when $t = \tau$, Equation 5.6 gives $v = 0.632v_t$. We can check that Equation 5.6 is a solution to Equation 5.5 by direct differentiation:

---

[1]There is also a *buoyant* force acting on the submerged object, and this force is constant and equal to the weight of the displaced fluid, as will be discussed in Chapter 12. This force will only change the apparent weight of the sphere by a constant factor, so we can ignore it here.

$$\frac{dv}{dt} = \frac{d}{dt}\left(\frac{mg}{b} - \frac{mg}{b}\,e^{-bt/m}\right) = -\frac{mg}{b}\,\frac{d}{dt}\,e^{-bt/m} = ge^{-bt/m}$$

Substituting this expression and Equation 5.6 into Equation 5.5 shows that our solution satisfies the differential equation.

Example 5.8   A Sphere Falling in Oil

A small sphere of mass 2.00 g is released from rest in a large vessel filled with oil. The sphere reaches a terminal speed of 5.00 cm/s. Determine the time constant τ and the time it takes the sphere to reach 90% of its terminal speed.

Solution   Because the terminal speed is given by $v_t = mg/b$, the coefficient b is

$$b = \frac{mg}{v_t} = \frac{(2.00\text{ g})(980\text{ cm/s}^2)}{5.00\text{ cm/s}} = 392\text{ g/s}$$

Therefore, the time τ is

$$\tau = \frac{m}{b} = \frac{2.00\text{ g}}{392\text{ g/s}} = 5.10 \times 10^{-3}\text{ s}$$

The speed of the sphere as a function of time is given by Equation 5.6. To find the time t it takes the sphere to reach a speed of $0.900v_t$, we set $v = 0.900v_t$ into the expression and solve for t:

$$0.900v_t = v_t(1 - e^{-t/\tau})$$

$$1 - e^{-t/\tau} = 0.900$$

$$e^{-t/\tau} = 0.100$$

$$-\frac{t}{\tau} = -2.30$$

$$t = 2.30\tau = 2.30(5.10 \times 10^{-3}\text{ s})$$

$$= 11.7 \times 10^{-3}\text{ s} = \boxed{11.7\text{ ms}}$$

EXERCISE 11   What is the sphere's speed through the oil at $t = 11.7$ ms? Compare this value with the speed the sphere would have if it were falling in a vacuum and so influenced only by gravity.   Answer   4.50 cm/s in oil versus 11.5 cm/s in free fall

Air Drag at High Speeds

For large objects moving at high speeds through air, such as airplanes, skydivers, and baseballs, the resistive force is approximately proportional to the square of the speed. In these situations, the magnitude of the resistive force can be expressed as

$$R = \tfrac{1}{2} D\rho A v^2 \qquad\qquad [5.7]$$

where ρ is the density of air, A is the cross-sectional area of the falling object measured in a plane perpendicular to its motion, and D is a dimensionless empirical quantity called the *drag coefficient*. The drag coefficient has a value of about 0.5 for spherical objects but can be as high as 2.0 for irregularly shaped objects.

Aerodynamic car. Streamlined bodies are used for sports cars and other vehicles to reduce air drag and increase fuel efficiency. *(© 1992 Dick Kelley)*

Consider an airplane in flight that experiences such a resistive force. Equation 5.7 shows that the force is proportional to the density of air and hence decreases with decreasing air density. Because air density decreases with increasing altitude, the resistive force on a jet airplane flying at a given speed must also decrease with increasing altitude. However, if at given altitude the plane's speed is doubled, the resistive force increases by a factor of four. In order to maintain this increased speed, the propulsive force must also increase by a factor of four.

Now let us analyze the motion of a mass in free fall subject to an upward air resistive force, the magnitude of which is $R = \frac{1}{2}D\rho Av^2$. Suppose a mass m is released from rest from the position $y = 0$, as in Figure 5.13. The mass experiences two external forces: the downward force of gravity, $m\mathbf{g}$, and the resistive force, $\mathbf{R}$, upward. (There is also an upward buoyant force that we neglect.) Hence, the magnitude of the net force is

$$F_{\text{net}} = mg - \tfrac{1}{2}D\rho Av^2 \qquad \text{[5.8]}$$

Substituting $F_{\text{net}} = ma$ into Equation 5.8, we find that the mass has a downward acceleration of magnitude

$$a = g - \left(\frac{D\rho A}{2m}\right)v^2 \qquad \text{[5.9]}$$

Again, we can calculate the terminal speed, v_t, using the fact that when the force of gravity is balanced by the resistive force, the net force is zero and therefore the acceleration is zero. Setting $a = 0$ in Equation 5.9 gives

$$g - \left(\frac{D\rho A}{2m}\right)v_t^2 = 0$$

$$v_t = \sqrt{\frac{2mg}{D\rho A}} \qquad \text{[5.10]}$$

Using this expression, we can determine how the terminal speed depends on the dimensions of the object. Suppose the object is a sphere of radius r. In this case, $A \propto r^2$ and $m \propto r^3$ (because the mass is proportional to the volume). Therefore, $v_t \propto \sqrt{r}$.

Table 5.2 lists the terminal speeds for several objects falling through air.

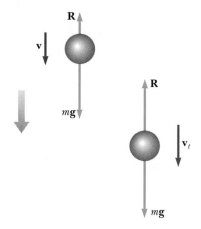

Figure 5.13   An object falling through air experiences a resistive drag force $\mathbf{R}$ and a gravitational force $m\mathbf{g}$. The object reaches terminal speed (on the right) when the net force is zero, that is, when $\mathbf{R} = -m\mathbf{g}$, or $R = mg$. Before this occurs, the acceleration varies with speed according to Equation 5.9.

TABLE 5.2   Terminal Speeds for Various Objects Falling Through Air

Object	Mass (kg)	Surface Area (m²)	v_t(m/s)[a]
Skydiver	75.0	0.70	60
Baseball (radius 3.7 cm)	0.145	4.2×10^{-3}	33
Golf ball (radius 2.1 cm)	0.046	1.4×10^{-3}	32
Hailstone (radius 0.50 cm)	4.8×10^{-4}	7.9×10^{-5}	14
Raindrop (radius 0.20 cm)	3.4×10^{-5}	1.3×10^{-5}	9

[a] The drag coefficient, D, is assumed to be 0.5 in each case.

Figure 5.14   (Conceptual Problem 5) By spreading their arms and legs and by keeping their bodies parallel to the ground, skydivers experience maximum air resistance resulting in a minimum terminal speed.   *(Guy Sauvage, Photo Researchers, Inc.)*

CONCEPTUAL PROBLEM 5

Consider a skydiver falling through air before reaching her terminal speed, as in Figure 5.14. As the speed of the skydiver increases, what happens to her acceleration? What is her acceleration once she reaches terminal speed?

5.5 · NUMERICAL MODELING IN PARTICLE DYNAMICS[2]

As we have seen in this and the preceding chapter, the study of dynamics of a particle focuses on describing the position, velocity, and acceleration as functions of time. Cause-and-effect relationships exist between these quantities: A velocity causes position to change, an acceleration causes velocity to change, and an acceleration is the direct result of applied forces. Therefore, the study of motion usually begins with an evaluation of the net force on the particle.

In this section, we confine our discussion to one-dimensional motion, so boldface notation will not be used for vector quantities. If a particle of mass m moves under the influence of a net force F, Newton's second law tells us that the acceleration of the particle is given by $a = F/m$. In general, we can then obtain the solution to a dynamics problem by using the following procedure:

1. Sum all the forces on the particle to get the net force F.
2. Use this force to determine the acceleration, using $a = F/m$.
3. Use this acceleration to determine the velocity from $dv/dt = a$.
4. Use this velocity to determine the position from $dx/dt = v$.

---

[2]The author is most grateful to Colonel James Head of the U.S. Air Force Academy for preparing this section.

Example 5.9 An Object Falling in a Vacuum

We can illustrate the procedure just described by considering a particle falling in a vacuum under the influence of the force of gravity, as in Figure 5.15.

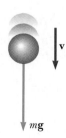

Figure 5.15 (Example 5.9) An object falling in vacuum under the influence of gravity.

Reasoning and Solution Applying Newton's second law, we set the sum of the external forces equal to the mass of the particle times its acceleration (taking upward to be the positive direction):

$$F = ma = -mg$$

Thus, $a = -g$, a constant. Because $dv/dt = a$, the resulting differential equation for velocity is $dv/dt = -g$, which may be integrated to give

$$v(t) = v_0 - gt$$

Then, because $dx/dt = v$, the position of the particle is obtained from another integration, which yields the well-known result

$$x(t) = x_0 + v_0 t - \tfrac{1}{2}gt^2$$

In this expression, x_0 and v_0 represent the position and speed of the particle at $t = 0$.

The procedure just described is straightforward for some physical situations, such as the one described in the previous example. In the real world, however, complications often arise that make analytical solutions for many practical situations difficult and perhaps beyond the mathematical abilities of most students taking introductory physics. For example, the force may depend on the position of the particle, as in cases in which the variation in gravitational acceleration with height must be taken into account. The force may also vary with velocity, as we have seen in cases of resistive forces caused by motion through a liquid or gas. The force may depend on both position and velocity, as in the case of an object falling through air where the resistive force depends on velocity and on height (air density). In rocket motion, the mass changes with time, so even if the force is constant, the acceleration is not.

Another complication arises because the equations relating acceleration, velocity, position, and time are differential equations, not algebraic equations. Differential equations are usually solved using integral calculus and other special techniques that introductory students may not have mastered.

So how does one proceed to solve real-world problems without advanced mathematics? One answer is to solve such problems on personal computers, using elementary numerical methods. The simplest of these is the Euler method, named after the Swiss mathematician Leonhard Euler (1707–1783).

The Euler Method

In the **Euler method** of solving differential equations, derivatives are approximated as finite differences. Considering a small increment of time, Δt, the relationship between speed and acceleration may be approximated as

$$a(t) = \frac{\Delta v}{\Delta t} = \frac{v(t + \Delta t) - v(t)}{\Delta t}$$

Then the speed of the particle at the end of the period Δt is approximately equal to the speed at the beginning of the period, plus the acceleration during the interval multiplied by Δt:

$$v(t + \Delta t) = v(t) + a(t)\,\Delta t \qquad \textbf{[5.11]}$$

Because the acceleration is a function of time, this estimate of $v(t + \Delta t)$ will be accurate only if the time interval Δt is short enough that the change in acceleration during it is very small (as will be discussed later).

The position can be found in the same manner:

$$v(t) = \frac{x(t + \Delta t) - x(t)}{\Delta t}$$

$$x(t + \Delta t) = x(t) + v(t)\,\Delta t \qquad \textbf{[5.12]}$$

It may be tempting to add the term $\frac{1}{2}a(\Delta t)^2$ to this result to make it look like the familiar kinematics equation, but this term is not included in the Euler method of integration because Δt is assumed to be so small that $(\Delta t)^2$ is nearly zero.

If the acceleration at any instant t is known, the particle's velocity and position at $(t + \Delta t)$ can be calculated from Equations 5.11 and 5.12. The calculation can then proceed in a series of finite steps to determine the velocity and position at any later time. The acceleration is determined by the net force acting on the object,

$$a(x, v, t) = \frac{F(x, v, t)}{m} \qquad \textbf{[5.13]}$$

which may depend explicitly on the position, velocity, or time.

It is convenient to set up the numerical solution to this kind of problem by numbering the steps and entering the calculations in a table. Table 5.3 illustrates how to do this in an orderly way. The equations provided in the table can be entered into a spreadsheet and the calculations performed row by row to determine the velocity, position, and acceleration as functions of time. The calculations can also be carried out using a program written in BASIC, PASCAL, or FORTRAN, or with commercially available mathematics packages for personal computers. Many small increments can be taken, and accurate results can usually be obtained with the help of a computer. Graphs of velocity versus time or position versus time can be displayed to help you visualize the motion.

TABLE 5.3   The Euler Method for Solving Dynamics Problems

Step	Time	Position	Velocity	Acceleration
0	t_0	x_0	v_0	$a_0 = F(x_0, v_0, t_0/m$
1	$t_1 = t_0 + \Delta t$	$x_1 = x_0 + v_0\,\Delta t$	$v_1 = v_0 + a_0\,\Delta t$	$a_1 = F(x_1, v_1, t_1)/m$
2	$t_2 = t_1 + \Delta t$	$x_2 = x_1 + v_1\,\Delta t$	$v_2 = v_1 + a_1\,\Delta t$	$a_2 = F(x_2, v_2, t_2)/m$
3	$t_3 = t_2 + \Delta t$	$x_3 = x_2 + v_2\,\Delta t$	$v_3 = v_2 + a_2\,\Delta t$	$a_3 = F(x_3, v_3, t_3)/m$
.	.	.	.	.
.	.	.	.	.
.	.	.	.	.
n	t_n	x_n	v_n	a_n

The Euler method has the advantage that the dynamics are not obscured—the fundamental relationships of acceleration to force, velocity to acceleration, and position to velocity are clearly evident. Indeed, these relationships form the heart of the calculations. There is no need to use advanced mathematics, and the basic physics governs the dynamics.

The Euler method is completely reliable for infinitesimally small time increments, but for practical reasons a finite increment size must be chosen. In order for the finite difference approximation of Equation 5.11 to be valid, the time increment must be small enough that the acceleration can be approximated as being constant during the increment. We can determine an appropriate size for the time increment by examining the particular problem that is being investigated. The criterion for the size of the time increment may need to be changed during the course of the motion. In practice, however, we usually choose a time increment appropriate to the initial conditions and use the same value throughout the calculations.

The size of the time increment influences the accuracy of the result, but unfortunately it is not easy to determine the accuracy of a solution by the Euler method without a knowledge of the correct analytical solution. One method of determining the accuracy of the numerical solution is to repeat the calculations with a smaller time increment and compare results. If the two calculations agree to a certain number of significant figures, you can assume that the results are correct to that precision.

5.6 · THE FUNDAMENTAL FORCES OF NATURE

We have described a variety of forces that are experienced in our everyday activities, such as the force of gravity that acts on all objects at or near the Earth's surface and the force of friction as one surface slides over another. Forces also act in the atomic and subatomic world. For example, atomic forces within the atom are responsible for holding its constituents together, and nuclear forces act on different parts of the nucleus to keep its parts from separating.

Until recently, physicists believed that there were four fundamental forces in nature: the gravitational force, the electromagnetic force, the strong nuclear force, and the weak nuclear force.

The Gravitational Force

The gravitational force is the mutual force of attraction between any two objects in the Universe. It is interesting and rather curious that, although the gravitational force can be very strong between macroscopic objects, it is the weakest of all the fundamental forces. For example, the gravitational force between the electron and proton in the hydrogen atom is only about 10^{-47} N, whereas the electromagnetic force between these same two particles is about 10^{-7} N.

Newton's law of gravitation states that every particle in the Universe attracts every other particle with a force that is directly proportional to the product of the two masses and inversely proportional to the square of the distance between them.

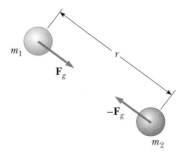

Figure 5.16 Two particles of masses m_1 and m_2 attract each other with a force of magnitude Gm_1m_2/r^2.

If the particles have masses m_1 and m_2 and are separated by a distance r, as in Figure 5.16, the magnitude of the gravitational force is

$$F_g = G\frac{m_1 m_2}{r^2}$$ [5.14]

where $G = 6.67 \times 10^{-11}$ N·m²/kg² is the universal gravitational constant.

The Electromagnetic Force

The electromagnetic force is the force that binds atoms and molecules in compounds to form ordinary matter. It is much stronger than the gravitational force. The force that causes a rubbed comb to attract bits of paper and the force that a magnet exerts on an iron nail are electromagnetic forces. Essentially all forces at work in our macroscopic world (apart from the gravitational force) are manifestations of the electromagnetic force. For example, friction forces, contact forces, tension forces, and forces in elongated springs are consequences of electromagnetic forces between charged particles in proximity.

The electromagnetic force involves two types of particles: those with positive charge and those with negative charge. (The properties of these two types of charges will be discussed shortly.) Unlike the gravitational force, which is always an attractive interaction, the electromagnetic force can be either attractive or repulsive, depending on the charges on the particles. In general, the electromagnetic force acts between two charged particles that are in relative motion.

When the charged particles are at rest relative to each other, we refer to the electromagnetic force between them as the electrostatic force. Coulomb's law expresses the magnitude of the electrostatic force F_e between two charged particles separated by a distance r:

$$F_e = k_e\frac{q_1 q_2}{r^2}$$ [5.15]

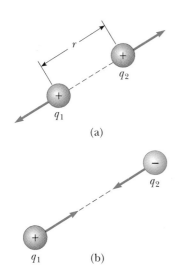

Figure 5.17 Two point charges separated by a distance r exert an electrostatic force on each other given by Coulomb's law. (a) When the charges are of the same sign, the charges repel each other. (b) When the charges are of opposite sign, the charges attract each other.

where q_1 and q_2 are the charges on the two particles, measured in coulombs (C), and k_e is the Coulomb constant, with the value $k_e = 8.99 \times 10^9$ N·m²/C². Note that the electrostatic force has the same mathematical form as Newton's law of gravity (Eq. 5.14); however, charge replaces mass, and the constants are different. The electrostatic force is attractive if the two charges have opposite signs and repulsive if the two charges have the same sign, as indicated in Figure 5.17. In other words, opposite charges attract each other, and like charges repel each other.

The smallest amount of isolated charge found (so far) in nature is the charge on an electron or proton. This fundamental unit of charge has the magnitude $e = 1.60 \times 10^{-19}$ C. Theories developed in the 1970s and 1980s propose that protons and neutrons are made up of smaller particles called quarks, which have charges of either $2e/3$ or $-e/3$ (discussed further in Chapter 31). Although experimental evidence has been found for such particles inside nuclear matter, free quarks have never been detected.

The most important properties of electric charges can be summarized as follows:

- Charges are scalar quantities; hence, their values are additive.
- The two kinds of charges that exist in nature are labeled positive and negative. Charges of like sign repel each other, and charges of opposite sign attract each other.
- The net charge on any object has discrete values; in other words, charge is said to be quantized. Any isolated elementary charged particle has a charge of e. For example, electrons have a charge $-e$ and protons have a charge $+e$. (Neutrons have no charge.) Because e is the fundamental unit of charge, the net charge of an object is Ne, where N is an integer. For macroscopic objects, N is very large and the quantization of charge is not noticed.
- Electric charge is always conserved. This means that in any kind of process— such as a collision event, a chemical reaction, or nuclear decay—the total charge of an isolated system remains constant.

- *Properties of electric charge*

The Strong Force

An atom, as we currently understand it, consists of an extremely dense, positively charged nucleus surrounded by a cloud of negatively charged electrons, with the electrons attracted to the nucleus by the Coulomb force. Because all nuclei except those of hydrogen are combinations of positively charged protons and neutral neutrons (collectively called nucleons), why does the repulsive electrostatic force between the protons not cause nuclei to break apart? Clearly, there must be an attractive force that overcomes the strong electrostatic repulsive force and is responsible for the stability of nuclei. This great force that binds the nucleons to form a nucleus is called the *strong force*. It should be noted that electrons and certain other particles are immune to the strong force. Unlike the gravitational and electromagnetic forces, which depend on distance in an inverse-square fashion, the strong force is extremely short-range; its strength decreases very rapidly outside the nucleus and is negligible for separations greater than approximately 10^{-14} m. For separations of approximately 10^{-15} m (a typical nuclear dimension), the strong force is about two orders of magnitude stronger than the electromagnetic force.

The Weak Force

The *weak force* is a short-range force that tends to produce instability in certain nuclei. It was first observed in naturally occurring radioactive substances and was later found to play a key role in most radioactive decay reactions. The weak force is about 10^{25} times stronger than the gravitational force and about 10^{12} times weaker than the electromagnetic force.

The Current View of Fundamental Forces

For years physicists have searched for a simplification scheme that would reduce the number of fundamental forces in nature. In 1967 physicists predicted that the electromagnetic force and the weak force, originally thought to be independent of each other and both fundamental, are in fact manifestations of one force, now called the *electroweak force*. The prediction was confirmed experimentally in 1984. We also now know that protons and neutrons are not fundamental particles but are composed of simpler particles called quarks. For example, the proton consists of

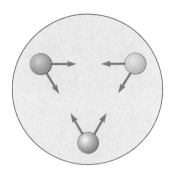

Figure 5.18   A proton is composed of three quarks that are bound together by a very strong force called the color force.

three quarks as shown in Figure 5.18. Quarks are bound together by a strong force called the **color force.**

Scientists believe that the fundamental forces of nature are closely related to the origin of the Universe. The Big Bang theory holds that the Universe began with a cataclysmic explosion 15 billion to 20 billion years ago. According to this theory, the first moments after the Big Bang saw such extremes of energy that all the fundamental forces were unified into one force. Physicists are continuing their search for connections among the known fundamental forces—connections that could eventually prove that the forces are all merely different forms of a single superforce. The recent success linking the electromagnetic and weak nuclear forces has spurred greater efforts at a unification scheme (yet to be proved) called the Grand Unified Theory. This fascinating subject continues to be at the forefront of physics.

5.7 • THE GRAVITATIONAL FIELD

When Newton first published his theory of gravitation, his contemporaries found it difficult to accept the concept of a force that one object could exert on another without anything happening in the space between them. They asked how it was possible for two masses to interact even though they were not in contact with each other. Although Newton himself could not answer this question, his theory was considered a success because it satisfactorily explained the motions of the planets.

An alternative approach is to think of the gravitational interaction as a two-step process. First, one object creates a gravitational field $\mathbf{g}$ throughout the space around it. Then a second object of mass m experiences a force $\mathbf{F}_g = m\mathbf{g}$. In other words, the field $\mathbf{g}$ exerts a force on the particle. Hence, the gravitational field is defined by

Definition of gravitational field •

$$\mathbf{g} \equiv \frac{\mathbf{F}_g}{m} \qquad\qquad [5.16]$$

That is, the gravitational field at a point in space equals the gravitational force that a test mass experiences at that point divided by the mass. As a consequence, if $\mathbf{g}$ is known at some point in space, a test particle of mass m experiences a gravitational force $m\mathbf{g}$ when placed at that point.

As an example, consider an object of mass m near the Earth's surface. The gravitational force on the object is directed toward the center of the Earth and has a magnitude mg. Thus we see that the gravitational field experienced by the object at some point has a magnitude equal to the acceleration of gravity at that point. Because the gravitational force on the object has a magnitude $GM_E m/r^2$ (where M_E is the mass of the Earth), the field $\mathbf{g}$ at a distance r from the center of the Earth is given by

Gravitational field of the •
Earth

$$\mathbf{g} = \frac{\mathbf{F}_g}{m} = -\frac{GM_E}{r^2}\,\hat{\mathbf{r}} \qquad\qquad [5.17]$$

where $\hat{\mathbf{r}}$ is a unit vector pointing radially outward from the Earth, and the minus sign indicates that the field points toward the center of the Earth, as shown in Figure

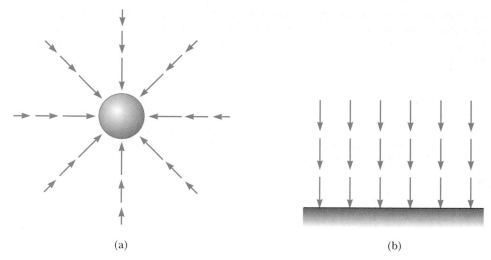

(a)   (b)

Figure 5.19   (a) The gravitational field vectors in the vicinity of a uniform spherical mass vary in both direction and magnitude. (b) The gravitational field vectors in a small region near the Earth's surface are uniform—that is, they match in both direction and magnitude.

5.19a. Note that the field vectors at different points surrounding the spherical mass vary in both direction and magnitude. In a small region near the Earth's surface, **g** is approximately constant and the downward field is uniform, as indicated in Figure 5.19b. This expression is valid at all points outside the Earth's surface, assuming that the Earth is spherical and that rotation can be neglected. At the Earth's surface, where $r = R_E$, **g** has a magnitude of 9.80 m/s².

The field concept is used in many other areas of physics. In fact, it was introduced by Michael Faraday (1791–1867) in his study of electromagnetism. In spite of its abstract nature, the field concept is particularly useful for describing electric and magnetic interactions.

Gravitational, electric, and magnetic fields are vector fields, because a vector is associated with each point in space. In the same way, a scalar field is one in which a scalar quantity is used to describe each point in space. For example, the variation in temperature over a given region can be described by a scalar temperature field.

SUMMARY

The maximum force of static friction, $f_{s, max}$, between a body and a surface is proportional to the normal force acting on the body. This maximum force occurs when the body is on the verge of slipping. In general, $f_s \leq \mu_s n$, where μ_s is the *coefficient of static friction* and **n** is the normal force. When a body slides over a rough surface, the *force of kinetic friction*, $\mathbf{f}_k$, is opposite the motion and is also proportional to the normal force. The magnitude of this force is given by $f_k = \mu_k n$, where μ_k is the *coefficient of kinetic friction*. Usually, $\mu_k < \mu_s$.

Newton's second law, applied to a particle moving in **uniform circular motion,** states that the net force in the radial direction must equal the product of the mass and the centripetal acceleration:

$$F_r = ma_r = \frac{mv^2}{r} \qquad\qquad [5.3]$$

A particle moving in nonuniform circular motion has both a centripetal acceleration and a tangential component of acceleration. In the case of a particle rotating in a vertical circle, the force of gravity provides the tangential acceleration and part or all of the centripetal acceleration.

A body moving through a liquid or gas experiences a **resistive force** that is velocity-dependent. This resistive force, which opposes the motion, generally increases with velocity. The force depends on the shape of the body and on the properties of the medium through which the body is moving. In the limiting case for a falling body, when the resistive force balances the weight ($a = 0$), the body reaches its **terminal speed**.

There are four fundamental forces in nature: the strong nuclear force, the electromagnetic force, the gravitational force, and the weak nuclear force.

CONCEPTUAL QUESTIONS

1. If you push on a heavy box that is at rest, it requires some force **F** to start its motion. However, once it is sliding, it requires a smaller force to maintain that motion. Why?

2. Suppose you are driving a car along a highway at a high speed. Why should you avoid slamming on your brakes if you want to stop in the shortest distance? That is, why should you keep the wheels turning as you brake?

3. Twenty people are playing tug-of-war, with ten people on a team. The teams are so evenly matched that neither team wins. After they give up the game, they notice that one of the team member's cars is mired in mud. They attach the tug-of-war rope to the bumper of the car and the 20 people try to pull the car from the mud. The friction force between the car and the mud is huge, the car does not move, and the rope breaks. Why did the rope break in this situation when it did not break when 20 people pulled on it in the tug-of-war?

4. What causes a rotary lawn sprinkler to turn?

5. It has been suggested that rotating cylinders about 10 mi in length and 5 mi in diameter be placed in space and used as colonies. The purpose of the rotation is to simulate gravity for the inhabitants. Explain this concept for producing an effective gravity.

6. Consider a rotating space station, spinning with just the right speed such that the centripetal acceleration on the inner surface is *g*. Thus, astronauts standing on this inner surface would feel pressed to the surface as if they were pressed into the floor because of the Earth's gravitational force. Suppose an astronaut in this station holds a ball above her head and "drops" it to the floor. Will the ball fall just like it would on the Earth?

7. A pail of water can be whirled in a vertical path such that none is spilled. Why does the water stay in, even when the pail is above your head?

8. How would you explain the force that pushes a rider toward the side of a car as the car rounds a corner?

9. Why does a pilot tend to black out when pulling out of a steep dive?

10. Suppose a race track is built on the Moon. For a given speed of a car around the track, would the roadway have to be banked at a steeper angle or a shallower angle compared to the same roadway on the Earth?

11. A skydiver in free fall reaches terminal speed. After the parachute is opened, what parameters change to decrease this terminal speed?

12. Consider a small raindrop and a large raindrop falling through the atmosphere. Compare their terminal speeds. What are their accelerations when they reach terminal speed?

13. On long journeys, jet aircraft usually fly at high altitudes of about 30 000 ft. What is the main advantage of flying at these altitudes from an economic viewpoint?

14. If someone told you that astronauts are weightless in orbit because they are beyond the pull of gravity, would you accept the statement? Explain.

PROBLEMS

Section 5.1   Forces of Friction

1. A 25.0-kg block is initially at rest on a horizontal surface. A horizontal force of 75.0 N is required to set the block in motion. After it is in motion, a horizontal force of 60.0 N is required to keep the block moving with constant speed. Find the coefficients of static and kinetic friction from this information.

2. A racing car accelerates uniformly from 0 to 80.0 mi/h in 8.00 s. The external force that accelerates the car is the frictional force between the tires and the road. If the tires do

not spin, determine the *minimum* coefficient of friction between the tires and the road.

3. A car is traveling at 50.0 mi/h on a horizontal highway. (a) If the coefficient of friction between road and tires on a rainy day is 0.100, what is the minimum distance in which the car will stop? (b) What is the stopping distance when the surface is dry and $\mu_s = 0.600$?

4. A woman at an airport is towing her 20.0-kg suitcase at constant speed by pulling on a strap at an angle of θ above the horizontal (Fig. P5.4). She pulls on the strap with a 35.0-N force, and the friction force on the suitcase is 20.0 N. (a) Draw a free-body diagram of the suitcase. (b) What angle does the strap make with the horizontal? (c) What normal force does the ground exert on the suitcase?

Figure P5.4

5. A 3.00-kg block is released from rest at the top of a 30.0° incline and slides a distance of 2.00 m down the incline in 1.50 s. Find (a) the magnitude of the acceleration of the block, (b) the coefficient of kinetic friction between block and plane, (c) the frictional force acting on the block, and (d) the speed of the block after it has slid 2.00 m.

6. In order to determine the coefficients of friction between rubber and various surfaces, a student uses a rubber eraser and an incline. In one experiment the eraser slips down the incline when the angle of inclination is 36.0° and then moves down the incline with constant speed when the angle is reduced to 30.0°. From these data, determine the coefficients of static and kinetic friction for this experiment.

7. A boy drags his 60.0-N sled at constant speed up a 15.0° hill. He does so by pulling with a 25.0-N force on a rope attached to the sled. If the rope is inclined at 35.0° to the horizontal, (a) what is the coefficient of kinetic friction between sled and snow? (b) At the top of the hill, he jumps on the sled and slides down the hill. What is the magnitude of his acceleration down the slope?

8. Determine the stopping distance for a skier with a speed of 20.0 m/s (Fig. P5.8). Assume $\mu_k = 0.180$ and $\theta = 5.00°$.

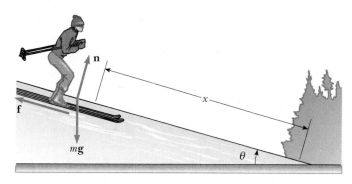

Figure P5.8

9. A 9.00-kg hanging weight is connected by a string over a pulley to a 5.00-kg block that is sliding on a flat table (Fig. P5.9). If the coefficient of kinetic friction is 0.200, find the tension in the string.

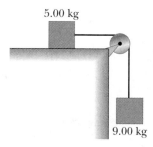

5.00 kg

9.00 kg

Figure P5.9

10. Three masses are connected on the table as shown in Figure P5.10. The table is rough and has a coefficient of sliding friction of 0.350. The three masses are $m_1 = 4.00$ kg, $m_2 = 1.00$ kg, and $m_3 = 2.00$ kg, respectively, and the pulleys are frictionless. (a) Draw free-body diagrams for each of the masses. (b) Determine the acceleration of each block and their directions. (c) Determine the tensions in the two cords.

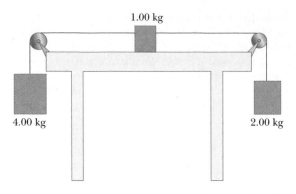

Figure P5.10

11. A mass $M = 2.20$ kg is accelerated across a rough surface by a rope passing over a pulley, as shown in Figure P5.11. The tension in the rope is 10.0 N and the pulley is 10.0 cm above the top of the block. The coefficient of kinetic friction is 0.400. (a) Determine the acceleration of the block when $x = 40.0$ cm. (b) Find the value of x at which the acceleration becomes zero.

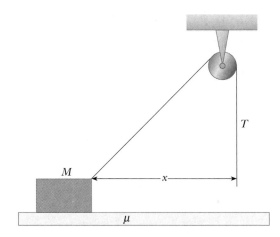

Figure P5.11

Section 5.2 Newton's Second Law Applied to Uniform Circular Motion

12. A toy car moving at constant speed completes one lap around a circular track (a distance of 200 m) in 25.0 s. (a) What is the average speed? (b) If the mass of the car is 1.50 kg, what is the magnitude of the force that keeps it moving in a circle?

13. A 3.00-kg mass attached to a light string rotates on a horizontal, frictionless table. The radius of the circle is 0.800 m, and the string can support a mass of 25.0 kg before breaking. What range of speeds can the mass have before the string breaks?

14. A 55.0-kg ice skater is moving at 4.00 m/s when she grabs the loose end of a rope the opposite end of which is tied to a pole. She then moves in a circle of radius 0.800 m around the pole. (a) Determine the force exerted by the rope on her arms. (b) Compare this force with her weight.

15. In the Bohr model of the hydrogen atom, the speed of the electron is approximately 2.20×10^6 m/s. Find (a) the centripetal acceleration of the electron and (b) the force acting on the electron as it revolves in a circular orbit of radius 0.530×10^{-10} m.

16. A crate of eggs is located in the middle of the flat bed of a pickup truck as the truck negotiates an unbanked curve in the road. The curve may be regarded as an arc of a circle of radius 35.0 m. If the coefficient of static friction between crate and truck is 0.600, what must be the maximum speed of the truck if the crate is not to slide?

17. A coin placed 30.0 cm from the center of a rotating, horizontal turntable slips when its speed is 50.0 cm/s. (a) What force provides the centripetal acceleration when the coin is stationary relative to the turntable? (b) What is the coefficient of static friction between coin and turntable?

18. A small turtle, appropriately named Dizzy, is placed on a horizontal, rotating turntable 20.0 cm from its center. Dizzy's mass is 50.0 g, and the coefficient of static friction between his feet and the turntable is 0.300. Find (a) the maximum number of revolutions per second the turntable can have if Dizzy is to remain stationary relative to the turntable and (b) Dizzy's speed and centripetal acceleration when he is on the verge of slipping.

19. Consider a conical pendulum with an 80.0-kg bob on a 10.0-m wire making an angle of 5.00° with the vertical (Fig. P5.19). Determine (a) the horizontal and vertical components of the force exerted by the wire on the pendulum, and (b) the centripetal acceleration of the bob.

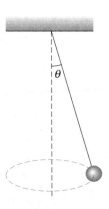

Figure P5.19

Section 5.3   Nonuniform Circular Motion

20. A 40.0-kg child sits in a swing supported by two chains, each 3.00 m long. If the tension in each chain at the lowest point is 350 N, find (a) the child's speed at the lowest point and (b) the force of the seat on the child at the lowest point. (Neglect the mass of the seat.)

21. Tarzan ($m = 85.0$ kg) tries to cross a river by swinging from a vine. The vine is 10.0 m long, and his speed at the bottom of the swing (as he just clears the water) is 8.00 m/s. Tarzan doesn't know that the vine has a breaking strength of 1000 N. Does he make it safely across the river?

22. At the Six Flags Great America amusement park in Gurnee, Illinois, there is a roller coaster that incorporates some of the latest design technology and some basic physics. Each vertical loop, instead of being circular, is shaped like a teardrop (Fig. P5.22). The cars ride on the inside of the loop at the top, and the speeds are high enough to ensure that the cars remain on the track. The biggest loop is 40.0 m high, with a maximum speed of 31.0 m/s (nearly 70 mph) at the bottom. Suppose the speed at the top is 13.0 m/s and the corresponding centripetal acceleration is $2g$. (a) What is the radius of the arc of the teardrop at the top? (b) If the total mass of the cars plus people is M, what force does the rail exert on it at the top? (c) Suppose the roller coaster had a loop of radius 20.0 m. If the cars have the same speed, 13.0 m/s at the top, what is the centripetal acceleration at the top? Comment on the normal force at the top in this situation.

Figure P5.22   *(Frank Cezus, FPG International)*

23. An 1800-kg car passes over a hump in a road that follows the arc of a circle of radius 42.0 m, as in Figure P5.23. (a) What force does the road exert on the car as the car passes the highest point of the hump if the car travels at 16.0 m/s? (b) What is the maximum speed the car can have as it passes this highest point before losing contact with the road?

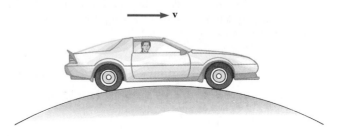

Figure P5.23

24. A car of mass m passes over a hump in a road that follows the arc of a circle of radius R, as in Figure P5.23. (a) What force does the road exert on the car as the car passes the highest point of the hump if the car travels at a speed v? (b) What is the maximum speed the car can have as it passes this highest point before losing contact with the road?

25. A pail of water is rotated in a vertical circle of radius 1.00 m. What is the minimum speed of the pail at the top of the circle if no water is to spill out?

Section 5.4   Motion in the Presence of Velocity-Dependent Resistive Forces

26. A sky diver of mass 80.0 kg jumps from a slow-moving aircraft and reaches a terminal speed of 50.0 m/s. (a) What is the acceleration of the sky diver when her speed is 30.0 m/s? What is the drag force on the diver when her speed is (b) 50.0 m/s? (c) 30.0 m/s?

27. A small piece of Styrofoam packing material is dropped from a height of 2.00 m above the ground. Until the terminal speed is reached, the magnitude of the acceleration is given by $a = g - cv$. After falling 0.500 m, it reaches terminal speed, and then the Styrofoam takes an extra 5.00 s to reach the ground. (a) What is the value of the constant c? (b) What is the acceleration at $t = 0$? (c) What is the acceleration when the speed is 0.150 m/s?

28. (a) Estimate the terminal speed of a wooden sphere (density 0.830 g/cm^3) moving in air if its radius is 8.00 cm. (b) From what height would a freely falling object reach this speed in the absence of air resistance? Note that the density of air is 1.20 kg/m^3.

29. A motorboat cuts its engine when its speed is 10.0 m/s and coasts to rest. The equation governing the motion of the motorboat during this period is $v = v_0 e^{-ct}$, where v is the

speed at time t, v_0 is the initial speed, and c is a constant. At $t = 20.0$ s, the speed is 5.00 m/s. (a) Find the constant c. (b) What is the speed at $t = 40.0$ s? (c) Differentiate the expression for $v(t)$ and thus show that the acceleration of the boat is proportional to the speed at any time.

30. Assume that the resistive force exerted on a speed skater is $f = -kmv^2$, where k is a constant and m is the skater's mass. The skater crosses the finish line of a straight-line race with speed v_f and then slows down by coasting without skating. Show that the skater's speed at any time t after crossing the finish line is $v(t) = v_f/(1 + ktv_f)$.

Section 5.5   Numerical Modeling in Particle Dynamics

31. A hailstone of mass 4.80×10^{-4} kg falls through the air and experiences a net force given by

$$F = -mg + Dv^2$$

where $D = 2.50 \times 10^{-5}$ kg/m. (a) Calculate the terminal speed of the hailstone. (b) Use Euler's method of numerical analysis to find the speed and position of the hailstone at 0.2-s intervals, taking the initial speed to be zero. Continue the calculation until the hailstone reaches 99% of terminal speed.

32. A 3.00-g leaf is dropped from a height of 2.00 m above the ground. Assume the net downward force on the leaf is $F = mg - bv$, where the drag factor is $b = 0.0300$ kg/s. (a) Calculate the terminal speed of the leaf. (b) Use Euler's method of numerical analysis to find the speed and position of the leaf, as functions of time, from the instant it is released until 99% of terminal speed is reached. (*Hint:* Try $\Delta t = 0.005$ s.)

33. A 50.0-kg parachutist jumps from an airplane and falls to Earth with a drag force proportional to the square of the speed, $R = Dv^2$. Take $D = 0.200$ kg/m (with the parachute closed) and $D = 20.0$ kg/m (with the chute open). (a) Determine the terminal speed of the parachutist in both configurations, before and after the chute is opened. (b) Set up a numerical analysis of the motion, and compute the speed and position as functions of time, assuming the jumper begins the descent at 1000 m above the ground and is in free fall for 10.0 s before opening the parachute. (*Hint:* When the parachute opens, a sudden large acceleration takes place; a smaller time step may be necessary in this region.)

34. Consider a 10.0-kg projectile launched with an initial speed of 100 m/s, at an angle of 35.0° elevation. The resistive force is $\mathbf{R} = -b\mathbf{v}$, where $b = 10.0$ kg/s. (a) Use a numerical method to determine the horizontal and vertical positions of the projectile as functions of time. (b) What is the range of this projectile? (c) Determine the elevation angle that gives the maximum range for the projectile. (*Hint:* Adjust the elevation angle by trial and error to find the greatest range.)

Section 5.6   The Fundamental Forces of Nature

35. Two identical isolated particles, each of mass 2.00 kg, are separated by a distance of 30.0 cm. What is the magnitude of the gravitational force of one particle on the other?

36. In a thundercloud there may be electric charges of $+40.0$ C near the top of the cloud and -40.0 C near the bottom of the cloud. These charges are separated by 2.00 km. What is the electric force on the top charge?

Section 5.7   The Gravitational Field

37. When a falling meteor is at a distance above the Earth's surface of 3.00 times the Earth's radius, what is its free-fall acceleration due to the gravitational force exerted on it?

Additional Problems

38. A crate of weight w is pushed by a force $\mathbf{F}$ on a horizontal floor. (a) If the coefficient of static friction is μ_s and $\mathbf{F}$ is directed at angle ϕ below the horizontal, show that the minimum value of F that will move the crate is given by

$$F = \frac{\mu_s w \sec \phi}{1 - \mu_s \tan \phi}$$

(b) Find the minimum value of F that can produce motion when $\mu_s = 0.400$, $w = 100$ N, and $\phi = 0°, 15.0°, 30.0°, 45.0°$, and 60.0°.

39. A 1.30-kg toaster is not plugged in. The coefficient of static friction between the toaster and a horizontal countertop is 0.350. To make the toaster start moving, you carelessly pull on its electric cord. (a) In order for the cord tension to be as small as possible, you should pull at what angle above the horizontal? (b) Then how large must the tension be?

40. A student builds and calibrates an accelerometer, which she uses to determine the speed of her car around a certain unbanked highway curve. The accelerometer is a plumb bob with a protractor that she attaches to the roof of her car. A friend riding in the car with her observes that the plumb bob hangs at an angle of 15.0° from the vertical when the car has a speed of 23.0 m/s. (a) What is the centripetal acceleration of the car rounding the curve? (b) What is the radius of the curve? (c) What is speed of the car if the plumb bob deflection is 9.00° while rounding the same curve?

41. A space station, in the form of a large wheel 120 m in diameter, rotates to provide an artificial gravity of 3.00 m/s² for persons situated at the outer rim. Find the rotational frequency of the wheel (in revolutions per minute) that will produce this effect.

42. An amusement park ride consists of a rotating circular platform 8.00 m in diameter from which 10.0-kg seats are suspended at the end of 2.50-m massless chains (Fig. P5.42). When the system rotates, the chains make an angle $\theta = 28.0°$ with the vertical. (a) What is the speed of each seat?

(b) Draw a free-body diagram of a 40.0-kg child riding in a seat and find the tension in the chain.

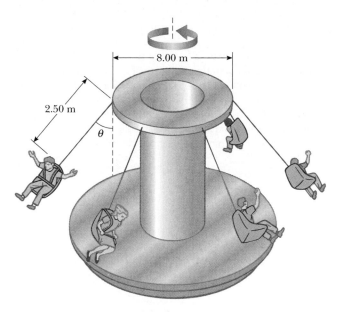

Figure P5.42

43. Because the Earth rotates about its axis, a point on the Equator experiences a centripetal acceleration of 0.0337 m/s², while a point at the poles experiences no centripetal acceleration. (a) Show that at the Equator the magnitude of the gravitational force on an object (its weight) must exceed the object's apparent weight. (b) What is the apparent weight at the Equator and at the poles of a person having a mass of 75.0 kg? (Assume the Earth is a uniform sphere and take $g = 9.800$ N/kg.)

44. The Earth rotates around its axis with a period of 24.0 h. Imagine that the rotational speed can be increased. If an object at the Equator is to have zero apparent weight, (a) what must the new period be? (b) By what factor would the speed of the object be increased when the planet is rotating at the higher speed? (*Hint:* See Problem 43 and note that the apparent weight of the object becomes zero when the normal force exerted on it is zero. Also, the distance traveled during one period is $2\pi R$, where R is the Earth's radius.)

45. An engineer wishes to design a curved exit ramp for a toll road in such a way that a car will not have to rely on friction to round the curve without skidding. Suppose that a typical car rounds the curve with a speed of 30 mi/h (13.4 m/s) and that the radius of the curve is 50.0 m. At what angle should the curve be banked? (See Fig. P5.45.)

46. A car rounds a banked curve as in Figure P5.45. The radius of curvature of the road is R, the banking angle is θ, and the coefficient of static friction is μ_s. (a) Determine the

range of speeds the car can have without slipping up or down the road. (b) Find the minimum value for μ_s such that the minimum speed is zero. (c) What is the range of speeds possible if $R = 100$ m, $\theta = 10.0°$, and $\mu_s = 0.100$ (slippery conditions)?

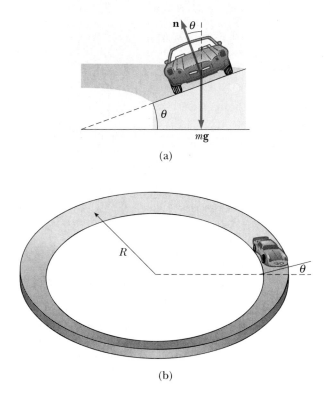

(a)

(b)

Figure P5.45

47. An amusement park ride consists of a large vertical cylinder that spins around its axis fast enough that any person inside is held up against the wall when the floor drops away (Fig. P5.47). The coefficient of static friction between person and wall is μ_s, and the radius of the cylinder is R. (a) Show that the maximum period of revolution necessary to keep the person from falling is $T = (4\pi^2 R\mu_s/g)^{1/2}$. (b) Obtain a numerical value for T if $R = 4.00$ m and $\mu_s = 0.400$. How many revolutions per minute does the cylinder make?

48. The expression $F = arv + br^2v^2$ gives the magnitude of the resistive force (in newtons) on a sphere of radius r (in meters) exerted by a stream of air with speed v (in meters per second), where a and b are constants with appropriate SI units. Their numerical values are $a = 3.10 \times 10^{-4}$ and $b = 0.870$. Using this formula, find the terminal speed for water droplets falling under their own weight in air, taking the following values for the drop radii: (a) 10.0 μm, (b) 100 μm, (c) 1.00 mm. Note that for (a) and (c) you can obtain accurate answers without solving a quadratic equa-

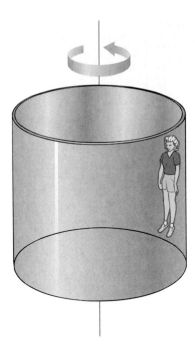

Figure P5.47

tion, by considering which of the two contributions to the air resistance is dominant and ignoring the lesser contribution.

49. A model airplane of mass 0.750 kg flies in a horizontal circle at the end of a 60.0-m control wire, with a speed of 35.0 m/s. Compute the tension in the wire if it makes a constant angle of 20.0° with the horizontal. The forces exerted on the airplane are the pull of the control wire, the force of gravity, and aerodynamic lift, which acts at 20.0° inward from the vertical as shown in Figure P5.49.

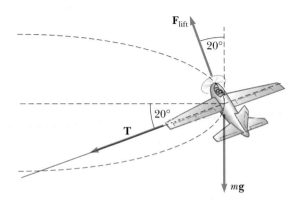

Figure P5.49

50. A 9.00-kg object starting from rest moves through a viscous medium and experiences a resistive force $\mathbf{R} = -b\mathbf{v}$, where $\mathbf{v}$ is the velocity of the object. If the object's speed reaches one half its terminal speed in 5.54 s, (a) determine the terminal speed. (b) At what time is the speed of the object three fourths the terminal speed? (c) How far has the object traveled in the first 5.54 s of motion?

51. Plaskett's binary system consists of two stars that revolve in a circular orbit around a center of mass midway between them. This means that the masses of the two stars are equal (Fig. P5.51). If the orbital velocity of each star is 220 km/s and the orbital period of each is 14.4 days, find the mass M of each star. (For comparison, the mass of our Sun is 1.99×10^{30} kg.)

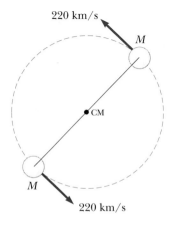

Figure P5.51

52. One block of mass 5.00 kg sits on top of a second rectangular block of mass 15.0 kg, which in turn is on a horizontal table. The coefficients of friction between the two blocks are $\mu_s = 0.300$ and $\mu_k = 0.100$. The coefficients of friction between the lower block and the rough table are $\mu_s = 0.500$ and $\mu_k = 0.400$. You apply a constant horizontal force to the lower block, just large enough to make this block start sliding out from between the upper block and the table. (a) Draw a free-body diagram of each block, naming the forces on each. (b) Determine the magnitude of each force on each block at the instant at which you have started pushing but motion has not yet started. In particular, what force must you apply? (c) Determine the acceleration you measure for each block.

53. Determine the order of magnitude of the gravitational force that you exert on another person 2 m away. In your solution state the quantities you estimate and the values you estimate for them.

Spreadsheet Problems

S1. A 0.142-kg baseball has a terminal speed of 42.5 m/s (95 mph). (a) If a baseball experiences a drag force of magnitude $R = Dv^2$, what is the value of D? (b) What is the magnitude of the drag force when the speed of the baseball is 36 m/s? (c) Set up a spreadsheet to determine the motion of a baseball thrown vertically upward at an initial speed of 36 m/s. What maximum height does the ball reach? How long is it in the air? What is its speed just before it hits the ground?

S2. A 10.0-kg projectile is launched with an initial speed of 150 m/s at an elevation angle of 35.0° with the horizontal. The resistive force acting on the projectile is $\mathbf{R} = -b\mathbf{v}$, where $b = 15$ kg/s. (a) Set up a spreadsheet to determine the horizontal and vertical positions of the projectile as functions of time. (b) Find the range of this projectile. (c) Determine the elevation angle that gives the maximum range for this projectile. (*Hint:* Adjust the elevation angle by trial and error to find the greatest range.)

S3. A professional golfer hits her 5-iron 155 m (170 yd). A 46-g golf ball experiences a drag force of magnitude $R = Dv^2$ and has a terminal speed of 44 m/s. (a) Calculate the drag coefficient D for the golf ball. (b) Set up a spreadsheet to calculate the trajectory of this shot. If the initial velocity of the ball makes an angle of 31° (the loft angle) with the horizontal, what initial speed must the ball have to reach the 155-m distance? (c) If this same golfer hits her 9-iron (47° loft) a distance of 119 m, what is the initial speed of the ball in this case? Discuss the differences in trajectories between the two shots.

ANSWERS TO CONCEPTUAL PROBLEMS

1. It is easier to attach the rope and pull. In this case, there is a component of your applied force which is upward. This reduces the normal force between the sled and the snow. In turn, this reduces the friction force between the sled and the snow, making it easier to move. If you push from behind, with a force with a downward component, the normal force is larger, the friction force is larger, and the sled is harder to move.

2. On the way up the ramp, the friction force and the component of the gravitational force on the book are both directed downhill. When the book returns back down the ramp, the component of the gravitational force is still downhill, but the friction force is uphill. Thus, there is a smaller net force when the book is sliding downhill, and according to Newton's second law, a smaller magnitude of acceleration. With a smaller acceleration, the velocity of the book at any point on the ramp is smaller when coming downhill than it was when going uphill. As a result, it takes longer to come down than to go up. The extreme case would be if the angle of the hill is such that once the book stops at the top of its motion, the maximum static friction force is larger than the component of the gravitational force, and the book "sticks" to the ramp. Then it takes *forever* to come down!

3. (a) Doubling the truck's mass would double the normal force the road exerts and double the backward force of kinetic friction. With twice the friction force acting on the mass, its acceleration would be the same and so would all other parameters of its motion, including distance. (b) With the same acceleration but half its original speed, the time required to stop would be halved, and moving with one-half the average speed for one-half the time, the truck would travel one-quarter the distance.

4. An object can move in a circle even if the total force on it is not perpendicular to its velocity, but then its speed will change. Resolve the total force into an inward radial component and a perpendicular tangential component. If the tangential force is forward, the object will speed up, and if the tangential force acts backward, it will slow down.

5. The forces exerted on the sky diver are the downward force of gravity, $m\mathbf{g}$, and an upward force of air resistance, $\mathbf{R}$, the magnitude of which is less than her weight before she reaches terminal speed. As her downward speed increases, the force of air resistance increases. The vector sum of the force of gravity and the force of air resistance gives a total force that decreases with time, so her acceleration decreases. Once she reaches terminal speed, the two forces balance each other, the net force is zero, and her acceleration is zero.

6

Work and Energy

Energy is present in the Universe in various forms, including mechanical energy, electromagnetic energy, chemical energy, thermal energy, and nuclear energy. When energy is changed from one form to another, its total amount remains the same. Conservation of energy says that if an isolated system loses energy in some form, it will gain an equal amount of energy in other forms.

In this chapter we shall concern ourselves only with the mechanical form of energy. We shall see that the concepts of work and energy can be applied to the dynamics of a mechanical system without resorting to Newton's laws. (However, it is important to note that the work–energy concepts are based on Newton's laws and therefore do not involve any new physical principles.) This energy approach to describing motion is especially useful when the force acting on a particle is not constant. Because the acceleration is not constant, we cannot apply the kinematic equations we developed in Chapter 2. Particles in nature are often subject to forces that vary with the particles' positions. These forces include gravitational forces and the force exerted on a body attached to a spring. We shall describe techniques for treating such systems with the help of an extremely important concept called the *work-kinetic energy theorem*, which is the central topic of this chapter.

We begin by defining work, a concept that provides a link between the concepts of force and energy.

These cyclists are working ▶ **hard and expending energy as they pedal uphill in Marin County, California.** *(David Madison, Tony Stone Images)*

6.1 · WORK DONE BY A CONSTANT FORCE

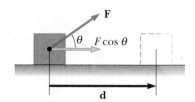

Figure 6.1 If an object undergoes a displacement **d**, the work done by the force **F** is $(F \cos \theta)\, d$.

Almost all of the terms we have used thus far—*velocity, acceleration, force,* and so on—have conveyed the same meaning in physics as they do in everyday life. Now, however, we encounter a term the meaning of which in physics is distinctly different from its everyday meaning. This new term is **work**. Consider a particle that undergoes a displacement **d** along a straight line while acted on by a constant force **F**, which makes an angle θ with **d**, as in Figure 6.1.

> The work W done by an agent exerting a constant force is the product of the component of the force along the direction of the displacement of the point of application of the force and the magnitude of the displacement:
>
> $$W \equiv Fd \cos \theta \qquad \text{[6.1]}$$

From this definition, we see that a force does no work on a particle if the particle does not move: If $d = 0$, Equation 6.1 gives $W = 0$. Also note from Equation 6.1 that the work done by a force is zero when the force is perpendicular to the displacement. That is, if $\theta = 90°$, then $\cos 90° = 0$, and $W = 0$. For example, in Figure 6.2, the work done by the normal force and the work done by the force of gravity during the horizontal displacement are zero for the same reason. In general, the particle may be moving with a constant or varying velocity under the influence of several forces. In that case, because work is a scalar quantity, the total work done as the particle undergoes some displacement is the algebraic sum of the work done by each of the forces.

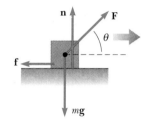

Figure 6.2 When an object is displaced horizontally, the normal force **n** and the force of gravity $m\mathbf{g}$ do no work.

The sign of the work also depends on the direction of **F** relative to **d**. The work done by the applied force is positive when the vector associated with the component $F \cos \theta$ is in the *same direction* as the displacement. For example, when an object is lifted, the work done by the applied force is positive because the lifting force is upward—that is, in the same direction as the displacement.

When the vector associated with the component $F \cos \theta$ is in the direction *opposite* the displacement, W is *negative*. In the case of the object being lifted, for instance, the work done by the gravitational force is negative. It is important to note that work is an energy transfer; if energy is transferred *to* the system, W is positive; if energy is transferred *from* the system, W is negative.

If an applied force **F** acts along the direction of the displacement, then $\theta = 0$ and $\cos 0 = 1$. In this case, Equation 6.1 gives

$$W = Fd \qquad \text{[6.2]}$$

Work is a scalar quantity, and its units are force multiplied by length. Therefore, the SI unit of work is the **newton·meter** (N·m). The newton·meter, when it refers to work or energy, is called the **joule** (J). The unit of work in the cgs system is the **dyne·cm**, also called the **erg**; the unit in the British engineering system is the **ft·lb**. These are summarized in Table 6.1. Note that $1\text{ J} = 10^7$ ergs.

TABLE 6.1   Units of Work in the Three Common Systems of Measurement

System	Unit of Work	Name of Combined Unit
SI	newton·meter (N·m)	joule (J)
cgs	dyne·centimeter (dyne·cm)	erg
British engineering (conventional)	foot·pound (ft·lb)	foot·pound (ft·lb)

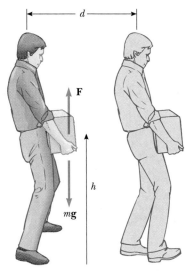

Figure 6.3   (Thinking Physics 1) A person lifts a cement block of mass *m* a vertical distance *h* and then walks horizontally a distance *d*.

Thinking Physics 1

A person slowly lifts a cement block of mass m a vertical height h and then walks horizontally a distance d while holding the block, as in Figure 6.3. Determine the work done by the person and by the force of gravity in this process.

Reasoning Assuming that the person lifts the block with a force of magnitude equal to the weight of the block, mg, the work done by the person during the vertical displacement is mgh, because the force in this case is in the direction of the displacement. The work done by the person during the horizontal displacement of the block is zero, because the applied force in this process is perpendicular to the displacement. Thus, the net work done by the person is mgh. The work done by the force of gravity during the vertical displacement of the block is $-mgh$, because this force is opposite the displacement. The work done by the force of gravity is zero during the horizontal displacement because this force is also perpendicular to the displacement. Hence, the net work done by the force of gravity is $-mgh$. The net work on the block is zero $(+mgh - mgh = 0)$.

Thinking Physics 2

A planet moves in a perfectly circular orbit around the Sun. What can be said about the work done by the gravitational force on the planet? Suppose the orbit is elliptical; what can be said now about the work done by the gravitational force?

Reasoning In the case of the circular orbit, the gravitational force is always perpendicular to the displacement of the planet as it moves around the orbit. Thus, there is no work done over any portion of the orbit, and no net work done in one complete orbit. In the case of the elliptical orbit, the Sun is at one focus of the ellipse. Thus, the force vector is not, in general, perpendicular to the displacement. While the planet is moving away from the Sun toward its farthest distance, called the aphelion, the force vector has a component opposite to the displacement. Thus, negative work is done on the planet and it slows down. As the planet passes aphelion and moves back toward its closest distance to the Sun, called the perihelion, there is a component of the gravitational force that is parallel to the displacement. The planet thus has positive work done on it and speeds up. The total negative work done on the outbound trip is canceled by the positive work done on the return trip. Thus, in the elliptical orbit, there is work done over any given portion of the orbit (except for portions symmetric about the aphelion or perihelion), but no net work is done over one complete orbit.

CONCEPTUAL PROBLEM 1

Consider a tug-of-war, in which the two teams pulling on the rope are evenly matched, so that no motion takes place. Is work done on the rope? On the pullers? On the ground? Is work done on anything?

CONCEPTUAL PROBLEM 2

A team of furniture movers wishes to load a truck using a ramp from the ground to the rear of the truck. One of the movers claims that less work would be required to load the truck if the length of the ramp were increased, reducing the angle of the ramp with respect to the horizontal. Is his claim valid? Explain.

CONCEPTUAL PROBLEM 3

Roads going up mountains are formed into *switchbacks*, with the road weaving back and forth along the face of the slope, such that there is only a gentle rise on any portion of the roadway. Does this require any less work to be done by an automobile climbing the mountain, compared to driving on a roadway that runs straight up the slope? Why are the switchbacks used?

Example 6.1   Mr. Clean

A man cleaning his apartment pulls a vacuum cleaner with a force of magnitude $F = 50$ N. The force makes an angle of 30° with the horizontal as shown in Figure 6.4. The vacuum cleaner is displaced 3.0 m to the right. Calculate the work done by the 50-N force.

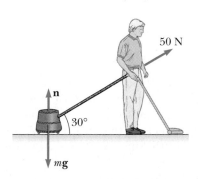

Figure 6.4   (Example 6.1) A vacuum cleaner being pulled at an angle of 30° with the horizontal.

Solution   Using the definition of work (Equation 6.1), we have

$$W_F = (F \cos \theta) d = (50 \text{ N})(\cos 30°)(3.0 \text{ m})$$
$$= 130 \text{ N} \cdot \text{m} = 130 \text{ J}$$

Note that the normal force, **n**, the force of gravity, $m\mathbf{g}$, and the upward component of the applied force, (50 N) sin 30°, do *no* work because they are perpendicular to the displacement.

EXERCISE 1   Find the work done by the man on the vacuum cleaner if he pulls it 3.0 m with a horizontal force of 32 N.
Answer   96 J

EXERCISE 2   If a person lifts a 20.0-kg bucket from a well and does 6.00 kJ of work, how deep is the well? Assume the speed of the bucket remains constant as it is lifted.
Answer   30.6 m

EXERCISE 3   A 65-kg woman climbs a flight of 20 stairs, each 23 cm high. How much work is done against the force of gravity in the process?   Answer   2.93 kJ

6.2 • THE SCALAR PRODUCT OF TWO VECTORS

We have defined work as a *scalar* quantity given by the product of the magnitude of the displacement and the component of a force in the direction of the displacement. It is convenient to express Equation 6.1 in terms of a **scalar product** of the two vectors **F** and **d**. We write this scalar product **F·d**. Because of the dot symbol, the scalar product is often called the *dot product*. Thus, we can express Equation 6.1 as a scalar product:

Work expressed as a scalar •
product

$$W = \mathbf{F \cdot d} = Fd \cos \theta \qquad [6.3]$$

In other words, **F·d** (read "F dot d") is a shorthand notation for $Fd \cos \theta$.

> In general, the scalar product of any two vectors **A** and **B** is a scalar quantity equal to the product of the magnitudes of the two vectors and the cosine of the angle θ between them:
>
> *Scalar product of any two* •
> *vectors **A** and **B***
>
> $$\mathbf{A \cdot B} \equiv AB \cos \theta \qquad [6.4]$$

where θ is the smaller angle between **A** and **B**, as in Figure 6.5. Note that **A** and **B** need not have the same units.

In Figure 6.5, $B \cos \theta$ is the projection of **B** onto **A**. Therefore, Equation 6.4 says that **A·B** is the product of the magnitude of **A** and the projection of **B** onto **A**.[1]

From Equation 6.4 we also see that the scalar product is *commutative*. That is,

The order of the scalar •
product can be reversed.

$$\mathbf{A \cdot B} = \mathbf{B \cdot A} \qquad [6.5]$$

Finally, the scalar product obeys the *distributive law of multiplication*, so that

$$\mathbf{A \cdot (B + C)} = \mathbf{A \cdot B} + \mathbf{A \cdot C} \qquad [6.6]$$

The dot product is simple to evaluate from Equation 6.4 when **A** is either perpendicular or parallel to **B**. If **A** is perpendicular to **B** ($\theta = 90°$), then **A·B** = 0. (The equality **A·B** = 0 also holds in the more trivial case when either **A** or **B** is zero.) If **A** and **B** point in the same direction ($\theta = 0°$), then **A·B** = AB. If **A** and **B** point in opposite directions ($\theta = 180°$), then **A·B** = $-AB$. The scalar product is negative when $90° < \theta < 180°$.

The unit vectors **i, j**, and **k**, which were defined in Chapter 1, lie in the positive *x, y*, and *z* directions, respectively, of a right-handed coordinate system. Therefore, it follows from the definition of **A·B** that the scalar products of these unit vectors are given by

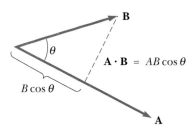

$$\mathbf{i \cdot i} = \mathbf{j \cdot j} = \mathbf{k \cdot k} = 1 \qquad [6.7]$$

Scalar products of unit •
vectors

$$\mathbf{i \cdot j} = \mathbf{i \cdot k} = \mathbf{j \cdot k} = 0 \qquad [6.8]$$

Two vectors **A** and **B** can be expressed in component form as

Figure 6.5 The scalar product **A·B** equals the magnitude of **A** multiplied by the projection of **B** onto **A**.

[1]This is equivalent to stating that **A·B** equals the product of the magnitude of **B** and the projection of **A** onto **B** or, vice versa, the product of the magnitude of **A** and the projection of **B** onto **A.**

$$\mathbf{A} = A_x\mathbf{i} + A_y\mathbf{j} + A_z\mathbf{k}$$

$$\mathbf{B} = B_x\mathbf{i} + B_y\mathbf{j} + B_z\mathbf{k}$$

Therefore, Equations 6.7 and 6.8 reduce the scalar product of **A** and **B** to

$$\mathbf{A}\cdot\mathbf{B} = A_xB_x + A_yB_y + A_zB_z \qquad\qquad \textbf{[6.9]}$$

In the special case where **A** = **B**, we see that

$$\mathbf{A}\cdot\mathbf{A} = A_x^2 + A_y^2 + A_z^2 = A^2$$

Example 6.2   The Scalar Product

The vectors **A** and **B** are given by **A** = 2**i** + 3**j** and **B** = −**i** + 2**j**.
(a) Determine the scalar product **A·B**.

Solution $\mathbf{A}\cdot\mathbf{B} = (2\mathbf{i} + 3\mathbf{j})\cdot(-\mathbf{i} + 2\mathbf{j})$

$$= -2\mathbf{i}\cdot\mathbf{i} + 2\mathbf{i}\cdot 2\mathbf{j} - 3\mathbf{j}\cdot\mathbf{i} + 3\mathbf{j}\cdot 2\mathbf{j}$$

$$= -2 + 6 = \boxed{4}$$

where we have used the facts that **i·i** = **j·j** = 1 and **i·j** = **j·i** = 0. The same result is obtained using Equation 6.9 directly, where $A_x = 2$, $A_y = 3$, $B_x = -1$, and $B_y = 2$.
   (b) Find the angle θ between **A** and **B**.

Solution   The magnitudes of **A** and **B** are given by

$$A = \sqrt{A_x^2 + A_y^2} = \sqrt{(2)^2 + (3)^2} = \sqrt{13}$$

$$B = \sqrt{B_x^2 + B_y^2} = \sqrt{(-1)^2 + (2)^2} = \sqrt{5}$$

Using Equation 6.4 and the result from (a) gives

$$\cos\theta = \frac{\mathbf{A}\cdot\mathbf{B}}{AB} = \frac{4}{\sqrt{13}\sqrt{5}} = \frac{4}{\sqrt{65}}$$

$$\theta = \cos^{-1}\frac{4}{8.06} = 60.2°$$

EXERCISE 4   For **A** = 4**i** + 3**j** and **B** = −**i** + 3**j**, find (a) **A·B** and (b) the angle between **A** and **B**.   Answer   (a) 5.00   (b) 71.6°

EXERCISE 5   As a particle moves from the origin to (3**i** − 4**j**) m, it is acted on by a force given by (4**i** − 5**j**) N. Calculate the work done by this force as the particle moves through the given displacement.   Answer   32 J

6.3 • WORK DONE BY A VARYING FORCE

Consider a particle being displaced along the x axis under the action of a varying force, as in Figure 6.6. The particle is displaced in the direction of increasing x from $x = x_i$ to $x = x_f$. In such a situation, we cannot use $W = (F\cos\theta)d$ to calculate the work done by the force, because this relationship applies only when **F** is constant in magnitude and direction. However, if we imagine that the particle undergoes a small displacement, Δx, shown in Figure 6.6a, then the x component of the force, F_x, is approximately constant over this interval and we can express the work done by the force for this small displacement as

$$W_1 = F_x\,\Delta x \qquad\qquad \textbf{[6.10]}$$

This quantity is just the area of the shaded rectangle in Figure 6.6a. If we imagine that the F_x versus x curve is divided into a large number of such intervals, then the

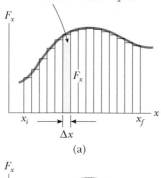

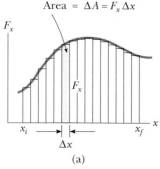

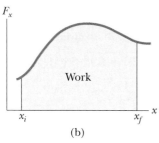

Figure 6.6 (a) The work done by the force F_x for the small displacement Δx is $F_x \Delta x$, which equals the area of the shaded rectangle. The total work done for the displacement from x_i to x_f is approximately equal to the sum of the areas of all the rectangles. (b) The work done by the variable force F_x as the particle moves from x_i to x_f is *exactly* equal to the area under this curve.

total work done for the displacement from x_i to x_f is approximately equal to the sum of a large number of such terms:

$$W \cong \sum_{x_i}^{x_f} F_x \, \Delta x$$

If the displacements Δx are allowed to approach zero, then the number of terms in the sum increases without limit, but the value of the sum approaches a definite value equal to the area under the curve bounded by F_x and the x axis as in Figure 6.6b. As you probably have learned in calculus, this limit of the sum is called an *integral* and is represented by

$$\lim_{\Delta x \to 0} \sum_{x_i}^{x_f} F_x \, \Delta x = \int_{x_i}^{x_f} F_x \, dx$$

The limits on the integral, $x = x_i$ to $x = x_f$, define what is called a **definite integral.** (An *indefinite integral* is the limit of a sum over an unspecified interval. Appendix B.7 gives a brief description of integration.) This definite integral is numerically equal to the area under the F_x versus x curve between x_i and x_f. Therefore, we can express the work done by F_x for the displacement of the particle from x_i to x_f as

$$W = \int_{x_i}^{x_f} F_x \, dx \qquad \text{[6.11]}$$

This equation reduces to Equation 6.1 when $F_x = F \cos \theta$ is constant.

If more than one force acts on a particle, the total work done is just the work done by the resultant force. If we express the resultant force in the x direction as ΣF_x, then the *net work* done as the particle moves from x_i to x_f is

$$W_{\text{net}} = \int_{x_i}^{x_f} \left(\sum F_x \right) dx \qquad \text{[6.12]}$$

Thinking Physics 3

Consider a single force that causes an object to follow a path. If the force is constant, under what conditions can the work done over the entire path be zero? If the force is varying, under what conditions can the work done over the entire path be zero?

Reasoning In the case of the constant force, there are several possibilities that will give a result of zero work. One possibility is that the constant value of the force is zero, in which case we have a trivial result of nothing pushing. The second possibility for zero work is that a nonzero force does act, but it does not succeed in moving the object, although this does not fit with the question, because it is stated that the object *does* move over a path. The third possibility is that the force is always perpendicular to the path, as in the case of a planet moving in a circular orbit around a gravitational-force center.

For the varying force, only one of the above possibilities is consistent with the question. The zero-force possibility is not available, because a force whose value is *always* zero is not a *varying* force. The second possibility, that of zero displacement, as is the case above, does not fit with the statement that the object does move through a path. The third possibility, in which the force is perpendicular to the displacement, will also give zero work in the case of a varying force.

Does the weight lifter do any work as he holds the weight on his shoulders? Does he do any work as he raises the weight?

We have additional possibilities in the case of the varying force that were not available in the constant force case. One possibility is that the varying force (or a constant force applied opposite the velocity) carries the object around a path such that the object ends up back where it started. In this case the net work is zero, because there was no net displacement. This is the case for a comet in an elliptical orbit around the Sun—no net work is done by the gravitational force during one complete orbit. Another possibility is that the relative direction of the displacement and the force reverse during the path in such a way that the area above the axis in the force-displacement graph exactly equals the area below the axis. This is the case for an automobile accelerating to some speed and then braking to a stop. The positive work done while the automobile is speeding up exactly equals the negative work done as the automobile slows back down (the equality of the positive and negative work is probably not obvious at this point, but will be after Section 6.4).

Example 6.3   Calculating Total Work Done from a Graph

A force acting on a particle varies with x, as shown in Figure 6.7. Calculate the work done by the force as the particle moves from $x = 0$ to $x = 6.0$ m.

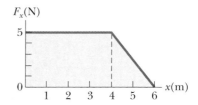

Solution   The work done by the force is equal to the area under the curve from $x = 0$ to $x = 6.0$ m. This area is equal to the area of the rectangular section from $x = 0$ to $x = 4.0$ m plus the area of the triangular section that extends from $x = 4.0$ m to $x = 6.0$ m. The area of the rectangle is $(4.0)(5.0)$ N·m $= 20$ J, and the area of the triangle is $\frac{1}{2}(2.0)(5.0)$ N·m $= 5.0$ J. Therefore, the total work done is 25 J.

Figure 6.7   (Example 6.3) The force acting on a particle is constant for the first 4.0 m of motion and then decreases linearly with x from $x = 4.0$ m to $x = 6.0$ m. The net work done by this force is the area under this curve.

Work Done by a Spring

A common physical system for which the force varies with position is shown in Figure 6.8. A block on a horizontal, frictionless surface is connected to a spring. If the spring is stretched or compressed a small distance from its unstretched, or equilibrium, configuration, the spring will exert a force on the block given by

$$F_s = -kx \qquad\qquad \text{[6.13]}$$ •   *Spring force*

where x is the displacement of the block from its unstretched ($x = 0$) position and k is a positive constant called the *force constant* of the spring. As mentioned in Chapter 4, Section 4.2, this force law for springs is known as **Hooke's law.** For many materials, Hooke's law can be very accurate, provided the displacement is not too large. The value of k is a measure of the stiffness of the spring. Stiff springs have large k values, and soft springs have small k values.

The negative sign in Equation 6.13 signifies that the force exerted by the spring is always directed *opposite* the displacement. For example, when $x > 0$, as in Figure 6.8a, the spring force is to the left, or negative. When $x < 0$, as in Figure 6.8c, the

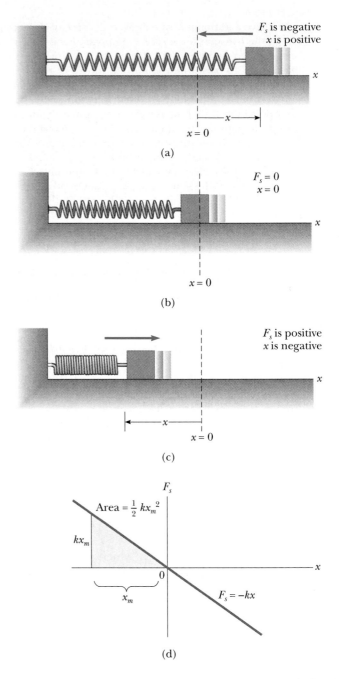

Figure 6.8   The force exerted by a spring on a block varies with the block's displacement from the equilibrium position $x = 0$. (a) When x is positive (stretched spring), the spring force is to the left. (b) When x is zero, the spring force is zero (natural length of the spring). (c) When x is negative (compressed spring), the spring force is to the right. (d) Graph of F_s versus x for the mass–spring system. The work done by the spring force as the block moves from $-x_m$ to 0 is the area of the shaded triangle, $\frac{1}{2}kx_m^2$.

spring force is to the right, or positive. Of course, when $x = 0$, as in Figure 6.8b, the spring is unstretched and $F_s = 0$. Because the spring force always acts toward the equilibrium position, it is sometimes called a *restoring force*. Once the block is displaced some distance x_m from equilibrium and then released, it moves from $-x_m$ through zero to $+x_m$. The details of the ensuing oscillating motion will be given in Chapter 12.

Suppose that the block is pushed to the left a distance x_m from equilibrium, as in Figure 6.8c, and then released. Let us calculate the *work done by the spring force* as the block moves from $x_i = -x_m$ to $x_f = 0$. Applying Equation 6.11 assuming the block may be treated as a particle, we get

$$W_s = \int_{x_i}^{x_f} F_s \, dx = \int_{-x_m}^{0} (-kx) \, dx = \tfrac{1}{2}kx_m^2 \qquad \textbf{[6.14]}$$

• *Work done by a spring*

where we have used the indefinite integral $\int x \, dx = x^2/2$. That is, the work done by the spring force is positive because the spring force is in the same direction as the displacement caused by the spring force (both are to the right). However, if we consider the work done by the spring force as the block moves from $x_i = 0$ to $x_f = x_m$, we find that $W_s = -\tfrac{1}{2}kx_m^2$, because for this part of the motion the displacement caused by the spring force is to the right and the spring force is to the left. Therefore, the *net* work done by the spring force as the block moves from $x_i = -x_m$ to $x_f = x_m$ is *zero*.

If we plot F_s versus x, as in Figure 6.8d, we arrive at the same results. Note that the work calculated in Equation 6.14 is equal to the area of the shaded triangle in Figure 6.8d, with base x_m and height kx_m. This area is $\tfrac{1}{2}kx_m^2$.

If the block undergoes an *arbitrary* displacement from $x = x_i$ to $x = x_f$, the work done by the spring force is

$$W_s = \int_{x_i}^{x_f} (-kx) \, dx = \tfrac{1}{2}kx_i^2 - \tfrac{1}{2}kx_f^2 \qquad \textbf{[6.15]}$$

From this equation we see that the work done by the spring force is zero for any motion that ends where it began ($x_i = x_f$). We shall make use of this important result in Chapter 7, where we describe the motion of this system in more detail.

Equations 6.14 and 6.15 describe the work done by the spring force exerted on the block. Now let us consider the work done on the spring by an *external agent* as the spring is stretched *very slowly* from $x_i = 0$ to $x_f = x_m$, as in Figure 6.9. This work can be easily calculated by noting that the *applied force*, $\mathbf{F}_{app}$, is equal to and opposite the spring force, $\mathbf{F}_s$, at any value of the displacement, so that $F_{app} = -(-kx) = kx$. Therefore, the work done on the spring by this applied force (the external agent) is

$$W_{F_{app}} = \int_0^{x_m} F_{app} \, dx = \int_0^{x_m} kx \, dx = \tfrac{1}{2}kx_m^2$$

You should note that this work is equal to the negative of the work done by the spring force for this displacement.

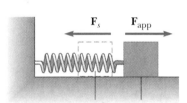

Figure 6.9   A block being pulled from $x = 0$ to $x = x_m$ on a frictionless surface by a force $\mathbf{F}_{app}$. If the process is carried out very slowly, the applied force is equal to and opposite the spring force at all times.

CONCEPTUAL PROBLEM 4

A spring is stretched by a distance x_m, resulting in work W being done by the external agent. The spring is then relaxed, cut in half, and one of the halves stretched by the same distance. How much work is done in this case?

Example 6.4   **Work Required to Stretch a Spring**

One end of a horizontal Hooke's-law spring ($k = 80$ N/m) is held fixed while an external force is applied to the free end, stretching it from $x_0 = 0$ to $x_1 = 4.0$ cm. (a) Find the work done by the external force.

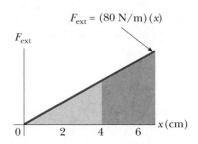

Figure 6.10 (Example 6.4) A graph of the external force required to stretch a spring that obeys Hooke's law versus the elongation of the spring.

Solution We place the zero reference of the coordinate axis at the free end of the unstretched spring (Fig. 6.10). The external force is $F_{ext} = (80 \text{ N/m})(x)$. The work done by F_{ext} is the area of the triangle from 0 to 4.0 cm:

$$W = \tfrac{1}{2}kx_1{}^2 = \tfrac{1}{2}(80 \text{ N/m})(0.040 \text{ m})^2 = \boxed{64 \text{ mJ}}$$

(b) Find the additional work done in stretching the spring from $x_1 = 4.0$ cm to $x_2 = 7.0$ cm.

Solution The work done in stretching the spring the additional amount, from $x_1 = 0.040$ m to $x_2 = 0.070$ m, is the darker-shaded area between these limits. Geometrically, it is

$$W = \tfrac{1}{2}kx_2{}^2 - \tfrac{1}{2}kx_1{}^2$$
$$= \tfrac{1}{2}(80 \text{ N/m})\,[(0.070 \text{ m})^2 - (0.040 \text{ m})^2] = 130 \text{ mJ}$$

Using calculus, we find that

$$W = \int_{x_1}^{x_2} F_{ext}\, dx = \int_{0.04\,\text{m}}^{0.07\,\text{m}} (80 \text{ N/m})\, dx$$

$$= \tfrac{1}{2}(80 \text{ N/m})(x^2)\Big|_{0.04\,\text{m}}^{0.07\,\text{m}}$$

$$W = \tfrac{1}{2}(80 \text{ N/m})\,[(0.070 \text{ m})^2 - (0.040 \text{ m})^2] = \boxed{130 \text{ mJ}}$$

EXERCISE 6 If an applied force varies with position according to $F_x = 3x^2 - 5$, where x is in meters, how much work is done by this force on an object that moves from $x = 4$ m to $x = 7$ m?   Answer   264 J

6.4 • KINETIC ENERGY AND THE WORK-KINETIC ENERGY THEOREM

Solutions using Newton's second law can be difficult if the forces in the problem are complex. An alternative approach that enables us to understand and solve such motion problems is to relate the speed of the particle to its displacement under the influence of some net force. As we shall see in this section, if the work done by the net force on a particle can be calculated for a given displacement, the change in the particle's speed will be easy to evaluate.

Figure 6.11 shows a particle of mass m moving to the right under the action of a constant net force **F**. Because the force is constant, we know from Newton's second law that the particle will move with a constant acceleration **a**. If the particle is displaced a distance d, the work done by the force **F** is

$$W_{net} = Fd = (ma)d \qquad \text{[6.16]}$$

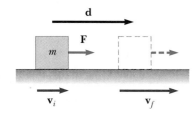

Figure 6.11 A particle undergoing a displacement and change in velocity under the action of a constant net force **F**.

In Chapter 2 (Eqs. 2.4 and 2.9) we found that the following relationships are valid when a particle undergoes constant acceleration:

$$d = \tfrac{1}{2}(v_i + v_f)t \qquad a = \frac{v_f - v_i}{t}$$

where v_i is the speed at $t = 0$ and v_f is the speed at time t. Substituting these expressions into Equation 6.16 gives

$$W_{net} = m\left(\frac{v_f - v_i}{t}\right)\tfrac{1}{2}(v_i + v_f)\,t$$

$$W_{net} = \tfrac{1}{2}mv_f^2 - \tfrac{1}{2}mv_i^2 \qquad \text{[6.17]}$$

The quantity $\tfrac{1}{2}mv^2$ represents the energy associated with the motion of a particle. It is so important that it has been given a special name—**kinetic energy.** The kinetic energy, K, of a particle of mass, m, moving with a speed, v, is defined as

$$K \equiv \tfrac{1}{2}mv^2 \qquad \text{[6.18]}$$

Kinetic energy is a scalar quantity and has the same units as work. For example, a 2.0-kg mass moving with a speed of 4.0 m/s has a kinetic energy of 32 J. It is often convenient to write Equation 6.17 in the form

$$W_{net} = K_f - K_i = \Delta K \qquad \text{[6.19]}$$

That is,

> the work done by the constant net force $\mathbf{F}_{net}$ in displacing a particle equals the change in kinetic energy of the particle.

Equation 6.19 is an important result known as the **work-kinetic energy theorem.** For convenience, it was derived under the assumption that the net force acting on the particle was constant. A more general derivation would show that this equation is also valid for a variable force.

The work-kinetic energy theorem also says that the speed of the particle will increase if the net work done on it is positive, because the final kinetic energy will be greater than the initial kinetic energy. The speed will decrease if the net work is negative, because the final kinetic energy will be less than the initial kinetic energy. The speed and kinetic energy of a particle change only if work is done on the particle by some external force.

Consider the relationship between the work done on a particle and the change in its kinetic energy as expressed by Equation 6.19. Because of this connection, we can also think of kinetic energy as the work the particle can do in coming to rest. For example, suppose a hammer is on the verge of striking a nail, as in Figure 6.12. The moving hammer has kinetic energy and can do work on the nail. The work done on the nail appears as the product Fd, where F is the average force exerted on the nail by the hammer and d is the distance the nail is driven into the wall. However, the hammer and the nail are not particles, so part of the kinetic energy of the hammer goes into warming the hammer and nail, and all of the work done on the nail goes into warming the nail and wall and locally deforming the wall.

• *Kinetic energy is energy associated with the motion of a particle.*

• *The work-kinetic energy theorem states that the work done on a particle equals the change in its kinetic energy.*

Figure 6.12 The hammer has kinetic energy associated with its motion and can do work on the nail, driving it into the wall.

CONCEPTUAL PROBLEM 5

You are working in a library, reshelving books. You lift a book from the floor to the top shelf. The kinetic energy of the book on the floor was zero, and the kinetic energy of the book on the top shelf is zero, so there is no change in kinetic energy. Yet you did some work in lifting the book. Is the work-kinetic energy theorem violated?

Situations Involving Kinetic Friction

When dealing with a force acting on an extended object, one must be careful in calculating the work done by that force because the displacement of the object is generally not equal to the displacement of the point of application of that force. Suppose that an object of mass m sliding on a horizontal surface is pulled with a constant horizontal external force **F** to the right and a kinetic frictional force **f** acts to the left, where $\mathbf{F} > \mathbf{f}$. In this case, the net force is to the right as in Figure 6.11, and the net work done on the object as it undergoes a displacement **d** to the right is

$$W_{\text{net}} = (\mathbf{F} - \mathbf{f}) \cdot \mathbf{d} = Fd - fd \qquad [6.20]$$

The quantity Fd is the work done on the object by the constant force **F**. The quantity $-fd$ is negative because the force of kinetic friction is opposite the displacement. The work done by kinetic friction depends on both the displacement of the object and on the details of the motion between the initial and final positions. In fact, the work done by kinetic friction on an extended object cannot be explicitly evaluated because friction forces and their individual displacements are complex.

Now suppose that a block moving on a horizontal surface and given an initial horizontal velocity $\mathbf{v}_i$ slides a distance d before reaching a final velocity $\mathbf{v}_f$ as in Figure 6.13. The external force that causes the block to undergo an acceleration in the negative x direction is the force of kinetic friction **f** acting to the left, opposite the motion. The initial kinetic energy of the block is $\frac{1}{2}mv_i^2$ and its final kinetic energy is $\frac{1}{2}mv_f^2$. The change in kinetic energy of the block is equal to $-fd$. This can be shown by applying Newton's second law to the block. (Newton's second law gives the acceleration of the center of mass of any object regardless of how or where the forces act.) Because the net force on the block in the x direction is the friction force, Newton's second law gives $-f = ma$. Multiplying both sides of this expression by d and using the expression $v_f^2 - v_i^2 = 2ad$ for motion under constant acceleration gives $-fd = (ma)\,d = \frac{1}{2}mv_f^2 - \frac{1}{2}mv_i^2$ or

$$\Delta K = -fd \qquad [6.21]$$

This result says that the change in kinetic energy of the block is equal to $-fd$, which corresponds to the energy dissipated by the force of kinetic friction. Part of this energy is transferred to internal thermal energy of the block, and part is transferred from the block to the surface.[2] In effect, the loss in kinetic energy of the block results in an increase in thermal energy of both the block and surface. For example, if the loss in kinetic energy of the block is 300 J, and 100 J appears as an increase in thermal energy of the block, then the remaining 200 J must have been transferred from the block to the surface.

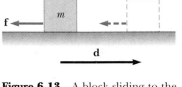

Figure 6.13 A block sliding to the right on a horizontal surface slows down in the presence of a force of kinetic friction acting to the left. The initial velocity of the block is $\mathbf{v}_i$, and its final velocity is $\mathbf{v}_f$. The normal force and force of gravity are not included in the diagram because they are perpendicular to the direction of motion and therefore do not influence the change in velocity of the block.

Thinking Physics 4

A car traveling at a speed v skids a distance d after its brakes lock. Estimate how far it will skid if its brakes lock when its initial speed is $2v$. What happens to the car's kinetic energy as it stops?

---

[2]For more details on energy transfer situations involving forces of kinetic friction, see B. A. Sherwood and W. H. Bernard, *American Journal of Physics*, 52:1001, 1984, and R. P. Bauman, *The Physics Teacher*, 30:264, 1992.

Reasoning   Let us assume that the force of kinetic friction between car and road surface is constant and the same in both cases. The net force times the displacement of the car is equal to its initial kinetic energy. If the speed is doubled as in this example, the kinetic energy of the car is quadrupled. For a given applied force (in this case, the frictional force), the distance traveled is four times as great when the initial speed is doubled, so the estimated distance it skids is $4d$. The kinetic energy of the car is changed into internal energy associated with the tires, brake pads, and road as they increase in temperature.

Thinking Physics 5

In most situations we have encountered in this chapter, frictional forces tend to reduce the kinetic energy of an object. However, frictional forces can sometimes increase an object's kinetic energy. Describe a few situations in which friction causes an increase in kinetic energy.

Reasoning   If a crate is located on the bed of a truck and the truck accelerates to the east, the static friction force exerted on the crate by the truck acts to the east to give the crate the same acceleration as the truck (assuming the crate doesn't slip). Another example is a car that accelerates because of the frictional forces exerted on the car's tires by the road. These forces act in the direction of the car's motion, and the sum of these forces causes an increase in the car's kinetic energy.

CONCEPTUAL PROBLEM 6

It is a known fact that more energy is expended by walking downstairs than by walking horizontally at the same speed. Why do you think this is so?

Example 6.5   A Block Pulled on a Frictionless Surface

A 6.0-kg block initially at rest is pulled to the right along a horizontal, frictionless surface by a constant, horizontal force of 12 N, as in Figure 6.14. Find the speed of the block after it has moved 3.0 m.

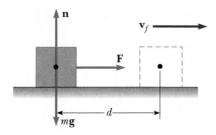

Figure 6.14   (Example 6.5)

Solution   The gravitational force acting on the block is balanced by the normal force, and neither of these forces does

work because the displacement is horizontal. Because there is no friction, the resultant external force is the 12-N force. The work done by this force is

$$W = Fd = (12 \text{ N})(3.0 \text{ m}) = 36 \text{ N·m} = 36 \text{ J}$$

Using the work-kinetic energy theorem and noting that the initial kinetic energy is zero, we get

$$W = K_f - K_i = \tfrac{1}{2}mv_f^2 - 0$$

$$v_f^2 = \frac{2W}{m} = \frac{2(36 \text{ J})}{6.0 \text{ kg}} = 12 \text{ m}^2/\text{s}^2$$

$$v_f = \boxed{3.5 \text{ m/s}}$$

EXERCISE 7   Find the acceleration of the block using the kinematic equation $v_f^2 = v_i^2 + 2ax$.
Answer $a = 2.0 \text{ m/s}^2$

Example 6.6   A Block Pulled on a Rough Surface

Find the final speed of the block described in Example 6.5 if the surface is rough and the coefficient of kinetic friction is 0.15.

Reasoning   In this case, we must use Equation 6.21 to calculate the change in kinetic energy, ΔK. The net force exerted on the block is the sum of the applied 12-N force and the frictional force, as in Figure 6.15. Because the frictional force is in the direction opposite the displacement, it must be subtracted.

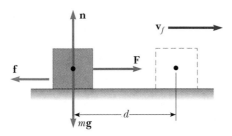

Figure 6.15   (Example 6.6)

Solution   The magnitude of the frictional force is $f = \mu_k n = \mu_k mg$. Therefore the net force acting on the block is

$$F_{net} = F - \mu_k mg = 12 \text{ N} - (0.15)(6.0 \text{ kg})(9.80 \text{ m/s}^2)$$
$$= 12 \text{ N} - 8.82 \text{ N} = 3.18 \text{ N}$$

Multiplying this constant force by the displacement and using Equation 6.21 gives

$$\Delta K = F_{net}d = (3.18 \text{ N})(3.0 \text{ m}) = 9.54 \text{ J} = \tfrac{1}{2}mv_f^2$$

using the information that $v_i = 0$. Therefore,

$$v_f^2 = \frac{2(9.54 \text{ J})}{6.0 \text{ kg}} = 3.18 \text{ m}^2/\text{s}^2$$

$$v_f = \boxed{1.8 \text{ m/s}}$$

EXERCISE 8   Find the acceleration of the block from Newton's second law.   Answer $a = 0.53 \text{ m/s}^2$

Example 6.7   A Mass-Spring System

A block of mass 1.6 kg is attached to a spring that has a force constant of 1.0×10^3 N/m as in Figure 6.16. The spring is compressed a distance of 2.0 cm, and the block is released from rest. (a) Calculate the speed of the block as it passes through the equilibrium position $x = 0$ if the surface is frictionless.

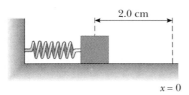

2.0 cm

$x = 0$

Figure 6.16   (Example 6.7)

Solution   We use Equation 6.14 to find the work done by the spring with $x_m = -2.0$ cm $= -2.0 \times 10^{-2}$ m:

$$W_s = \tfrac{1}{2}kx_m^2 = \tfrac{1}{2}(1.0 \times 10^3 \text{ N/m})(-2.0 \times 10^{-2} \text{ m})^2 = 0.20 \text{ J}$$

Using the work-kinetic energy theorem with $v_i = 0$ gives

$$W_s = \tfrac{1}{2}mv_f^2 - \tfrac{1}{2}mv_i^2$$

$$0.20 \text{ J} = \tfrac{1}{2}(1.6 \text{ kg})v_f^2 - 0$$

$$v_f^2 = \frac{0.40 \text{ J}}{1.6 \text{ kg}} = 0.25 \text{ m}^2/\text{s}^2$$

$$v_f = \boxed{0.50 \text{ m/s}}$$

(b) Calculate the speed of the block as it passes through the equilibrium position if a constant frictional force of 4.0 N retards its motion.

Solution   We use Equation 6.21 to calculate the kinetic energy lost due to friction and add this to the kinetic energy found in the absence of friction. Considering only the frictional force, the kinetic energy lost due to friction is

$$-fd = -(4.0 \text{ N})(2.0 \times 10^{-2} \text{ m}) = -0.08 \text{ J}$$

The final kinetic energy, without this loss, was found in part (a) to be 0.20 J. Therefore, the final kinetic energy in the presence of friction is

$$K_f = 0.20 \text{ J} - 0.08 \text{ J} = 0.12 \text{ J} = \tfrac{1}{2}mv_f^2$$

$$\tfrac{1}{2}(1.6 \text{ kg})v_f^2 = 0.12 \text{ J}$$

$$v_f^2 = \frac{0.24 \text{ J}}{1.6 \text{ kg}} = 0.15 \text{ m}^2/\text{s}^2$$

$$v_f = \boxed{0.39 \text{ m/s}}$$

Note that this value for v_f is less than that obtained in the frictionless case. Does this result surprise you?

EXERCISE 9 A 100-kg sled is dragged by a team of dogs a distance of 2.00 km over a horizontal surface at a constant velocity. If the coefficient of friction between the sled and snow is 0.150, find (a) the work done by the team of dogs and (b) the energy lost due to friction. Answer (a) 2.94×10^5 J (b) All of this energy is dissipated due to friction.

EXERCISE 10 A 15-kg block is dragged over a horizontal surface by a constant force of 70 N acting at an angle of 20° above the horizontal. The block is displaced 5.0 m, and the coefficient of kinetic friction between the block and surface is 0.30. Find the work done by (a) the 70-N force, (b) the normal force, and (c) the force of gravity. (d) Calculate the energy lost due to friction. Answer (a) 330 J (b) 0 (c) 0 (d) -185 J

6.5 • POWER

From a practical viewpoint, it is interesting to know not only the work done on an object but also the rate at which the work is being done. The time rate at which work is done, or the time rate of energy transfer, is called **power**.

If an external force is applied to an object (which we assume acts as a particle), and if the work done by this force is W in the time interval Δt, then the **average power** during this interval is defined as

$$\overline{P} \equiv \frac{W}{\Delta t} \qquad \text{[6.22]}$$

• *Average power*

The work done on the object contributes to increasing the energy of the object. A more general definition of power is the *time rate of energy transfer*. The **instantaneous power,** P, is the limiting value of the average power as Δt approaches zero:

$$P \equiv \lim_{\Delta t \to 0} \frac{W}{\Delta t} = \frac{dW}{dt} \qquad \text{[6.23]}$$

• *Instantaneous power*

where we have represented the infinitesimal value of the work done by dW (even though it is not a change and therefore not a differential). We know from Equation 6.3 that $dW = \mathbf{F} \cdot d\mathbf{x}$. Therefore, the instantaneous power can be written

$$P = \frac{dW}{dt} = \mathbf{F} \cdot \frac{d\mathbf{x}}{dt} = \mathbf{F} \cdot \mathbf{v} \qquad \text{[6.24]}$$

• *The watt*

where we have used the fact that $\mathbf{v} = d\mathbf{x}/dt$.

The SI unit of power is joules per second (J/s), also called a *watt* (W) (after James Watt):

$$1 \text{ W} = 1 \text{ J/s} = 1 \text{ kg} \cdot \text{m}^2/\text{s}^3$$

The upright symbol W for watt should not be confused with the italic symbol W for work.

The unit of power in the British engineering system is the horsepower (hp):

$$1 \text{ hp} \equiv 550 \text{ ft} \cdot \text{lb/s} = 746 \text{ W}$$

A new unit of energy (or work) can now be defined in terms of the unit of power. One kilowatt hour (kWh) is the energy converted or consumed in 1 h at the constant rate of 1 kW. The numerical value of 1 kWh is

$$1 \text{ kWh} = (10^3 \text{ W})(3600 \text{ s}) = 3.60 \times 10^6 \text{ J}$$

It is important to realize that a kilowatt hour is a unit of energy, not power. When you pay your electric bill, you are buying energy, and the amount of electricity used by an appliance is usually expressed in kilowatt hours. For example, an electric bulb rated at 100 W would "consume" 3.60×10^5 J of energy in 1 h.

Thinking Physics 6

A light bulb is described by some individuals as "having 60 watts." What's wrong with this phrase?

Reasoning   The number given represents the power of the light bulb, which is the rate at which energy passes through it, entering as electrical energy from the power source. It is not a value that the light bulb "possesses," however. It is only the approximate power rating of the bulb when it is connected to a normal 120-volt household supply of electricity. If the bulb were connected to a 12-volt supply, energy would enter the bulb at a rate other than 60 watts. Thus, the 60 watts is not an intrinsic property of the bulb, such as density is for a given solid material.

CONCEPTUAL PROBLEM 7

An older model car accelerates from 0 to speed v in 10 seconds. A newer, more powerful sports car accelerates from 0 to $2v$ in the same time period. What is the ratio of powers expended by the two cars? Consider the energy coming from the engine to appear only as kinetic energy of the car.

Example 6.8   **Power Delivered by an Elevator Motor**

An elevator has a mass of 1000 kg and carries a maximum load of 800 kg. A constant frictional force of 4000 N retards its motion upward, as in Figure 6.17. (a) What must be the minimum power delivered by the motor to lift the elevator at a constant speed of 3.00 m/s?

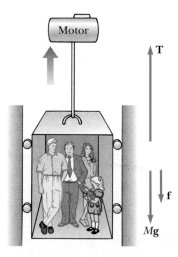

Figure 6.17   (Example 6.8) A motor exerts an upward force, equal in magnitude to the tension, T, on the elevator. A frictional force **f** and the force of gravity $M\mathbf{g}$ act downward.

Solution   The motor must supply the force **T** that pulls the elevator upward. From Newton's second law and from the fact that $a = 0$ because v is constant, we get

$$T - f - Mg = 0$$

where M is the *total* mass (elevator plus load), equal to 1800 kg. Therefore,

$$\begin{aligned}
T &= f + Mg \\
&= 4.00 \times 10^3 \text{ N} + (1.80 \times 10^3 \text{ kg})(9.80 \text{ m/s}^2) \\
&= 2.16 \times 10^4 \text{ N}
\end{aligned}$$

Using Equation 6.24 and the fact that **T** is in the same direction as **v** gives

$$\begin{aligned}
P &= \mathbf{T} \cdot \mathbf{v} = Tv \\
&= (2.16 \times 10^4 \text{ N})(3.00 \text{ m/s}) = 6.48 \times 10^4 \text{ W} \\
&= \boxed{64.8 \text{ kW}}
\end{aligned}$$

(b) What power must the motor deliver at any instant if it is designed to provide an upward acceleration of 1.00 m/s²?

Solution   Applying Newton's second law to the elevator gives

$$T - f - Mg = Ma$$

$$
\begin{aligned}
T &= M(a + g) + f \\
&= (1.80 \times 10^3 \text{ kg})(1.00 + 9.80) \text{ m/s}^2 \\
&\quad + 4.00 \times 10^3 \text{ N} \\
&= 2.34 \times 10^4 \text{ N}
\end{aligned}
$$

Therefore, using Equation 6.24, for the required power we get

$$P = Tv = \boxed{(2.34 \times 10^4 \, v) \text{ W}}$$

where v is the instantaneous speed of the elevator in meters per second. Hence, the power required increases with increasing speed.

EXERCISE 11   A 65-kg athlete runs a distance of 600 m up a mountainside that is inclined at 20° to the horizontal. He performs this feat in 80 s. Assuming that air resistance is negligible, (a) how much work is done against gravity, and (b) what is his average power output during the run?   Answer   (a) 1.3×10^5 J   (b) 1.6 kW

SUMMARY

The **work** done by a *constant* force **F** acting on a particle is defined as the product of the component of the force in the direction of the particle's displacement and the magnitude of the displacement. If **F** makes an angle θ with the displacement **d**, the work done by **F** is

$$W \equiv Fd \cos \theta \qquad \qquad \textbf{[6.1]}$$

The **scalar**, or dot **product** of any two vectors **A** and **B** is defined by the relationship

$$\mathbf{A} \cdot \mathbf{B} \equiv AB \cos \theta \qquad \qquad \textbf{[6.5]}$$

where the result is a scalar quantity and θ is the angle between the directions of the two vectors. The scalar product obeys the commutative and distributive laws.

The **work** done by a *varying* force acting on a particle moving along the x axis from x_i to x_f is

$$W \equiv \int_{x_i}^{x_f} F_x \, dx \qquad \qquad \textbf{[6.11]}$$

where F_x is the component of force in the x direction. If there are several forces acting on the particle, the net work done by all forces is the sum of the individual amounts of work done by each force.

The **kinetic energy** of a particle of mass m moving with a speed v (where v is small compared with the speed of light) is

$$K \equiv \tfrac{1}{2}mv^2 \qquad \qquad \textbf{[6.18]}$$

The **work-kinetic energy theorem** states that the net work done on a particle by external forces equals the change in kinetic energy of the particle:

$$W_{\text{net}} = K_f - K_i = \tfrac{1}{2}mv_f^2 - \tfrac{1}{2}mv_i^2 \qquad \qquad \textbf{[6.19]}$$

Average power is the time rate of doing work:

$$\overline{P} \equiv \frac{W}{\Delta t} \qquad \qquad \textbf{[6.22]}$$

If an agent applies a force **F** to an object moving with a velocity **v**, the **instantaneous power** delivered by that agent is

$$P \equiv \frac{dW}{dt} = \mathbf{F} \cdot \mathbf{v} \qquad \qquad \textbf{[6.24]}$$

CONCEPTUAL QUESTIONS

1. Explain why the work done by the force of sliding friction is negative when an object undergoes a displacement on a rough surface.

2. When a particle rotates in a circle, a force acts on it directed toward the center of rotation. Why is it that this force does no work on the particle?

3. When a punter kicks a football, is he doing any work on the ball while his toe is in contact with it? Is he doing any work on the ball after it loses contact with his toe? Are any forces doing work on the ball while it is in flight?

4. Cite two examples in which a force is exerted on an object without doing any work on the object.

5. As a simple pendulum swings back and forth, the forces acting on the suspended mass are the force of gravity, the tension in the supporting cord, and air resistance. (a) Which of these forces, if any, does no work on the pendulum? (b) Which of these forces does negative work at all times during its motion? (c) Describe the work done by the force of gravity while the pendulum is swinging.

6. If the dot product of two vectors is positive, does this imply that the vectors must have positive rectangular components?

7. A hockey player pushes a puck with his hockey stick over frictionless ice. If the puck starts from rest and ends up moving at speed v, has the hockey player performed any work in his reference frame? An ant is on the puck and hangs on tightly while the puck is accelerated. In the reference frame of the ant, has any work been done? Is the work-kinetic energy theorem satisfied for each observer?

8. Can kinetic energy be negative? Explain.

9. One bullet has twice the mass of a second bullet. If both are fired so that they have the same speed, which has more kinetic energy? What is the ratio of the kinetic energies of the two bullets?

10. If the speed of a particle is doubled, what happens to its kinetic energy?

11. What can be said about the speed of a particle if the net work done on it is zero?

12. Can the average power ever equal the instantaneous power? Explain.

13. In Example 6.8, does the required power increase or decrease as the force of friction is reduced?

14. Sometimes physics words are used in popular literature in interesting ways. For example, consider a description of a rock falling from the top of a cliff as "gathering force as it falls to the beach below." What does the phrase "gathering force" mean and can you repair this phrase?

PROBLEMS

Section 6.1   Work Done by a Constant Force

1. A tugboat exerts a constant force of 5000 N on a ship moving at constant speed through a harbor. How much work does the tugboat do on the ship in a distance of 3.00 km?

2. A shopper in a supermarket pushes a cart with a force of 35.0 N directed at an angle of 25.0° downward from the horizontal. Find the work done by the shopper as he moves down a 50.0-m length of aisle.

3. A raindrop ($m = 3.35 \times 10^{-5}$ kg) falls vertically at constant speed under the influence of gravity and air resistance. After the drop has fallen 100 m, what is the work done (a) by gravity and (b) by air resistance?

4. A block of mass 2.50 kg is pushed 2.20 m along a frictionless horizontal table by a constant 16.0-N force directed 25.0° below the horizontal. Determine the work done by (a) the applied force, (b) the normal force exerted by the table, (c) the force of gravity, and (d) the net force on the block.

5. Batman, whose mass is 80.0 kg, is holding on to the free end of a 12.0-m rope, the other end of which is fixed to a tree limb directly above. He is able to get the rope in motion as only Batman knows how, eventually getting it to swing enough that he can reach a ledge when the rope makes a 60.0° angle with the vertical. How much work was done by him against the force of gravity in this maneuver?

Section 6.2   The Scalar Product of Two Vectors

6. Vector **A** has a magnitude of 5.00 units, and **B** has a magnitude of 9.00 units. The two vectors make an angle of 50.0° with each other. Find **A·B**.

7. Vector **A** extends from the origin to a point having polar coordinates (7, 70°), and vector **B** extends from the origin to a point having polar coordinates (4, 130°). Find **A·B**.

8. Given two arbitrary vectors **A** and **B**, show that $\mathbf{A \cdot B} = A_x B_x + A_y B_y + A_z B_z$. (*Hint:* Write **A** and **B** in unit vector form and use Equations 6.7 and 6.8.)

9. A force $\mathbf{F} = (6\mathbf{i} - 2\mathbf{j})$ N acts on a particle that undergoes a displacement $\mathbf{d} = (3\mathbf{i} + \mathbf{j})$ m. Find (a) the work done by the force on the particle and (b) the angle between **F** and **d**.

10. For $\mathbf{A} = 3\mathbf{i} + \mathbf{j} - \mathbf{k}$, $\mathbf{B} = -\mathbf{i} + 2\mathbf{j} + 5\mathbf{k}$, and $\mathbf{C} = 2\mathbf{j} - 3\mathbf{k}$, find $\mathbf{C \cdot (A - B)}$.

11. Using the definition of the scalar product, find the angles between (a) $\mathbf{A} = 3\mathbf{i} - 2\mathbf{j}$ and $\mathbf{B} = 4\mathbf{i} - 4\mathbf{j}$; (b) $\mathbf{A} = -2\mathbf{i} +$

4**j** and **B** = 3**i** − 4**j** + 2**k**; (c) **A** = **i** − 2**j** + 2**k** and **B** = 3**j** + 4**k**.

Section 6.3 Work Done by a Varying Force

12. The force acting on a particle varies as in Figure P6.12. Find the work done by the force as the particle moves (a) from $x = 0$ to $x = 8.00$ m, (b) from $x = 8.00$ m to $x = 10.0$ m, and (c) from $x = 0$ to $x = 10.0$ m.

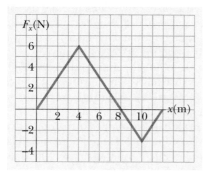

Figure P6.12

13. A particle is subject to a force F_x that varies with position as in Figure P6.13. Find the work done by the force on the body as it moves (a) from $x = 0$ to $x = 5.00$ m, (b) from $x = 5.00$ m to $x = 10.0$ m, and (c) from $x = 10.0$ m to $x = 15.0$ m. (d) What is the total work done by the force over the distance $x = 0$ to $x = 15.0$ m?

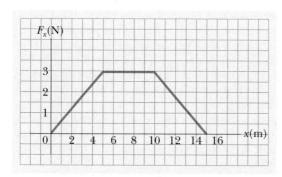

Figure P6.13

14. The force acting on a particle is $F_x = (8x − 16)$ N, where x is in meters. (a) Make a plot of this force versus x from $x = 0$ to $x = 3.00$ m. (b) From your graph, find the net work done by this force as the particle moves from $x = 0$ to $x = 3.00$ m.

15. A force **F** = (4.0x**i** + 3.0y**j**) N acts on an object as the object moves in the x direction from the origin to $x = 5.00$ m. Find the work done on the object by the force.

16. When a 4.00-kg mass is hung vertically on a certain light spring that obeys Hooke's law, the spring stretches 2.50 cm. If the 4.00-kg mass is removed, (a) how far will the spring stretch if a 1.50-kg mass is hung on it, and (b) how much work must an external agent do to stretch the same spring 4.00 cm from its unstretched position?

17. An archer pulls her bow string back 0.400 m by exerting a force that increases uniformly from 0 to 230 N. (a) What is the equivalent spring constant of the bow? (b) How much work is done in pulling the bow?

18. A 100-g bullet is fired from a rifle having a barrel 0.600 m long. Assuming the origin is placed where the bullet begins to move, the force (in newtons) exerted on the bullet by the expanding gas is $15\,000 + 10\,000x − 25\,000x^2$, where x is in meters. (a) Determine the work done by the gas on the bullet as the bullet travels the length of the barrel. (b) If the barrel is 1.00 m long, how much work is done, and how does this value compare to the work calculated in (a)?

19. If it takes 4.00 J of work to stretch a Hooke's-law spring 10.0 cm from its unstressed length, determine the extra work required to stretch it an additional 10.0 cm.

20. If it takes work W to stretch a Hooke's-law spring a distance d from its unstressed length, determine the extra work required to stretch it an additional distance d.

Section 6.4 Kinetic Energy and the Work-Kinetic Energy Theorem

21. A 0.600-kg particle has a speed of 2.00 m/s at point A and kinetic energy of 7.50 J at point B. What is (a) its kinetic energy at A? (b) its speed at B? (c) the total work done on the particle as it moves from A to B?

22. A 0.300-kg ball has a speed of 15.0 m/s. (a) What is its kinetic energy? (b) If its speed is doubled, what is its kinetic energy?

23. A 3.00-kg mass has an initial velocity $\mathbf{v}_0 = (6.00\mathbf{i} − 2.00\mathbf{j})$ m/s. (a) What is its kinetic energy at this time? (b) Find the total work done on the object if its velocity changes to $(8.00\mathbf{i} + 4.00\mathbf{j})$ m/s. (*Hint:* Remember that $v^2 = \mathbf{v} \cdot \mathbf{v}$.)

24. You can think of the work-kinetic energy theorem as a second theory of motion, parallel to Newton's laws in describing how outside influences affect the motion of an object. In this problem work out parts (a) and (b) separately from parts (c) and (d) to compare the predictions of the two theories. In a rifle barrel, a 15.0-g bullet is accelerated from rest to a speed of 780 m/s. (a) Find the work done on it. (b) If the rifle barrel is 72.0 cm long, find the magnitude of the average total force that acted on it, as $F = W/(d \cos \theta)$. (c) Find the constant acceleration of a bullet that starts from rest and gains speed 780 m/s over a distance

of 72.0 cm. (d) If the bullet has mass 15.0 g, find the total force that acted on it as $\Sigma F = ma$.

25. A cart loaded with bricks has a total mass of 18.0 kg and is pulled at constant speed by a rope. The rope is inclined at 20.0° above the horizontal, and the cart moves a distance of 20.0 m on a horizontal surface. The coefficient of kinetic friction between the cart and surface is 0.500. (a) What is the tension of the rope? (b) How much work is done on the cart by the rope? (c) What is the energy lost due to friction?

26. A 40.0-kg box initially at rest is pushed 5.00 m along a rough, horizontal floor with a constant applied horizontal force of 130 N. If the coefficient of friction between box and floor is 0.300, find (a) the work done by the applied force, (b) the energy lost due to friction, (c) the change in kinetic energy of the box, and (d) the final speed of the box.

27. A sled of mass m is given a kick on a frozen pond, imparting to it an initial speed $v_i = 2.00$ m/s. The coefficient of kinetic friction between sled and ice is $\mu_k = 0.100$. Use energy considerations to find the distance the sled moves before stopping.

28. A block of mass 12.0 kg slides from rest down a frictionless 35.0° incline and is stopped by a strong spring with $k = 3.00 \times 10^4$ N/m. The block slides 3.00 m from the point of release to the point where it comes to rest against the spring. When the block comes to rest, how far has the spring been compressed?

29. A crate of mass 10.0 kg is pulled up a rough incline with an initial speed of 1.50 m/s. The pulling force is 100 N parallel to the incline, which makes an angle of 20.0° with the horizontal. The coefficient of kinetic friction is 0.400, and the crate is pulled 5.00 m. (a) How much work is done by gravity? (b) How much energy is lost due to friction? (c) How much work is done by the 100-N force? (d) What is the change in kinetic energy of the crate? (e) What is the speed of the crate after being pulled 5.00 m?

30. A picture tube in a certain television set is 36.0 cm long. The electrical force accelerates an electron in the tube from rest to 1.00% of the speed of light over this distance. Determine (a) the kinetic energy of the electron as it strikes the screen at the end of the tube, (b) the magnitude of the average electrical force acting on the electron over this distance, (c) the magnitude of the average acceleration of the electron over this distance, and (d) the time of flight.

31. A time-varying net force acting on a 4.00-kg object causes the object to have a displacement given by $x = 2.0t - 3.0t^2 + 1.0t^3$, where x is in meters and t is in seconds. Find the work done on the object in the first 3.00 s of motion.

32. A 5.00-kg steel ball is dropped onto a copper plate from a height of 10.0 m. If the ball leaves a dent 0.320 cm deep, what is the average force exerted on the ball by the plate during the impact?

Section 6.5   Power

33. A 700-N Marine in basic training climbs a 10.0-m vertical rope at a constant speed in 8.00 s. What is her power output?

34. Make an order-of-magnitude estimate of the power your car engine puts into speeding the car up to highway speed. In your solution state the physical quantities you take as data and the values you measure or estimate for them. The mass of the vehicle is given in the owner's manual. If not a car, consider a bus or truck that you specify.

35. If a certain horse can maintain 1.00 hp of output for 2.00 h, how many 70.0-kg bundles of shingles can that horse hoist (via some pulley arrangement) to the roof of a house 8.00 m tall, assuming 70.0% efficiency?

36. A certain automobile engine delivers 30.0 hp (2.24×10^4 W) to its wheels when moving at a constant speed of 27.0 m/s ($\approx$60 mi/h). What is the resistive force acting on the automobile at that speed?

37. A skier of mass 70.0 kg is pulled up a slope by a motor-driven cable. (a) How much work is required to pull him a distance of 60.0 m up a 30.0° slope (assumed frictionless) at a constant speed of 2.00 m/s? (b) A motor of what power is required to perform this task?

38. A 650-kg elevator starts from rest. It moves upward for 3.00 s with constant acceleration until it reaches its cruising speed of 1.75 m/s. (a) What is the average power of the elevator motor during this period? (b) How does this power compare with its power when it moves at its cruising speed?

Additional Problems

39. A baseball outfielder throws a 0.150-kg baseball at a speed of 40.0 m/s and an initial angle of 30.0°. What is the kinetic energy of the baseball at the highest point of the trajectory?

40. While running, a person dissipates about 0.600 J of mechanical energy per step per kilogram of body mass. If a 60.0-kg runner dissipates a power of 70.0 W during a race, how fast is the person running? Assume a running step is 1.50 m long.

41. A particle of mass m moves with a constant acceleration $\mathbf{a}$. If the initial position vector and velocity of the particle are $\mathbf{r}_0$ and $\mathbf{v}_0$, respectively, show that its speed v at any time satisfies the equation

$$v^2 = v_0{}^2 + 2\mathbf{a} \cdot (\mathbf{r} - \mathbf{r}_0)$$

where $\mathbf{r}$ is the position vector of the particle at that same time.

42. The direction of an arbitrary vector $\mathbf{A}$ can be completely specified with the angles α, β, and γ that the vector makes with the x, y, and z axes, respectively. If $\mathbf{A} = A_x\mathbf{i} + A_y\mathbf{j} + A_z\mathbf{k}$, (a) find expressions for $\cos \alpha$, $\cos \beta$, and $\cos \gamma$ (these

are known as *direction cosines*), and (b) show that these angles satisfy the relation $\cos^2 \alpha + \cos^2 \beta + \cos^2 \gamma = 1$. (*Hint:* Take the scalar product of **A** with **i**, **j**, and **k** separately.)

43. A 4.00-kg particle moves along the *x* axis. Its position varies with time according to $x = t + 2.0t^3$, where *x* is in meters and *t* is in seconds. Find (a) the kinetic energy at any time *t*, (b) the acceleration of the particle and the force acting on it at time *t*, (c) the power being delivered to the particle at time *t*, and (d) the work done on the particle in the interval $t = 0$ to $t = 2.00$ s.

44. When a spring is stretched beyond its elastic limit, the restoring force satisfies the equation $F = -kx + \beta x^3$. If $k = 10.0$ N/m and $\beta = 100$ N/m^3, calculate the work done by this force when the spring is stretched 0.100 m.

45. A 2100-kg pile driver is used to drive a steel I-beam into the ground. The pile driver falls 5.00 m before contacting the beam, and it drives the beam 12.0 cm into the ground before coming to rest. Using energy considerations, calculate the average force the beam exerts on the pile driver while the pile driver is brought to rest after each hit.

46. A cyclist and her bicycle have a combined mass of 75.0 kg. She coasts down a road inclined at 2.00° with the horizontal at 4.00 m/s and down a road inclined at 4.00° at 8.00 m/s. She then holds on to a moving vehicle and coasts on a level road. What power must the vehicle expend to maintain her speed at 3.00 m/s? Assume that the force of air resistance is proportional to her speed and assume that other frictional forces remain constant.

47. A 200-g block is pressed against a spring of force constant 1.40 kN/m until the block compresses the spring 10.0 cm. The spring rests at the bottom of a ramp inclined at 60.0° to the horizontal. Use energy considerations to determine how far up the incline the block moves before it stops if (a) there is no friction between block and ramp and if (b) the coefficient of kinetic friction is 0.400.

48. A 0.400-kg particle slides on a horizontal circular track 1.50 m in radius. It is given an initial speed of 8.00 m/s. After one revolution, its speed drops to 6.00 m/s because of friction. (a) Find the energy lost due to friction in one revolution. (b) Calculate the coefficient of kinetic friction. (c) What is the total number of revolutions the particle makes before stopping?

49. The ball launcher in a pinball machine has a spring that has a force constant of 1.20 N/cm (Fig. P6.49). The surface on which the ball moves is inclined 10.0° with respect to the horizontal. If the spring is initially compressed 5.00 cm, find the launching speed of a 100-g ball when the plunger

is released. Friction and the mass of the plunger are negligible.

50. In diatomic molecules, the constituent atoms exert attractive forces on each other at large distances and repulsive forces at short distances. For many molecules, the Lennard-Jones law is a good approximation to the magnitude of these forces:

$$ F = F_0 \left[2 \left(\frac{\sigma}{r} \right)^{13} - \left(\frac{\sigma}{r} \right)^7 \right] $$

where *r* is the center-to-center distance between the atoms in the molecule, σ is a length parameter, and F_0 is the force when $r = \sigma$. For an oxygen molecule, the values of F_0 and σ are $F_0 = 9.60 \times 10^{-11}$ N and $\sigma = 3.50 \times 10^{-10}$ m. Determine the work done by this force from $r = 4.00 \times 10^{-10}$ m to $r = 9.00 \times 10^{-10}$ m.

51. Suppose a car is modeled as a cylinder with cross-sectional area *A* moving with a speed *v*, as in Figure P6.51. In a time Δt, a column of air of mass Δm must be moved a distance $v \, \Delta t$ and given a kinetic energy $\frac{1}{2}(\Delta m) v^2$. Using this model, show that the power loss due to air resistance is $\frac{1}{2}\rho A v^3$ and the resistive force is $\frac{1}{2}\rho A v^2$, where ρ is the density of air.

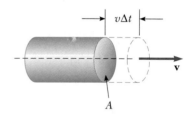

$v\Delta t$

v

A

Figure P6.51

Spreadsheet Problems

S1. When different weights are hung on a spring, the spring stretches to different lengths as shown in the table that follows. (a) Use a spreadsheet to plot the length of the spring versus applied force. Use the spreadsheet's least-squares fitting features to determine the straight line that best fits the data. (You may not want to use all the data points.) (b) From the slope of the least-squares fit line, find the spring constant *k*. (c) If the spring is extended to 105 mm, what force does it exert on the suspended weight?

Figure P6.49

10.0°

F(N)	*L*(mm)	*F*(N)	*L*(mm)
2.0	15	14	112
4.0	32	16	126
6.0	49	18	149
8.0	64	20	175
10	79	22	190
12	98		

S2. A 0.178-kg particle moves along the x axis from $x = 12.8$ m to $x = 23.7$ m under the influence of a force

$$F = \frac{375}{x^3 + 3.75x}$$

where F is in newtons and x is in meters. Use numerical integration and set up a spreadsheet to determine the total work done by this force during this displacement. Your calculations should have an accuracy of at least 2%.

ANSWERS TO CONCEPTUAL PROBLEMS

1. Since there is no motion taking place, the rope experiences no displacement. Thus, no work is done on it. For the same reason, no work is being done on the pullers or the ground. Work is only being done within the bodies of the pullers. For example, the heart of each puller is applying forces on the blood to move it through the body.

2. Less force will be necessary with a longer ramp, but the force must act over a longer distance to do the same amount of work. Suppose the refrigerator is rolled up the ramp at constant speed. The normal force does no work because it acts at 90° to the motion. The work by gravity is just the weight of the refrigerator times the vertical height through which it is displaced times (cos 180°), or $Wg = -mgh$. Therefore, the movers must do work mgh on the refrigerator, however long the ramp.

3. If we ignore the effects of rolling friction on the tires of the car, the same amount of work would be done in driving up the switchback and driving straight up the mountain, since the weight of the car is moved upward against gravity by the same vertical distance in each case. If we include friction, there is more work done in driving the switchback, since the distance over which the friction force acts is much longer. So why do we use the switchback? The answer lies in the force required, not the work. The force from the engine required to follow a gentle rise is much smaller than that required to drive straight up the hill. Roadways running straight uphill would require redesigning engines so as to be able to apply much larger forces. This is similar to the ease with which heavy objects can be rolled up ramps into moving trucks, compared to lifting the object straight up from the ground.

4. If the spring is cut in half and stretched by the same distance, each adjacent pair of coils must now be twice as far apart as in the original stretched spring. Thus, the force applied by the spring is twice as great for any given extension. This leads to an increase in the spring constant by a factor of 2. Thus, if the cut spring is stretched by the same amount, it must require twice as much work.

5. We focus on the book as the system. You indeed did some work in applying an upward force to lift the book. But there is another force on the book—gravity. The work done by gravity is negative, since the force of gravity is opposite to the displacement. Assuming that you lifted the book slowly, the upward force that you applied is approximately the same in magnitude as the weight of the book. Thus, the positive work done by you and the negative work done by gravity cancel. Thus, there is no net work performed, and no net change in the kinetic energy—the work-kinetic energy theorem is satisfied. If we consider the book and the Earth to be the system, then there is net work done by you on this system. This work appears in the system as potential energy, which will be investigated in Chapter 7.

6. When walking downstairs, a person has to stop the motion of the entire body at each step. This requires work to be done by the muscles. The work done by the muscles is negative, since the forces applied are opposite to the displacements representing the movement of the body. Negative work represents a transfer of energy out of the muscles (from the store of energy from the food eaten by the walker). When walking horizontally, the body can be kept in motion with an almost constant horizontal speed—only the feet and legs start and stop. Thus, more energy is transferred from the body in bringing the entire body to a stop in the trip downstairs than in walking.

7. Since the time periods are the same for both cars, we need only to compare the work done. Since the sports car is moving twice as fast as the older car at the end of the time interval, it has four times the kinetic energy. Thus, according to the work-kinetic energy theorem, four times as much work was done, and the engine must have expended four times the power.

7

Potential Energy and Conservation of Energy

In Chapter 6 we introduced the concept of kinetic energy, which is associated with the motion of an object. In this chapter we introduce another form of mechanical energy, called *potential energy*, that is associated with the position or configuration of a system. The potential energy of a system can be thought of as stored energy that can be converted to kinetic energy or other forms of energy.

The potential energy concept can be used only with a special class of forces called *conservative forces*. When only internal conservative forces, such as gravitational or spring forces, act within a system, the kinetic energy gained (or lost) by the system as its members change their relative positions is compensated by an equal loss (or gain) in potential energy.

◀ Twin Falls on the Island of Kauai, Hawaii. The gravitational potential energy of the water-Earth system when the water is at the top of the falls is converted to kinetic energy of the water at the bottom. In many locations, this mechanical energy is used to produce electrical energy. *(Bruce Byers, FPG)*

165

7.1 • POTENTIAL ENERGY

In Chapter 6 we saw that an object with kinetic energy can do work on another object, as illustrated by the moving hammer driving a nail into a wall. Now we shall see that an object can also do work because of the energy resulting from its *position* in space.

As an object falls in a gravitational field, the field exerts a force on the object, does work on it, and thereby changes its kinetic energy. Consider a brick dropped from rest directly above a nail in a board that is lying horizontally on the ground. When the brick is released, it falls toward the ground, gaining speed and therefore gaining kinetic energy. The potential energy of the brick–Earth system is converted into kinetic energy as the brick falls. When the brick reaches the ground, it does work on the nail, driving it into the board. The potential energy of a system consisting of any object and the Earth is called **gravitational potential energy.**

Gravitational potential •
energy

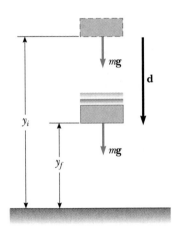

Figure 7.1 The work done by the gravitational force as the block falls from y_i to y_f is equal to $mgy_i - mgy_f$.

Let us now derive an expression for the gravitational potential energy associated with an object at a given location in space. To do this, consider a block of mass m at an initial height y_i above the ground, as in Figure 7.1. With air resistance neglected as the block falls, the only force that does work on it is the gravitational force, $m\mathbf{g}$. The work done by the gravitational force as the block undergoes a downward displacement $\mathbf{d}$ is given by the product of the downward force $m\mathbf{g}$ and the displacement, or

$$W_g = (m\mathbf{g}) \cdot \mathbf{d} = (-mg\mathbf{j}) \cdot [(y_f - y_i)\mathbf{j}] = mgy_i - mgy_f$$

We can represent the quantity mgy to be the gravitational potential energy, U_g:

$$U_g = mgy \qquad\qquad [7.1]$$

In this representation, the gravitational potential energy associated with an object at any point in space is the product of the object's weight and its vertical coordinate. The origin of the coordinate system could be located at the surface of the Earth or at any other convenient point.

If we substitute U for the mgy terms in the expression for W_g, we have

$$W_g = U_i - U_f \qquad\qquad [7.2]$$

From this result, we see that the work done on any object by the force of gravity —that is, the energy transferred to the object from the gravitational field—is equal to the initial value of the potential energy minus the final value of potential energy.

The units of gravitational potential energy are the same as those of work. That is, potential energy may be expressed in joules, ergs, or foot-pounds. Potential energy, like work and kinetic energy, is a scalar quantity.

Note that the gravitational potential energy depends only on the vertical height of the object above the surface of the Earth. From this result, we see that the same amount of work is done on an object if it falls vertically to the Earth as if it starts at the same point and slides down a frictionless incline to the Earth. Also note that Equation 7.1 is valid only for objects near the surface of the Earth, where $\mathbf{g}$ is approximately constant.

In working problems involving gravitational potential energy, it is always necessary to choose a location at which to set the gravitational potential energy equal to zero. The choice of zero level is completely arbitrary, because the important quantity is the *difference* in potential energy, and this difference is independent of the choice of zero level.

It is often convenient to choose the surface of the Earth as the reference position for zero potential energy, but this is not essential. Often, the statement of the problem suggests a convenient level to use.

Thinking Physics 1

Elevators that are lifted by a cable have a counterweight, as shown in Figure 7.2. The elevator and the counterweight are connected by a cable that passes over a drive wheel. The counterweight moves in the direction opposite to that of the elevator. What is the purpose of the counterweight?

Reasoning Without the counterweight, the elevator motor would need to lift the weight of the elevator and the people riding it. This would represent a large increase in gravitational potential energy, which means a large expenditure of energy from the motor. With the counterweight, as the elevator moves upward, the counterweight moves down. Thus, the center of mass of the combined system of the elevator and counterweight moves only over a short distance near the middle region of the vertical shaft. This results in a much smaller variation of potential energy of the system than without the counterweight. In the particular case of the elevator and occupants having exactly the same mass as the counterweight, the center of mass is midway between them and never moves as the elevator changes height. In this case, once the elevator is in motion, the motor is no longer required—the system will continue to move as described by Newton's first law (as long as we ignore friction). The motor is needed again to stop the elevator at the desired floor. Thus, the presence of the counterweight reduces the energy requirements on the motor—instead of having to pull the elevator through its entire motion, it is needed only to start and stop the elevator, to overcome friction, and to account for any difference in mass between the counterweight and the elevator, associated with the number of passengers on the elevator.

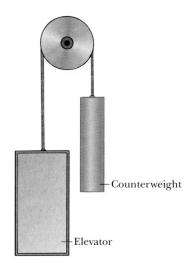

Figure 7.2   (Thinking Physics 1)

EXERCISE 1   What is the gravitational potential energy, relative to the ground, of a 0.15-kg baseball at the top of a 100−m-tall building?   Answer   147 J

7.2 • CONSERVATIVE AND NONCONSERVATIVE FORCES

Forces found in nature can be divided into two categories: conservative and nonconservative. We shall describe the properties of conservative and nonconservative forces separately in this section.

Conservative Forces

A force is conservative if the work it does on an object moving between any two points is independent of the path taken by the object. The work done by a conservative force depends only on the initial and final coordinates of the object. A conservative force can also be defined in a second way. **A force is conservative if the work it does on an object moving through any closed path is zero.**

• *Definition of a conservative force*

The force of gravity is conservative. As we learned in the preceding section, the work done by the gravitational force on an object moving between any two points near the Earth's surface is

$$W_g = mgy_i - mgy_f$$

From this, we see that W_g depends only on the initial and final coordinates of the object and hence is independent of path. Furthermore, W_g is zero when the object moves over any closed path (where $y_i = y_f$).

We can associate a potential energy function with any conservative force. In the preceding section, the potential energy function associated with the gravitational force was found to be

$$U_g = mgy$$

Potential energy functions can be defined only for conservative forces. In general, the work, W_c, done on an object by a conservative force is given by the initial value of the potential energy associated with the object minus the final value:

$$W_c = U_i - U_f \qquad \text{[7.3]}$$

Another example of a conservative force is the force of a spring on an object attached to the spring, where the spring force is given by $F_s = -kx$. As we learned in Chapter 6 (Eq. 6.15), the work done by the spring force is

$$W_s = \tfrac{1}{2}kx_i^2 - \tfrac{1}{2}kx_f^2$$

where the initial and final coordinates of the object are measured from its equilibrium position, $x = 0$. Again we see that W_s depends only on the initial and final coordinates of the object and is zero for any closed path. Hence, the spring force is conservative. The **elastic potential energy** function associated with the spring force is defined by

Potential energy stored in a •
spring

$$U_s \equiv \tfrac{1}{2}kx^2 \qquad \text{[7.4]}$$

where x is measured from the uncompressed position of the spring. The elastic potential energy can be thought of as the energy stored in the deformed spring (one that is either compressed or stretched from its equilibrium position). To visualize this, consider Figure 7.3a, which shows an undeformed spring on a frictionless, horizontal surface. When the block is pushed against the spring (Fig. 7.3b), compressing the spring a distance x, the elastic potential energy stored in the spring is $kx^2/2$. When the block is released, the spring snaps back to its original length and the stored elastic potential energy is transformed into kinetic energy of the block (Fig. 7.3c). The elastic potential energy stored in the spring is zero whenever the spring is undeformed ($x = 0$). Energy is stored in the spring only when the spring is either stretched or compressed. Furthermore, the elastic potential energy is a maximum when the spring has reached its maximum compression or extension (that is, when $|x|$ is a maximum). Finally, because the elastic potential energy is proportional to x^2, we see that U_s is always positive in a deformed spring. If the spring and object are taken together as the system, then no work is done as the spring changes length because the forces are internal.

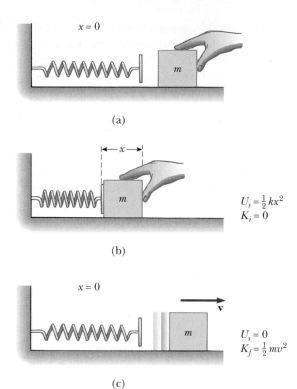

(a)

$U_s = \frac{1}{2} kx^2$
$K_i = 0$

(b)

Figure 7.3 A block of mass m on a frictionless horizontal surface is pushed against a spring and then released. If x is the compression in the spring as in (b), the elastic potential energy stored in the spring is $\frac{1}{2} kx^2$. This energy is transferred to the block in the form of kinetic energy as in (c).

$U_s = 0$
$K_f = \frac{1}{2} mv^2$

(c)

Nonconservative Forces

A force is called **nonconservative** if it leads to dissipation of mechanical energy. For example, if you move an object on a horizontal surface, returning to the same location and same state of motion, but you found that it was necessary to do a net amount of work on the object, then something must have dissipated that energy transferred to the object. That dissipative force is recognized as friction between the surface and the object. Friction is a dissipative, or nonconservative, force. By contrast, if the object is lifted, work is required, but that energy is recovered when the object is lowered. The gravitational force is a nondissipative or conservative force.

 Suppose you displace a book between two points on a table. If the book is displaced in a straight line along the blue path between points A and B in Figure 7.4, the loss in mechanical energy due to the friction force f is simply $-fd$, where d is the distance between the two points. However, if the book is moved along any other path between the two points, the loss in mechanical energy due to friction is greater (in absolute magnitude) than $-fd$. For example, the loss in mechanical energy due to friction along the red semicircular path in Figure 7.4 is equal to $-f(\pi d/2)$, where d is the diameter of the circle.

• *Definition of a nonconservative force*

Figure 7.4 The decrease in mechanical energy due to the force of friction depends on the path taken as the book is moved from A to B, and hence friction is a nonconservative force. The decrease in mechanical energy is greater along the red path compared to the blue path.

7.3 • CONSERVATIVE FORCES AND POTENTIAL ENERGY

In the preceding section we found that the work done on a particle by a conservative force does not depend on the path taken by the particle and is independent of the particle's velocity. The work done is a function only of the particle's initial and final coordinates. As a consequence, we are able to define a **potential energy function** *U* such that the work done on a particle equals the decrease in the potential energy of the particle. The work done by a conservative force **F** as the particle moves along the *x* axis can be expressed as[1]

Work done by a conservative force •

$$W_c = \int_{x_i}^{x_f} F_x \, dx = -\Delta U = U_i - U_f \qquad [7.5]$$

That is, **the work done by a conservative force equals the negative of the change in the potential energy associated with that force,** where the change in the potential energy is defined as $\Delta U = U_f - U_i$. For example, the work, W_c, done by the gravitational field on an object, as the object is lowered in the field, is $U_i - U_f$, which is positive, showing that energy has been transferred from the gravitational field to the object. This energy may appear as kinetic energy of the falling object or may be transferred to something else. We can also express Equation 7.5 as

Change in potential energy •

$$\Delta U = U_f - U_i = -\int_{x_i}^{x_f} F_x \, dx \qquad [7.6]$$

where F_x is the component of **F** in the direction of the displacement: **F** is the force exerted by the field on the object. Therefore ΔU is negative when F_x and dx are in the same direction, as when an object is lowered in a gravitational field or a spring pushes an object toward equilibrium.

The term *potential energy* implies that an object has the potential, or capability, of either gaining kinetic energy or doing work when released from some point under the influence of gravity. It is often convenient to establish some particular location, x_i, to be a reference point and measure all potential energy differences with respect to that point. We can then define the potential energy function as

$$U_f = -\int_{x_i}^{x_f} F_x \, dx + U_i \qquad [7.7]$$

Furthermore, as we discussed earlier, the value of U_i is often taken to be zero at some arbitrary reference point. It really doesn't matter what value we assign to U_i, because any value only shifts U_f by a constant, and it is only the *change* in potential energy that is physically meaningful. If the conservative force is known as a function of position, we can use Equation 7.7 to calculate the change in potential energy of a body as it moves from x_i to x_f.

The amount of mechanical energy dissipated by a nonconservative force depends on the path as an object moves from one position to another and can also depend on the object's speed or on other quantities. Because the work done by a nonconservative force is not simply a function of the initial and final coordinates, we conclude that there is no potential energy function associated with a nonconservative force.

---

[1]For a general displacement, the work done in two or three dimensions also equals $U_i - U_f$, where $U = U(x, y, z)$. We write this formally as $W = \int_i^f \mathbf{F} \cdot d\mathbf{s} = U_i - U_f$.

7.4 • CONSERVATION OF MECHANICAL ENERGY

An object held at some height h above the floor has no kinetic energy, but, as we learned earlier, there is an associated gravitational potential energy equal to mgh, relative to the floor if the gravitational field is included as part of the system. If the object is dropped, it falls to the floor; as it falls, its speed and thus its kinetic energy increase while the potential energy decreases. If factors such as air resistance are ignored, whatever potential energy the object loses as it moves downward appears as kinetic energy. In other words, the sum of the kinetic and potential energies, called the *mechanical energy E,* remains constant in time. This is an example of principle of **conservation of mechanical energy.** For the case of an object in free-fall, this principle tells us that any increase (or decrease) in potential energy is accompanied by an equal decrease (or increase) in kinetic energy.

Because the total mechanical energy E is defined as the sum of the kinetic and potential energies, we can write

$$E \equiv K + U \qquad [7.8]$$

• *Total mechanical energy*

Therefore, we can apply conservation of mechanical energy in the form $E_i = E_f$, or

$$K_i + U_i = K_f + U_f \qquad [7.9]$$

• *Conservation of mechanical energy*

We can state this in a more formal way: **conservation of mechanical energy requires that the total mechanical energy of a system remains constant in any isolated system of objects which interact only through conservative forces.** It is important to note that Equation 7.9 is valid *provided* no energy is added to or removed from the system. Furthermore, there must be no nonconservative forces within the system.

• *A formal statement of conservation of energy*

Because mechanical energy E remains constant with time, $dE/dt = 0$. Taking the derivative of Equation 7.8 with respect to time,

$$\frac{dE}{dt} = 0 = \frac{dK}{dt} + \frac{dU}{dt} \qquad [7.10]$$

Because $K = \frac{1}{2}mv^2$, then

$$\frac{dK}{dt} = \frac{d}{dt}\left(\tfrac{1}{2}mv^2\right) = mv\frac{dv}{dt} = mva = F_x v$$

Applying the chain rule (see Appendix B.6) to dU/dt, we have

$$\frac{dU}{dt} = \frac{dU}{dx}\frac{dx}{dt} = \left(\frac{dU}{dx}\right)v$$

Substituting these expressions for dK/dt and dU/dt into Equation 7.10 gives

$$F_x v + \left(\frac{dU}{dx}\right)v = 0$$

$$F_x = -\frac{dU}{dx} \qquad [7.11]$$

• *Relation between a conservative force and potential energy*

That is, **the conservative internal force acting between parts of a system equals the negative derivative of the potential energy associated with that system.**

We can easily check this relationship for the two instances already discussed. In the case of the deformed spring, $U_s = \frac{1}{2}kx^2$, and therefore

$$F_s = -\frac{dU_s}{dx} = -\frac{d}{dx}\left(\tfrac{1}{2}kx^2\right) = -kx$$

which corresponds to the restoring force exerted by the spring. In the case of an object located a distance y above some reference point, the gravitational potential energy function is given by $U_g = mgy$, and it follows from Equation 7.11 that $F_g = -mg$.

We now see that U is an important function; from it can be derived the conservative force acting in any system. Furthermore, Equation 7.11 should clarify the fact that adding a constant to the potential energy is unimportant, because the location of the reference point is arbitrary.

Equation 7.11 can also be written in the form $dU = -F\,dx$, which, when integrated between the initial and final position values, gives

$$U_f - U_i = -\int_{x_i}^{x_f} F\,dx \qquad\qquad [7.12]$$

This result, which is identical to Equation 7.6, tells us that if the conservative force F acting on an object within a system is known as a function of x, we can calculate the *difference* in the potential energy associated with the object between the initial and final positions.

If more than one conservative force acts on the object, then a potential energy function is associated with *each* force. In such a case, we can apply the law of conservation of energy for the system as

Conservation of mechanical • energy

$$K_i + \sum U_i = K_f + \sum U_f \qquad\qquad [7.13]$$

where the number of terms in the sums equals the number of conservative forces present. For example, if a mass connected to a spring oscillates vertically, two conservative forces act on it: the spring force and the force of gravity. (We will discuss this situation later in a worked example.)

If the force of gravity is the *only* force acting on a body, then the total mechanical energy of the body is constant. Therefore, the law of conservation of energy for a freely falling body can be written

Conservation of mechanical • energy for a freely falling body

$$\tfrac{1}{2}mv_i^2 + mgy_i = \tfrac{1}{2}mv_f^2 + mgy_f \qquad\qquad [7.14]$$

Thinking Physics 2

You have graduated from college and are designing roller coasters for a living. You design a roller coaster in which a car is pulled to the top of a hill of height h and then, starting from a momentary rest, rolls freely down the hill and upward toward the peak of the next hill, which is at height $1.1h$. Will you have a long career in this business?

Reasoning Your career will probably not be long, because this roller coaster will not work! At the top of the first hill, the roller coaster train has no kinetic energy, and gravitational potential energy associated with a height h. If it were to reach the top of the next hill, it would have higher potential energy, that associated with height $1.1h$.

This would violate the principle of conservation of mechanical energy. If this coaster were actually built, the car would move upward on the second hill to a height h (ignoring the effects of friction), stop short of the peak, and then start rolling backward, becoming trapped between the two peaks.

CONCEPTUAL PROBLEM 1

Discuss the energy transformations that occur during the pole vault event pictured in the multiflash photograph shown in Figure 7.5. Ignore rotational motion.

Figure 7.5 (Conceptual Problem 1) Multiflash photograph of a pole vault event. How many forms of energy can you identify in this picture? (*©Harold E. Edgerton, Courtesy of Palm Press, Inc.*)

CONCEPTUAL PROBLEM 2

Three identical balls are thrown from the top of a building, all with the same initial speed. The first ball is thrown horizontally, the second at some angle above the horizontal, and the third at some angle below the horizontal as in Figure 7.6. Neglecting air resistance, describe their motions and compare the speeds of the balls as they reach the ground.

Figure 7.6 (Conceptual Problem 2) Three identical balls are thrown with the same initial speed from the top of a building.

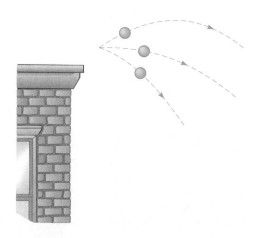

Example 7.1   Ball in Free-Fall

A ball of mass m is dropped from a height h above the ground, as in Figure 7.7. (a) Neglecting air resistance, determine the speed of the ball when it is at a height y above the ground.

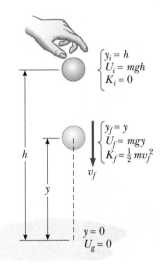

Figure 7.7   (Example 7.1) A ball is dropped from a height h above the ground. Initially, the total energy is gravitational potential energy, equal to mgh relative to the ground. At the elevation y, the total energy is the sum of kinetic and potential energies.

Reasoning   Because the ball is in free-fall, the only force acting on it is the gravitational force. Therefore, we can use the principle of conservation of mechanical energy. Initially, the ball has potential energy and no kinetic energy. As it falls, its total energy (the sum of kinetic and potential energies) remains constant and equal to its initial potential energy.

Solution   When the ball is released from rest at a height h above the ground, its kinetic energy is $K_i = 0$ and its potential energy is $U_i = mgh$, where the y coordinate is measured from ground level. When the ball is at a distance y above the ground, its kinetic energy is $K_f = \frac{1}{2}mv_f^2$ and its potential energy relative to the ground is $U_f = mgy$. Applying Equation 7.9, we get

$$K_i + U_i = K_f + U_f$$

$$0 + mgh = \tfrac{1}{2}mv_f^2 + mgy$$

$$v_f^2 = 2g(h - y)$$

$$v_f = \sqrt{2g(h - y)}$$

(b) Determine the speed of the ball at y if it is given an initial speed v_i at the initial altitude h.

Solution   In this case, the initial energy includes kinetic energy equal to $\frac{1}{2}mv_i^2$, and Equation 7.14 gives

$$\tfrac{1}{2}mv_i^2 + mgh = \tfrac{1}{2}mv_f^2 + mgy$$

$$v_f^2 = v_i^2 + 2g(h - y)$$

$$v_f = \sqrt{v_i^2 + 2g(h - y)}$$

This result is consistent with the expression from kinematics, $v_y^2 = v_{y0}^2 - 2g(y - y_0)$, where $y_0 = h$. Furthermore, this result is valid even if the initial velocity is at an angle to the horizontal (the projectile situation).

Example 7.2   One Way to Lift an Object

Two blocks are connected by a massless cord that passes over a frictionless pulley and a frictionless peg, as in Figure 7.8. One end of the cord is attached to a mass $m_1 = 3.00$ kg that is a distance $R = 1.20$ m from the peg. The other end of the cord is connected to a block of mass $m_2 = 6.00$ kg resting on a table. From what angle θ (measured from the vertical) must the 3.00-kg mass be released in order to just begin to lift the 6.00-kg block off the table?

Reasoning   It is necessary to use several concepts to solve this problem. First, we use conservation of energy to find the speed of the 3.00-kg mass at the bottom of the circular path as a function of θ and the radius of the path. Next, we apply Newton's second law to the 3.00-kg mass at the bottom of its path to find the tension as a function of the given parameters. Finally, we note that the 6.00-kg block lifts off the table when

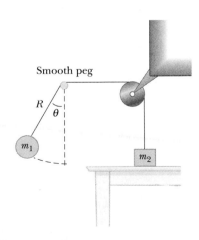

Figure 7.8   (Example 7.2)

the upward force exerted on it by the cord exceeds the force of gravity acting on the block. This procedure enables us to find the required angle.

Solution   Applying conservation of energy to the 3.00-kg mass gives

$$K_i + U_i = K_f + U_f$$

$$(1) \quad 0 + m_1 g y_i = \tfrac{1}{2} m_1 v^2 + 0$$

where v is the speed of the 3.00-kg mass at the bottom of its path. (Note that $K_i = 0$, because the 3.00-kg mass starts from rest and $U_f = 0$, because the bottom of the circle is the zero level of potential energy.) From the geometry in Figure 7.8, we see that $y_i = R - R \cos \theta = R(1 - \cos \theta)$. Using this relation in (1) gives

$$(2) \quad v^2 = 2gR(1 - \cos \theta)$$

Now we apply Newton's second law to the 3.00-kg mass when it is at the bottom of the circular path:

$$T - m_1 g = m_1 \frac{v^2}{R}$$

$$(3) \quad T = m_1 g + m_1 \frac{v^2}{R}$$

This same force is transmitted to the 6.00-kg block, and if it is to be just lifted off the table, the normal force on it becomes zero, and we require that $T = m_2 g$. Using this condition, together with (2) and (3), gives

$$m_2 g = m_1 g + m_1 \frac{2gR(1 - \cos \theta)}{R}$$

Solving for θ, and substituting in the given parameters, we get

$$\cos \theta = \frac{3m_1 - m_2}{2m_1} = \frac{3(3.00 \text{ kg}) - 6.00 \text{ kg}}{2(3.00 \text{ kg})} = \frac{1}{2}$$

$$\theta = 60.0°$$

EXERCISE 2   If the initial angle is $\theta = 40.0°$, find the speed of the 3.00-kg mass and the tension in the cord when the 3.00-kg mass is at the bottom of its circular path.
Answer   2.35 m/s; 43.2 N

EXERCISE 3   A 4.0-kg particle moves along the *x* axis under the influence of a single conservative force. If the work done on the particle is 80.0 J as it moves from the point *x* = 2.0 m to *x* = 5.0 m, find (a) the change in its kinetic energy, (b) the change in its potential energy, and (c) its speed at *x* = 5.0 m if it starts at rest at *x* = 2.0 m.   Answer   (a) 80.0 J
(b) −80.0 J   (c) 6.32 m/s

EXERCISE 4   A rocket is launched at an angle of 53° to the horizontal from an altitude *h* with a speed v_0. Use energy methods to find its speed when its altitude is *h*/2.
Answer $v = \sqrt{v_0^2 + gh}$

7.5 · WORK DONE BY NONCONSERVATIVE FORCES

As we have seen, if the forces acting on a system are conservative, the mechanical energy of the system remains constant. However, if some of the forces acting on the system are not conservative, the mechanical energy of the system does not remain constant. Let us examine two types of nonconservative forces: an applied force and the force of kinetic friction.

Work Done by an Applied Force

When you lift a book through some distance by applying a force to it, the force you apply does work W_{app} on the book, and the force of gravity does work W_g on the book. If we treat the book as a particle, then the net work done on the book is related to the change in its kinetic energy through the work-kinetic energy theorem given by Equation 6.19:

$$W_{app} + W_g = \Delta K \qquad\qquad \textbf{[7.15]}$$

Because the force of gravity is conservative, we can use Equation 7.2 to express the work done by the force of gravity in terms of the change in gravitational potential energy or $W_g = -\Delta U$. Substituting this into Equation 7.15 gives

$$W_{app} = \Delta K + \Delta U \qquad \text{[7.16]}$$

Note that the right side of this equation is the change in the mechanical energy of the book–Earth system. This result says that your applied force transfers energy to the system in the form of kinetic energy of the book and to the gravitational potential energy of the book–Earth system. Thus, we conclude that if an object is part of a system, then **an applied force can transfer energy into or out of the system.**

Situations Involving Kinetic Friction

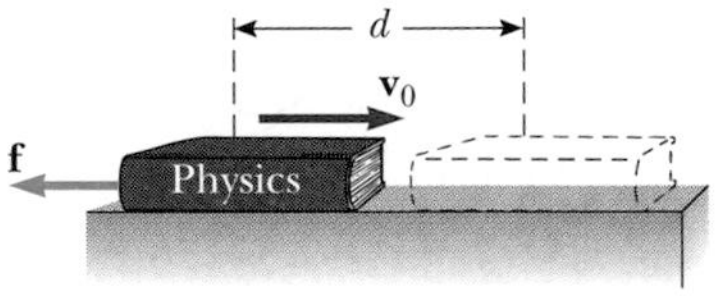

Figure 7.9 A book that is given an initial velocity $\mathbf{v}_0$ on a horizontal surface comes to rest due to the force of kinetic friction after traveling a distance d.

Kinetic friction is an example of a nonconservative force. If a book is given some initial velocity on a horizontal surface that is not frictionless, as in Figure 7.9, the force of kinetic friction acting on the book opposes its motion and negative work is done by the friction force. The book slows down and eventually stops after undergoing a displacement **d**. The force of friction reduces the kinetic energy of the book by transferring energy to thermal (internal) energy of the book and part of the horizontal surface. Note that only part of the book's kinetic energy is transferred to thermal energy in the book. The rest is transferred as thermal energy from the book to the surface.

As the book in Figure 7.9 moves through a distance d, the only force that does work is the force of kinetic friction. In this situation, we can apply Equation 7.16, where $\Delta U = 0$ and note that the force of kinetic friction is opposite the displacement. Thus, we see that the amount by which the force of friction decreases the kinetic energy is $-f_k d$, or

$$\Delta K = -f_k d \qquad \text{[7.17]}$$

If the book moves on an *incline* that is not frictionless, then a change in gravitational potential energy also occurs, and $-f_k d$ is the amount by which the force of kinetic friction reduces the mechanical energy $E = K + U$ of the system. In such cases,

$$\Delta K + \Delta U = -f_k d \qquad \text{[7.18]}$$

CONCEPTUAL PROBLEM 3

A rock is thrown vertically in the air. Ignoring air friction, it takes the same time to go up as it does to come down. Is this still true in the presence of air friction?

CONCEPTUAL PROBLEM 4

A driver brings an automobile to a stop. If the brakes lock, so that the car skids, where is the energy, and in what form is it, after the car stops? Answer the same question for the case in which the brakes do not lock, but the wheels continue to turn.

PROBLEM-SOLVING STRATEGIES • Conservation of Energy

Many problems in physics can be solved using the principle of conservation of energy. The following procedure should be used when you apply this principle:

1. Define your system, which may consist of more than one object and may or may not include fields, springs, or other sources of potential energy. Choose instants to call the initial and final points.

2. Select a reference position for the zero point of potential energy (both gravitational and spring), and use this throughout your analysis. If there is more than one conservative force, write an expression for the potential energy associated with each force.

3. Determine whether any nonconservative forces are present. Remember that if friction or air resistance is present, mechanical energy *is not constant*.

4. If mechanical energy is *constant*, you can write the total initial energy, E_i, at some point as the sum of the kinetic and potential energy at that point. Then write an expression for the total final energy, $E_f = K_f + U_f$, at the final point that is of interest. Because mechanical energy is *constant*, you can equate the two total energies and solve for the quantity that is unknown.

5. If nonconservative forces are present (and thus mechanical energy is not *constant*), first write expressions for the total initial and total final energies. In this case, the difference between the total final mechanical energy and the total initial mechanical energy equals the energy transferred to or from the system by the nonconservative forces. It is convenient to think of a general work-energy theorem as a combination of Equations 7.16 and 7.18, written as

$$K_i + U_i + W_{app} - f_k d = K_f + U_f$$

Example 7.3   Crate Sliding Down a Ramp

A 3.00-kg crate slides down a ramp at a loading dock. The ramp is 1.00 m in length, and inclined at an angle of 30.0°, as shown in Figure 7.10. The crate starts from rest at the top, experiences a constant frictional force of magnitude 5.00 N, and continues to move a short distance on the flat floor. Use energy methods to determine the speed of the crate when it reaches the bottom of the ramp.

Solution   Because $v_i = 0$, the initial kinetic energy is zero. If the y coordinate is measured from the bottom of the ramp, then $y_i = 0.500$ m. Therefore, the total mechanical energy of the crate–Earth system at the top is all gravitational potential energy:

$$U_i = mgy_i = (3.00 \text{ kg}) \left(9.80 \frac{\text{m}}{\text{s}^2} \right) (0.5000 \text{ m}) = 14.7 \text{ J}$$

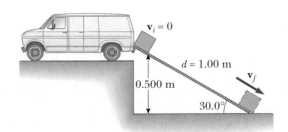

Figure 7.10   (Example 7.3) A crate slides down a ramp under the influence of gravity. The potential energy of the system decreases, and the kinetic energy of the crate increases.

When the crate reaches the bottom, the gravitational potential energy of the system is *zero*, because the elevation of the crate is $y_f = 0$. Therefore, the total mechanical energy at the bottom is all kinetic energy,

$$K_f = \tfrac{1}{2}mv_f^2$$

However, we cannot say that $U_i = K_f$ in this case, because there is an external nonconservative force that reduces the mechanical energy of the system: the force of kinetic friction. In this case, $\Delta E = -f_k d$ where d is the displacement along the ramp. (Remember that the forces normal to the ramp do no work on the crate because they are perpendicular to the displacement.) With $f_k = 5.00$ N and $d = 1.00$ m, we have

$$\Delta E = -f_k d = (-5.00 \text{ N})(1.00 \text{ m}) = -5.00 \text{ J}$$

This says that the decrease in mechanical energy is due to the presence of the force of kinetic friction that opposes the motion. Because $\Delta E = \tfrac{1}{2}mv_f^2 - mgy_i$ in this situation, Equation 7.18 gives

$$-f_k d = \tfrac{1}{2}mv_f^2 - mgy_i$$
$$\tfrac{1}{2}mv_f^2 = 14.7 \text{ J} - 5.00 \text{ J} = 9.70 \text{ J}$$
$$v_f^2 = \frac{19.4 \text{ J}}{3.00 \text{ kg}} = 6.47 \text{ m}^2/\text{s}^2$$
$$v_f = \boxed{2.54 \text{ m/s}}$$

EXERCISE 5 Use Newton's second law to find the acceleration of the crate along the ramp and the equations of kinematics to determine the final speed of the crate. Answer   3.23 m/s²; 2.54 m/s

EXERCISE 6 If the ramp is assumed to be frictionless, find the final speed of the crate and its acceleration along the ramp.   Answer   3.13 m/s; 4.90 m/s²

Example 7.4   Motion on a Curved Track

A child of mass m takes a ride on an irregularly curved slide of height $h = 6.00$ m, as in Figure 7.11. The child starts from rest at the top. (a) Determine the speed of the child at the bottom, assuming no friction is present.

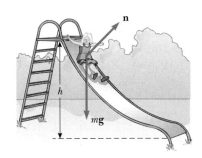

Figure 7.11   (Example 7.4) If the slide is frictionless, the speed of the child at the bottom depends only on the height of the slide.

Reasoning   The normal force, **n**, does no work on the child, because this force is always perpendicular to each element of the displacement. Furthermore, because there is no friction, mechanical energy is constant—that is, $K + U = $ constant.

Solution   If we measure the y coordinate from the bottom of the slide, then $y_i = h$, $y_f = 0$, and we get

$$K_i + U_i = K_f + U_f$$
$$0 + mgh = \tfrac{1}{2}mv_f^2 + 0$$
$$v_f = \sqrt{2gh}$$

Note that the result is the same as it would be if the child fell vertically through a distance h! In this example, $h = 6.00$ m, giving

$$v_f = \sqrt{2gh} = \sqrt{2\left(9.80\,\frac{\text{m}}{\text{s}^2}\right)(6.00 \text{ m})} = \boxed{10.8 \text{ m/s}}$$

(b) If a frictional force acts on the child, how much mechanical energy is dissipated by this force? Assume that $v_f = 8.00$ m/s and $m = 20.0$ kg.

Solution   In this case, $\Delta E \neq 0$ and mechanical energy is *not* constant. We can use Equation 7.18 to find the loss of mechanical energy due to friction, assuming the final speed at the bottom is known:

$$\Delta E = E_f - E_i = \tfrac{1}{2}mv_f^2 - mgh$$
$$\Delta E = \tfrac{1}{2}(20.0 \text{ kg})(8.00 \text{ m/s})^2 - (20.0 \text{ kg})\left(9.80\,\frac{\text{m}}{\text{s}^2}\right)(6.00 \text{ m})$$
$$= \boxed{-536 \text{ J}}$$

Again, ΔE is negative because friction reduces the mechanical energy of the system. Note, however, that because the slide is curved, the normal force changes in magnitude and direction during the motion. Therefore, the frictional force, which is proportional to n, also changes during the motion. Do you think it would be possible to determine μ from these data?

Example 7.5   Mass–Spring Collision

A mass of 0.80 kg is given an initial velocity $v_i = 1.2$ m/s to the right and collides with a light spring of force constant $k = 50$ N/m, as in Figure 7.12. (a) If the surface is frictionless, calculate the initial maximum compression of the spring after the collision.

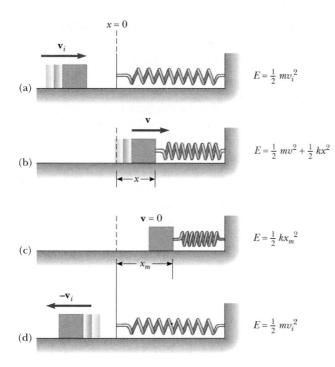

$E = \frac{1}{2} mv_i^2$

$E = \frac{1}{2} mv^2 + \frac{1}{2} kx^2$

$E = \frac{1}{2} kx_m^2$

$E = \frac{1}{2} mv_i^2$

Figure 7.12   (Example 7.12) A block sliding on a smooth, horizontal surface collides with a light spring. (a) Initially the mechanical energy is all kinetic energy. (b) The mechanical energy is the sum of the kinetic energy of the block and the elastic potential energy in the spring. (c) The energy is entirely potential energy. (d) The energy is converted back to the kinetic energy of the block. The total energy remains constant throughout the motion.

Reasoning   Before the collision, the mass has kinetic energy and the spring is uncompressed, so the energy stored in the spring is zero. Thus, the total energy of the system (mass plus spring) before the collision is $\frac{1}{2} mv_i^2$. After the collision, and when the spring is fully compressed, the mass is momentarily at rest and has zero kinetic energy, and the energy stored in the spring has its maximum value, $\frac{1}{2} kx_f^2$. The total mechanical energy of the system is constant because no nonconservative forces act on the system.

Solution   Because mechanical energy is conserved, the kinetic energy of the mass before the collision must equal the maximum energy stored in the spring when it is fully compressed, or

$$\tfrac{1}{2}mv_i^2 = \tfrac{1}{2}kx_f^2$$

$$x_f = \sqrt{\frac{m}{k}}\, v_i = \sqrt{\frac{0.80 \text{ kg}}{50 \text{ N/m}}}\,(1.2 \text{ m/s}) = 0.15 \text{ m}$$

(b) If a constant force of kinetic friction acts between block and surface with $\mu_k = 0.50$ and if the speed of the block just as it collides with the spring is $v_i = 1.2$ m/s, what is the maximum compression in the spring?

Solution   In this case, mechanical energy is *not* conserved because of friction. The magnitude of the frictional force is

$$f_k = \mu_k n = \mu_k mg = 0.50(0.80 \text{ kg})\left(9.80\,\frac{\text{m}}{\text{s}^2}\right) = 3.9 \text{ N}$$

Therefore, the decrease in kinetic energy due to friction as the block is displaced from $x_i = 0$ to $x_f = x$ is

$$\Delta K = -f_k x = (-3.92x) \text{ J}$$

Substituting this into Equation 7.16 gives

$$\Delta K = (0 + \tfrac{1}{2}kx^2) - (\tfrac{1}{2}mv_i^2 + 0)$$

$$-3.92x = \tfrac{50}{2}x^2 - \tfrac{1}{2}(0.80)(1.2)^2$$

$$25x^2 + 3.92x - 0.576 = 0$$

Solving the quadratic equation for x gives $x = 0.092$ m and $x = -0.25$ m. The physically meaningful root is $x = 0.092$ m = 9.2 cm. The negative root is meaningless because the block must be to the right of the origin when it comes to rest. Note that 9.2 cm is less than the distance obtained in the frictionless case (a). This result is what we expect because friction retards the motion of the system.

EXERCISE 7   A 3.0-kg block starts at a height $h = 60$ cm on a plane that has an inclination angle of 30°. On reaching the bottom, the block slides along a horizontal surface. If the coefficient of friction on both surfaces is $\mu_k = 0.20$, how far does the block slide on the horizontal surface before coming to rest?   Answer   1.96 m

EXERCISE 8 A child starts from rest at the top of a slide of height 4.0 m. (a) What is her speed at the bottom if the incline is frictionless? (b) If she reaches the bottom with a speed of 6.0 m/s, what percentage of her total energy at the top of the slide was lost as a result of friction? Answer (a) 8.85 m/s (b) 54.1%

7.6 • CONSERVATION OF ENERGY IN GENERAL

We have seen that the total mechanical energy of a system is constant when only conservative internal forces act within the system. Furthermore, we were able to associate a potential energy function with each conservative force. However, mechanical energy is not constant when nonconservative forces, such as friction, are present.

Whenever the mechanical energy of a system decreases, the energy does not disappear. Instead, the mechanical energy is transformed into other forms of energy. In the study of thermodynamics we shall find that mechanical energy can be transformed into internal energy of the system. For example, when a block slides over a surface, part of the mechanical energy is transformed into internal energy stored in the block and the surface, as evidenced by a measurable increase in the block's temperature. We shall see that, on a submicroscopic scale, this internal energy is associated with the vibration of atoms about their equilibrium positions. The energy acquired by the atoms is unevenly and randomly distributed among the atoms. This disorderly internal form of energy is often called **thermal energy.** Such internal atomic motion has kinetic and potential energy, and so one can say that frictional forces arise fundamentally from conservative atomic forces. Therefore, if we include this increase in the internal energy of the system in our energy expression, the total energy is conserved.

This is just one example of how you can analyze an isolated system and always find that its total energy does not change, as long as you account for all forms of energy. That is, **energy can never be created or destroyed. Energy may be transformed from one form to another, but the total energy of an isolated system is always constant.** From a universal point of view, we can say that **the total energy of the Universe is constant:** If one part of the Universe gains energy in some form, another part must lose an equal amount of energy. No violation of this principle has been found.

Total energy is always • conserved.

Other examples of energy transformations include the energy carried by sound waves resulting from the collision of two objects, the energy radiated by an accelerating charge in the form of electromagnetic waves (a radio antenna), and the elaborate sequence of energy conversions in a thermonuclear reaction.

In subsequent chapters we shall see that the energy concept, and especially transformations of energy, unite the branches of physics. The subjects of mechanics, thermodynamics, and electromagnetism cannot really be separated. In practical terms, all mechanical and electronic devices rely on energy transformations.

Thinking Physics 3

An automobile with kinetic energy and potential energy carried within the gasoline strikes a tree and comes to rest. The total mechanical energy in the system (the car) is now less than before. How did the energy leave the system, and in what form is energy left in the system?

Reasoning There are a number of ways that energy transferred out of the system. We shall describe the main mechanisms. The automobile did some *work* on the tree during the collision, causing the tree to become warmer and deform. There was a large crash during the collision, representing transfer of energy by *sound*. If the gasoline tank leaks after the collision, the automobile is losing energy by means of *mass transfer*. After the car has come to rest, there is no more kinetic energy. There is some *potential energy* left in any gas remaining within the tank. There is also more *internal energy* in the automobile, because its temperature is likely to be higher after the collision than before.

CONCEPTUAL PROBLEM 5

A toaster is turned on. Discuss the forms of energy and energy transfer occurring in the coils of the toaster.

CONCEPTUAL PROBLEM 6

Soft steel can be made red hot by continued hammering on it. This doesn't work as well for hard steel. Why is there this difference?

7.7 • GRAVITATIONAL POTENTIAL ENERGY REVISITED

Earlier in this chapter we introduced the concept of gravitational potential energy, that is, the energy associated with the position of a particle. We emphasized the fact that the gravitational potential energy function, $U = mgy$, is valid only when a particle is near the Earth's surface. Because the gravitational force between two particles varies as $1/r^2$, it follows that the correct potential energy function of the system depends on the amount of separation between the particles.

Consider a particle of mass m moving between two points P and Q above the Earth's surface, as in Figure 7.13. The gravitational force acting on m is

$$\mathbf{F}_g = -\frac{GM_em}{r^2}\,\hat{\mathbf{r}} \qquad [7.19]$$

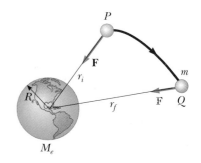

Figure 7.13 As a particle of mass m moves from P to Q above the Earth's surface, the potential energy changes according to Equation 7.20.

where $\hat{\mathbf{r}}$ is a unit vector directed from the Earth to the particle and the negative sign indicates that the force is attractive. This expression shows that the gravitational force depends only on the polar coordinate r. Furthermore, the gravitational force is conservative. Because the change in potential energy associated with a given displacement of the particle is defined as the negative of the work done by the conservative gravitational force during that displacement, Equation 7.12 gives

$$U_f - U_i = -\int_{r_i}^{r_f} F(r)\,dr = GM_em\int_{r_i}^{r_f}\frac{dr}{r^2} = GM_em\left[-\frac{1}{r}\right]_{r_i}^{r_f}$$

or

$$U_f - U_i = -GM_em\left(\frac{1}{r_f} - \frac{1}{r_i}\right) \qquad [7.20]$$

As always, the choice of a reference point for the potential energy is completely arbitrary. It is customary to locate the reference point where the force is zero. Taking $U_i = 0$ at $r_i = \infty$, we obtain the important result

Gravitational potential •
energy r > R_e

$$U(r) = -\frac{GM_e m}{r} \qquad [7.21]$$

This equation applies to the Earth–particle system separated by a distance r, provided that $r > R_e$. The result is not valid for particles moving inside the Earth, where $r < R_e$. Because of our choice of U_i, the function $U(r)$ is always negative (Fig. 7.14).

Although Equation 7.21 was derived for the particle–Earth system, it can be applied to *any* two particles. That is, the gravitational potential energy associated with *any pair* of particles of masses m_1 and m_2 separated by a distance r is

$$U_g = -\frac{Gm_1 m_2}{r} \qquad [7.22]$$

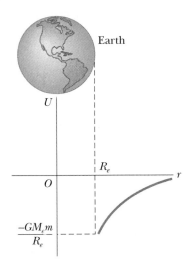

This expression also applies to larger objects *if they are spherically symmetric*, as first shown by Newton using integral calculus. Equation 7.22 shows that the gravitational potential energy for any pair of particles varies as $1/r$ (whereas the force between them varies as $1/r^2$). Furthermore, the potential energy is *negative*, because the force is attractive, and we have taken the potential energy as zero when the particle separation is infinity. Because the force between the particles is attractive, we know that an external agent must do positive work to increase the separation between the two particles. The work done by the external agent produces an increase in the potential energy as the two particles are separated. That is, U_g becomes less negative as r increases. (Note that part of the work done can also produce a change in kinetic energy of the system. That is, if the work done in separating the particles exceeds the increase in potential energy, the excess energy is accounted for by the increase in kinetic energy of the system.) When the two particles are separated by a distance r, an external agent would have to supply an energy *at least* equal to $+Gm_1 m_2/r$ in order to separate the particles by an infinite distance.

Figure 7.14 Graph of the gravitational potential energy, U_g, versus r for a particle above the Earth's surface. The potential energy of the system goes to zero as r approaches ∞.

It is convenient to think of the absolute value of the gravitational potential energy defined this way as the **binding energy** of the system. If the external agent supplies an energy *greater than* the binding energy, $Gm_1 m_2/r$, the additional energy of the system is in the form of kinetic energy when the particles are at an infinite separation.

We can extend this concept to three or more particles. In this case, the total potential energy of the system is the sum over all *pairs* of particles.[2] Each pair contributes a term of the form given by Equation 7.22. For example, if the system contains three particles, as in Figure 7.15, we find that

$$U_{\text{total}} = U_{12} + U_{13} + U_{23} = -G\left(\frac{m_1 m_2}{r_{12}} + \frac{m_1 m_3}{r_{13}} + \frac{m_2 m_3}{r_{23}}\right) \qquad [7.23]$$

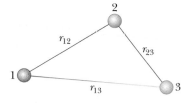

Figure 7.15 Diagram of three interacting particles.

The absolute value of U_{total} represents the work needed to separate the particles by an infinite distance. If the system consists of four particles, there are six terms in the sum, corresponding to the six distinct pairs of interaction forces. The total mechanical energy of this system of four particles includes ten terms: four kinetic energy terms (one for each particle) and six potential energy terms.

[2]The fact that one can add potential energy terms for all pairs of particles stems from the experimental fact that gravitational forces obey the superposition principle. That is, if $\Sigma\mathbf{F} = \mathbf{F}_{12} + \mathbf{F}_{13} + \mathbf{F}_{23} + \ldots$, then there exists a potential energy term for each interaction $\mathbf{F}_{ij}$.

Thinking Physics 4

Why is the Sun hot?

Reasoning The Sun was formed when a cloud of gas and dust coalesced, due to gravitational attraction, into a massive astronomical object. Before this occurred, the particles were widely scattered, representing a large amount of gravitational potential energy. As the particles came together to form the Sun, the gravitational potential energy decreased. According to the principle of conservation of energy, this potential energy must be transformed to another form. The form to which it transformed was internal energy, representing an increase in temperature. If enough particles come together, the temperature can rise to a point at which nuclear fusion occurs, and the object becomes a star. If there are not enough particles, the temperature rises, but not to a point at which fusion occurs. The object is a planet if it is in orbit around a star. Jupiter is an example of a large planet that might have been a star if more particles were available to be collected.

Example 7.6   The Change in Potential Energy

A particle of mass m is displaced through a small vertical distance Δy near the Earth's surface. Let us show that the general expression for the change in gravitational potential energy given by Equation 7.20 reduces to the familiar relationship $\Delta U_g = mg\,\Delta y$.

Solution   We can express Equation 7.20 in the form

$$\Delta U_g = -GM_e m \left(\frac{1}{r_f} - \frac{1}{r_i}\right) = GM_e m \left(\frac{r_f - r_i}{r_i r_f}\right)$$

If both the initial and final positions of the particle are close to the Earth's surface, then $r_f - r_i = \Delta y$ and $r_i r_f \approx R_e^2$. (Recall that r is measured from the center of the Earth.) Therefore, the *change* in potential energy becomes

$$\Delta U_g \approx \frac{GM_e m}{R_e^2}\,\Delta y = mg\,\Delta y$$

where we have used the fact that $g = GM_e/R_e^2$. Keep in mind that the reference point is arbitrary, because it is the *change* in potential energy that is meaningful.

EXERCISE 9 A satellite of the Earth has a mass of 100 kg and its altitude is 2.0×10^6 m. (a) What is the gravitational potential energy of the satellite–Earth system? (b) What is the magnitude of the force on the satellite?   Answer   (a) -4.8×10^9 J   (b) 570 N

7.8 • ENERGY DIAGRAMS AND STABILITY OF EQUILIBRIUM   O P T I O N A L

The motion of a system can often be understood qualitatively through an analysis of the system's potential energy curve. Consider the potential energy function for the mass–spring system, given by $U_s = \frac{1}{2}kx^2$. This function is plotted versus x in Figure 7.16a. The spring force is related to U through Equation 7.11:

$$F_s = -\frac{dU_s}{dx} = -kx$$

That is, the force is equal to the negative of the *slope* of the U-versus-x curve. When the mass is placed at rest at the equilibrium position ($x = 0$), where $F = 0$, it will remain there unless some external force acts on it. If the spring is stretched from equilibrium, x is positive and the slope dU/dx is positive; therefore, F_s is negative and the mass accelerates back toward $x = 0$. If the spring is compressed, x is negative

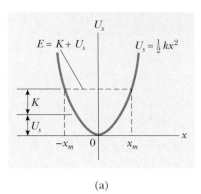

(a)

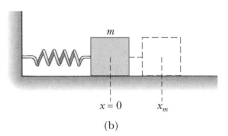

(b)

Figure 7.16   (a) The potential energy as a function of x for the mass–spring system shown in (b). The mass oscillates between the turning points, which have the coordinates $x = \pm x_m$. Note that the restoring force of the spring always acts toward $x = 0$, the position of stable equilibrium.

and the slope is negative; therefore, F_s is positive and again the mass accelerates toward $x = 0$.

Stable equilibrium •

From this analysis we conclude that the $x = 0$ position is one of **stable equilibrium.** That is, any movement away from this position results in a force that is directed back toward $x = 0$. In general, **positions of stable equilibrium correspond to those points for which $U(x)$ has a relative minimum value.**

From Figure 7.16 we see that if the mass is given an initial displacement x_m and released from rest, its total energy initially is the potential energy stored in the spring, given by $\frac{1}{2}kx_m^2$. As motion commences, the system acquires kinetic energy and loses an equal amount of potential energy. Because the total energy must remain constant, the mass oscillates between the two points $x = \pm x_m$, called the *turning points.* In fact, because no energy loss (no friction) takes place, the mass oscillates between $-x_m$ and $+x_m$ forever. (We shall discuss these oscillations further in Chapter 12). From an energy viewpoint, the energy of the system cannot exceed $\frac{1}{2}kx_m^2$; therefore, the mass must stop at these points and, because of the spring force, accelerate toward $x = 0$.

Another simple mechanical system that has a position of stable equilibrium is a ball rolling around in the bottom of a spherical bowl. If the ball is displaced from its lowest position, it always tends to return to that position when released.

Now consider an example in which the U-versus-x curve is as shown in Figure 7.17. In this case, $F_x = 0$ at $x = 0$, and so the particle is in equilibrium at this point. However, this is a position of **unstable equilibrium** for the following reason. Suppose the particle is displaced to the *right* ($x > 0$). Because the slope is negative for $x > 0$, $F_x = -dU/dx$ is positive and the particle accelerates away from $x = 0$. Now suppose the particle is displaced to the left ($x < 0$). In this case the force is *negative,* because the slope is positive for $x < 0$, and the particle again accelerates away from the equilibrium position. The $x = 0$ position in this situation is called a position of *unstable equilibrium* because, for any displacement from this point, the force pushes

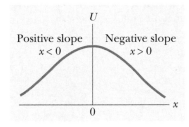

Figure 7.17   A plot of U versus x for a particle that has a position of unstable equilibrium, located at $x = 0$. For any finite displacement of the particle, the force on the particle is directed away from $x = 0$.

Unstable equilibrium •

the particle farther away from equilibrium. In fact, the force pushes the particle toward a position of lower potential energy. A ball placed on the top of an inverted spherical bowl is obviously in a position of unstable equilibrium. If the ball is displaced slightly from the top and released, it will surely roll off the bowl. In general, **positions of unstable equilibrium correspond to those points for which $U(x)$ has a relative maximum value.**[3]

Finally, a situation may arise where U is constant over some region, and hence $F = 0$. This is called a position of **neutral equilibrium.** Small displacements from this position produce neither restoring nor disrupting forces. A ball lying on a flat horizontal surface is an example of an object in neutral equilibrium.

- *Neutral equilibrium*

SUMMARY

If a particle of mass m is elevated a distance y near the Earth's surface, the **gravitational potential energy** of the particle–Earth system can be represented as

$$U_g = mgy \qquad \text{[7.1]}$$

The **elastic potential energy** stored in a spring of force constant k is

$$U_s \equiv \tfrac{1}{2}kx^2 \qquad \text{[7.4]}$$

A force is **conservative** if the work it does on a particle is independent of the path the particle takes between two given points. Alternatively, a force is conservative if the work it does is zero when the particle moves through an arbitrary closed path and returns to its initial position. A force that does not meet these criteria is said to be **nonconservative.**

A **potential energy** function U can be associated only with a conservative force. If a conservative force **F** acts on a particle that moves along the x axis from x_i to x_f, *the change in the potential energy equals the negative of the work done by that force:*

$$U_f - U_i = -\int_{x_i}^{x_f} F_x\, dx \qquad \text{[7.6]}$$

The **total mechanical energy of a system** is defined as the sum of the kinetic energy and potential energy:

$$E \equiv K + U \qquad \text{[7.8]}$$

The principle of **conservation of mechanical energy** states that if no external forces do work on the system, and there are no nonconservative forces, the total mechanical energy is constant:

$$K_i + U_i = K_f + U_f \qquad \text{[7.9]}$$

If some of the forces acting on a system are not conservative, the mechanical energy of the system does not remain constant. If an object is part of a system, the work done by an applied force is equal to the change in mechanical energy of the system:

$$W_{\text{app}} = \Delta K + \Delta U \qquad \text{[7.16]}$$

An applied force transfers energy to or from the system in the form of changes in kinetic energy and gravitational potential energy.

The gravitational force is conservative, and therefore a potential energy function can be

[3]You can test mathematically whether an extreme of U is stable or unstable by examining the sign of d^2U/dx^2. A positive sign gives stable equilibrium and a negative sign gives unstable equilibrium.

defined. The **gravitational potential energy** associated with two particles separated by a distance r is

$$U_g = -\frac{Gm_1 m_2}{r} \qquad [7.22]$$

where U_g is taken to be zero at $r = \infty$. The total gravitational potential energy for a system of particles is the sum of energies for all pairs of particles, with each pair represented by a term of the form given by Equation 7.22.

CONCEPTUAL QUESTIONS

1. One person drops a ball from the top of a building, while another person at the bottom observes its motion. Will these two people agree on the value of the gravitational potential energy? On the change in potential energy? On its kinetic energy?

2. Discuss the production and dissipation of mechanical energy in (a) lifting a weight, (b) holding the weight up, and (c) lowering the weight slowly. Include the muscles in your discussion.

3. A skier prepares to take off down a snow-covered hill. Is it correct to say that gravity provides the energy for the trip down the hill?

4. You walk leisurely up a flight of stairs. You then return to the bottom and run up the same set of stairs as fast as you can. Have you performed a different amount of work in these two cases?

5. Many mountain roads are built so that they spiral around the mountain rather than go straight up the slope. Discuss this design from the viewpoint of energy and power.

6. In an earthquake, a large amount of energy is "released," and spreads outward, potentially causing severe damage. In what form does this energy exist before the earthquake, and by what energy transfer mechanism does it travel from the focus?

7. You ride a bicycle. In what sense is your bicycle solar-powered?

8. Can the gravitational potential energy of a system ever have a negative value? Explain.

9. A bowling ball is suspended from the ceiling of a lecture hall by a strong cord. The bowling ball is drawn away from its equilibrium position and released from rest at the tip of the demonstrator's nose. If the demonstrator remains stationary, explain why she will not be struck by the ball on its return swing. Would the demonstrator be safe if she pushed the ball as she released it?

10. A pile driver is a device used to drive objects into the Earth by repeatedly dropping a heavy weight on them. By how much does the energy of a pile driver increase when the weight it drops is doubled? (Assume the weight is dropped from the same height each time.)

11. Our body muscles exert forces when we lift, push, run, jump, and so forth. Are these forces conservative?

12. When nonconservative forces act on a system, does the total mechanical energy remain constant?

13. A block is connected to a spring that is suspended from the ceiling. If the block is set in motion and air resistance is neglected, describe the energy transformations that occur within the system consisting of the block and spring.

14. What would the curve of U versus x look like if a particle were in a region of neutral equilibrium?

15. A ball rolls on a horizontal surface. Is the ball in stable, unstable, or neutral equilibrium?

16. Discuss the energy transformations that occur during the operation of an automobile.

17. A ball is thrown straight up into the air. At what position is its kinetic energy a maximum? At what position is the gravitational potential energy a maximum?

PROBLEMS

Section 7.1 Potential Energy

Section 7.2 Conservative and Nonconservative Forces

1. A 1000-kg roller coaster is initially at the top of a rise, at point A. It then moves 135 ft, at an angle of 40.0° below the horizontal, to a lower point, B. (a) Choose point B to be the zero level for gravitational potential energy, and find the potential energy at points A and B and the difference in potential energy, $U_A - U_B$, between these points. (b) Repeat part (a), setting the zero reference level at point A.

2. A 2.00-kg ball is attached to the bottom end of a 1.00-m–long string hanging from the ceiling of a room. The height of the room is 3.00 m. What is the gravitational potential energy relative to (a) the ceiling, (b) the floor, and (c) a point at the same elevation as the ball?

3. A 40.0-N child is in a swing that is attached to ropes

2.00 m long. Find the gravitational potential energy of the system relative to the child's lowest position when (a) the ropes are horizontal, (b) the ropes make a 30.0° angle with the vertical, and (c) the child is at the bottom of the circular arc.

4. A 4.00-kg particle moves from the origin to the position having coordinates $x = 5.00$ m and $y = 5.00$ m under the influence of gravity acting in the negative y direction (Fig. P7.4). Using Equation 6.3, calculate the work done by gravity in going from O to C along (a) OAC, (b) OBC, (c) OC. Your results should all be identical. Why?

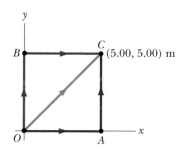

Figure P7.4

Section 7.3   Conservative Forces and Potential Energy

Section 7.4   Conservation of Mechanical Energy

5. A single conservative force $F_x = (2.0x + 4.0)$ N acts on a 5.00-kg particle, where x is in meters. As the particle moves along the x axis from $x = 1.00$ m to $x = 5.00$ m, calculate (a) the work done by this force, (b) the change in the potential energy of the system, and (c) the kinetic energy of the particle at $x = 5.00$ m if its speed at $x = 1.00$ m is 3.00 m/s.

6. A single constant force $\mathbf{F} = (3.0\mathbf{i} + 5.0\mathbf{j})$ N acts on a 4.00-kg particle. (a) Calculate the work done by this force if the particle moves from the origin to the point having the vector position $\mathbf{r} = (2.0\mathbf{i} - 3.0\mathbf{j})$ m. Does this result depend on the path? Explain. (b) What is the speed of the particle at $\mathbf{r}$ if its speed at the origin is 4.00 m/s? (c) What is the change in potential energy?

7. At time t_i, the kinetic energy of a particle is 30.0 J and the potential energy is 10.0 J. At some later time t_f, its kinetic energy is 18.0 J. (a) If only conservative forces act on the particle, what are the potential energy and the total energy at time t_f? (b) If the potential energy at time t_f is 5.00 J, are there any nonconservative forces acting on the particle? Explain.

8. A particle of mass 0.500 kg is shot from P as shown in Figure P7.8 with an initial velocity $\mathbf{v}_0$ having a horizontal component of 30.0 m/s. The particle rises to a maximum height of 20.0 m above P. Using conservation of energy, determine (a) the vertical component of $\mathbf{v}_0$, (b) the work done by the gravitational force on the particle during its motion from P

to B, and (c) the horizontal and the vertical components of the velocity vector when the particle reaches B.

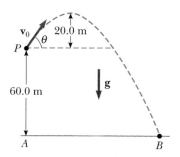

Figure P7.8

9. A bead slides without friction around a loop-the-loop (Fig. P7.9). If the bead is released from a height $h = 3.50R$, what is its speed at point A? How large is the normal force on it if its mass is 5.00 g?

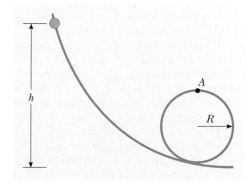

Figure P7.9

10. A simple 2.00-m–long pendulum is released from rest when the support string is at an angle of 25.0° from the vertical. What is the speed of the suspended mass at the bottom of the swing?

11. Two masses are connected by a light string passing over a light frictionless pulley as shown in Figure P7.11. The 5.00-kg mass is released from rest at a height of 4.00 m above the ground. Using the law of conservation of energy, (a) determine the speed of the 3.00-kg mass just as the 5.00-kg mass hits the ground. (b) Find the maximum height to which the 3.00-kg mass rises above the ground.

12. Two masses are connected by a light string passing over a light frictionless pulley as in Figure P7.11. The mass m_1 is released from rest at height h. Using the law of conservation of energy, (a) determine the speed of m_2 just as m_1 hits the ground. (b) Find the maximum height to which m_2 rises.

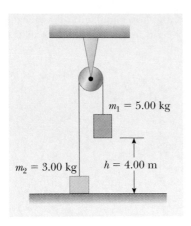

Figure P7.11

13. Dave Johnson, the bronze medalist at the 1992 Olympic decathlon in Barcelona, leaves the ground at the high jump with a vertical velocity of 6.00 m/s. How far does his center of gravity move up as he makes the jump?

14. A 0.400-kg ball is thrown into the air and reaches a maximum altitude of 20.0 m. Taking its initial position as the point of zero potential energy and using energy methods, find (a) its initial speed, (b) its total mechanical energy, and (c) the ratio of its kinetic energy to the potential energy when its altitude is 10.0 m.

15. A 20.0-kg cannon ball is fired from a cannon with the muzzle speed of 1000 m/s at an angle of 37.0° with the horizontal. A second ball is fired at an angle of 90.0°. Use the conservation of mechanical energy to find (a) the maximum height reached by each ball and (b) the total mechanical energy at the maximum height for each ball.

16. A 2.00-kg ball is attached to a 10-lb (44.5 N) fishing line. The ball is released from rest at the horizontal position ($\theta = 90.0°$). At what angle, θ, (measured from the vertical) will the fishing line break?

17. A child starts from rest and slides down the frictionless slide

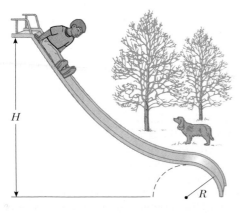

Figure P7.17

shown in Figure P7.17. In terms of R and H, at what height h will he lose contact with the section of radius R?

Section 7.5   Work Done by Nonconservative Forces

18. A 70.0-kg diver steps off a 10.0-m tower and drops straight down into the water. If he comes to rest 5.00 m beneath the surface of the water, determine the average resistance force exerted on the diver by the water.

19. A force F_x, shown as a function of distance in Figure P7.19, acts on a 5.00-kg mass. If the particle starts from rest at $x = 0$ m, determine the speed of the particle at $x = $ 2.00 m, 4.00 m, and 6.00 m.

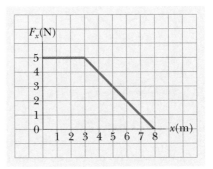

Figure P7.19

20. A softball pitcher swings in her hand a ball of mass 0.250 kg around a vertical circular path of radius 60.0 cm before releasing it. The pitcher maintains a component of force on the ball of constant magnitude 30.0 N in the direction of motion around the complete path. The speed of the ball at the top of the circle is 15.0 m/s. If the ball is released at the bottom of the circle, what is its speed on release?

21. The coefficient of friction between the 3.00-kg mass and surface in Figure P7.21 is 0.400. The system starts from rest. What is the speed of the 5.00-kg mass when it has fallen 1.50 m?

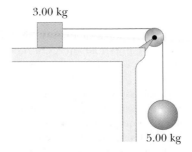

Figure P7.21

22. A 2000-kg car starts from rest at the top of a 5.00–m-long driveway that is sloped at an angle of 20.0° with the horizontal. If an average friction force of 4000 N impedes the motion of the car, find the speed of the car at the bottom of the driveway.

23. A 5.00-kg block is set into motion up an inclined plane with an initial speed of 8.00 m/s (Fig. P7.23). The block comes to rest after traveling 3.00 m along the plane, which is inclined at an angle of 30.0° to the horizontal. Determine (a) the change in the block's kinetic energy, (b) the change in potential energy, and (c) the frictional force exerted on it (assumed to be constant). (d) What is the coefficient of kinetic friction?

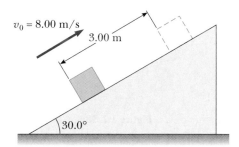

Figure P7.23

24. A parachutist of mass 50.0 kg jumps out of an airplane at a height of 1000 m and lands on the ground with a speed of 5.00 m/s. How much mechanical energy was lost to air friction during this jump?

25. A 80.0-kg skydiver jumps out of an airplane at an altitude of 1000 m and opens the parachute at an altitude of 200.0 m. (a) Assuming that the total retarding force on the diver is constant at 50.0 N with the parachute closed and constant at 3600 N with the parachute open, what is the speed of the diver when he lands on the ground? (b) Do you think the skydiver will get hurt? Explain. (c) At what height should the parachute be opened so that the final speed of the skydiver when he hits the ground is 5.00 m/s? (d) How realistic is the assumption that the total retarding force is constant? Explain.

26. A toy gun uses a spring to project a 5.30-g soft rubber ball. The spring is originally compressed by 5.00 cm and has stiffness constant 8.00 N/m. When it is fired, the ball moves 15.0 cm through the barrel of the gun, and there is a constant frictional force of 0.0320 N between barrel and ball. (a) With what speed does the projectile leave the barrel of the gun? (b) Just where does the ball have maximum speed? (c) What is this maximum speed?

27. A block slides down a curved frictionless track and then up an inclined plane, as in Figure P7.27. The coefficient of kinetic friction between block and incline is μ_k. Use energy methods to show that the maximum height reached by the block is

$$y_{max} = \frac{h}{1 + \mu_k \cot \theta}$$

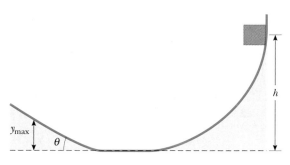

Figure P7.27

28. A 1.50-kg mass is held 1.20 m above a relaxed massless spring with a spring constant of 320 N/m. The mass is dropped onto the spring. (a) How far does it compress the spring? (b) How far does it compress the spring if the same experiment is performed on the moon where $g = 1.63$ m/s²? (c) Repeat part (a), but this time assume a constant air-resistance force of 0.700 N acts on the mass during its motion.

29. A skier starts from rest at the top of a hill that is inclined at an angle of 10.5° to the horizontal. The hill is 200 m long, and the coefficient of friction between the snow and the skis is 0.0750. At the bottom of the hill the snow is level and the coefficient of friction is unchanged. How far does the skier move along the horizontal portion of the snow before coming to rest?

Section 7.7 Gravitational Potential Energy Revisited

(Assume $U = 0$ at $r = \infty$.)

30. How much energy is required to move a 1000-kg mass from the Earth's surface to an altitude twice the Earth's radius?

31. After our Sun exhausts its nuclear fuel, its ultimate fate is possibly to collapse to a *white-dwarf* state, in which it has approximately the mass of the Sun but the radius of the Earth. Calculate (a) the average density of the white dwarf, (b) the free-fall acceleration at its surface, and (c) the gravitational potential energy of a 1.00-kg object at its surface. (Take $U_g = 0$ at infinity.)

32. At the Earth's surface a projectile is launched straight up at a speed of 10.0 km/s. To what height will it rise? Ignore air resistance.

33. A system consists of three particles, each of mass 5.00 g, located at the corners of an equilateral triangle with sides of 30.0 cm. (a) Calculate the potential energy of the system.

(b) If the particles are released simultaneously, where will they collide?

Section 7.8   Energy Diagrams and Stability of Equilibrium (Optional)

34. A right circular cone can be balanced on a horizontal surface in three different ways. Sketch these three equilibrium configurations and identify them as positions of stable, unstable, or neutral equilibrium.

35. For the potential energy curve shown in Figure P7.35, (a) determine whether the force F_x is positive, negative, or zero at the five points indicated. (b) Indicate points of stable, unstable, and neutral equilibrium. (c) Sketch the curve F_x versus x from $x = 0$ to $x = 8$ m.

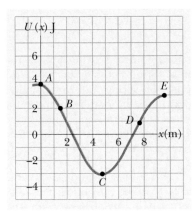

Figure P7.35

36. A hollow pipe has one or two weights attached to its inner surface as shown in Figure P7.36. Characterize each configuration as being stable, unstable, or neutral equilibrium and explain each of your choices.

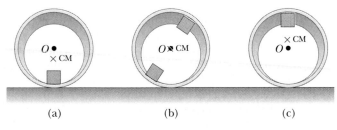

(a)   (b)   (c)

Figure P7.36

37. The potential energy of a two-particle system separated by a distance r is given by $U(r) = A/r$, where A is a constant. Find the radial force $\mathbf{F}_r$.

Additional Problems

38. A 200-g particle is released from rest at point A along the horizontal diameter on the inside of a frictionless, hemispherical bowl of radius $R = 30.0$ cm (Fig. P7.38). Calculate (a) the gravitational potential energy at point A relative to point B, (b) the kinetic energy of the particle at point B, (c) its speed at point B, and (d) its kinetic energy and the potential energy at point C.

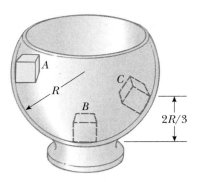

Figure P7.38

39. The particle described in Problem 38 (Fig. P7.38) is released from rest at A, and the surface of the bowl is rough. The speed of the particle at B is 1.50 m/s. (a) What is its kinetic energy at B? (b) How much mechanical energy is lost due to friction as the particle moves from A to B? (c) Is it possible to determine μ from these results in any simple manner? Explain.

40. Make an order-of-magnitude estimate of your power output. Let the work you do be climbing stairs. In your solution state the physical quantities you take as data and the values you measure or estimate for them. Do you consider your peak power or your sustainable power?

41. A 10.0-kg block is released from point A in Figure P7.41. The track is frictionless except for the portion BC, of length 6.00 m. The block travels down the track, hits a spring of force constant $k = 2250$ N/m, and compresses it 0.300 m from its equilibrium position before coming to rest momentarily. Determine the coefficient of kinetic friction between surface BC and block.

42. A child's pogo stick (Fig. P7.42) stores energy in a spring ($k = 2.50 \times 10^4$ N/m). At position A ($x_1 = -0.100$ m), the spring compression is a maximum and the child is momentarily at rest. At position B ($x = 0$), the spring is relaxed and the child is moving upward. At position C, the child is again momentarily at rest at the top of the jump. Assuming that the combined mass of child and pogo stick is 25.0 kg, (a) calculate the total energy of the system if both potential energies are zero at $x = 0$, (b) determine x_2, (c) calculate the speed of the child at $x = 0$, (d) determine the

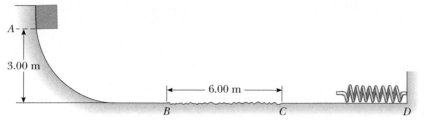

Figure P7.41

value of x for which the kinetic energy of the system is a maximum, and (e) obtain the child's maximum upward speed.

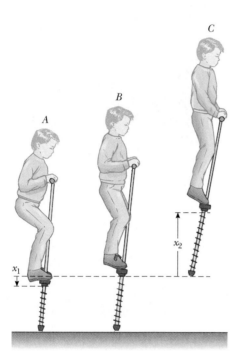

Figure P7.42

43. A 2.00-kg block situated on a rough incline is connected to a spring of negligible mass having a spring constant of 100 N/m (Fig. P7.43). The block is released from rest when the spring is unstretched, and the pulley is frictionless. The block moves 20.0 cm down the incline before coming to rest. Find the coefficient of kinetic friction between block and incline.

44. Suppose the incline is frictionless for the system described in Problem 43 (Fig. P7.43). The block is released from rest with the spring initially unstretched. (a) How far does it move down the incline before coming to rest? (b) What is its acceleration at its lowest point? Is the acceleration constant? (c) Describe the energy transformations that occur during the descent.

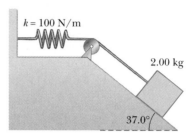

Figure P7.43

45. A 20.0-kg block is connected to a 30.0-kg block by a string that passes over a frictionless pulley. The 30.0-kg block is connected to a spring that has negligible mass and a force constant of 250 N/m, as in Figure P7.45. The spring is unstretched when the system is as shown in the figure, and the incline is frictionless. The 20.0-kg block is pulled 20.0 cm down the incline (so that the 30.0-kg block is 40.0 cm above the floor) and released from rest. Find the speed of each block when the 30.0-kg block is again 20.0 cm above the floor (that is, when the spring is unstretched).

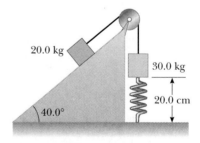

Figure P7.45

46. A potential energy function for a system is given by $U(x) = -x^3 + 2x^2 + 3x$. (a) Determine the force F_x as a function of x. (b) For what values of x is the force equal to zero? (c) Plot $U(x)$ versus x and F_x versus x, and indicate points of stable and unstable equilibrium.

47. A block of mass 0.500 kg is pushed against a horizontal spring of negligible mass, compressing the spring a distance of Δx (Fig. P7.47). The spring constant is 450 N/m. When

released, the block travels along a frictionless, horizontal surface to point B, the bottom of a vertical circular track of radius $R = 1.00$ m, and continues to move up the track. The speed of the block at the bottom of the track is $v_B = 12.0$ m/s, and the block experiences an average frictional force of magnitude 7.00 N while sliding up the track. (a) What is Δx? (b) What is the speed of the block at the top of the track? (c) Does the block reach the top of the track, or does it fall off before reaching the top?

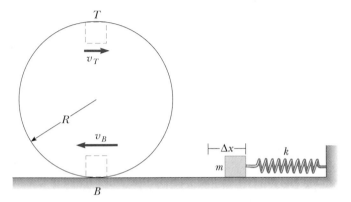

Figure P7.47

48. A 1.00-kg mass slides to the right on a surface having a coefficient of kinetic friction $\mu_k = 0.250$ (Fig. P7.48). The mass has a speed of $v_i = 3.00$ m/s when contact is made with a spring that has a spring constant $k = 50.0$ N/m. The

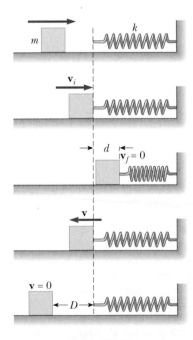

Figure P7.48

mass comes to rest after the spring has been compressed a distance d. The mass is then forced toward the left by the spring and continues to move in that direction beyond the unstretched position. Finally the mass comes to rest a distance D to the left of the unstretched spring. Find (a) the compressed distance d, (b) the speed v at the unstretched position when the system is moving to the left, and (c) the distance D where the mass comes to rest.

49. In the dangerous sport of bungee jumping, a student jumps from a balloon with a specially designed elastic cord attached to his ankles, as shown in the photograph. The unstretched length of the cord is 25.0 m, the student weighs 700 N, and the balloon is 36.0 m above the surface of a river below. Calculate the required force constant of the cord if the student is to stop safely 4.00 m above the river.

Bungee jumping (Problem 49).   *(Gamma)*

50. A uniform chain of length 8.00 m initially lies stretched out on a horizontal table. (a) If the coefficient of static friction between chain and table is 0.600, show that the chain will begin to slide off the table if at least 3.00 m of it hangs over the edge of the table. (b) Determine the speed of the chain as all of it leaves the table, given that the coefficient of kinetic friction between chain and table is 0.400.

51. An object of mass m is suspended from the top of a cart by a string of length L as in Figure P7.51a. The cart and object are initially moving to the right at constant speed v_0. The cart comes to rest after colliding and sticking to a bumper as in Figure P7.51b, and the suspended object swings through an angle θ. (a) Show that $v_0 = \sqrt{2gL(1 - \cos\theta)}$. (b) If $L = 1.20$ m and $\theta = 35.0°$, find the initial speed of the cart. (*Hint:* The force exerted by the string on the object does no work on the object.)

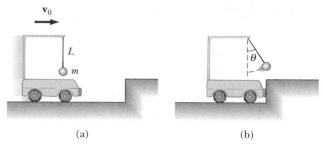

Figure P7.51

Spreadsheet Problems

S1. The potential energy function of a particle is

$$U(x) = \tfrac{1}{2}kx^2 + bx^3 + c$$

where $k = 300$ N/m, $b = -12.0$ N/m^2, and $c = -1000$ J. Use Spreadsheet 7.1 to plot the function from $x =$ -5.00 m to $x = +15.0$ m. Describe the general motion of a particle under the influence of this potential energy function. How does the motion change when the initial energy of the object is increased?

S2. The potential energy function associated with the force between two atoms is modeled by the Lennard-Jones potential

$$U(x) = 4\epsilon \left[\left(\frac{\sigma}{x} \right)^{12} - \left(\frac{\sigma}{x} \right)^{6} \right]$$

In this model, there are two adjustable parameters, σ and ϵ, that are determined from experiments (σ is a range parameter and ϵ is the well depth). Modify Spreadsheet 7.1 to plot $U(x)$ versus x for $\sigma = 0.263$ nm and $\epsilon = 1.51 \times 10^{-22}$ J. Numerically integrate $U(x)$ to find the force F_x.

S3. Using Spreadsheet 7.1, numerically differentiate the potential energy function given in Problem S1. Find the equilibrium points. Are they stable or unstable?

ANSWERS TO CONCEPTUAL PROBLEMS

1. As an athlete runs, chemical energy in the body is converted mostly into thermal energy but also into kinetic energy. The kinetic energy of the athlete just before ascending becomes elastic potential energy stored in the bent pole at the bottom, then gravitational potential energy in the elevated body of the vaulter, and then kinetic energy as he falls. When the vaulter lands on the pad, this energy turns into additional thermal energy, mostly in the pad.

2. The first and third balls speed up after they are thrown, while the second ball first slows down and then speeds up after reaching its peak. The paths of all three are portions of parabolas. The three take different times to reach the ground. However, all have the same impact speed because all start with the same kinetic energy and undergo the same change in gravitational potential energy. In other words, $E_{total} = \tfrac{1}{2}mv^2 + mgy$ is the same for all three balls.

3. Air friction is a nonconservative force, so it will result in a transformation of mechanical energy to internal energy. On the way up, the rock slows down from its projection speed of v to zero. On the way down, it speeds up to v_f. Because of the air friction, the mechanical energy of the system as the rock lands must be less than that when it was projected. The potential energy of the rock-Earth system is the same at the initial and final points, since the two points are at the same height. Thus, the final kinetic energy must be less than the initial kinetic energy. In turn, the final speed, v_f, must be less than the initial speed, v. Finally, since the trip on the way down went from zero velocity to a speed lower than v, the average velocity on the way down must be less than that on the way up, and it must have taken longer for the rock to come down than to go up.

4. If the brakes lock, then the car skids over the roadway. The friction force between the roadway and the tires brings the car to a stop. In this case, the energy is transformed to internal energy in the tires and is also laid out as internal energy along the skid path of the tires. If the brakes do not lock, then the major friction force is between the brake pads and the discs (or the brake shoes and the drums), so that the energy appears as internal energy within the brake structure of the car.

5. The coils change their *internal energy*, since the temperature of the coils rises. The energy transfer mechanism for energy coming into the coils is *electrical transmission*, through the wire plugged into the wall. Energy is transferring out of the coils by *electromagnetic radiation*, since the coils are hot and glowing. There is also some transfer of thermal energy from the hot surfaces of the coils into the air. After a short warm-up period, the coils will stabilize in temperature, and the internal energy will no longer change. In this situation, the energy input and output will be balanced.

6. The rise in temperature of the soft steel is an example of transferring energy into a system by work and having it appear as an increase in the internal energy in the system. This works well for the soft steel because it is *soft*. This softness results in a deformation of the steel under the blow of the hammer. Thus, there is a displacement of the point of application of the force from the hammer, and work is done. With the hard steel, there is much less deformation. Thus, there is less displacement of the point of application of the force, and less work done. The soft steel is thus more efficient at absorbing energy from the hammer by means of work, and its temperature rises more rapidly.

8

Momentum and Collisions

Consider what happens when a golf ball is struck by a club. The ball is given a very large initial velocity as a result of the collision; as a consequence, it is able to travel more than 100 m through the air. The ball experiences a large change in velocity and a correspondingly large acceleration. Furthermore, because the ball experiences this acceleration over a short time interval, the average force on it during the collision is very large. As Newton's third law predicts, the club experiences a reaction force that is equal to and opposite the force on the ball. This reaction force produces a change in the velocity of the club. Because the club is much more massive than the ball, however, the change in velocity of the club is much less than the change in velocity of the ball.

One of the main objectives of this chapter is to enable you to understand and analyze such events. As a first step, we shall introduce the concept of *momentum*, a term that is used in describing objects in motion. For example, a massive football player is often said to have a great deal of momentum as he runs down the field. A much less massive player, such as a halfback, can have equal or greater momentum if his speed is greater than that of the more massive player. This follows from the fact that momentum is defined as the product of mass and velocity.

The concept of momentum leads us to a second conservation law, that

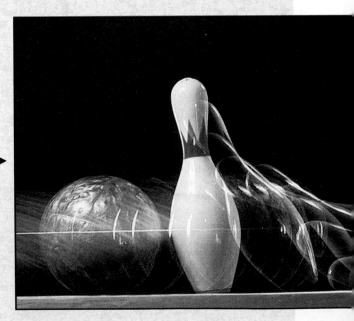

▶ **As a result of the collision between the bowling ball and pin, part of the ball's momentum is transferred to the pin. As a result, the pin acquires momentum and kinetic energy, and the ball loses momentum and kinetic energy. However, the total momentum of the system (ball and pin) remains constant.** *(Ben Rose/The Image Bank)*

of conservation of momentum. This law is especially useful for treating problems that involve collisions between objects.

8.1 • LINEAR MOMENTUM AND ITS CONSERVATION

The **linear momentum** of a particle of mass m moving with a velocity $\mathbf{v}$ is defined to be the product of the mass and velocity:[1]

$$\mathbf{p} \equiv m\mathbf{v} \qquad [8.1]$$

• *Definition of linear momentum of a particle*

Because momentum equals the product of a scalar, m, and a vector, $\mathbf{v}$, it is a vector quantity. Its direction is along $\mathbf{v}$, and it has dimensions of ML/T. In the SI system, momentum has the units kg·m/s.

If a particle is moving in an arbitrary direction in three-dimensional space, $\mathbf{p}$ will have three components, and Equation 8.1 is equivalent to the component equations

$$p_x = mv_x \qquad p_y = mv_y \qquad p_z = mv_z \qquad [8.2]$$

As you can see from its definition, the concept of momentum provides a quantitative distinction between heavy and light particles moving at the same velocity. For example, the momentum of a bowling ball moving at 10 m/s is much greater than that of a tennis ball moving at the same speed. Newton called the product $m\mathbf{v}$ *quantity of motion*, perhaps a more graphic description than *momentum*, which comes from the Latin word for movement.

By using Newton's second law of motion, we can relate the linear momentum of a particle to the resultant force acting on the particle. In Chapter 4 we learned that Newton's second law can be written as $\Sigma\mathbf{F} = m\mathbf{a}$. However, this form applies only when the mass of the system remains constant. In situations in which the mass is changing with time, one must use an alternative statement of Newton's second law: **The time rate of change of momentum of a particle is equal to the resultant force acting on the particle.**

$$\Sigma\mathbf{F} = \frac{d\mathbf{p}}{dt} \qquad [8.3]$$

• *Newton's second law for a particle*

From Equation 8.3 we see that if the resultant force is zero, the time derivative of the momentum is zero, and therefore the momentum of any object must be constant. In other words, the linear momentum of an object is *constant* when $\Sigma\mathbf{F} = 0$. Of course, if the particle is *isolated* (that is, if it does not interact with its environment), then by necessity $\Sigma\mathbf{F} = 0$ and $\mathbf{p}$ remains unchanged.

In a certain sense, conservation of linear momentum is just another way of stating Newton's first law. If an object is in motion, its linear momentum and, as a consequence, its velocity do not change unless an external force acts on the system. A good example is a rocket with its engines disengaged, which coasts through space far from any gravitational sources.

---

[1]This expression is nonrelativistic and is valid only when $v \ll c$, where c is the speed of light.

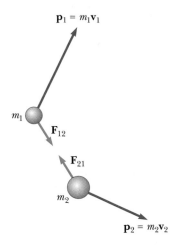

Figure 8.1 At some instant, the momentum of m_1 is $\mathbf{p}_1 = m_1\mathbf{v}_1$ and the momentum of m_2 is $\mathbf{p}_2 = m_2\mathbf{v}_2$. Note that $\mathbf{F}_{12} = -\mathbf{F}_{21}$.

Conservation of Linear Momentum for a Two-Particle System

To see how to apply the principle of *conservation of linear momentum,* consider a system of two particles that can interact with each other but are isolated from their surroundings (Fig. 8.1). That is, the particles may exert a force on each other, but no *external* forces are present. It is important to note the impact of Newton's third law on this analysis. Recall from Chapter 4 that Newton's third law states that the forces on these two particles are equal in magnitude and opposite in direction; that is, *forces always occur in pairs.* Thus, if an *internal* force (say a gravitational force) acts on particle 1, then there must be a second *internal* force, equal in magnitude but opposite in direction, that acts on particle 2.

Suppose that at some instant, the momentum of particle 1 is $\mathbf{p}_1$ and the momentum of particle 2 is $\mathbf{p}_2$. Applying Newton's second law to each particle, we can write

$$\mathbf{F}_{12} = \frac{d\mathbf{p}_1}{dt} \quad \text{and} \quad \mathbf{F}_{21} = \frac{d\mathbf{p}_2}{dt}$$

where $\mathbf{F}_{12}$ is the force exerted on particle 1 by particle 2 and $\mathbf{F}_{21}$ is the force exerted on particle 2 by particle 1. (These forces could be gravitational forces, or they could have some other origin. The source of the forces isn't important for the present discussion.) Newton's third law tells us that $\mathbf{F}_{12}$ and $\mathbf{F}_{21}$ are equal in magnitude and opposite in direction. That is, they form an action–reaction pair, and $\mathbf{F}_{12} = -\mathbf{F}_{21}$. We can also express this condition as

$$\mathbf{F}_{12} + \mathbf{F}_{21} = 0 \quad \text{or as} \quad \frac{d\mathbf{p}_1}{dt} + \frac{d\mathbf{p}_2}{dt} = \frac{d}{dt}(\mathbf{p}_1 + \mathbf{p}_2) = 0$$

Because the time derivative of the total momentum, $\mathbf{p}_{\text{tot}} = \mathbf{p}_1 + \mathbf{p}_2$, is *zero,* we conclude that the *total* momentum, $\mathbf{p}_{\text{tot}}$, must remain constant:

$$\mathbf{p}_{\text{tot}} = \mathbf{p}_1 + \mathbf{p}_2 = \text{constant} \qquad \text{[8.4]}$$

or, equivalently,

$$\mathbf{p}_{1i} + \mathbf{p}_{2i} = \mathbf{p}_{1f} + \mathbf{p}_{2f} \qquad \text{[8.5]}$$

where $\mathbf{p}_{1i}$ and $\mathbf{p}_{2i}$ are initial values and $\mathbf{p}_{1f}$ and $\mathbf{p}_{2f}$ are final values of the momentum during a time period, dt, over which the reaction pair interacts. Equation 8.5 in component form says that the total momenta in the x, y, and z directions are all *independently conserved;* that is,

$$p_{ix} = p_{fx} \qquad p_{iy} = p_{fy} \qquad p_{iz} = p_{fz} \qquad \text{[8.6]}$$

This result is known as the law of **conservation of linear momentum.** It is considered to be one of the most important laws of mechanics. We can state it as follows:

Conservation of momentum •

> Whenever two isolated particles interact with each other, their total momentum remains constant.

That is, **the total momentum of an isolated system at all times equals its initial total momentum.**

We can also describe the law of conservation of momentum in another way. Because we require that the system be isolated, no external forces are present, and the total momentum of the system remains constant. Therefore, momentum conservation is an alternative and more general statement of Newton's third law.

Notice that we have made no statement concerning the nature of the forces acting on the system. The only requirement was that the forces must be *internal* to the system. Thus, momentum is constant for a two-particle system *regardless* of the nature of the internal forces. One can use a similar and equivalent argument to show that the law of conservation of momentum also applies to a system of many particles.

Thinking Physics 1

A baseball is projected into the air at an upward angle to the ground. As it moves through its trajectory, its velocity and, therefore, its momentum constantly changes. Is this a violation of conservation of momentum?

Reasoning The principle of conservation of momentum states that the momentum of a particle or a system of particles is conserved *in the absence of external forces*. A ball projected through the air is subject to the external force of gravity, so we would not expect its momentum to be conserved. If we consider this in a little more detail, the gravitational force is in the vertical direction, so it is actually only the vertical component of the momentum that changes due to this force. In the horizontal direction, there is no force (ignoring air friction), so the horizontal component of the momentum is conserved. In Chapter 3, we used the same idea in stating that the horizontal component of the *velocity* of a projectile remains constant.

If we consider the baseball and the Earth as a system of particles, then the gravitational force is an internal force to this system. The momentum of the ball–Earth system remains unchanged. The downward impulse due to the thrower's feet gives the Earth an initial motion. As the ball rises and falls, the Earth initially sinks but then rises (although imperceptibly) due to the upward gravitational force from the ball, so that the total momentum of the system remains unchanged.

Example 8.1 The Recoiling Pitching Machine

A baseball player uses a pitching machine to help him improve his batting average. He places the 50-kg machine on a frozen pond as in Figure 8.2. The machine fires a 0.15-kg baseball horizontally with a velocity of $36\mathbf{i}$ m/s. What is the recoil velocity of the machine?

Reasoning We take the system to consist of the baseball and the pitching machine. Because of the force of gravity and the normal force, the system is not really isolated. However, both of these forces are directed perpendicularly to the motion of the system. Therefore, momentum is constant in the x direction because there are no external forces in this direction (assuming the surface is frictionless).

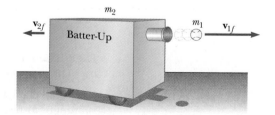

Figure 8.2 (Example 8.1) When the baseball is fired horizontally to the right, the pitching machine recoils to the left. The total momentum of the system before and after firing is zero.

Solution   The total momentum of the system before firing is zero ($m_1\mathbf{v}_{1i} + m_2\mathbf{v}_{2i} = 0$). Therefore, the total momentum after firing must be zero; that is,

$$m_1\mathbf{v}_{1f} + m_2\mathbf{v}_{2f} = 0$$

With $m_1 = 0.15$ kg, $\mathbf{v}_{1i} = 36\mathbf{i}$ m/s, and $m_2 = 50$ kg, solving for $\mathbf{v}_{2f}$, we find the recoil velocity of the pitching machine to be

$$\mathbf{v}_{2f} = -\frac{m_1}{m_2}\mathbf{v}_{1f} = -\left(\frac{0.15 \text{ kg}}{50 \text{ kg}}\right)(36\mathbf{i} \text{ m/s}) = -0.11\mathbf{i} \text{ m/s}$$

The negative sign for $\mathbf{v}_{2f}$ indicates that the pitching machine is moving to the left after firing, in the direction opposite the direction of motion of the baseball. In the words of Newton's third law, for every force (to the left) on the pitching machine, there is an equal but opposite force (to the right) on the ball. Because the pitching machine is much more massive than the ball, the acceleration and consequent speed of the pitching machine are much smaller than the acceleration and speed of the ball.

Example 8.2   Decay of the Kaon at Rest

A meson is a nuclear particle that is more massive than an electron but less massive than a proton or neutron. One type of meson, called the neutral kaon (K^0), decays into a pair of charged pions (π^+ and π^-) that are oppositely charged but equal in mass, as in Figure 8.3. A pion is a particle associated with the strong nuclear force that binds the protons and neutrons together in the nucleus. Assuming the kaon is initially at rest, prove that after the decay, the two pions must have momenta that are equal in magnitude and opposite in direction.

After decay

Figure 8.3   (Example 8.2) A kaon at rest decays spontaneously into a pair of oppositely charged pions. The pions move apart with momenta of equal magnitudes but opposite directions.

Solution   The decay of the kaon, represented in Figure 8.3, can be written

$$K^0 \longrightarrow \pi^+ + \pi^-$$

If we let $\mathbf{p}_+$ be the momentum of the positive pion and $\mathbf{p}_-$ be the momentum of the negative pion after the decay, then the final momentum of the system can be written

$$\mathbf{p}_f = \mathbf{p}_+ + \mathbf{p}_-$$

Because the kaon is at rest before the decay, we know that $\mathbf{p}_i = 0$. Furthermore, because momentum is conserved, $\mathbf{p}_i = \mathbf{p}_f = 0$, so that $\mathbf{p}_+ + \mathbf{p}_- = 0$, or

$$\mathbf{p}_+ = -\mathbf{p}_-$$

Thus, we see that the two momentum vectors of the pions are equal in magnitude and opposite in direction. Furthermore, because the pions have equal mass, they will have equal and opposite velocities.

EXERCISE 1   The momentum of a 1250-kg car is equal to the momentum of a 5000-kg truck traveling at a speed of 10 m/s. What is the speed of the car?   Answer   40.0 m/s

EXERCISE 2   A 1500-kg car moving with a speed of 15 m/s collides with a utility pole and is brought to rest in 0.3 s. Find the average force exerted on the car during the collision. Answer 7.5×10^4 N

EXERCISE 3   A 60-kg boy and a 40-kg girl, both wearing skates, face each other at rest. The girl pushes the boy, sending him eastward with a speed of 4 m/s. Describe the subsequent motion of the girl. (Neglect friction.)   Answer   She moves westward at 6 m/s.

8.2 · IMPULSE AND MOMENTUM

As we have seen, the momentum of a particle changes if a net force acts on the particle. Let us assume that a force $\mathbf{F}$ acts on a particle and that this force may vary with time. According to Newton's second law, $\mathbf{F} = d\mathbf{p}/dt$, or

$$d\mathbf{p} = \mathbf{F}\, dt \qquad [8.7]$$

We can integrate this expression to find the change in the momentum of a particle as it changes from $\mathbf{p}_i$ at time t_i to $\mathbf{p}_f$ at time t_f. Integrating Equation 8.7 gives

$$\Delta\mathbf{p} = \mathbf{p}_f - \mathbf{p}_i = \int_{t_i}^{t_f} \mathbf{F}\, dt \qquad [8.8]$$

The integral of a force over the time for which it acts is called the **impulse** of the force. The impulse of the force $\mathbf{F}$ for the time interval $\Delta t = t_f - t_i$ is a vector defined by

$$\mathbf{I} \equiv \int_{t_i}^{t_f} \mathbf{F}\, dt = \Delta\mathbf{p} \qquad [8.9]$$

• *Impulse of a force*

• *Impulse-momentum theorem*

That is, the **impulse** of the force $\mathbf{F}$ equals the change in the momentum of any object.[2] This statement, known as the **impulse-momentum theorem**, is equivalent to Newton's second law. From this definition we see that impulse is a vector quantity having a magnitude equal to the area under the force–time curve, as described in Figure 8.4. In this figure it is assumed that the force varies in time in the general manner shown and is nonzero in the time interval $\Delta t = t_f - t_i$. The direction of the impulse vector is the same as the direction of the change in momentum. Impulse has the dimensions of momentum, ML/T. Note that impulse is *not* a property of the particle itself; rather, it is a measure of the degree to which an external force changes the momentum of the particle. Therefore, when we say that an impulse is given to a particle, it is implied that momentum is transferred from an external agent to that particle.

Because the force can generally vary in time as in Figure 8.4a, it is convenient to define a time-averaged force $\overline{\mathbf{F}}$, given by

$$\overline{\mathbf{F}} \equiv \frac{1}{\Delta t} \int_{t_i}^{t_f} \mathbf{F}\, dt \qquad [8.10]$$

where $\Delta t = t_f - t_i$. Therefore, we can express Equation 8.9 as

$$\mathbf{I} = \Delta\mathbf{p} = \overline{\mathbf{F}}\,\Delta t \qquad [8.11]$$

This average force, described in Figure 8.4b, can be thought of as the constant force that would give the same impulse to the particle in the time interval Δt as the actual time-varying force gives over this same interval.

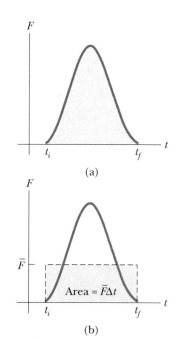

Figure 8.4 (a) A force acting on a particle may vary in time. The impulse is the area under the force-versus-time curve. (b) The average force (horizontal dashed line) gives the same impulse to the particle in the time Δt as the time-varying force described in (a).

[2]In general, if several forces act on a particle, the total impulse injected into the particle by the forces acting on it constitutes the change in momentum of the particle.

In principle, if **F** is known as a function of time, the impulse can be calculated from Equation 8.9. The calculation becomes especially simple if the force acting on the particle is constant. In this case, $\overline{\mathbf{F}} = \mathbf{F}$ and Equation 8.11 becomes

$$\mathbf{I} = \Delta\mathbf{p} = \mathbf{F}\,\Delta t \qquad\qquad [8.12]$$

In many physical situations, we shall use what is called the **impulse approximation: We assume that one of the forces exerted on a particle acts for a short time but is much greater than any other force present.** This approximation is especially useful in treating collisions, where the duration of the collision is very short. When this approximation is made, we refer to the force as an *impulsive force.* For example, when a baseball is struck with a bat, the duration of the collision is about 0.01 s, and the average force the bat exerts on the ball in this time is typically several thousand newtons. This is much greater than the force of gravity, so the impulse approximation is justified. It is important to remember that $\mathbf{p}_i$ and $\mathbf{p}_f$ represent the momenta *immediately* before and after the collision, respectively. Therefore, in the impulse approximation very little motion of the particle takes place during the collision.

Thinking Physics 2

A race car travels rapidly around a circular race track at constant speed. For a given portion of the track, what is the direction of the impulse vector? Once the car returns to the starting point, there is no net change in momentum. Thus, according to the impulse-momentum theorem, there must be no impulse. Yet a nonzero force acted over the time interval for one rotation around the track. How can the impulse be zero?

Reasoning For the movement around a short portion of the track, the direction of the impulse vector is the same as that of the change in momentum. The direction of the vector representing the change in momentum is the same as the direction of the vector representing the change in velocity. From our discussion of circular motion, we know that the direction of the vector representing the change in velocity is that of the acceleration vector, which is toward the center of the circle, because this is uniform circular motion. Thus, the impulse vector is directed toward the center of the circle.

If the car makes one rotation around the track, there is indeed no net change in momentum, so there must be zero net impulse. Keeping in mind that the force on the car causing the circular motion is the friction between the tires and the roadway, and that this force is always directed toward the center of the circle, we can argue the zero value of the total impulse. For every location of the car on the track, there is another point of its motion diametrically opposed across the circle, at which its force vector will be directed in the opposite direction. Thus, as we add up the vector impulses **F** *dt* around the circle, they will cancel in pairs for a net impulse of zero.

CONCEPTUAL PROBLEM 1

In boxing matches of the nineteenth century, bare fists were used. In modern boxing, fighters wear padded gloves. How does this better protect the brain of the boxer from injury? Boxers often "roll with the punch." How does this maneuver protect their health?

Example 8.3   How Good Are the Bumpers?

In a particular crash test, an automobile of mass 1500 kg collides with a wall, as in Figure 8.5. The initial and final velocities of the automobile are $\mathbf{v}_i = -15.0\mathbf{i}$ m/s and $\mathbf{v}_f = 2.6\mathbf{i}$ m/s. If the collision lasts for 0.150 s, find the impulse due to the collision and the average force exerted on the automobile.

Solution   The initial and final momenta of the automobile are

$$\mathbf{p}_i = m\mathbf{v}_i = (1500 \text{ kg})(-15.0\mathbf{i} \text{ m/s}) = -2.25 \times 10^4\mathbf{i} \text{ kg} \cdot \text{m/s}$$

$$\mathbf{p}_f = m\mathbf{v}_f = (1500 \text{ kg})(2.6\mathbf{i} \text{ m/s}) = 0.39 \times 10^4\mathbf{i} \text{ kg} \cdot \text{m/s}$$

Hence, the impulse is

$$\mathbf{I} = \Delta\mathbf{p} = \mathbf{p}_f - \mathbf{p}_i$$
$$= 0.39 \times 10^4\mathbf{i} \text{ kg} \cdot \text{m/s} - (-2.25 \times 10^4\mathbf{i} \text{ kg} \cdot \text{m/s})$$
$$\mathbf{I} = \boxed{2.64 \times 10^4\mathbf{i} \text{ kg} \cdot \text{m/s}}$$

The average force exerted on the automobile is

$$\overline{\mathbf{F}} = \frac{\Delta\mathbf{p}}{\Delta t} = \frac{2.64 \times 10^4\mathbf{i} \text{ kg} \cdot \text{m/s}}{0.150 \text{ s}} = \boxed{1.76 \times 10^5\mathbf{i} \text{ N}}$$

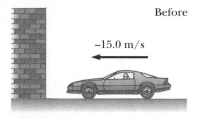

Before

−15.0 m/s

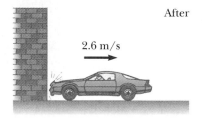

After

2.6 m/s

Figure 8.5   (Example 8.3)

EXERCISE 4   A child bounces a superball on the sidewalk. The linear impulse delivered by the sidewalk to the ball is 2.00 N·s during the 1/800 s of contact. What is the magnitude of the average force exerted on the ball by the sidewalk?   Answer   1.60 kN upward

EXERCISE 5   A 0.15-kg baseball is thrown with a speed of 40 m/s. It is hit straight back at the pitcher with a speed of 50 m/s. (a) What is the impulse delivered to the baseball? (b) Find the average force exerted by the bat on the ball if the two are in contact for 2.0×10^{-3} s.   Answer   (a) 13.5 kg·m/s   (b) 6.75 kN

8.3 · COLLISIONS

In this section we use the law of conservation of momentum to describe what happens when two objects collide. **The force due to the collision is assumed to be much larger than any external forces present.**

A collision may be the result of physical contact between two objects, as described in Figure 8.6a. This is a common observation when two macroscopic objects, such as two billiard balls or a baseball and a bat, collide. The notion of what we mean by *collision* must be generalized because "contact" on a submicroscopic scale is ill-defined and hence meaningless. More accurately, contact forces between two bodies arise from the electrostatic interaction of the electrons in the surface atoms of the bodies.

To understand this distinction between macroscopic and microscopic collisions, consider the collision of a proton with an alpha particle (the nucleus of the helium atom), such as occurs in Figure 8.6b. Because the two particles are positively charged, they repel each other.

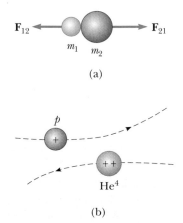

Figure 8.6   (a) The collision between two objects as the result of direct contact. (b) The "collision" between two charged particles.

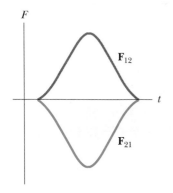

Figure 8.7   The impulse force as a function of time for the two colliding particles described in Figure 8.6a. Note that $\mathbf{F}_{12} = -\mathbf{F}_{21}$.

When two particles of masses m_1 and m_2 collide, the collision forces may vary in time in a complicated way (Fig. 8.7). If $\mathbf{F}_{12}$ is the force on m_1 due to m_2, then the change in momentum of m_1 due to the collision is given by Equation 8.8:

$$\Delta\mathbf{p}_1 = \int_{t_i}^{t_f} \mathbf{F}_{12}\, dt$$

Likewise, if $\mathbf{F}_{21}$ is the force exerted on m_2 by m_1, the change in momentum of m_2 is

$$\Delta\mathbf{p}_2 = \int_{t_i}^{t_f} \mathbf{F}_{21}\, dt$$

Newton's third law states that the force on m_1 by m_2 is equal to and opposite the force exerted on m_2 by m_1; that is, $\mathbf{F}_{12} = -\mathbf{F}_{21}$. (This is described graphically in Fig. 8.7.) Hence, we conclude that

$$\Delta\mathbf{p}_1 = \Delta\mathbf{p}_2$$
$$\Delta\mathbf{p}_1 + \Delta\mathbf{p}_2 = 0$$

Because the total momentum of the system is $\mathbf{p} = \mathbf{p}_1 + \mathbf{p}_2$, we conclude that the *change* in the momentum of the system due to the collision is zero:

$$\mathbf{p} = \mathbf{p}_1 + \mathbf{p}_2 = \text{constant}$$

This is precisely what we expect if no external forces are acting on the system (Section 8.2), or if the external forces are assumed to act for too short a time to make a significant change in momentum. However, the result is also valid if we consider the motion just before and just after the collision. Because the forces due to the collision are internal to the system, they do not change the total momentum of the system. Therefore, for any type of collision, the total momentum of a system just before the collision equals the total momentum of a system just after the collision.

CONCEPTUAL PROBLEM 2

You are watching a movie about a superhero and notice that the superhero hovers in the air and throws a piano at some bad guys while remaining stationary in the air. What's wrong with this scenario?

Example 8.4   Carry Collision Insurance

An 1800-kg car stopped at a traffic light is struck from the rear by a 900-kg car and the two become entangled. If the smaller car was moving at 20 m/s before the collision, what is the speed of the entangled mass after the collision?

Reasoning   The total momentum of the system (the two cars) before the collision equals the total momentum of the system after the collision because momentum is conserved.

Solution   The magnitude of the total momentum of the system before the collision is equal to that of the smaller car because the larger car is initially at rest:

$$p_i = m_i v_i = (900 \text{ kg})(20 \text{ m/s}) = 1.80 \times 10^4 \text{ kg·m/s}$$

After the collision, the mass that moves is the sum of the masses of the cars. The magnitude of the momentum of the combination is

$$p_f = (m_1 + m_2)v_f = (2700 \text{ kg})v_f$$

Equating the momentum before to the momentum after and solving for v_f, the speed of the combined mass, we have

$$v_f = \frac{p_i}{m_1 + m_2} = \frac{1.80 \times 10^4 \text{ kg·m/s}}{2700 \text{ kg}} = 6.67 \text{ m/s}$$

8.4 · ELASTIC AND INELASTIC COLLISIONS IN ONE DIMENSION

As we have seen, momentum is conserved in any type of collision. Kinetic energy, however, is generally *not* constant in a collision because some of it is converted to thermal energy, to internal elastic potential energy when the objects are deformed, and to rotational kinetic energy.

We define an inelastic collision as one in which total kinetic energy is not constant (even though momentum is constant). The collision of a rubber ball with a hard surface is inelastic, because some of the kinetic energy of the ball is lost when it is deformed while in contact with the surface. When two objects collide and stick together after a collision, some kinetic energy is lost, and the collision is called **perfectly inelastic.** For example, if two vehicles collide and become entangled, as in Example 8.4, they move with some common velocity after the perfectly inelastic collision. If a meteorite collides with the Earth, it becomes buried and the collision is perfectly inelastic.

An elastic collision is defined as one in which total kinetic energy is constant (as well as momentum). Real collisions in the macroscopic world, such as those between billiard balls, are only approximately elastic because some deformation and loss of kinetic energy takes place. Billiard-ball collisions and the collisions of air molecules with the walls of a container at ordinary temperatures are highly elastic. Truly elastic collisions do occur, however, between atomic and subatomic particles. Elastic and perfectly inelastic collisions are *limiting* cases; most collisions fall in a category between them.

In the remainder of this section we treat collisions in one dimension and consider the two extreme cases—perfectly inelastic and elastic collisions. The important distinction between these two types of collisions is that **momentum is conserved in all cases, but kinetic energy is constant only in elastic collisions.**

* *Inelastic collision*

* *Elastic collision*

Perfectly Inelastic Collisions

Consider two objects of masses m_1 and m_2 moving with initial velocities $\mathbf{v}_{1i}$ and $\mathbf{v}_{2i}$ along a straight line, as in Figure 8.8. If the two objects collide head-on, stick together, and move with some common velocity $\mathbf{v}_f$ after the collision, the collision is perfectly inelastic. Because the total momentum before the collision equals the total momentum of the composite system after the collision, we have

$$m_1\mathbf{v}_{1i} + m_2\mathbf{v}_{2i} = (m_1 + m_2)\mathbf{v}_f \qquad [8.13]$$

$$\mathbf{v}_f = \frac{m_1\mathbf{v}_{1i} + m_2\mathbf{v}_{2i}}{m_1 + m_2} \qquad [8.14]$$

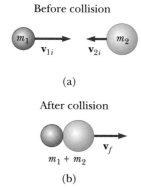

Before collision

(a)

After collision

(b)

Figure 8.8 A perfectly inelastic head-on collision between two particles.

Elastic Collisions

Now consider two particles that undergo an elastic head-on collision (Fig. 8.9). In this case, both momentum and kinetic energy are constant; therefore, we can write

$$m_1\mathbf{v}_{1i} + m_2\mathbf{v}_{2i} = m_1\mathbf{v}_{1f} + m_2\mathbf{v}_{2f} \qquad [8.15]$$

$$\tfrac{1}{2}m_1 v_{1i}^2 + \tfrac{1}{2}m_2 v_{2i}^2 = \tfrac{1}{2}m_1 v_{1f}^2 + \tfrac{1}{2}m_2 v_{2f}^2 \qquad [8.16]$$

Before collision

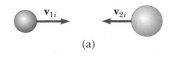

(a)

After collision

(b)

Figure 8.9 An elastic head-on collision between two particles.

where v is positive if a particle moves to the right and negative if it moves to the left.

In a typical problem involving elastic collisions, there are two unknown quantities, and Equations 8.15 and 8.16 can be solved simultaneously to find them. An alternative approach, employing a little mathematical manipulation of Equation 8.16, often simplifies this process. To see this, let us cancel the factor of $\frac{1}{2}$ in Equation 8.16 and rewrite it as

$$m_1(v_{1i}^2 - v_{1f}^2) = m_2(v_{2f}^2 - v_{2i}^2)$$

Here we have moved the terms containing m_1 to one side of the equation and those containing m_2 to the other. Next, let us factor both sides:

$$m_1(v_{1i} - v_{1f})(v_{1i} + v_{1f}) = m_2(v_{2f} - v_{2i})(v_{2f} + v_{2i}) \qquad \textbf{[8.17]}$$

We now separate the terms containing m_1 and m_2 in the equation for the conservation of momentum (Eq. 8.15) to get

$$m_1(v_{1i} - v_{1f}) = m_2(v_{2f} - v_{2i}) \qquad \textbf{[8.18]}$$

To obtain our final result, we divide Equation 8.17 by Equation 8.18 and get

$$v_{1i} + v_{1f} = v_{2f} + v_{2i}$$

$$v_{1i} - v_{2i} = -(v_{1f} - v_{2f}) \qquad \textbf{[8.19]}$$

This equation, in combination with the condition for conservation of momentum, can be used to solve problems dealing with elastic collisions. According to Equation 8.19, the relative speed of the two objects before the collision, $v_{1i} - v_{2i}$, equals the negative of their relative speed after the collision, $-(v_{1f} - v_{2f})$.

Suppose that the masses and the initial velocities of both particles are known. Equations 8.15 and 8.16 can be solved for the final speeds in terms of the initial speeds, because there are two equations and two unknowns.

*Elastic collision: relations •
between final and initial
velocities*

$$v_{1f} = \left(\frac{m_1 - m_2}{m_1 + m_2}\right)v_{1i} + \left(\frac{2m_2}{m_1 + m_2}\right)v_{2i} \qquad \textbf{[8.20]}$$

$$v_{2f} = \left(\frac{2m_1}{m_1 + m_2}\right)v_{1i} + \left(\frac{m_2 - m_1}{m_1 + m_2}\right)v_{2i} \qquad \textbf{[8.21]}$$

It is important to remember that the appropriate signs for v_{1i} and v_{2i} must be included in Equations 8.20 and 8.21. For example, if m_2 is moving to the left initially, as in Figure 8.9, then v_{2i} is negative.

Let us consider some special cases. If $m_1 = m_2$, then $v_{1f} = v_{2i}$ and $v_{2f} = v_{1i}$. That is, the particles exchange speeds if they have equal masses. This is what one observes in head-on billiard ball collisions.

If m_2 is initially at rest, $v_{2i} = 0$, and Equations 8.20 and 8.21 become

$$v_{1f} = \left(\frac{m_1 - m_2}{m_1 + m_2}\right)v_{1i} \qquad \textbf{[8.22]}$$

$$v_{2f} = \left(\frac{2m_1}{m_1 + m_2}\right)v_{1i} \qquad \textbf{[8.23]}$$

If m_1 is very large compared with m_2, we see from Equations 8.22 and 8.23 that $v_{1f} \approx v_{1i}$ and $v_{2f} \approx 2v_{1i}$. That is, when a very heavy particle collides head-on with a very light one initially at rest, the heavy particle continues its motion unaltered after the collision, and the light particle rebounds with a speed equal to about twice the initial speed of the heavy particle. An example of such a collision would be that of a moving heavy atom, such as uranium, with a light atom, such as hydrogen.

If m_2 is much larger than m_1, and m_2 is initially at rest, then we find from Equations 8.22 and 8.23 that $v_{1f} \approx -v_{1i}$ and $v_{2f} \approx 0$. That is, when a very light particle collides head-on with a very heavy particle initially at rest, the velocity of the light particle is reversed, and the heavy particle remains approximately at rest. For example, imagine what happens when a marble hits a stationary bowling ball.

CONCEPTUAL PROBLEM 3

A great deal of expense is involved in repairing automobiles after collisions. It seems that much of this money could be saved if cars were built from hard rubber, so that they made approximately elastic collisions and simply bounced off each other without denting. Why are cars not made of hard rubber?

CONCEPTUAL PROBLEM 4

In perfectly inelastic collisions between two objects, under what conditions is all of the original kinetic energy transformed to forms other than kinetic?

The photograph of a car crash test (an inelastic collision) illustrates that much of the car's initial kinetic energy is transformed into the energy it took to damage the vehicle. Why do safety belts and air bags help prevent serious injury in such collisions? *(Courtesy of General Motors)*

Example 8.5 Slowing Down Neutrons by Collisions

In a nuclear reactor, neutrons are produced when a $^{235}_{92}$U atom splits in a process called fission. These neutrons are moving at about 10^7 m/s and must be slowed down to about 10^3 m/s before they take part in another fission event. They are slowed down by being passed through a solid or liquid material called a *moderator*. The slowing-down process involves elastic collisions. Let us show that a neutron can lose most of its kinetic energy if it collides elastically with a moderator containing light nuclei, such as deuterium (in "heavy water," D_2O) or carbon (in graphite).

Reasoning Because momentum and energy are constant, Equations 8.22 and 8.23 can be applied to the head-on collision of a neutron with the moderator nucleus.

Solution Let us assume that the moderator nucleus of mass m_m is at rest initially and that the neutron of mass m_n and initial speed v_{ni} collides head-on with it. The initial kinetic energy of the neutron is

$$K_{ni} = \tfrac{1}{2}m_n v_{ni}^2$$

After the collision, the neutron has a kinetic energy $\tfrac{1}{2}m_n v_{nf}^2$, where v_{nf} is given by Equation 8.22.

$$K_{nf} = \tfrac{1}{2}m_n v_{nf}^2 = \frac{m_n}{2}\left(\frac{m_n - m_m}{m_n + m_m}\right)^2 v_{ni}^2$$

Therefore, the fraction of the total kinetic energy possessed by the neutron after the collision is

$$(1) \quad f_n = \frac{K_{nf}}{K_{ni}} = \left(\frac{m_n - m_m}{m_n + m_m}\right)^2$$

From this result, we see that the final kinetic energy of the neutron is small when m_m is close to m_n and is zero when $m_n = m_m$.

We can calculate the kinetic energy of the moderator nucleus after the collision using Equation 8.23:

$$K_{mf} = \tfrac{1}{2}m_m v_{mf}^2 = \frac{2m_n^2 m_m}{(m_n + m_m)^2} v_{ni}^2$$

Hence, the fraction of the total kinetic energy transferred to the moderator nucleus is

$$(2) \quad f_m = \frac{K_{mf}}{K_{ni}} = \frac{4m_n m_m}{(m_n + m_m)^2}$$

Because the total energy is constant, (2) can also be obtained from (1) with the condition that $f_n + f_m = 1$, so that $f_m = 1 - f_n$.

Suppose that heavy water is used for the moderator. For collisions of the neutrons with deuterium nuclei in D_2O ($m_m = 2m_n$), $f_n = 1/9$ and $f_m = 8/9$. That is, 89% of the neutron's kinetic energy is transferred to the deuterium nucleus. In practice, the moderator efficiency is reduced because head-on collisions are very unlikely to occur.

How do the results differ with graphite as the moderator?

Example 8.6   A Two-Body Collision with Spring

A block of mass $m_1 = 1.60$ kg, initially moving to the right with a speed of 4.00 m/s on a frictionless horizontal track, collides with a spring attached to a second block of mass $m_2 = 2.10$ kg, moving to the left with a speed of 2.50 m/s, as in Figure 8.10a. The spring has a spring constant of 600 N/m. (a) At the instant when m_1 is moving to the right with a speed of 3.00 m/s, as in Figure 8.10b, determine the speed of m_2.

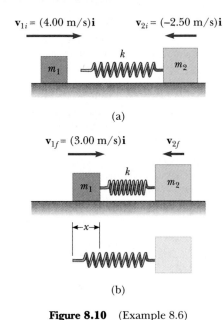

$\mathbf{v}_{1i} = (4.00$ m/s$)\mathbf{i}$ $\mathbf{v}_{2i} = (-2.50$ m/s$)\mathbf{i}$

k

m_1 m_2

(a)

$\mathbf{v}_{1f} = (3.00$ m/s$)\mathbf{i}$ $\mathbf{v}_{2f}$

k

m_1 m_2

$\leftarrow x \rightarrow$

(b)

Figure 8.10   (Example 8.6)

Solution First, note that the initial velocity of m_2 is -2.50 m/s because its direction is to the left. Because the total momentum is conserved, we have

$$m_1 v_{1i} + m_2 v_{2i} = m_1 v_{1f} + m_2 v_{2f}$$

$(1.60$ kg$)(4.00$ m/s$) + (2.10$ kg$)(-2.50$ m/s$)$
$$= (1.60$ kg$)(3.00$ m/s$) + (2.10$ kg$)v_{2f}$$

$$v_{2f} = -1.74 \text{ m/s}$$

The negative value for v_{2f} means that m_2 is still moving to the left at the instant we are considering.

(b) Determine the distance the spring is compressed at that instant.

Solution To determine the compression in the spring, *x*, shown in Figure 8.10b, we use conservation of energy, because there are no friction or other nonconservative forces acting on the system. Thus, we have

$$\tfrac{1}{2}m_1 v_{1i}^2 + \tfrac{1}{2}m_2 v_{2i}^2 = \tfrac{1}{2}m_1 v_{1f}^2 + \tfrac{1}{2}m_2 v_{2f}^2 + \tfrac{1}{2}kx^2$$

Substituting the given values and the result to part (a) into this expression gives

$$x = 0.173 \text{ m}$$

EXERCISE 6   Find the velocity of m_1 and the compression in the spring at the instant that m_2 is at rest.
Answer   0.719 m/s to the right; 0.251 m

EXERCISE 7   A 2.5-kg mass, moving initially with a speed of 10 m/s, makes a perfectly inelastic head-on collision with a 5.0-kg mass that is initially at rest. (a) Find the final velocity of the composite particle.   (b) How much kinetic energy is lost in the collision?
Answer   (a) 3.33 m/s   (b) 83.4 J

8.5 • TWO-DIMENSIONAL COLLISIONS

In Section 8.1 we showed that the total momentum of a system of particles is constant when the system is isolated (that is, when no external forces act on the system). For a general collision of two particles in three-dimensional space, the conservation-of-momentum law implies that the total momentum in each direction is constant. However, an important subset of collisions takes place in a plane. The game of billiards is a familiar example involving multiple collisions of objects moving on a two-dimensional surface. For such two-dimensional collisions, we obtain two component equations for the conservation of momentum:

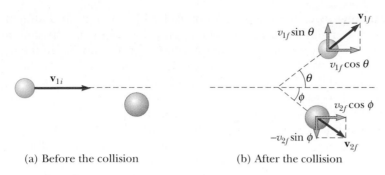

(a) Before the collision   (b) After the collision

Figure 8.11   An elastic glancing collision between two particles.

$$m_1 v_{1ix} + m_2 v_{2ix} = m_1 v_{1fx} + m_2 v_{2fx}$$

$$m_1 v_{1iy} + m_2 v_{2iy} = m_1 v_{1fy} + m_2 v_{2fy}$$

Consider a two-dimensional problem in which a particle of mass m_1 collides elastically with a particle of mass m_2 that is initially at rest, as in Figure 8.11. After the collision, m_1 moves at an angle θ with respect to the horizontal, and m_2 moves at an angle ϕ with respect to the horizontal. This is called a *glancing* collision. Applying the law of conservation of momentum in component form, and noting that the total y component of momentum is zero, we get

x component: $m_1 v_{1i} + 0 = m_1 v_{1f} \cos \theta + m_2 v_{2f} \cos \phi$ **[8.24]**   • *Conservation of momentum*

y component: $0 + 0 = m_1 v_{1f} \sin \theta - m_2 v_{2f} \sin \phi$ **[8.25]**

We now have two independent equations. So long as only two of the preceding quantities are unknown, we can solve the problem completely.

Because the collision is elastic, we can write a third equation for kinetic energy, in the form

$$\tfrac{1}{2} m_1 v_{1i}^2 = \tfrac{1}{2} m_1 v_{1f}^2 + \tfrac{1}{2} m_2 v_{2f}^2$$ **[8.26]**   • *Conservation of energy*

If we know the initial velocity, v_{1i}, and the masses, we are left with four unknowns. Because we have only three equations, one of the four remaining quantities (v_{1f}, v_{2f}, θ, or ϕ) must be given to determine the motion after the collision from conservation principles alone.

If the collision is inelastic, kinetic energy is *not* constant, and Equation 8.26 does *not* apply.

PROBLEM-SOLVING STRATEGY • **Collisions**

The following procedure is recommended when dealing with problems involving collisions between two objects.

1. Set up a coordinate system and define your velocities with respect to that system. It is convenient to have the x axis coincide with one of the initial velocities.
2. In your sketch of the coordinate system, draw and label all velocity vectors and include all the given information. Write expressions for the x and y

components of the momentum of each object before and after the collision. Remember to include the appropriate signs for the components of the velocity vectors. For example, if an object is moving in the negative x direction, its x component of velocity must be taken to be negative. It is essential that you pay careful attention to signs.

3. Write expressions for the *total* momentum in the x direction *before* and *after* the collision, and equate the two. Repeat this procedure for the total momentum in the y direction. These steps follow from the fact that, because the momentum is conserved in any collision, the total momentum in any direction must be constant. It is important to emphasize that it is the momentum of the *system* (the two colliding objects) that is constant, not the momenta of the individual objects.

4. If the collision is inelastic, kinetic energy is *not* constant, and additional information is probably required. If the collision is perfectly inelastic, the final velocities of the two objects are equal. Proceed to solve the momentum equations for the unknown quantities.

5. If the collision is elastic, kinetic energy is constant, and you can equate the total kinetic energy before the collision to the total kinetic energy after the collision. This provides an additional relationship between the velocities.

Example 8.7   Proton–Proton Collision

A proton collides in an elastic fashion with another proton that is initially at rest. The incoming proton has an initial speed of 3.5×10^5 m/s and makes a glancing collision with the second proton, as in Figure 8.11. (At close separations, the protons exert a repulsive electrostatic force on each other.) After the collision, one proton moves off at an angle of $37°$ to the original direction of motion, and the second deflects at an angle of ϕ to the same axis. Find the final speeds of the two protons and the angle ϕ.

Solution   Both momentum and kinetic energy are constant in this glancing elastic collision. Because $m_1 = m_2$, $\theta = 37°$, and we are given $v_{1i} = 3.5 \times 10^5$ m/s, Equations 8.24, 8.25, and 8.26 become

$$v_{1f}\cos 37° + v_{2f}\cos \phi = 3.5 \times 10^5$$

$$v_{1f}\sin 37° - v_{2f}\sin \phi = 0$$

$$v_{1f}^2 + v_{2f}^2 = (3.5 \times 10^5)^2$$

Simultaneous solution of these three equations with three unknowns gives

$$v_{1f} = 2.80 \times 10^5 \text{ m/s} \qquad v_{2f} = 2.11 \times 10^5 \text{ m/s}$$

$$\phi = 53.0°$$

It is interesting to note that $\theta + \phi = 90°$. This result is *not* accidental. **Whenever two equal masses collide elastically in a glancing collision and one of them is initially at rest, their final velocities are *always* at right angles to each other.**

Example 8.8   Collision at an Intersection

A 1500-kg car traveling east with a speed of 25.0 m/s collides at an intersection with a 2500-kg van traveling north at a speed of 20.0 m/s, as shown in Figure 8.12. Find the direction and magnitude of the velocity of the wreckage after the collision, assuming that the vehicles undergo a perfectly inelastic collision (that is, they stick together).

Solution   Let us choose east to be along the positive x direction and north to be along the positive y direction, as in Figure 8.12. Before the collision, the only object having momen-

tum in the x direction is the car. Thus, the magnitude of the total initial momentum of the system (car plus van) in the x direction is

$$\sum p_{xi} = (1500 \text{ kg})(25.0 \text{ m/s}) = 3.75 \times 10^4 \text{ kg·m/s}$$

Now let us assume that the wreckage moves at an angle θ and speed v after the collision. The magnitude of the total momentum in the x direction after the collision is

$$\sum p_{xf} = (4000 \text{ kg})v \cos \theta$$

Because the total momentum in the x direction is constant, we can equate these two equations to get

(1) (3.75×10^4) kg·m/s = $(4000$ kg$)v \cos \theta$

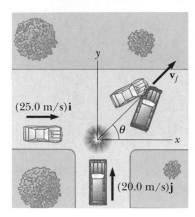

Figure 8.12 (Example 8.8) Top view of an eastbound car colliding with a northbound van.

In a similar way, the total initial momentum of the system in the y direction is that of the van, the magnitude of which is equal to $(2500$ kg$)(20.0$ m/s$)$. Applying conservation of momentum to the y direction, we have

$$\sum p_{yi} = \sum p_{yf}$$

$$(2500 \text{ kg})(20.0 \text{ m/s}) = (4000 \text{ kg})v \sin \theta$$

(2) 5.00×10^4 kg·m/s = $(4000$ kg$)v \sin \theta$

If we divide (2) by (1), we get

$$\tan \theta = \frac{5.00 \times 10^4}{3.75 \times 10^4} = 1.33$$

$$\theta = 53.1°$$

When this angle is substituted into (2), the value of v is

$$v = \frac{5.00 \times 10^4 \text{ kg·m/s}}{(4000 \text{ kg}) \sin 53°} = 15.6 \text{ m/s}$$

EXERCISE 8 A 3.00-kg mass with an initial velocity of $5.00\mathbf{i}$ m/s collides with and sticks to a 2.00-kg mass with an initial velocity of $-3.00\mathbf{j}$ m/s. Find the final velocity of the composite mass.   Answer $(3.00\mathbf{i} - 1.20\mathbf{j})$ m/s

EXERCISE 9 A bird and a bee are approaching each other at right angles. The bird has a mass of 0.125 kg and a speed of 0.600 m/s, and the bee has a mass of 5.00×10^{-3} kg and a speed of 15.0 m/s. If the bird catches the bee, what is the new speed of the bird? Answer   0.816 m/s

8.6 • THE CENTER OF MASS

In this section we describe the overall motion of a mechanical system in terms of a special point called the **center of mass** of the system. The mechanical system can be either a system of particles or any object. We shall see that the center of mass accelerates as if all of the system's mass were concentrated at that point and all external forces acted there. Furthermore, if the resultant external force on the system is **F** and the total mass of the system is M, the center of mass moves with an acceleration given by $\mathbf{a} = \mathbf{F}/M$. That is, the system moves as if the resultant external force were applied to a single particle of mass M located at the center of mass. This result was implicitly assumed in earlier chapters, because nearly all examples referred to the motion of extended objects.

Consider a system consisting of a pair of particles connected by a light, rigid rod (Fig. 8.13). The center of mass is located somewhere on the rod and is closer to the larger mass. If a single force is applied at some point on the rod that is closer to the smaller mass than to the larger one, the system rotates clockwise (Fig. 8.13a). If the force is applied at a point on the rod closer to the larger mass, the system rotates counterclockwise (Fig. 8.13b). If the force is applied at the center of mass,

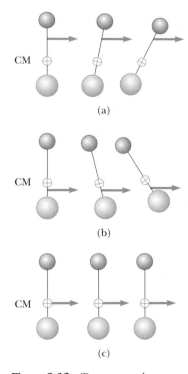

Figure 8.13 Two unequal masses are connected by a light, rigid rod. (a) The system rotates clockwise when a force is applied between the smaller mass and the center of mass. (b) The system rotates counterclockwise when a force is applied between the larger mass and the center of mass. (c) The system moves in the direction of the force without rotating when a force is applied at the center of mass.

This multiflash photograph shows that as the diver executes a back flip, his center of mass follows a parabolic path, the same path that a particle would follow. *(© the Harold E. Edgerton 1992 Trust. Courtesy of Palm Press, Inc.)*

the system moves in the direction of **F** without rotating (Fig. 8.13c). Thus, the center of mass can be easily located.

The position of the center of mass of a system can be described as being the *average position* of the system's mass. For example, the center of mass of the pair of particles described in Figure 8.14 is located on the x axis, somewhere between the particles. The x coordinate of the center of mass in this case is

$$x_{CM} \equiv \frac{m_1 x_1 + m_2 x_2}{m_1 + m_2} \qquad [8.27]$$

For example, if $x_1 = 0$, $x_2 = d$, and $m_2 = 2m_1$, we find that $x_{CM} = \frac{2}{3}d$. That is, the center of mass lies closer to the more massive particle. If the two masses are equal, the center of mass lies midway between the particles.

We can extend the center of mass concept to a system of many particles in three dimensions. The x coordinate of the center of mass of n particles is defined to be

$$x_{CM} \equiv \frac{m_1 x_1 + m_2 x_2 + m_3 x_3 + \ldots + m_n x_n}{m_1 + m_2 + m_3 + \ldots + m_n} = \frac{\Sigma m_i x_i}{\Sigma m_i} \qquad [8.28]$$

where x_i is the x coordinate of the ith particle and Σm_i is the *total mass* of the system. For convenience, we shall express the total mass as $M = \Sigma m_i$, where the sum runs over all n particles. The y and z coordinates of the center of mass are similarly defined by the equations

$$y_{CM} \equiv \frac{\Sigma m_i y_i}{M} \qquad \text{and} \qquad z_{CM} \equiv \frac{\Sigma m_i z_i}{M} \qquad [8.29]$$

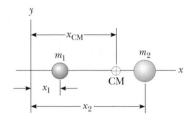

Figure 8.14 The center of mass of two particles on the x axis is located at x_{CM}, a point between the particles, closer to the larger mass.

The center of mass can also be located by its position vector, $\mathbf{r}_{CM}$. The rectangular coordinates of this vector are x_{CM}, y_{CM}, and z_{CM}, defined in Equations 8.28 and 8.29. Therefore,

$$\mathbf{r}_{CM} = x_{CM}\mathbf{i} + y_{CM}\mathbf{j} + z_{CM}\mathbf{k} = \frac{\sum m_i x_i \mathbf{i} + \sum m_i y_i \mathbf{j} + \sum m_i z_i \mathbf{k}}{M}$$

$$\mathbf{r}_{CM} \equiv \frac{\sum m_i \mathbf{r}_i}{M} \qquad \text{[8.30]}$$

where $\mathbf{r}_i$ is the position vector of the *i*th particle, defined by

$$\mathbf{r}_i \equiv x_i\mathbf{i} + y_i\mathbf{j} + z_i\mathbf{k}$$

The center of mass of a homogeneous, symmetric body must lie on an axis of symmetry. For example, the center of mass of a homogeneous rod must lie midway between the ends of the rod. The center of mass of a homogeneous sphere or a homogeneous cube must lie at the geometric center of the object. One can experimentally determine the center of mass of an irregularly shaped object, such as a wrench, by suspending the wrench from two different points (Fig. 8.15). The wrench is first hung from point *A*, and a vertical line, *AB*, is drawn (which can be established with a plumb bob) when the wrench is in equilibrium. The wrench is then hung from point *C*, and a second vertical line, *CD*, is drawn. The center of mass coincides with the intersection of these two lines. In fact, if the wrench is hung freely from any point, the vertical line through that point will pass through the center of mass.

Because a rigid body is a continuous distribution of mass, each portion is acted on by the force of gravity. The net effect of all of these forces is equivalent to the effect of a single force, $M\mathbf{g}$, acting through a special point called the **center of gravity.** If $\mathbf{g}$ is constant over the mass distribution, then the center of gravity coincides with the center of mass. If a rigid body is pivoted at its center of gravity, it will be balanced in any orientation.

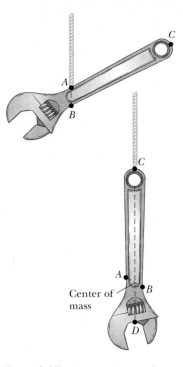

Figure 8.15 An experimental technique for determining the center of mass of a wrench. The wrench is hung freely from two different pivots, *A* and *C*. The intersection of the two vertical lines *AB* and *CD* locates the center of mass.

Example 8.9 The Center of Mass of Three Particles

A system consists of three particles located at the corners of a right triangle as in Figure 8.16. Find the center of mass of the system.

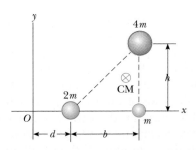

Figure 8.16 (Example 8.9) Locating the center of mass for a system of three particles.

Solution Using the basic defining equations for the coordinates of the center of mass, and noting that $z_{CM} = 0$, we get

$$x_{CM} = \frac{\sum m_i x_i}{M} = \frac{2md + m(d + b) + 4m(d + b)}{7m} = d + \tfrac{5}{7}b$$

$$y_{CM} = \frac{\sum m_i y_i}{M} = \frac{2m(0) + m(0) + 4mh}{7m} = \tfrac{4}{7}h$$

Therefore, we can express the position vector to the center of mass measured from the origin as

$$\mathbf{r}_{CM} = x_{CM}\mathbf{i} + y_{CM}\mathbf{j} + z_{CM}\mathbf{k} = \left(d + \tfrac{5}{7}b\right)\mathbf{i} + \tfrac{4}{7}h\mathbf{j}$$

8.7 • MOTION OF A SYSTEM OF PARTICLES

We can begin to understand the physical significance and utility of the center of mass concept by taking the time derivative of the position vector of the center of mass $\mathbf{r}_{CM}$, given by Equation 8.30. Assuming that M remains constant—that is, no particles enter or leave the system—we get the following expression for the **velocity of the center of mass:**

Velocity of the center of mass •

$$\mathbf{v}_{CM} = \frac{d\mathbf{r}_{CM}}{dt} = \frac{1}{M}\sum m_i \frac{d\mathbf{r}_i}{dt} = \frac{\sum m_i \mathbf{v}_i}{M} \qquad [8.31]$$

where $\mathbf{v}_i$ is the velocity of the ith particle. Rearranging Equation 8.31 gives

Total momentum of a system •
of particles

$$M\mathbf{v}_{CM} = \sum m_i \mathbf{v}_i = \sum \mathbf{p}_i = \mathbf{p}_{tot} \qquad [8.32]$$

This result tells us that **the total momentum of the system equals the total mass multiplied by the velocity of the center of mass**—in other words, the total momentum of a single particle of mass M moving with a velocity $\mathbf{v}_{CM}$.

If we now differentiate Equation 8.32 with respect to time, we get the **acceleration of the center of mass:**

Acceleration of the center of •
mass for a system of particles

$$\mathbf{a}_{CM} = \frac{d\mathbf{v}_{CM}}{dt} = \frac{1}{M}\sum m_i \frac{d\mathbf{v}_i}{dt} = \frac{1}{M}\sum m_i \mathbf{a}_i \qquad [8.33]$$

Rearranging this expression and using Newton's second law, we get

$$M\mathbf{a}_{CM} = \sum m_i \mathbf{a}_i = \sum \mathbf{F}_i \qquad [8.34]$$

where $\mathbf{F}_i$ is the force on particle i.

The forces on any particle in the system may include both external and internal forces. However, by Newton's third law, the force exerted by particle 1 on particle 2, for example, is equal to and opposite the force exerted by particle 2 on particle 1. When we sum over all internal forces in Equation 8.34, they cancel in pairs and the net force on the system is due *only* to external forces. Thus, we can write Equation 8.34 in the form

Newton's second law for a •
system of particles

$$\sum \mathbf{F}_{ext} = M\mathbf{a}_{CM} = \frac{d\mathbf{p}_{tot}}{dt} \qquad [8.35]$$

That is, the resultant external force on the system of particles equals the total mass of the system multiplied by the acceleration of the center of mass. If we compare this to Newton's second law for a single particle, we see that the center of mass moves like an imaginary particle of mass M under the influence of the resultant external force on the system. In the absence of external forces, the center of mass moves with uniform velocity, as in the case of the rotating wrench in Figure 8.17. If the resultant force acts along a line through the center of mass of an extended body such as the wrench, the body is accelerated without rotation, and its kinetic energy is associated entirely with its linear motion. If the resultant force does not act through the center of mass, the body will be rotationally accelerated, and the body will acquire a rotational kinetic energy in addition to the kinetic energy of its linear motion. The linear acceleration of the center of mass is the same in either case, as given by Equation 8.35.

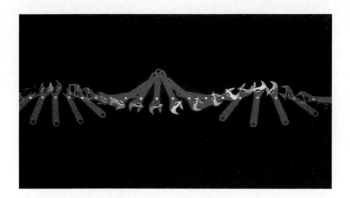

Figure 8.17 Multiflash photograph of a wrench moving on a horizontal surface. The center of mass of the wrench moves in a straight line as the wrench rotates about this point as marked with a white dot. *(Richard Megna, Fundamental Photographs)*

Finally, we see that if the resultant external force is zero, then from Equation 8.35 it follows that

$$\frac{d\mathbf{p}_{tot}}{dt} = M\mathbf{a}_{CM} = 0$$

so that

$$\mathbf{p}_{tot} = M\mathbf{v}_{CM} = \text{constant} \qquad \left(\text{when } \sum \mathbf{F}_{ext} = 0\right) \qquad [8.36]$$

That is, the total linear momentum of a system of particles is constant if there are no external forces acting on the system. It follows that, for an *isolated* system of particles, both the total momentum and the velocity of the center of mass are constant in time. The law of conservation of momentum that was derived in Section 8.1 for a two-particle system is thus generalized to a many-particle system.

Suppose an isolated system consisting of two or more members is at rest. The center of mass of such a system remains at rest unless acted on by an external force. For example, consider a system, initially at rest, made up of a swimmer and a raft. When the swimmer dives off the raft, the center of mass of the system remains at rest (if we neglect friction between raft and water). Furthermore, the horizontal momentum of the diver is equal in magnitude to the momentum of the raft, but opposite in direction.

As another example, suppose an unstable atom initially at rest suddenly decays into two fragments of masses M_1 and M_2, with velocities $\mathbf{v}_1$ and $\mathbf{v}_2$, respectively. Because the total momentum of the system before the decay is zero, the total momentum of the system after the decay must also be zero. Therefore, we see that $M_1\mathbf{v}_1 + M_2\mathbf{v}_2 = 0$. If the velocity of one of the fragments after the decay is known, the recoil velocity of the other fragment can be calculated.

Thinking Physics 3

A boy stands at one end of a floating canoe that is stationary relative to the shore (Fig. 8.18). He then walks to the opposite end of the canoe, away from the shore. Does the canoe move?

Reasoning Yes, the canoe moves toward the shore. Neglecting friction between the canoe and water, there are no horizontal forces acting on the system consisting of the boy and canoe. Therefore, the center of mass of the system remains fixed relative to the shore (or any stationary point). As the boy moves away from the shore, the canoe

Figure 8.18   (Thinking Physics 3)

must move toward the shore such that the center of mass of the system remains constant. An alternative explanation is that the momentum of the system remains constant if friction is neglected. As the boy acquires a momentum away from the shore, the canoe must acquire an equal momentum toward the shore such that the total momentum of the system is always zero.

CONCEPTUAL PROBLEM 5

Suppose you tranquilize a polar bear on a glacier as part of a research effort. How might you be able to estimate the weight of the polar bear using a measuring tape, a rope, and a knowledge of your own weight?

CONCEPTUAL PROBLEM 6

A projectile is fired into the air and suddenly explodes into several fragments (Fig. 8.19). What can be said about the motion of the center of mass of the fragments after the explosion?

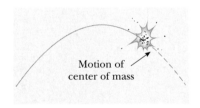

Motion of
center of mass

Figure 8.19   (Conceptual Problem 6)

EXERCISE 10   A 5.0-kg particle moves along the x axis with a velocity of 3.0 m/s. A 2.0-kg particle moves along the x axis with a velocity of -2.5 m/s. Find (a) the velocity of the center of mass and (b) the total momentum of the system.   Answer   (a) 1.4 m/s to the right (b) 10 kg·m/s to the right

EXERCISE 11   A 2.0-kg particle has a velocity $(2.0\mathbf{i} - 3.0\mathbf{j})$ m/s, and a 3.0-kg particle has a velocity $(1.0\mathbf{i} + 6.0\mathbf{j})$ m/s. Find (a) the velocity of the center of mass and (b) the total momentum of the system.   Answer   (a) $(1.4\mathbf{i} + 2.4\mathbf{j})$ m/s   (b) $(7.0\mathbf{i} + 12\mathbf{j})$ kg·m/s

OPTIONAL

8.8 • ROCKET PROPULSION

When ordinary vehicles, such as automobiles and locomotives, are propelled, the driving force for the motion is one of friction. In the case of the automobile, the driving force is the force exerted by the road on the car. A locomotive "pushes" against the tracks; hence, the driving force is the reaction force exerted by the tracks on the locomotive. However, a rocket moving in space has no road or tracks to "push" against. Therefore, the source of the propulsion of a rocket must be different. **The operation of a rocket depends on the law of conservation of**

Figure 8.20 Lift-off of the space shuttle *Columbia.* Enormous thrust is generated by the shuttle's liquid-fueled engines, aided by the two solid-fuel boosters. Many physical principles from mechanics, thermodynamics, and electricity and magnetism are involved in such a launch. *(Courtesy of NASA)*

momentum as applied to a system of particles, where the system is the rocket plus its ejected exhaust gases.

The propulsion of a rocket (Fig. 8.20) can be understood by first considering the mechanical system consisting of a machine gun mounted on a cart on wheels. As the machine gun is fired, each bullet receives a momentum $m\mathbf{v}$ in some direction, where $\mathbf{v}$ is the velocity of the bullet relative to the gun. For each bullet that is fired, the gun and cart must receive a compensating momentum in the opposite direction (neglecting the momentum imparted to the escaping gas in the barrel of the gun). That is, the reaction force exerted by the bullet on the gun accelerates the cart and gun. If there are n bullets fired each second, then the average force exerted on the gun is $\mathbf{F}_{av} = nm\mathbf{v}$.

In a similar manner, as a rocket moves in free space (a vacuum), **its momentum changes when some of its mass is released in the form of ejected gases. Because the ejected gases acquire some momentum, the rocket receives a compensating momentum in the opposite direction.** Therefore, **the rocket is accelerated as a result of the "push," or thrust, from the exhaust gases.** In free space, the center of mass of the system moves uniformly, independent of the propulsion process. It is interesting to note that the rocket and machine gun represent cases of the *inverse* of an inelastic collision—that is, momentum is conserved, but the kinetic energy of the system is *increased* (at the expense of potential energy carried in the fuel).

Suppose that at some time t, the momentum of the rocket plus the fuel is $(M + \Delta m)v$ (Fig. 8.21a). At some short time later, Δt, the rocket ejects fuel of mass Δm and the rocket's speed therefore increases to $v + \Delta v$ (Fig. 8.21b). If the fuel is ejected with a velocity v_e *relative to the rocket*, then the speed of the fuel relative to a stationary frame of reference is $v - v_e$. Thus, if we equate the total initial momentum of the system with the total final momentum, we get

$$(M + \Delta m)v = M(v + \Delta v) + \Delta m(v - v_e)$$

Simplifying this expression gives

$$M \Delta v = \Delta m(v_e)$$

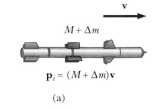

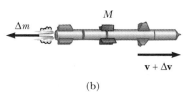

Figure 8.21 Rocket propulsion. (a) The initial mass of the rocket is $M + \Delta m$ at a time t, and its speed is v. (b) At a time $t + \Delta t$, the rocket's mass has reduced to M, and an amount of fuel Δm has been ejected. The rocket's speed increases by an amount Δv.

We can also arrive at this result by considering the system in the center-of-mass frame of reference—that is, a frame the velocity of which equals the center-of-mass velocity. In this frame, the total momentum is zero; therefore, if the rocket gains a momentum $M \Delta v$ by ejecting some fuel, the exhaust gases obtain a momentum $\Delta m(v_e)$ in the *opposite* direction, and so $M \Delta v - \Delta m(v_e) = 0$. If we now take the limit as Δt goes to zero, the $\Delta v \rightarrow dv$ and $\Delta m \rightarrow dm$. Furthermore, the increase in the exhaust mass, dm, corresponds to an equal decrease in the rocket mass, so that $dm = -dM$. Note that dM is given a negative sign because it represents a decrease in mass. Using this fact, we get

$$M \, dv = - v_e \, dM \qquad \text{[8.37]}$$

Integrating this equation and taking the initial mass of the rocket plus fuel to be M_i and the final mass of the rocket plus its remaining fuel to be M_f we get

$$\int_{v_i}^{v_f} dv = - v_e \int_{M_i}^{M_f} \frac{dM}{M}$$

$$v_f - v_i = v_e \ln\!\left(\frac{M_i}{M_f}\right) \qquad \text{[8.38]}$$

Basic expression for rocket • propulsion

This is the basic expression of rocket propulsion. First, it tells us that the increase in speed is proportional to the exhaust speed, v_e. Therefore, the exhaust velocity should be very high. Second, the increase in speed is proportional to the logarithm of the ratio M_i/M_f. Therefore, this ratio should be as large as possible, which means that the rocket should carry as much fuel as possible.

The **thrust** on the rocket is the force exerted on the rocket by the ejected exhaust gases. We can obtain an expression for the thrust from Equation 8.37:

$$\text{Thrust} = M \frac{dv}{dt} = \left| v_e \frac{dM}{dt} \right| \qquad \text{[8.39]}$$

Here we see that the thrust increases as the exhaust speed increases and as the rate of change of mass (burn rate) increases.

Thinking Physics 4

When Robert Goddard proposed the possibility of rocket-propelled vehicles, the *New York Times* agreed that such vehicles would be useful and successful within the Earth's atmosphere ("Topics of the Times," *New York Times,* Jan. 13, 1920, p. 12). But the *Times* balked at the idea of using such a rocket in the vacuum of space, noting that, ". . . its flight would be neither accelerated nor maintained by the explosion of the charges it then might have left. To claim that it would be is to deny a fundamental law of dynamics . . . That Professor Goddard, with his 'chair' in Clark College and the countenancing of the Smithsonian Institution, does not know the relation of action to reaction, and of the need to have something better than a vacuum against which to react—to say that would be absurd."

What did the writer of these passages overlook?

Reasoning The writer was making a common mistake in believing that a rocket works by expelling gases that *push* on something, propelling the rocket forward. With this belief, it is impossible to see how a rocket fired in empty space *would* work.

Gases do not need to push on anything—it is the act of expelling the gases itself that pushes the rocket forward. This can be argued from Newton's third law—the rocket pushes the gases backward, resulting in the gases pushing the rocket forward. It can also be argued from conservation of momentum—as the gases gain momentum in one direction, the rocket must gain momentum in the opposite direction to conserve the original momentum of the rocket–gas system.

The *New York Times* did publish a retraction 49 years later ("A Correction," *New York Times,* July 17, 1969, p. 43), at the same time that the Apollo astronauts were on their way to the Moon. It appeared on a page with two other articles, one titled, "Fundamentals of Space Travel," and the other, "Spacecraft, Like Squid, Maneuver by 'Squirts.'" The following passages were part of these articles: ". . . an editorial feature of the *New York Times* dismissed the notion that a rocket could function in a vacuum and commented on the ideas of Robert H. Goddard. . . . Further investigation and experimentation have confirmed the findings of Isaac Newton in the 17th century, and it is now definitely established that a rocket can function in a vacuum as well as in an atmosphere. The *Times* regrets the error."

Example 8.10   A Rocket in Space

A rocket moving in free space has a speed of 3.0×10^3 m/s relative to Earth. Its engines are turned on, and exhaust is ejected in a direction opposite the rocket's motion at a speed of 5.0×10^3 m/s relative to the rocket. (a) What is the speed of the rocket relative to Earth once its mass is reduced to one half its mass before ignition?

Solution   Applying Equation 8.38, we get

$$v_f = v_i + v_e \ln \left(\frac{M_i}{M_f} \right)$$

$$= 3.0 \times 10^3 + 5.0 \times 10^3 \ln \left(\frac{M_i}{0.5M_i} \right)$$

$$= 6.5 \times 10^3 \text{ m/s}$$

(b) What is the thrust on the rocket if it burns fuel at the rate of 50 kg/s?

Solution

$$\text{Thrust} = \left| v_e \frac{dM}{dt} \right| = \left(5.0 \times 10^3 \, \frac{\text{m}}{\text{s}} \right) \left(50 \, \frac{\text{kg}}{\text{s}} \right)$$

$$= 2.5 \times 10^5 \text{ N}$$

EXERCISE 12   A rocket engine consumes 80 kg of fuel per second. If the exhaust speed is 2.5×10^3 m/s, calculate the thrust on the rocket.   Answer   200 kN

SUMMARY

The **linear momentum** of any object of mass m moving with a velocity $\mathbf{v}$ is

$$\mathbf{p} \equiv m\mathbf{v} \qquad [8.1]$$

Conservation of momentum applied to two interacting objects states that if the two objects form an isolated system, their total momentum is constant regardless of the nature of the force between them. Therefore, the total momentum of the system at all times equals its initial total momentum:

$$\mathbf{p}_{1i} + \mathbf{p}_{2i} = \mathbf{p}_{1f} + \mathbf{p}_{2f} \qquad [8.5]$$

The **impulse** of a force $\mathbf{F}$ on any object is equal to the change in the momentum of the object and is given by

$$\mathbf{I} \equiv \int_{t_i}^{t_f} \mathbf{F} \, dt = \Delta\mathbf{p} = \mathbf{p}_f - \mathbf{p}_i \qquad [8.9]$$

This is known as the **impulse-momentum theorem.**

Impulsive forces are forces that are very strong compared with other forces on the system. They usually act for a short time, as in the case of collisions.

When two bodies collide, the total momentum of the system before the collision always equals the total momentum after the collision, regardless of the nature of the collision. An **inelastic collision** is one for which kinetic energy is not constant, but momentum is. A perfectly inelastic collision is one in which the colliding bodies stick together after the collision. An **elastic collision** is one in which both momentum and kinetic energy are constant.

In a two- or three-dimensional collision, the components of momentum in each of the directions are conserved independently.

The **vector position of the center of mass of a system of particles** is defined as

$$\mathbf{r}_{CM} \equiv \frac{\Sigma m_i \mathbf{r}_i}{M} \qquad \qquad [8.30]$$

where $M = \Sigma m_i$ is the total mass of the system and $\mathbf{r}_i$ is the vector position of the ith particle.

The **velocity of the center of mass for a system of particles** is

$$\mathbf{v}_{CM} = \frac{\Sigma m_i \mathbf{v}_i}{M} \qquad \qquad [8.31]$$

The total momentum of a system of particles equals the total mass multiplied by the velocity of the center of mass—that is, $\mathbf{p}_{tot} = M\mathbf{v}_{CM}$.

Newton's second law applied to a system of particles is

$$\Sigma \mathbf{F}_{ext} = M\mathbf{a}_{CM} = \frac{d\mathbf{p}_{tot}}{dt} \qquad \qquad [8.35]$$

where $\mathbf{a}_{CM}$ is the acceleration of the center of mass and the sum is over all external forces. Therefore, the center of mass moves like an imaginary particle of mass M under the influence of the resultant external force on the system. It follows from Equation 8.35 that the total momentum of the system is constant if there are no external forces acting on it.

CONCEPTUAL QUESTIONS

1. Does a large force always produce a larger impulse on a body than a smaller force does? Explain.

2. If two objects collide and one is initially at rest, is it possible for both to be at rest after the collision? Is it possible for one to be at rest after the collision? Explain.

3. Explain how linear momentum is conserved when a ball bounces from a floor.

4. Consider a perfectly inelastic collision between a car and a large truck. Which vehicle loses more kinetic energy as a result of the collision?

5. Your physical education teacher, who knows something about physics, throws you a tennis ball at a certain velocity, and you catch it. You are now given the following choice: The teacher can throw you a medicine ball, which is much more massive than the tennis ball, with the same velocity as the tennis ball, the same momentum, or the same kinetic energy. Which choice would you make in order to make the easiest catch, and why?

6. Can the center of mass of a body lie outside the body? If so, give examples.

7. Is the center of mass of a mountain higher or lower than the center of gravity?

8. In golf, novice players are often advised to be sure to "follow through" with their swing. Why does this make the ball travel a longer distance? What about a chip shot? There is very little follow-through here—why?

9. A sharpshooter fires a rifle while standing with the butt of the gun against his shoulder. If the forward momentum of a bullet is the same as the backward momentum of the gun, why isn't it as dangerous to be hit by the gun as by the bullet?

10. A pole-vaulter falls from a height of 6.0 m onto a foam rubber pad. Can you calculate his speed just before he reaches the pad? Could you calculate the force exerted on him due to the collision? Explain.

11. Does the center of mass of a rocket in free space accelerate? Explain. Can the speed of a rocket exceed the exhaust speed of the fuel? Explain.

12. An airbag is inflated when a collision occurs, which protects the passenger (the dummy, in this case) from serious injury (Fig. Q8.12). Why does the airbag soften the blow? Discuss the physics involved in this dramatic photograph.

13. A large bedsheet is held vertically by two students. A third student, who happens to be the star pitcher on the baseball team, throws a raw egg at the sheet. Explain why the egg does not break when it hits the sheet, regardless of its initial speed. (If you try this, make sure the pitcher hits the sheet near its center, and do not allow the egg to fall on the floor after being caught in the sheet.)

Figure Q8.12 *(Courtesy of Saab)*

PROBLEMS

Section 8.1 Linear Momentum and Its Conservation

1. A 3.00-kg particle has a velocity of $(3.00\mathbf{i} - 4.00\mathbf{j})$ m/s. Find its x and y components of momentum and the magnitude of its total momentum.

2. A 0.100-kg ball is thrown straight up into the air with an initial speed of 15.0 m/s. Find the momentum of the ball (a) at its maximum height and (b) halfway up to its maximum height.

3. A 40.0-kg child standing on a frozen pond throws a 0.500-kg stone to the east with a speed of 5.00 m/s. Neglecting friction between child and ice, find the recoil velocity of the child.

4. A pitcher claims she can throw a baseball with as much momentum as a 3.00-g bullet moving with a speed of 1500 m/s. A baseball has a mass of 0.145 kg. What must be its speed if the pitcher's claim is valid?

5. Two blocks of masses M and $3M$ are placed on a horizontal, frictionless surface. A light spring is attached to one of them, and the blocks are pushed together with the spring between them (Fig. P8.5). A cord holding them together is burned, after which the block of mass $3M$ moves to the right with a speed of 2.00 m/s. (a) What is the speed of the block of mass M? (b) Find the original elastic potential energy in the spring if $M = 0.350$ kg.

Section 8.2 Impulse and Momentum

6. A car is stopped for a traffic signal. When the light turns green, the car accelerates, increasing its speed from 0 to 5.20 m/s in 0.832 s. What linear impulse and average force does a 70.0-kg passenger in the car experience?

7. An estimated force–time curve for a baseball struck by a bat

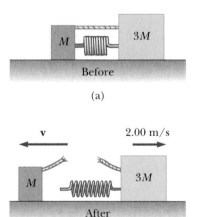

Figure P8.5

is shown in Figure P8.7. From this curve, determine (a) the impulse delivered to the ball, (b) the average force exerted on the ball, and (c) the peak force exerted on the ball.

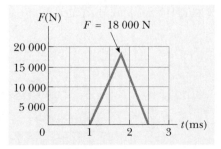

Figure P8.7

8. A tennis player receives a shot with the ball (0.0600 kg) traveling horizontally at 50.0 m/s and returns the shot with the ball traveling horizontally at 40.0 m/s in the opposite direction. (a) What is the impulse delivered to the ball by the racket? (b) What work does the racket do on the ball?

9. A 3.00-kg steel ball strikes a wall with a speed of 10.0 m/s at an angle of 60.0° with the surface. It bounces off with the same speed and angle (Fig. P8.9). If the ball is in contact with the wall for 0.200 s, what is the average force exerted on the ball by the wall?

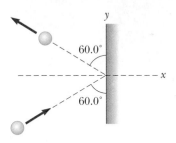

Figure P8.9

10. In a slow-pitch softball game, a 0.200-kg softball crossed the plate at 15.0 m/s at an angle of 45.0° below the horizontal. The ball was hit at 40.0 m/s, 30.0° above the horizontal. (a) Determine the impulse applied to the ball. (b) If the force on the ball increased linearly for 4.00 ms, held constant for 20.0 ms, then decreased to 0 linearly in another 4.00 ms, find the maximum force on the ball.

11. A machine gun fires 35.0 g bullets at a speed of 750.0 m/s. If the gun can fire 200 bullets/min, what is the average force the shooter must exert to keep the gun from moving?

Section 8.3   Collisions

Section 8.4   Elastic and Inelastic Collisions in One Dimension

12. High-speed stroboscopic photographs show that the head of a golf club of mass 200 g is traveling at 55.0 m/s just before it strikes a 46.0-g golf ball at rest on a tee. After the collision, the club head travels (in the same direction) at 40.0 m/s. Find the speed of the golf ball just after impact.

13. A 10.0-g bullet is stopped in a block of wood ($m = 5.00$ kg). The speed of the bullet-plus-wood combination immediately after the collision is 0.600 m/s. What was the original speed of the bullet?

14. A 75.0-kg ice skater, moving at 10.0 m/s, crashes into a stationary skater of equal mass. After the collision, the two skaters move as a unit at 5.00 m/s. The average force a skater can experience without breaking a bone is 4500 N. If the impact time is 0.100 s, does a bone break?

15. As shown in Figure P8.15, a bullet of mass m and speed v passes completely through a pendulum bob of mass M. The bullet emerges with a speed of $v/2$. The pendulum bob is suspended by a stiff rod of length ℓ and negligible mass. What is the minimum value of v such that the pendulum bob will barely swing through a complete vertical circle?

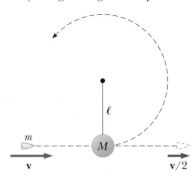

Figure P8.15

16. Gayle runs at a speed of 4.00 m/s and dives on a sled, initially at rest on the top of a frictionless snow-covered hill. After she has descended a vertical distance of 5.00 m, her brother, who is initially at rest, hops on her back and together they continue down the hill. What is their speed at the bottom of the hill if the total vertical drop is 15.0 m? Gayle's mass is 50.0 kg, the sled has a mass of 5.00 kg, and her brother has a mass of 30.0 kg.

17. A 1200-kg car traveling initially with a speed of 25.0 m/s in an easterly direction crashes into the rear end of a 9000-kg truck moving in the same direction at 20.0 m/s (Fig. P8.17). The velocity of the car right after the collision is 18.0 m/s to the east. (a) What is the velocity of the truck right after the collision? (b) How much mechanical energy is lost in the collision? Account for this loss in energy.

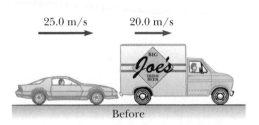
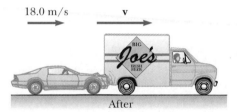

Figure P8.17

18. A railroad car of mass 2.50×10^4 kg, moving with a speed of 4.00 m/s, collides and couples with three other coupled railroad cars, each of the same mass as the single car and moving in the same direction with an initial speed of 2.00 m/s. (a) What is the speed of the four cars after the collision? (b) How much energy is lost in the collision?

19. A neutron in a reactor makes an elastic head-on collision with the nucleus of a carbon atom initially at rest. (a) What fraction of the neutron's kinetic energy is transferred to the carbon nucleus? (b) If the initial kinetic energy of the neutron is 1.60×10^{-13} J, find its final kinetic energy and the kinetic energy of the carbon nucleus after the collision. (The mass of the carbon nucleus is about 12.0 times the mass of the neutron.)

20. A 7.00-kg bowling ball collides head-on with a 2.00-kg bowling pin. The pin flies forward with a speed of 3.00 m/s. If the ball continues forward with a speed of 1.80 m/s, what was the initial speed of the ball? Ignore rotation of the ball.

21. A 12.0-g bullet is fired into a 100-g wooden block initially at rest on a horizontal surface. After impact, the block slides 7.50 m before coming to rest. If the coefficient of friction between block and surface is 0.650, what was the speed of the bullet immediately before impact?

22. A bullet of mass m_1 is fired into a wooden block of mass m_2 initially at rest on a horizontal surface. After impact, the block slides a distance d before coming to rest. If the coefficient of friction between block and surface is μ, what was the speed of the bullet immediately before impact?

23. Consider a frictionless track *ABC* as shown in Figure P8.23. A block of mass $m_1 = 5.00$ kg is released from *A*. It makes a head-on elastic collision with a block of mass $m_2 = 10.0$ kg at *B*, initially at rest. Calculate the maximum height to which m_1 rises after the collision.

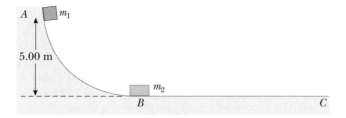

Figure P8.23

24. A 7.00-g bullet, when fired from a gun into a 1.00-kg block of wood held in a vise, penetrates the block to a depth of 8.00 cm. This block of wood is placed on a frictionless horizontal surface, and a 7.00-g bullet is fired from the gun into the block. To what depth will the bullet penetrate the block in this case?

Section 8.5 Two-Dimensional Collisions

25. A 90.0-kg fullback running east with a speed of 5.00 m/s is tackled by a 95.0-kg opponent running north with a speed of 3.00 m/s. If the collision is perfectly inelastic, (a) calculate the speed and direction of the players just after the tackle and (b) determine the energy lost as a result of the collision. Account for the missing energy.

26. The mass of the blue puck in Figure P8.26 is 20.0% greater than the mass of the green one. Before colliding, the pucks approach each other with equal and opposite momenta, and the green puck has an initial speed of 10.0 m/s. Find the speeds of the pucks after the collision if half the kinetic energy is lost during the collision.

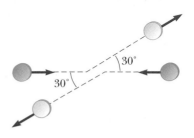

Figure P8.26

27. A billiard ball moving at 5.00 m/s strikes a stationary ball of the same mass. After the collision, the first ball moves at 4.33 m/s, at an angle of 30.0° with respect to the original line of motion. Assuming an elastic collision, find the struck ball's velocity after the collision.

28. A proton, moving with a velocity of $v_0\mathbf{i}$, collides elastically with another proton that is initially at rest. If the two protons have the same speed after the collision, find (a) the speed of each proton after the collision in terms of v_0 and (b) the direction of the velocity vectors after the collision.

29. A 0.300-kg puck, initially at rest on a horizontal, frictionless surface, is struck by a 0.200-kg puck moving initially along the *x* axis with a speed of 2.00 m/s. After the collision, the 0.200-kg puck has a speed of 1.00 m/s at an angle of $\theta = 53.0°$ to the positive *x* axis (Fig. 8.11). (a) Determine the velocity of the 0.300-kg puck after the collision. (b) Find the fraction of kinetic energy lost in the collision.

30. During the battle of Gettysburg, the gunfire was so intense that several bullets collided in mid-air and fused together. Assume a 5.00-g Union musket ball was moving to the right at a speed of 250 m/s, 20.0° above the horizontal, and that a 3.00-g Confederate ball was moving to the left at a speed of 280 m/s, 15.0° above the horizontal. Immediately after they fuse together, what is their velocity?

31. An unstable nucleus of mass 17.0×10^{-27} kg initially at rest disintegrates into three particles. One of the particles, of mass 5.00×10^{-27} kg, moves along the *y* axis with a velocity of 6.00×10^6 m/s. Another particle, of mass 8.40×10^{-27} kg, moves along the *x* axis with a speed of $4.00 \times$

10^6 m/s. Find (a) the velocity of the third particle and (b) the total energy given off in the process.

32. At an intersection, a 1500-kg car, traveling east with a speed of 20.0 m/s, collides with a 2500-kg van traveling south at a speed of 15.0 m/s. The vehicles undergo a perfectly inelastic collision, and the wreckage slides 6.00 m before coming to rest. Find the magnitude and direction of the constant force that decelerated them.

Section 8.6   The Center of Mass

33. Four objects are situated along the y axis as follows: a 2.00-kg object is at $+3.00$ m, a 3.00-kg object is at $+2.50$ m, a 2.50-kg object is at the origin, and a 4.00-kg object is at -0.500 m. Where is the center of mass of these objects?

34. A uniform carpenter's square has the shape of an L, as in Figure P8.34. Locate the center of mass relative to an origin at the lower left corner. (*Hint:* Note that the mass of each rectangular part is proportional to its area.)

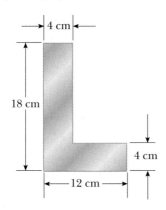

Figure P8.34

35. A uniform piece of sheet steel is shaped as in Figure P8.35. Compute the x and y coordinates of the center of mass of the piece.

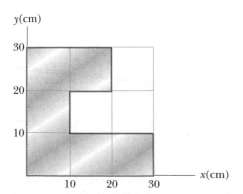

Figure P8.35

36. The mass of the Earth is 5.98×10^{24} kg, and the mass of the Moon is 7.36×10^{22} kg. The distance of separation, measured between their centers, is 3.84×10^8 m. Locate the center of mass of the Earth–Moon system as measured from the center of the Earth.

37. A water molecule consists of an oxygen atom with two hydrogen atoms bound to it (Fig. P8.37). The angle between the two bonds is 106°. If the bonds are 0.100 nm long, where is the center of mass of the molecule?

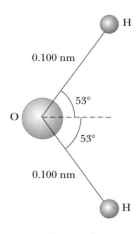

Figure P8.37

Section 8.7   Motion of a System of Particles

38. Consider a system of two particles in the xy plane: $m_1 = 2.00$ kg is at $\mathbf{r}_1 = (1.00\mathbf{i} + 2.00\mathbf{j})$ m and has velocity $(3.00\mathbf{i} + 0.500\mathbf{j})$ m/s; $m_2 = 3.00$ kg is at $\mathbf{r}_2 = (-4.00\mathbf{i} - 3.00\mathbf{j})$ m and has velocity $(3.00 \mathbf{i} - 2.00\mathbf{j})$ m/s. (a) Plot these particles on a grid or graph paper. Draw their position vectors and show their velocities. (b) Find the position of the center of mass of the system and mark it on the grid. (c) Determine the velocity of the center of mass and also show it on the diagram. (d) What is the total linear momentum of the system?

39. Romeo (77.0 kg) entertains Juliet (55.0 kg) by playing his guitar from the rear of their boat in still water, 2.70 m away from Juliet in the front of the boat. After the serenade, Juliet carefully moves to the rear of the boat (away from shore) to plant a kiss on Romeo's cheek. How far does the 80.0-kg boat move toward the shore it is facing?

40. A ball of mass 0.200 kg has velocity $1.50\mathbf{i}$ m/s; a ball of mass 0.300 kg has velocity $-0.400\mathbf{i}$ m/s. Then they meet in a head-on elastic collision. (a) Find their velocities thereafter. (b) Find the velocity of their center of mass before the collision and afterward.

Section 8.8   Rocket Propulsion (Optional)

41. The first stage of a Saturn V space vehicle consumes fuel at the rate of 1.50×10^4 kg/s, with an exhaust speed of

2.60×10^3 m/s. (a) Calculate the thrust produced by these engines. (b) Find the initial acceleration of the vehicle on the launch pad if its initial mass is 3.00×10^6 kg. [*Hint:* You must include the force of gravity to solve part (b).]

42. A large rocket with an exhaust speed of $v_e = 3000$ m/s develops a thrust of 24.0 million newtons. (a) How much mass is being blasted out of the rocket exhaust per second? (b) What is the maximum speed the rocket can attain if it starts from rest in a force-free environment with $v_e = 3.00$ km/s and if 90.0% of its initial mass is fuel?

43. A rocket for use in deep space is to have the capability of accelerating a payload (including the rocket frame and engine) of 3.00 metric tons from rest to a final speed of 10 000 m/s with an engine and fuel designed to produce an exhaust speed of 2000 m/s. (a) How much fuel plus oxidizer is required? (b) If a different fuel and engine design could give an exhaust speed of 5000 m/s, what amount of fuel and oxidizer would be required for the same task?

Additional Problems

44. A golf ball ($m = 46.0$ g) is struck a blow that makes an angle of 45.0° with the horizontal. The drive lands 200 m away on a flat fairway. If the golf club and ball are in contact for 7.00 ms, what is the average force of impact? (Neglect air resistance.)

45. An 8.00-g bullet is fired into a 2.50-kg block initially at rest at the edge of a frictionless table of height 1.00 m (Fig. P8.45). The bullet remains in the block, and after impact the block lands 2.00 m from the bottom of the table. Determine the initial speed of the bullet.

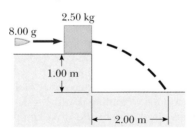

Figure P8.45

46. A spacecraft is stationary in deep space when its rocket engine is ignited for a 100-s "burn." Hot gases are ejected at a constant rate of 150 kg/s, with a speed of 3000 m/s relative to the spacecraft. The initial mass of the spacecraft (plus fuel and oxidizer) is 25 000 kg. Determine the thrust of the rocket engine and the initial acceleration in units of *g*. What is the final speed of the spacecraft?

47. A small block of mass $m_1 = 0.500$ kg is released from rest at the top of a curved-shaped frictionless wedge of mass $m_2 = 3.00$ kg, which sits on a frictionless horizontal surface as in Figure P8.47a. When the block leaves the wedge, its velocity is measured to be 4.00 m/s to the right, as in Figure

P8.47b. (a) What is the velocity of the wedge after the block reaches the horizontal surface? (b) What is the height, *h*, of the wedge?

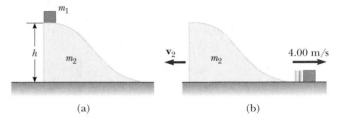

(a)   (b)

Figure P8.47

48. Tarzan, whose mass is 80.0 kg, swings from a 3.00-m vine that is horizontal when he starts. At the bottom of his arc, he picks up 60.0-kg Jane in a perfectly inelastic collision. What is the height of the highest tree limb they can reach on their upward swing?

49. A 80.0-kg astronaut is working on the engines of her ship, which is drifting through space with a constant velocity. The astronaut, wishing to get a better view of the Universe, pushes against the ship and much later finds herself 30.0 m behind the ship. Without a thruster, the only way to return to the ship is to throw her 0.500-kg wrench directly away from the ship. If she throws the wrench with a speed of 20.0 m/s relative to the ship, how long does it take the astronaut to reach the ship?

50. A 40.0-kg child stands at one end of a 70.0-kg boat that is 4.00 m in length (Fig. P8.50). The boat is initially 3.00 m from the pier. The child notices a turtle on a rock at the far end of the boat and proceeds to walk to that end to catch the turtle. Neglecting friction between boat and water, (a) describe the subsequent motion of the system (child plus boat). (b) Where is the child *relative to the pier* when he reaches the far end of the boat? (c) Will he catch the turtle? (Assume he can reach out 1.00 m from the end of the boat.)

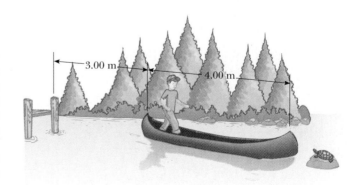

Figure P8.50

51. A 5.00-g bullet moving with an initial speed of 400 m/s is fired into and passes through a 1.00-kg block, as in Figure P8.51. The block, initially at rest on a frictionless, horizontal surface, is connected to a spring of force constant 900 N/m. If the block moves 5.00 cm to the right after impact, find (a) the speed at which the bullet emerges from the block and (b) the energy lost in the collision.

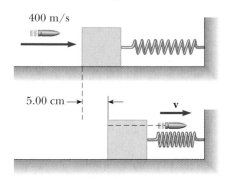

Figure P8.51

52. A student performs a ballistic pendulum experiment using the apparatus shown in Figure P8.52a. A ball of mass

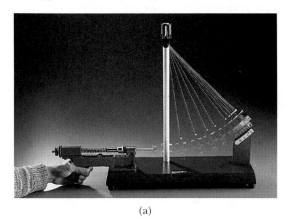

(a)

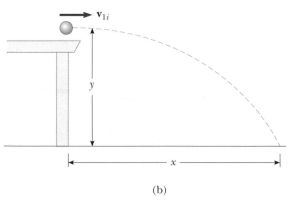

(b)

Figure P8.52 (a) and (b)   *(Photo courtesy of CENCO)*

68.8 g is fired into a pendulum of mass 263 g hanging from a low friction pivot. The ball is caught inside the pendulum, which then swings up through a height 8.68 cm. (a) Determine the initial speed v_{1i} of the ball. (b) The second part of her experiment is to obtain v_{1i} by firing the same ball horizontally (with the pendulum removed from the path), by measuring its horizontal displacement, x, and vertical displacement, y (Fig. P8.52b). Show that the initial speed of the ball is related to x and y through the relation

$$v_{1i} = \frac{x}{\sqrt{2y/g}}$$

What numerical value does she obtain for v_{1i} based on her measured values of $x = 257$ cm and $y = 85.3$ cm? What factors might account for the difference in this value compared to that obtained in part (a)?

53. Two particles of masses m and $3m$ are moving toward each other along the x axis with the same initial speeds v_0. Mass m is traveling to the left, and mass $3m$ is traveling to the right. They undergo a head-on elastic collision. Find the final speeds of the particles.

54. Two particles of masses m and $3m$ are moving toward each other along the x axis with the same initial speeds v_0. Mass m is traveling to the left, and mass $3m$ is traveling to the right. They undergo an elastic glancing collision such that mass m is moving downward after the collision at right angles from its initial direction. (a) Find the final speeds of the two masses. (b) What is the angle θ at which the mass $3m$ is scattered?

55. How fast can you set the Earth moving? In particular, when you jump straight up as high as you can, you give the Earth a maximum recoil speed of what order of magnitude? Visualize the Earth as a perfectly rigid object. In your solution state the physical quantities you take as data and the values you measure or estimate for them.

56. There are (one can say) three co-equal theories of motion: Newton's second law, holding that the total force on an object causes its acceleration; the work-kinetic energy theorem, holding that the total work on an object causes its change in kinetic energy; and the impulse-momentum theorem, holding that the total impulse on an object causes its change in momentum. In this problem you compare the predictions of the three theories in one particular case. A 3.00-kg object has velocity $7.00\mathbf{j}$ m/s. Then a total force $12.0\mathbf{i}$ N acts on the object for 5.00 s. (a) Calculate its final velocity from the impulse-momentum theorem. (b) Calculate its acceleration from $\mathbf{a} = (\mathbf{v}_f - \mathbf{v}_0)/t$. (c) Calculate its acceleration from $\mathbf{a} = \Sigma\mathbf{F}/m$. (d) Find the object's vector displacement from $\mathbf{s} = \mathbf{v}_0 t + \frac{1}{2}\mathbf{a}t^2$. (e) Find the work done on the object from $W = \mathbf{F}\cdot\mathbf{s}$. (f) Find the final kinetic energy from $\frac{1}{2}mv_f^2 = \frac{1}{2}m\mathbf{v}_f\cdot\mathbf{v}_f$. (g) Find the final kinetic energy from $\frac{1}{2}mv_0^2 + W$.

🔲 Spreadsheet Problems

S1. A rocket has an initial mass of 20 000 kg, of which 20% is the payload. The rocket burns fuel at a rate of 200 kg/s and exhausts gas at a relative speed of 2.00 km/s. Its acceleration dv/dt is determined from the equation of motion

$$M\frac{dv}{dt} = v_e\left|\frac{dM}{dt}\right| + F_{ext}$$

Assume that there are no external forces and that the rocket's initial velocity is zero. When $F_{ext} = 0$, the rocket's velocity is $v(t) = v_e \ln(M_i/M)$, where M_i is the rocket's initial mass. Spreadsheet 8.1 calculates the acceleration and velocity as a function of time. (a) Using the given parameters, find the maximum acceleration and velocity. (b) At what time is the velocity equal to half its maximum value? Why isn't this time half of the burn time?

S2. Modify Spreadsheet 8.1 to calculate the distance traveled by the rocket in Problem S1. Add a new column to the spreadsheet to find the new position x_{i+1}. Estimate the new position of the rocket by

$$x_{i+1} = x_i + \tfrac{1}{2}(v_{i+1} + v_i)\Delta t$$

where x_i is the old position, v_i is the old velocity, and v_{i+1} is the new velocity.

S3. If the rocket in the previous problem burns its fuel twice as fast (400 kg/s) while maintaining the same exhaust speed, how far does it travel before it runs out of fuel? The velocity reached by the rocket is the same in both cases; that is

$$v_f - v_i = v_e \ln\left(\frac{M_i}{M_f}\right)$$

Is there an advantage in burning the fuel faster? Is there a disadvantage? (*Hint:* It may be useful to plot the acceleration, velocities, and position for the same time intervals in both cases. In which case does the rocket travel farther? In which case does the rocket have to withstand the largest acceleration? By how much?)

ANSWERS TO CONCEPTUAL PROBLEMS

1. The brain is immersed in a cushioning fluid inside the skull. If the head is struck suddenly by a bare fist, there is a rapid acceleration of the skull. The brain matches this acceleration only because of the large impulsive force of the skull on the brain. This large and sudden force can severely injure the brain. Padded gloves extend the time over which the force is applied to the head. Thus, for a given impulse, the gloves provide a longer time interval than the bare fist, decreasing the average force. Because the average force is decreased, the acceleration of the skull is decreased, reducing (but not eliminating) the chance for brain injury. The same argument can be made for "rolling with the punch." If the head is held steady while being struck, the time interval over which the force is applied is relatively short and the average force is large. If the head is allowed to move in the same direction of the punch, the time interval is lengthened, and the average force reduced.

2. The scenario described violates the principle of conservation of momentum. If the superhero, initially at rest, throws the piano in one direction, he must recoil in the other direction, in order to maintain the zero momentum of the system.

3. Making cars out of hard rubber would save on automobile repair bills, but the results for the drivers and passengers would be disastrous. Imagine two identical cars traveling toward each other at identical speeds, and making a head-on collision. In the approximation that the collision is elastic, the cars bounce off each other and leave the collision with their velocities reversed in direction. Thus, the inhabitants of the cars would suffer a large change in momentum in a short period of time. As a result, the force on the inhabitants would become very large, causing significant bodily injury. For real cars in the above situation, the cars would come to rest during the collision, as the front sections of the cars crumpled. Thus, the change in momentum is only half what it was in the elastic case, and the change occurs over the longer time period during which the cars crumpled. The resulting smaller force on the inhabitants will result in fewer injuries.

4. If all of the kinetic energy is transformed into other forms, then there is no motion of either of the objects after the collision. If nothing is moving, there is no momentum. Thus, if the final momentum of the system is zero, then the initial momentum of the system must have been zero. The required condition, then, is that the two objects are approaching each other so as to make a head-on collision, with equal, but oppositely directed, momenta.

5. Tie one end of the rope around the bear. Lay out the tape measure on the ice between the bear's original position and yours, as you hold the opposite end of the rope. Take off your spiked shoes and pull on the rope. Both you and the bear will slide over the ice until you meet. From the tape observe how far you have slid, x_y, and how far the bear has slid, x_b. The point where you meet the bear is the constant location of the center of mass of the system (bear plus you),

so you can determine the mass of the bear from $m_b x_b = m_y x_y$. (Unfortunately, if the bear now wakes up you cannot get back to your spiked shoes.)

6. Neglecting air resistance, the only external force on the projectile is the force of gravity. Thus, the projectile follows a parabolic path. If the projectile did not explode, it would continue to move along the parabolic path indicated by the broken line in Figure 8.19. Because the forces due to the explosion are internal, they do not affect the motion of the center of mass. Thus, after the explosion, and until the first fragment hits the ground, the center of mass of the fragments follows the same parabolic path the projectile would have followed if there had been no explosion.

9

Relativity

Most of our everyday experiences and observations have to do with objects that move at speeds much less than that of light. Newtonian mechanics and early ideas on space and time were formulated to describe the motion of such objects. This formalism is successful in describing a wide range of phenomena that occur at low speeds. It fails, however, when applied to particles the speeds of which approach that of light. The predictions of Newtonian theory can be tested experimentally at high speeds by accelerating electrons or other charged particles. For example, it is possible to accelerate an electron to a speed of $0.99c$ (where c is the speed of light) by using particle accelerators. According to Newtonian mechanics, if the energy transferred to an electron is increased by a factor of 4, the electron speed should jump to $1.98c$. However, experiments show that the speed of the electron—as well as the speeds of all other particles in the Universe—always remains less than the speed of light, regardless of the particle energy. Because it places no upper limit on speed, New-

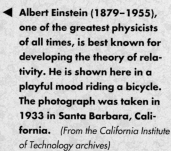

◀ **Albert Einstein (1879–1955), one of the greatest physicists of all times, is best known for developing the theory of relativity. He is shown here in a playful mood riding a bicycle. The photograph was taken in 1933 in Santa Barbara, California.** *(From the California Institute of Technology archives)*

"Imagination is more important than knowledge."
Albert Einstein

tonian mechanics is contrary to modern experimental results and is clearly a limited theory.

In 1905, at the age of 26, Einstein published his special theory of relativity. Einstein wrote,

> The relativity theory arose from necessity, from serious and deep contradictions in the old theory from which there seemed no escape. The strength of the new theory lies in the consistency and simplicity with which it solves all these difficulties, using only a few very convincing assumptions. . . . [1]

Although Einstein made many important contributions to science, special relativity alone represents one of the greatest intellectual achievements of the twentieth century. With special relativity, experimental observations can be correctly predicted over the range of speeds from $v = 0$ to speeds approaching the speed of light. Newtonian mechanics, which was accepted for more than 200 years, is in fact a special case of special relativity. This chapter gives an introduction to special relativity, with emphasis on some of its consequences.

Special relativity covers phenomena such as the slowing down of clocks and the contraction of lengths in moving reference frames as measured by a laboratory observer. We also discuss the relativistic forms of momentum and energy, and some consequences of the famous mass–energy formula, $E = mc^2$.

In addition to its well known and essential role in theoretical physics, special relativity has practical applications, including the design of accelerators and other devices that use high-speed particles. These devices will not work if designed according to nonrelativistic principles. We shall have occasion to use special relativity in some subsequent chapters of this textbook, most often presenting only the outcome of relativistic effects.

9.1 • THE PRINCIPLE OF NEWTONIAN RELATIVITY

To describe a physical event, it is necessary to establish a frame of reference. You should recall from Chapter 5 that Newton's laws are valid in all inertial frames of reference. Because an inertial frame is defined as one in which Newton's first law is valid, it can be said that **an inertial system is one in which a free body experiences no acceleration.** Furthermore, any system moving with constant velocity with respect to an inertial system must also be an inertial system.

Inertial frame of reference •

There is no preferred frame. This means that the results of an experiment performed in a vehicle moving with uniform velocity will be identical to the results of the same experiment performed in a stationary vehicle. The formal statement of this result is called the principle of **Newtonian relativity:**

> The laws of mechanics must be the same in all inertial frames of reference.

Let us consider an observation that illustrates the equivalence of the laws of mechanics in different inertial frames. Consider a pickup truck moving with a constant velocity, as in Figure 9.1a. If a passenger in the truck throws a ball straight up in the air, the passenger observes that the ball moves in a vertical path. The motion

[1] A. Einstein and L. Infeld, *The Evolution of Physics*, New York, Simon and Schuster, 1961.

(a)   (b)

Figure 9.1 (a) The observer in the truck sees the ball move in a vertical path when thrown upward. (b) The Earth observer views the path of the ball to be a parabola.

of the ball appears to be precisely the same as if the ball were thrown by a person at rest on Earth. The law of gravity and the equations of kinematics are obeyed whether the truck is at rest or in uniform motion. Now consider the same experiment viewed by an observer at rest on Earth. This stationary observer sees the path of the ball as a parabola, as in Figure 9.1b. Furthermore, according to this observer, the ball has a horizontal component of velocity equal to the velocity of the truck. Although the two observers see different velocities and rates of energy change, they see the same forces and agree on the validity of Newton's laws as well as on principles like conservation of energy and conservation of momentum. All differences between the two views stem from the relative motion of one frame with respect to the other. That is, the notion of absolute motion through space plays no role in what is seen.

Suppose that some physical phenomenon that we call an *event* occurs in an inertial system. The event's location and time of occurrence can be specified by the coordinates (x, y, z, t). We would like to be able to transform these coordinates from one inertial system to another moving with uniform relative velocity. This is accomplished by using what is called a *Galilean transformation,* which owes its origin to Galileo.

Consider two inertial systems S and S' (Fig. 9.2). The system S' moves with a constant velocity **v** along the xx' axes, where **v** is measured relative to S. We assume that an event occurs at the point P and that the origins of S and S' coincide at $t = 0$. An observer in S describes the event with space–time coordinates (x, y, z, t), while an observer in S' uses (x', y', z', t') to describe the same event. As we can see from Figure 9.2, these coordinates are related by the equations

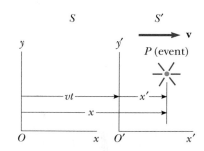

Figure 9.2 An event occurs at a point P. The event is observed by two observers in inertial frames S and S', where S' moves with a velocity **v** relative to S.

$$x' = x - vt$$

$$y' = y$$

$$z' = z \qquad\qquad\qquad \text{[9.1]}$$

$$t' = t$$

• *Galilean transformation of coordinates*

These equations constitute what is known as a **Galilean transformation of coordinates.** Note that time is assumed to be the same in both inertial systems. That is, within the framework of classical mechanics, all clocks run at the same rate, regardless of their velocity, so that the time at which an event occurs for an observer in *S* is the same as the time for the same event in *S'*. As a consequence, the time interval between two successive events should be the same for both observers. Although this assumption may seem obvious, it turns out to be incorrect when treating situations in which *v* is comparable to the speed of light. This point of equal time intervals represents one of the most profound differences between Newtonian concepts and Einstein's special relativity.

Now suppose two events are separated by a distance *dx* and a time interval *dt* as measured by an observer in *S*. It follows from Equation 9.1 that the corresponding distance *dx'* measured by an observer in *S'* is $dx' = dx - vdt$, where *dx* is the distance between the two events measured by an observer in *S*. Because $dt = dt'$, we find that

$$\frac{dx'}{dt'} = \frac{dx}{dt} - v$$

or

Galilean addition law for velocities

$$u_x' = u_x - v \qquad [9.2]$$

where u_x and u_x' are the instantaneous velocities of the object relative to *S* and *S'*, respectively. This result, which is called the **Galilean addition law for velocities** (or Galilean velocity transformation), is used in everyday observations and is consistent with our intuitive notion of time and space. As we will soon see, however, it leads to serious contradictions when applied to electromagnetic waves.

9.2 • THE MICHELSON–MORLEY EXPERIMENT

Many experiments similar to the one described in the preceding section show us that the laws of mechanics are the same in all inertial frames of reference. When similar inquiries are made into the laws of other branches of physics, the results are contradictory. In particular, the laws of electricity and magnetism are found to depend on the frame of reference used. It might be argued that it is these laws that are wrong, but that is difficult to accept because the laws are in total agreement with all known experimental results. The Michelson–Morley experiment was an attempt to explain this dilemma.

The experiment stemmed from a misconception early physicists had concerning the manner in which light propagates. The properties of mechanical waves, such as water and sound waves, were well known, and all of these waves require a medium to support the disturbances. In the nineteenth century, physicists thought that electromagnetic waves also required a medium through which to propagate. They proposed that such a medium existed, and they named it the **luminiferous ether.** This ether was assumed to be present everywhere, even in a vacuum, and to have the unusual property of being a massless but rigid medium (a strange concept indeed!). In addition, it was found that the troublesome laws of electricity and magnetism took on their simplest forms in a frame of reference *at rest* with respect to the luminiferous ether; this frame was called the **absolute frame.** In any refer-

Albert A. Michelson (1852–1931)

A German American physicist, Michelson spent much of his life making accurate measurements of the speed of light. In 1907 he was the first American to be awarded the Nobel prize, which he received for his work in optics. His most famous experiment, conducted with Edward Morley in 1887, implied that it was impossible to measure the absolute velocity of the Earth with respect to the ether. *(AIP Emilio Segrè Visual Archives, Michelson Collection)*

ence frame moving with respect to it, the laws of electricity and magnetism would remain valid but would have to be modified.

As a result of the importance attached to this absolute frame, experimental proof of its existence became of considerable interest in physics. However, all attempts to detect the presence of the ether (and hence the absolute frame) proved futile. The most famous experiment designed to show the presence of the ether was performed in 1887 by A. A. Michelson (1852–1931) and E. W. Morley (1838–1923). The objective was to determine the velocity of the Earth with respect to the ether, and the experimental tool used was a device called the interferometer, shown in Figure 9.3.

Suppose one arm of an interferometer ($M_0 M_2$ in Fig. 9.3) were aligned along the direction of the motion of the Earth through space. The Earth moving through the ether would be equivalent to the ether flowing past the Earth in the opposite direction. This "ether wind" blowing in the direction opposite the Earth's motion should cause the speed of light, as measured in the Earth's frame of reference, to be $c - v$ as the light approaches mirror M_2 in Figure 9.3, and $c + v$ after reflection, where c is the speed of light in the ether frame and v is the speed of the Earth through space and hence the speed of the ether wind. The incident and reflected beams of light would recombine, and a pattern would form as the two beams interfere with each other.

During the experiment, the interference pattern was observed while the interferometer was rotated through an angle of 90°. The idea was that this rotation would change the speed of the ether wind along the direction of the arms of the interferometer, and as a consequence the interference pattern would shift slightly but measurably. Measurements failed to show any change in the pattern. The Michelson–Morley experiment was repeated by other researchers under varying conditions and at different locations, but the results were always the same: *no interference-pattern shift of the magnitude required was ever observed.*[2]

The negative result of the Michelson–Morley experiment not only contradicted the ether hypothesis; it also meant that it was impossible to measure the absolute (orbital) velocity of the Earth with respect to the ether frame. From a theoretical viewpoint, this meant that it was impossible to find the absolute frame. However, as we shall see in the next section, Einstein offered a postulate that places a different interpretation on the negative result. In later years, when more was known about the nature of light, the idea of an ether that permeates all space was relegated to the ash heap of worn-out concepts. **Light is now understood to be an electromagnetic wave that requires no medium for its propagation.** As a result, an ether in which light could travel became an unnecessary construct.

Modern versions of the Michelson–Morley experiment have compared the frequencies of resonant laser cavities of identical length, oriented at right angles to each other. Most recently, Doppler shift experiments using gamma rays emitted by a radioactive sample of ^{57}Fe have placed an upper limit of about 5 cm/s on ether wind velocity. These results have shown quite conclusively that the motion of the Earth has no effect on the speed of light!

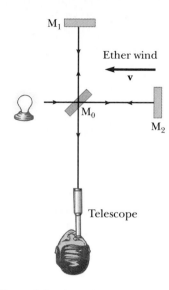

Figure 9.3 According to the ether wind theory, the speed of light should be $c - v$ as the beam approaches mirror M_2 and $c + v$ after reflection.

[2]From an Earth-observer's point of view, changes in the Earth's speed and direction of motion in the course of a year are viewed as ether wind shifts. Even if the speed of the Earth with respect to the ether were zero at some time, 6 months later the speed of the Earth would be 60 km/s with respect to the ether and a clear fringe shift should be found. None has ever been observed, however.

9.3 • EINSTEIN'S PRINCIPLE OF RELATIVITY

In the previous section we noted the impossibility of measuring the speed of the ether with respect to the Earth and the failure of the Galilean addition law for velocities in the case of light. Albert Einstein proposed a theory that boldly removed these difficulties and at the same time completely altered our notion of space and time.[3] Einstein based special relativity on two postulates:

The postulates of special • relativity

> 1. **The Principle of Relativity:** All the laws of physics are the same in all inertial reference frames.
> 2. **The Constancy of the Speed of Light:** The speed of light in vacuum has the same value, $c = 3.00 \times 10^8$ m/s, in all inertial frames, regardless of the velocity of the observer or the velocity of the source emitting the light.

The first postulate asserts that *all* the laws of physics, those dealing with mechanics, electricity and magnetism, optics, thermodynamics, and so on, are the same in all reference frames moving with constant velocity relative to each other. This postulate is a sweeping generalization of the principle of Newtonian relativity that only refers to the laws of mechanics. From an experimental point of view, Einstein's principle of relativity means that any kind of experiment (measuring the speed of light, for example), performed in a laboratory at rest must give the same result when performed in a laboratory moving at constant velocity past the first one. Hence no preferred inertial reference frame exists and it is impossible to detect absolute motion.

Note that postulate 2, the principle of the constancy of the speed of light, is required by postulate 1: If the speed of light were not the same in all inertial frames, it would be possible to distinguish between inertial frames and a preferred, absolute frame could be identified in contradiction to postulate 1. Postulate 2 also eliminates the problem of measuring the speed of the ether by denying the existence of the ether and boldly asserting that light always moves with speed c relative to all inertial observers.

Although the Michelson–Morley experiment was performed before Einstein published his work on relativity, it is not clear whether or not Einstein was aware of the details of the experiment. Nonetheless, the null result of the experiment can be readily understood within the framework of Einstein's theory. According to his principle of relativity, the premises of the Michelson–Morley experiment were incorrect. In the process of trying to explain the expected results, we stated that when light traveled against the ether wind its speed was $c - v$, in accordance with the Galilean addition law for velocities. However, if the state of motion of the observer or of the source has no influence on the value found for the speed of light, one will always measure the value to be c. Likewise, the light makes the return trip after reflection from the mirror at speed c, not at speed $c + v$. Thus, the motion of the Earth does not influence the interference pattern observed in the Michelson–Morley experiment, and a null result should be expected.

[3]A. Einstein, "On the Electrodynamics of Moving Bodies," *Ann. Physik* 17:891, 1905. For an English translation of this article and other publications by Einstein, see the book by H. Lorentz, A. Einstein, H. Minkowski, and H. Weyl, *The Principle of Relativity*, New York, Dover, 1958.

If we accept special relativity, we must conclude that relative motion is unimportant when measuring the speed of light. At the same time, we must alter our common-sense notion of space and time and be prepared for some bizarre consequences.

Thinking Physics 1

Imagine a very powerful lighthouse, with a rotating beacon. Imagine also drawing a horizontal circle around the lighthouse, with the lighthouse at the center. Along the circumference of the circle, the light beam lights up a portion of the circle and the lit portion of the circle moves around the circle at a certain tangential speed. If we now imagine a circle twice as big in radius, the tangential speed of the lit portion is faster, because it must travel a larger circumference in the time of one rotation of the light source. Imagine that we continue to make the circle larger and larger, eventually moving it out into space. The tangential speed of the lit portion will keep increasing. Is it possible that the tangential speed could become larger than the speed of light? Would this violate a principle of special relativity?

Reasoning   For a large enough circle, it is possible that the tangential speed of the lit portion of the circle could be larger than the speed of light. This does not violate a principle of special relativity, however, because no matter or information is traveling faster than the speed of light.

CONCEPTUAL PROBLEM 1

You are in a speedboat on a lake. You see ahead of you a wavefront, caused by the previous passage of another boat, moving away from you. You accelerate, catch up with, and pass the wavefront. Is this scenario possible if you are in a rocket and you detect a wavefront of *light* ahead of you?

9.4 • CONSEQUENCES OF SPECIAL RELATIVITY

Before we discuss the consequences of special relativity, we must first understand how an observer located in an inertial reference frame describes an event. As mentioned earlier, an event is an occurrence describable by three space coordinates and one time coordinate. Different observers in different inertial frames usually describe the same event with different space–time coordinates.

The reference frame used to describe an event consists of a coordinate grid and a set of synchronized clocks located at the grid intersections as shown in Figure 9.4 in two dimensions. The clocks can be synchronized in many ways with the help of light signals. For example, suppose the observer is located at the origin with his master clock and sends out a pulse of light at $t = 0$. The light pulse takes a time r/c to reach a second clock located a distance r from the origin. Hence, the second clock is synchronized with the clock at the origin if the second clock reads a time r/c at the instant the pulse reaches it. This procedure of synchronization assumes that the speed of light has the same value in all directions and in all inertial frames. Furthermore, the procedure concerns an event recorded by an observer in a specific inertial reference frame. An observer in some other inertial frame would assign different space–time coordinates to events being observed by using another coordinate grid and another array of clocks.

Albert Einstein (1879–1955)

One of the greatest physicists of all times, Einstein was born in Ulm, Germany. In 1905, at the age of 26, he published four scientific papers that revolutionized physics. Two of these papers were concerned with what is now considered his most important contribution of all, the special theory of relativity. In 1916, Einstein published his work on the general theory of relativity. The most dramatic prediction of this theory is the degree to which light is deflected by a gravitational field. Measurements made by astronomers on bright stars in the vicinity of the eclipsed Sun in 1919 confirmed Einstein's prediction, and Einstein suddenly became a world celebrity. Einstein was deeply disturbed by the development of quantum mechanics in the 1920s despite his own role as a scientific revolutionary. In particular, he could never accept the probabilistic view of events in nature that is a central feature of quantum theory. The last few decades of his life were devoted to an unsuccessful search for a unified theory that would combine gravitation and electromagnetism into one picture.

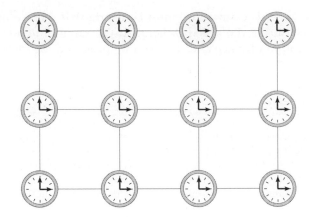

Figure 9.4   In relativity, we use a reference frame consisting of a coordinate grid and a set of synchronized clocks.

Almost everyone who has dabbled even superficially with science is aware of some of the startling predictions that arise because of Einstein's approach to relative motion. As we examine some of the consequences of relativity in the following three sections, we see that they conflict with our basic notions of space and time. We restrict our discussion to the concepts of length, time, and simultaneity, which are quite different in relativistic mechanics than in Newtonian mechanics. For example, **the distance between two points and the time interval between two events depend on the frame of reference in which they are measured.** That is, **there is no such thing as absolute length or absolute time in relativity.** Furthermore, **events at different locations that occur simultaneously in one frame are not simultaneous in another frame moving uniformly past the first.**

Simultaneity and the Relativity of Time

A basic premise of Newtonian mechanics is that a universal time scale exists that is the same for all observers. In fact, Newton wrote, "Absolute, true, and mathematical time, of itself, and from its own nature, flows equably without relation to anything external." Thus, Newton and his followers simply took simultaneity for granted. In his development of special relativity, Einstein abandoned this assumption. According to Einstein, **a time interval measurement depends on the reference frame in which the measurement is made.**

Einstein devised the following thought experiment to illustrate this point. A boxcar moves with uniform velocity, and two lightning bolts strike its ends, as in Figure 9.5a, leaving marks on the boxcar and on the ground. The marks on the boxcar are labeled A' and B', and those on the ground are labeled A and B. An observer at O' moving with the boxcar is midway between A' and B', and a ground observer at O is midway between A and B. The events recorded by the observers are the light signals from the lightning bolts.

The two light signals reach observer O at the same time, as indicated in Figure 9.5b. This observer realizes that the light signals have traveled at the same speed over equal distances, and so rightly concludes that the events at A and B occurred simultaneously. Now consider the same events as viewed by the observer on the boxcar at O'. The lightning strikes as A' passes A, O' passes O, and B' passes B. By

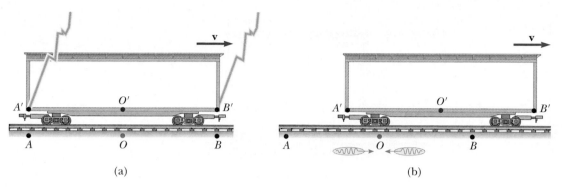

(a)   (b)

Figure 9.5 Two lightning bolts strike the end of a moving boxcar. (a) The events appear to be simultaneous to the stationary observer at *O*, who is midway between *A* and *B*. (b) The events do not appear to be simultaneous to the observer at *O'*, who claims that the front of the train is struck before the rear.

the time the light has reached observer *O*, observer *O'* has moved as indicated in Figure 9.5b. Thus, the light signal from *B'* has already swept past *O'*, and the light from *A'* has not yet reached *O'*. According to Einstein, **observer O' must find that light travels at the same speed as that measured by observer O.** Therefore, the observer *O'* concludes that the lightning struck the front of the boxcar before it struck the back. This thought experiment clearly demonstrates that the two events, which appear to be simultaneous to observer *O*, do not appear to be simultaneous to observer *O'*. In other words,

> two events that are simultaneous in one reference frame are in general not simultaneous in a second frame moving relative to the first. That is, simultaneity is not an absolute concept but one that depends on the state of motion of the observer.

At this point, you might wonder which observer is right concerning the two events. The answer is that *both are correct*, because the principle of relativity states that **there is no preferred inertial frame of reference.** Although the two observers reach different conclusions, both are correct in their own reference frame because the concept of simultaneity is not absolute. This, in fact, is the central point of relativity—any uniformly moving frame of reference can be used to describe events and do physics. However, observers in different inertial frames of reference always measure different time intervals with their clocks and different distances with their meter sticks. Nevertheless, all observers agree on the forms of the laws of physics in their respective frames, because these laws must be the same for all observers in uniform motion. It is the alteration of time and space that allows the laws of physics (including Maxwell's equations) to be the same for all observers in uniform motion.

Time Dilation

The fact that observers in different inertial frames always measure different time intervals between a pair of events can be illustrated by considering a vehicle moving to the right with a speed *v*, as in Figure 9.6a. A mirror is fixed to the ceiling of the

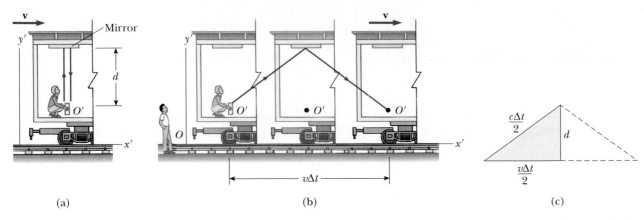

Figure 9.6 (a) A mirror is fixed to a moving vehicle, and a light pulse leaves O' at rest in the vehicle. (b) Relative to a stationary observer on Earth, the mirror and O' move with a speed v. Note that the distance the pulse travels is greater than $2d$ as measured by the stationary observer. (c) The right triangle for calculating the relationship between Δt and $\Delta t'$.

vehicle and observer O' at rest in this system holds a laser a distance d below the mirror. At some instant, the laser emits a pulse of light directed towards the mirror (event 1), and at some later time after reflecting from the mirror, the pulse arrives back at the laser (event 2). Observer O' carries a clock C' that she uses to measure the time interval Δt_p between these two events. Because the light pulse has a speed c, the time it takes the pulse to travel from O' to the mirror and back to O' can be found from the definition of speed:

$$\Delta t_p = \frac{\text{distance traveled}}{\text{speed}} = \frac{2d}{c} \qquad \text{[9.3]}$$

This time interval Δt_p measured by O', who is at rest in the moving vehicle, requires only a single clock C' located at the same place in this frame.

Now consider the same pair of events as viewed by observer O in a second frame, as in Figure 9.6b. According to this observer, the mirror and laser are moving to the right with a speed v. The sequence of events appears entirely different as viewed by this observer. By the time the light from the laser reaches the mirror, the mirror has moved a distance $v\,\Delta t/2$, where Δt is the time it takes the light to travel from O' to the mirror and back to O' as measured by observer O. In other words, the second observer concludes that, because of the motion of the vehicle, if the light is to hit the mirror, it must leave the laser at an angle with respect to the vertical direction. Comparing Figures 9.6a and 9.6b, we see that the light must travel farther in the second frame than in the first frame.

According to the second postulate of special relativity, both observers must measure c for the speed of light. Because the light travels farther in the second frame, it follows that the time interval Δt measured by the observer in the second frame is longer than the time interval Δt_p measured by the observer in the first frame. To obtain a relationship between these two time intervals, it is convenient to use the right triangle shown in Figure 9.6c. The Pythagorean theorem gives

$$\left(\frac{c\,\Delta t}{2}\right)^2 = \left(\frac{v\,\Delta t}{2}\right)^2 + d^2$$

Solving for Δt gives

$$\Delta t = \frac{2d}{\sqrt{c^2 - v^2}} = \frac{2d}{c\sqrt{1 - \dfrac{v^2}{c^2}}} \qquad \text{[9.4]}$$

Because $\Delta t_p = 2d/c$, we can express Equation 9.4 as

$$\Delta t = \frac{\Delta t_p}{\sqrt{1 - \dfrac{v^2}{c^2}}} = \gamma \, \Delta t_p \qquad \text{[9.5]} \qquad \bullet \textit{ Time dilation}$$

where $\gamma = (1 - v^2/c^2)^{-1/2}$. This result says that **the time interval Δt measured by an observer moving with respect to the clock is longer than the time interval Δt_p measured by an observer at rest with respect to the clock** because γ is always greater than unity. That is, $\Delta t > \Delta t_p$. This effect is known as **time dilation.**

The time interval Δt_p in Equation 9.5 is called the **proper time.** In general, proper time is defined as **the time interval between two events measured by an observer who sees the events occur at the same point in space.** In our case, observer O' measures the proper time. That is, **proper time is always the time interval measured with a single clock at rest in the frame in which the events take place at the same position.** $\bullet \textit{ Proper time}$

Because the time between ticks of a moving clock, $\gamma(2d/c)$, is observed to be longer than the time between ticks of an identical clock at rest, $2d/c$, it is often said, "*A moving clock runs slower than a clock at rest by a factor γ.*" This is true for ordinary mechanical clocks as well as for the light clock just described. In fact, we can generalize these results by stating that **all physical processes, including chemical and biological ones, slow down relative to a stationary clock when those processes occur in a moving frame.** For example, the heartbeat of an astronaut moving through space would keep time with a clock inside the spaceship. Both the astronaut's clock and heartbeat are slowed down relative to a stationary clock. The astronaut would not have any sensation of life slowing down in the spaceship. For the astronaut, it is the clock on Earth and the companions at Mission Control that are moving and therefore keep a slow time.

Time dilation is a verifiable phenomenon. For example, muons are unstable elementary particles that have a charge equal to that of the electron and a mass 207 times that of the electron. Muons can be produced by the collision of cosmic radiation with atoms high in the atmosphere. Slow-moving muons have a lifetime of only 2.2 μs in the laboratory. If we take 2.2 μs as the average lifetime of a muon and assume that its speed is close to the speed of light, we find that these particles can travel a distance of only approximately 600 m before they decay (Fig. 9.7a). Hence, they cannot reach the Earth from the upper atmosphere where they are produced. However, experiments show that a large number of muons *do* reach the Earth. The phenomenon of time dilation explains this effect. Relative to an observer on Earth, the muons have a lifetime equal to $\gamma\tau$, where $\tau = 2.2$ μs is the lifetime in a frame of reference traveling with the muons. For example, for $v = 0.99c$, $\gamma \approx 7.1$ and $\gamma\tau \approx 16$ μs. Hence, the average distance traveled as measured by an observer on Earth is $\gamma v\tau \approx 4800$ m, as indicated in Figure 9.7b.

In 1976, at the laboratory of the European Council for Nuclear Research in

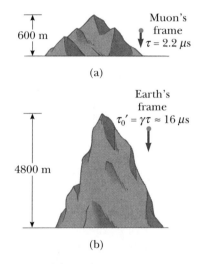

Figure 9.7 (a) Muons traveling with a speed of $0.99c$ travel only approximately 600 m as measured in the muons' reference frame, where their proper lifetime is about 2.2 μs. (b) The muons travel approximately 4800 m as measured by an observer on Earth. Because of time dilation, the muons' lifetime is longer as measured by the Earth observer.

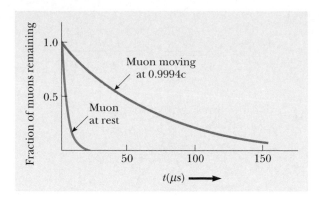

Figure 9.8   Decay curves for muons traveling at a speed of 0.9994c and for muons at rest.

Geneva, muons injected into a large storage ring reached speeds of approximately 0.9994c. Electrons produced by the decaying muons were detected by counters around the ring, enabling scientists to measure the decay rate and hence the muon lifetime. The lifetime of the moving muons was measured to be approximately 30 times as long as that of the stationary muon (Fig. 9.8), in agreement with the prediction of relativity to within 2 parts in 1000.

The results of an experiment reported by Hafele and Keating provided direct evidence of time dilation.[4] The experiment involved the use of very stable cesium-beam atomic clocks. Time intervals measured with four such clocks in jet flight were compared with time intervals measured by reference clocks located at the U.S. Naval Observatory. To compare these results with the theory, many factors had to be considered, including periods of acceleration and deceleration relative to the Earth, variations in direction of travel, and the weaker gravitational field experienced by the flying clocks compared to the Earth-based clock. Their results were in good agreement with the predictions of special relativity and can be explained in terms of the relative motion between the Earth's rotation and the jet aircraft. In their paper, Hafele and Keating report the following: "Relative to the atomic time scale of the U.S. Naval Observatory, the flying clocks lost 59 ± 10 ns during the eastward trip and gained 273 ± 7 ns during the westward trip. . . . These results provide an unambiguous empirical resolution of the famous clock paradox with macroscopic clocks."

Thinking Physics 2

Suppose a student explains time dilation with the following argument: If you start running at 0.99c away from a clock at 12:00, you would not see the time change, because the light from the clock representing 12:01 would never reach you. What is the flaw in this argument?

Reasoning The inference in this argument is that the velocity of light relative to the runner is approximately *zero*—"the light . . . would never reach you." This is a Galilean point of view, in which the relative velocity is a simple subtraction of running

[4]J. C. Hafele and R. E. Keating, "Around the World Atomic Clocks: Relativistic Time Gains Observed," *Science,* July 14, 1972, p. 168.

velocity from the light velocity. From the point of view of special relativity, one of the fundamental postulates is that the speed of light is the same for all observers, *including one running away from the light source at the speed of light.* Thus, the light from 12:01 will move toward the runner at the speed of light.

Example 9.1 What Is the Period of the Pendulum?

The period of a pendulum is measured to be 3.0 s in the rest frame of the pendulum's pivot. What is the period when measured by an observer moving at a speed of 0.95c relative to the pendulum?

Reasoning Instead of the observer moving at 0.95c, we can take the equivalent point of view that the observer is at rest and that the pendulum is moving at 0.95c past the stationary observer. Hence the pendulum is an example of a moving clock.

Solution The proper time is 3.0 s. Because a moving clock runs slower than a stationary clock by γ, Equation 9.5 gives

$$T = \gamma T_p = \frac{1}{\sqrt{1 - \dfrac{(0.95c)^2}{c^2}}} T_p = (3.2)(3.0 \text{ s}) = 9.6 \text{ s}$$

That is, a moving pendulum slows down or takes longer to complete a period compared to one at rest.

EXERCISE 1 Muons move in circular orbits at a speed of 0.9994c in a storage ring of radius 500 m. If a muon at rest decays into other particles after 2.2 μs (proper time), how many trips around the storage ring do we expect the muons to make before they decay?
Answer 6.1 revolutions

Length Contraction

The measured distance between two points also depends on the frame of reference. The **proper length** of an object is defined as **the length of the object measured by someone who is at rest relative to the object.** The length of an object measured by someone in a reference frame that is moving with respect to the object is always less than the proper length. This effect is known as **length contraction.**

• *Proper length*

Consider a spaceship traveling with a speed v from one star to another. There are two observers, one on Earth and the other in the spaceship. The observer at rest on Earth (and also assumed to be at rest with respect to the two stars) measures the distance between the stars to be L_p, the proper length. According to this observer, the time it takes the spaceship to complete the voyage is $\Delta t = L_p / v$. What does an observer in the moving spaceship measure for the distance between the stars? Because of time dilation, the space traveler measures a smaller time of travel: $\Delta t_p = \Delta t / \gamma$. The space traveler claims to be at rest and sees the destination star moving toward the spaceship with speed v. Because the space traveler reaches the star in the time Δt_p, she or he concludes that the distance, L, between the stars is shorter than L_p. This distance measured by the space traveler is

$$L = v \, \Delta t_p = v \frac{\Delta t}{\gamma}$$

Because $L_p = v \, \Delta t$, we see that $L = L_p / \gamma$ or

$$L = L_p \left(1 - \frac{v^2}{c^2} \right)^{1/2}$$

[9.6] • *Length contraction*

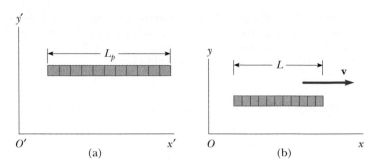

Figure 9.9 (a) A stick as viewed by an observer in a frame attached to the stick (i.e., both have the same velocity). (b) The stick as seen by an observer in a frame in which the stick has a velocity **v** relative to the frame. The length is *shorter* than the proper length, L_p, by a factor $(1 - v^2/c^2)^{1/2}$.

where $(1 - v^2/c^2)^{1/2}$ is a factor less than one. This result may be interpreted as follows:

> A moving observer measures the length L of an object (along the direction of motion) to be shorter than the length L_p measured by an observer at rest with respect to the object (the proper length).

Note that **length contraction takes place only along the direction of motion.** For example, suppose a stick moves past an Earth observer with speed v as in Figure 9.9. The length of the stick as measured by an observer in a frame attached to the stick is the proper length, L_p, as in Figure 9.9a. The length of the stick, L, measured by the Earth observer is shorter than L_p by the factor $(1 - v^2/c^2)^{1/2}$. Furthermore, length contraction is a symmetrical effect: If the stick is at rest on Earth, an observer in the moving frame would also measure its length to be shorter by the same factor $(1 - v^2/c^2)^{1/2}$.

It is important to emphasize that proper length and proper time are measured in different reference frames. As an example of this point, let us return to the decaying muons moving at speeds close to the speed of light. An observer in the muon's reference frame would measure the proper lifetime, whereas an Earth-based observer would measure the proper height of the mountain in Figure 9.7. In the muon's reference frame, there is no time dilation, but the distance of travel is observed to be shorter when measured in this frame. Likewise, in the Earth-based observer's reference frame, there is time dilation, but the distance of travel is measured to be the proper height of the mountain. Thus, when calculations on the muon are performed in both frames, the outcome of the experiment in one frame is the same as the outcome in the other frame!

Example 9.2   A Voyage to Sirius

An astronaut takes a trip to Sirius, located 8 lightyears from Earth. The astronaut measures the time of the one-way journey to be 6 y. If the spaceship moved at a constant speed of $0.8c$, how can the 8-lightyear distance be reconciled with the 6-y duration measured by the astronaut?

Reasoning The 8 lightyears (ly) represents the proper length (distance) from Earth to Sirius measured by an observer seeing both nearly at rest. The astronaut sees Sirius approaching her at $0.8c$ but also sees the distance contracted to

$$\frac{8 \text{ ly}}{\gamma} = (8 \text{ ly})\sqrt{1 - \frac{v^2}{c^2}} = (8 \text{ ly})\sqrt{1 - \frac{(0.8c)^2}{c^2}} = 5 \text{ ly}$$

So the travel time measured on the astronaut's clock is

$$\Delta t = \frac{d}{v} = \frac{5 \text{ ly}}{0.8c} = 6 \text{ y}$$

Example 9.3 The Triangular Spaceship

A spaceship in the form of a triangle flies by an observer with a speed of $0.95c$ along the x-direction. When the ship is at rest relative to the observer (Fig. 9.10a), the distances x and y are measured to be 52 m and 25 m, respectively. What is the shape of the ship as seen by an observer at rest when the ship is in motion along the direction shown in Figure 9.10b?

Solution   The observer sees the horizontal length of the ship to be contracted to a length

$$L = L_p \sqrt{1 - \frac{v^2}{c^2}} = (52 \text{ m}) \sqrt{1 - \frac{(0.95c)^2}{c^2}} = \boxed{16 \text{ m}}$$

The 25-m vertical height is unchanged because it is perpendicular to the direction of relative motion between observer and spaceship. Figure 9.10b represents the shape of the spaceship as seen by the observer at rest.

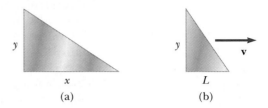

Figure 9.10   (Example 9.3) (a) When the spaceship is at rest, its shape is as shown. (b) The spaceship appears to look like this when it moves to the right with a speed v. Note that only its x dimension is contracted in this case.

EXERCISE 2   A spacecraft moves at $0.9c$. If its length is L_0 when measured from inside the spacecraft, what is its length measured by a ground observer?   Answer $0.436L_0$

EXERCISE 3   A spaceship is measured to be 120 m long while at rest relative to an observer. If this spaceship now flies by the observer with a speed $0.99c$, what length does the observer measure?   Answer   17 m

The Twins Paradox

OPTIONAL

An intriguing consequence of time dilation is the so-called twins paradox. Consider an experiment involving a set of twins named Speedo and Goslo who are, say, 20 years old. The twins carry with them identical clocks that have been synchronized (Fig. 9.11). Speedo, the more adventuresome of the two, sets out on an epic journey

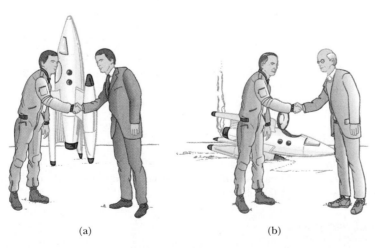

Figure 9.11   (a) As the twins depart, they are the same age. (b) When Speedo returns from his journey to Planet X, he is younger than his twin Goslo who remained on Earth.

to Planet X, located 10 lightyears from Earth. Furthermore, his spaceship is capable of reaching a speed of $0.500c$ relative to the inertial frame of his twin brother on Earth. After reaching Planet X, Speedo becomes homesick and immediately returns to Earth at the same high speed he had attained on the outbound journey. On his return, Speedo is shocked to discover that many things have changed in his absence. To Speedo, the most significant change to have occurred is that his twin brother Goslo has aged 40 years and is now 60 years old. Speedo, on the other hand, has aged by only 34.6 years.

At this point, it is fair to raise the following question—which twin is the traveler and which twin is really younger as a result of this experiment? From Goslo's frame of reference, he was at rest while his brother traveled at a high speed. From Speedo's perspective, it is he who is at rest while Goslo is on the high-speed space journey. According to Speedo, it is Goslo and the Earth that have raced away on a 17.3-year journey and then headed back for another 17.3 years. This leads to an apparent contradiction. Which twin has developed signs of excess aging?

To resolve this apparent paradox, recall that relative to any chosen inertial frame (like that of the stay-at-home twin), clocks in a moving object's frame move more slowly. However, the trip situation is not symmetrical. Speedo, the space traveler, must experience a series of accelerations during his journey. As a result, his speed is not always uniform and as a consequence Speedo is not in a single inertial frame. He cannot be regarded as always being at rest and Goslo to be in uniform motion because he changes his state of motion at least once. Therefore there is no paradox.

The conclusion that Speedo is in a noninertial frame is inescapable. The time required to accelerate and decelerate Speedo's spaceship may be made very small using large rockets, so that Speedo can claim that he spends most of his time traveling to Planet X at $0.500c$ in an inertial frame. However, Speedo must slow down, reverse his motion, and return to Earth in an altogether different inertial frame. At the very best, Speedo is in two different inertial frames during his journey. Only Goslo, who is in a single inertial frame, can apply the simple time dilation formula to Speedo's trip. Thus, Goslo finds that instead of aging 40 years, Speedo ages only $(1 - v^2/c^2)^{1/2}(40 \text{ years}) = 34.6 \text{ years}$. However, Speedo spends 17.3 years traveling to Planet X and 17.3 years returning, for a total travel time of 34.6 years, in agreement with our earlier statement.

CONCEPTUAL PROBLEM 2

You are packing for a trip to another star, to which you will be traveling at $0.99c$. Should you buy smaller sizes of your clothing, since you will be skinnier on your trip? Can you sleep in a smaller cabin than usual, since you will be shorter when you lie down?

CONCEPTUAL PROBLEM 3

Suppose astronauts were paid according to the time spent traveling in space. After a long voyage traveling at a speed near that of light, a crew of astronauts return to Earth and open their pay envelopes. What will their reaction be?

CONCEPTUAL PROBLEM 4

You are observing a rocket moving away from you. You notice that it is measured to be shorter than when it was at rest on the ground next to you, and, through the rocket window, you can see a clock. You observe that the passage of time on the clock is measured to be slower than that of the watch on your wrist. What if the rocket turns around and comes toward you? Will it appear to be *longer* and will the rocketbound clock move *faster*?

9.5 • THE LORENTZ TRANSFORMATION EQUATIONS

We have seen that the Galilean transformation is not valid when v approaches the speed of light. In this section, we state the correct transformation equations that apply for all speeds in the range $0 \leq v < c$.

Suppose an event that occurs at some point P is reported by two observers, one at rest in a frame S and another in a frame S' that is moving to the right with speed v as in Figure 9.12. The observer in S reports the event with space–time coordinates (x, y, z, t), and the observer in S' reports the same event using the coordinates (x', y', z', t'). We would like to find a relationship between these coordinates that is valid for all speeds. In Section 9.1, we found that the Galilean transformation of coordinates, given by Equation 9.1, does not agree with experiment at speeds comparable to the speed of light.

The equations that are valid from $v = 0$ to $v = c$ and enable us to transform coordinates from S to S' are given by the **Lorentz transformation equations:**

$$x' = \gamma(x - vt)$$
$$y' = y$$
$$z' = z$$
$$t' = \gamma\left(t - \frac{v}{c^2}x\right)$$

Lorentz transformation for **[9.7]**
$S \rightarrow S'$

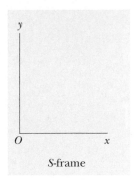

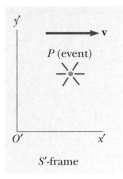

Figure 9.12 Representation of an event that occurs at some point P as observed by an observer at rest in the S frame and another in the S' frame, which is moving to the right with a speed v.

These transformation equations, known as the Lorentz transformation, were developed by Hendrik A. Lorentz (1853–1928) in 1890 in connection with electromagnetism. However, it was Einstein who recognized their physical significance and took the bold step of interpreting them within the framework of special relativity.

We see that the value for t' assigned to an event by an observer standing at O' depends both on the time t and on the coordinate x as measured by an observer at O. This is consistent with the notion that an event is characterized by four space–time coordinates (x, y, z, t). In other words, in relativity, space and time are not separate concepts but rather are closely interwoven with each other. This is unlike the case of the Galilean transformation in which $t = t'$.

If we wish to transform coordinates in the S' frame to coordinates in the S frame, we simply replace v by $-v$ and interchange the primed and unprimed coordinates in Equation 9.7:

$$x = \gamma(x' + vt')$$
$$y = y'$$
$$z = z'$$
$$t = \gamma\left(t' + \frac{v}{c^2}x'\right)$$

Inverse Lorentz **[9.8]**
transformation for $S' \rightarrow S$

When $v \ll c$, the Lorentz transformation should reduce to the Galilean transformation. To check this, note that as $v \rightarrow 0$, $v/c^2 \ll 1$ and $v^2/c^2 \ll 1$, so that

$\gamma = 1$ and Equation 9.7 reduces in this limit to the Galilean coordinate transformation equations

$$x' = x - vt \qquad y' = y \qquad z' = z \qquad t' = t$$

In many situations, we would like to know the difference in coordinates between two events or the time interval between two events as seen by observers at O and O'. This can be accomplished by writing the Lorentz equations in a form suitable for describing pairs of events. From Equations 9.7 and 9.8, we can express the differences between the four variables x, x', t, and t' in the form

$$\Delta x' = \gamma(\Delta x - v\,\Delta t)$$

$$\Delta t' = \gamma\left(\Delta t - \frac{v}{c^2}\,\Delta x\right) \qquad S \rightarrow S' \qquad\qquad \textbf{[9.9]}$$

$$\Delta x = \gamma(\Delta x' + v\,\Delta t')$$

$$\Delta t = \gamma\left(\Delta t' + \frac{v}{c^2}\,\Delta x'\right) \qquad S' \rightarrow S \qquad\qquad \textbf{[9.10]}$$

where $\Delta x' = x_2' - x_1'$ and $\Delta t' = t_2' - t_1'$ are the differences measured by the observer at O', and $\Delta x = x_2 - x_1$ and $\Delta t = t_2 - t_1$ are the differences measured by the observer at O. We have not included the expressions for relating the y and z coordinates because they are unaffected by motion along the x direction.[5]

Example 9.4   Simultaneity and Time Dilation Revisited

Use the Lorentz transformation equations in difference form to show that (a) simultaneity is not an absolute concept and (b) moving clocks run slower than stationary clocks.

Solution   (a) Suppose that two events are simultaneous according to a moving observer at O', so that $\Delta t' = 0$. From the expression for Δt given in Equation 9.10, we see that in this case, $\Delta t = \gamma v\,\Delta x'/c^2$. That is, the time interval for the same two events as measured by an observer at O is nonzero, and so they do not appear to be simultaneous in O.

(b) Suppose that an observer at O' finds that two events occur at the same place ($\Delta x' = 0$), but at different times ($\Delta t' \neq 0$). In this situation, the expression for Δt given in Equation 9.10 becomes $\Delta t = \gamma\,\Delta t'$. This is the equation for time dilation found earlier, Equation 9.5, where $\Delta t' = \Delta t$ is the proper time measured by the single clock located at O'.

EXERCISE 4   Use the Lorentz transformation equations in difference form to confirm that $L = L_p/\gamma$.

Lorentz Velocity Transformation

Let us now derive the Lorentz velocity transformation, which is the relativistic counterpart of the Galilean velocity transformation. Once again S is our stationary frame of reference and S' is our frame of reference that moves at a speed v relative to S. Suppose that an object is observed in the S' frame with an instantaneous speed u_x' measured in S' given by

$$u_x' = \frac{dx'}{dt'} \qquad\qquad \textbf{[9.11]}$$

---

[5]Although motion along x does not change y and z coordinates, it does change velocity components along y and z.

Using Equations 9.7, we have

$$dx' = \gamma(dx - v\,dt) \quad \text{and} \quad dt' = \gamma\left(dt - \frac{v}{c^2}\,dx\right)$$

Substituting these values into Equation 9.11 gives

$$u'_x = \frac{dx'}{dt'} = \frac{dx - v\,dt}{dt - \dfrac{v}{c^2}\,dx} = \frac{\dfrac{dx}{dt} - v}{1 - \dfrac{v}{c^2}\dfrac{dx}{dt}}$$

But dx/dt is just the velocity component u_x of the object measured in S, and so this expression becomes

$$u'_x = \frac{u_x - v}{1 - \dfrac{u_x v}{c^2}}$$

[9.12] • *Lorentz velocity transformation for $S \to S'$*

In a similar way, if the object has velocity components along y and z, the components in S' are

$$u'_y = \frac{u_y}{\gamma\left(1 - \dfrac{u_x v}{c^2}\right)} \quad \text{and} \quad u'_z = \frac{u_z}{\gamma\left(1 - \dfrac{u_x v}{c^2}\right)}$$

[9.13]

When u_x and v are both much smaller than c (the nonrelativistic case), the denominator of Equation 9.12 approaches unity and so $u'_x \approx u_x - v$. This corresponds to the Galilean velocity transformations. In the other extreme, when $u_x = c$, Equation 9.12 becomes

$$u'_x = \frac{c - v}{1 - \dfrac{cv}{c^2}} = \frac{c\left(1 - \dfrac{v}{c}\right)}{1 - \dfrac{v}{c}} = c$$

The speed of light is the speed limit of the Universe.

From this result, we see that an object whose speed approaches c relative to an observer in S also has a speed approaching c relative to an observer in S'—independent of the relative motion of S and S'. Note that this conclusion is consistent with Einstein's second postulate—namely, that the speed of light must be c relative to all inertial frames of reference. Furthermore, the speed of an object can never exceed c. That is, the speed of light is the ultimate speed. We return to this point later when we consider the energy of a particle.

To obtain u_x in terms of u'_x, we replace v by $-v$ in Equation 9.12 and interchange the roles of u_x and u'_x:

$$u_x = \frac{u'_x + v}{1 + \dfrac{u'_x v}{c^2}}$$

[9.14] • *Inverse Lorentz velocity transformation for $S' \to S$*

Example 9.5   Relative Velocity of Spaceships

Two spaceships A and B are moving directly toward each other, as in Figure 9.13. An observer on Earth measures the speed of A to be 0.750c and the speed of B to be 0.850c. Find the velocity of B with respect to A.

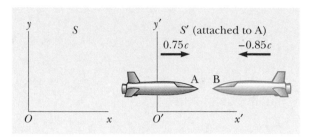

Figure 9.13   (Example 9.5) Two spaceships A and B move in opposite directions. The speed of B relative to A is *less* than c and is obtained by using the relativistic velocity transformation.

Solution   This problem can be solved by taking the S' frame as being attached to A, so that $v = 0.750c$ relative to the Earth observer (the S frame). Spaceship B can be considered as an object moving with a velocity $u_x = -0.850c$ relative to the Earth observer. Hence, the velocity of B with respect to A can be obtained using Equation 9.12:

$$u_x' = \frac{u_x - v}{1 - \frac{u_x v}{c^2}} = \frac{-0.850c - 0.750c}{1 - \frac{(-0.850c)(0.750c)}{c^2}} = -0.980c$$

The negative sign indicates that spaceship B is moving in the negative x direction as observed by A. Note that the result is less than c. That is, a body whose speed is less than c in one frame of reference must have a speed less than c in any other frame. (If the Galilean velocity transformation were used in this example, we would find that $u_x' = u_x - v = -0.850c - 0.750c = -1.60c$, which is greater than c. The Galilean transformation does not work in relativistic situations.)

Example 9.6   Relativistic Leaders of the Pack

Two motorcycle pack leaders named David and Emily are racing at relativistic speeds along perpendicular paths, as in Figure 9.14. How fast does Emily recede over David's right shoulder as seen by David?

Solution   Figure 9.14 represents the situation as seen by a police officer at rest in frame S, who observes the following:

David: $u_x = 0.75c$ $u_y = 0$

Emily: $u_x = 0$ $u_y = -0.90c$

To get Emily's speed of recession as seen by David, we take

S' to move along with David and we calculate u_x' and u_y' for Emily using Equations 9.12 and 9.13:

$$u_x' = \frac{u_x - v}{1 - \frac{u_x v}{c^2}} = \frac{0 - 0.75c}{1 - \frac{(0)(0.75c)}{c^2}} = -0.75c$$

$$u_y' = \frac{u_y}{\gamma\left(1 - \frac{u_x v}{c^2}\right)} = \frac{\sqrt{1 - \frac{(0.75c)^2}{c^2}}(-0.90c)}{\left(1 - \frac{(0)(0.75c)}{c^2}\right)} = -0.60c$$

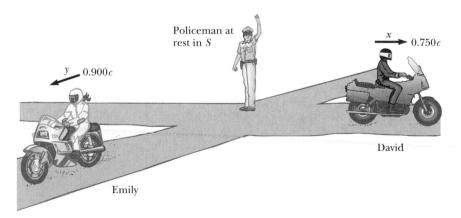

Figure 9.14   (Example 9.6) David moves to the east with a speed 0.750c relative to the policeman, while Emily travels south at a speed 0.900c.

Thus, the speed of Emily as observed by David is

$$u' = \sqrt{(u'_x)^2 + (u'_y)^2} = \sqrt{(-0.75c)^2 + (-0.60c)^2} = 0.96c$$

Note that this speed is less than c as required by special relativity.

EXERCISE 5   Calculate the classical speed of recession for Emily as observed by David using a Galilean transformation.
Answer $1.2c$

EXERCISE 6   A Klingon spaceship moves away from the Earth at a speed of $0.80c$. The starship *Enterprise* pursues at a speed of $0.90c$ relative to the Earth. Observers on Earth see the *Enterprise* overtaking the Klingon ship at a relative speed of $0.10c$. With what speed is the *Enterprise* overtaking the Klingon ship as seen by the crew of the *Enterprise*?
Answer $0.36c$

9.6 • RELATIVISTIC MOMENTUM AND THE RELATIVISTIC FORM OF NEWTON'S LAWS

We have seen that in order to properly describe the motion of particles within the framework of special relativity, the Galilean transformation must be replaced by the Lorentz transformation. Because the laws of physics must remain unchanged under the Lorentz transformation, we must generalize Newton's laws and the definitions of momentum and energy to conform to the Lorentz transform and the principle of relativity. These generalized definitions should reduce to the classical (nonrelativistic) definitions for $v \ll c$.

First, recall that the law of conservation of momentum states that when two bodies collide, the total momentum remains constant, assuming the bodies are isolated. Suppose the collision is described in a reference frame S in which momentum is conserved. If the velocities in a second reference frame S' are calculated using the Lorentz transformation and the classical definition of momentum, $\mathbf{p} = m\mathbf{u}$, it is found that momentum is *not* conserved in the second reference frame. (We use the symbol $\mathbf{u}$ for particle velocity rather than $\mathbf{v}$, which is used for the relative velocity of two reference frames.) However, because the laws of physics are the same in all inertial frames, the momentum must be conserved in all systems. In view of this condition and assuming the Lorentz transformation is correct, we must modify the definition of momentum to satisfy the following conditions:

- $\mathbf{p}$ must be conserved in all collisions
- $\mathbf{p}$ must approach the classical value $m\mathbf{u}$ as $\mathbf{u} \to 0$

The correct relativistic equation for momentum that satisfies these conditions is

$$\mathbf{p} \equiv \frac{m\mathbf{u}}{\sqrt{1 - \dfrac{u^2}{c^2}}} \qquad [9.15]$$

• *Definition of relativistic momentum*

where $\mathbf{u}$ is the velocity of the particle. When u is much less than c, the denominator of Equation 9.15 approaches unity, so that $\mathbf{p}$ approaches $m\mathbf{u}$. Therefore, the relativistic equation for **p** reduces to the classical expression when u is small compared with c. Because it is simpler, Equation 9.15 is often written as

$$\mathbf{p} = \gamma m \mathbf{u} \qquad [9.16]$$

where $\gamma = (1 - u^2/c^2)^{-1/2}$. Note that γ has the same functional form as the γ in the Lorentz transformation. The transformation is from that of the particle to the frame of the observer moving at speed u relative to the particle.

The relativistic force **F** on a particle whose momentum is **p** is defined as

$$\mathbf{F} \equiv \frac{d\mathbf{p}}{dt} \qquad [9.17]$$

where **p** is given by Equation 9.15. This expression is reasonable because it preserves classical mechanics in the limit of low velocities and requires conservation of momentum for an isolated system ($\mathbf{F} = 0$) both relativistically and classically.

It is left as an end-of-chapter problem (Problem 47) to show that the acceleration **a** of a particle decreases under the action of a constant force, in which case $a \propto (1 - u^2/c^2)^{3/2}$. From this formula note that as the particle's speed approaches c, the acceleration caused by any finite force approaches zero. Hence, it is impossible to accelerate a particle from rest to a speed $v \geq c$.

Example 9.7   Momentum of an Electron

An electron, which has a mass of 9.11×10^{-31} kg, moves with a speed of $0.750c$. Find its relativistic momentum and compare this with the momentum calculated from the classical expression.

Solution   Using Equation 9.15 with $u = 0.75c$, we have

$$p = \frac{mu}{\sqrt{1 - \dfrac{u^2}{c^2}}}$$

$$p = \frac{(9.11 \times 10^{-31} \text{ kg})(0.750 \times 3.00 \times 10^8 \text{ m/s})}{\sqrt{1 - \dfrac{(0.750c)^2}{c^2}}}$$

$$= 3.10 \times 10^{-22} \text{ kg} \cdot \text{m/s}$$

The incorrect classical expression gives

$$\text{Momentum} = m_e u = 2.05 \times 10^{-22} \text{ kg} \cdot \text{m/s}$$

Hence, the correct relativistic result is 50% greater than the classical result!

9.7 • RELATIVISTIC ENERGY

We have seen that the definition of momentum and the laws of motion require generalization to make them compatible with the principle of relativity. This implies that the definition of kinetic energy must also be modified.

To derive the relativistic form of the work–kinetic energy theorem, let us start with the definition of the work done on a particle by a force F and use the definition of relativistic force, Equation 9.17:

$$W = \int_{x_1}^{x_2} F\, dx = \int_{x_1}^{x_2} \frac{dp}{dt}\, dx \qquad [9.18]$$

for force and motion both along the x axis. In order to perform this integration, and find the work done on a particle, or the relativistic kinetic energy as a function of u, we first evaluate dp/dt:

$$\frac{dp}{dt} = \frac{d}{dt}\frac{mu}{\sqrt{1 - \dfrac{u^2}{c^2}}} = \frac{m(du/dt)}{\left(1 - \dfrac{u^2}{c^2}\right)^{3/2}}$$

Substituting this expression for dp/dt and $dx = u\,dt$ into Equation 9.18 gives

$$W = \int_0^t \frac{m(du/dt)\,u\,dt}{\left(1 - \dfrac{u^2}{c^2}\right)^{3/2}} = m\int_0^u \frac{u}{\left(1 - \dfrac{u^2}{c^2}\right)^{3/2}}\,du$$

where we have assumed that the particle is accelerated from rest to some final speed u. Evaluating the integral, we find that

$$W = \frac{mc^2}{\sqrt{1 - \dfrac{u^2}{c^2}}} - mc^2 \qquad\qquad \textbf{[9.19]}$$

Recall from Chapter 6 that the work done by a force acting on a particle equals the change in kinetic energy of the particle. Because the initial kinetic energy is zero, we conclude that the work W is equivalent to the relativistic kinetic energy K:

$$K = \frac{mc^2}{\sqrt{1 - \dfrac{u^2}{c^2}}} - mc^2 = \gamma mc^2 - mc^2 \qquad\qquad \textbf{[9.20]}$$ • *Relativistic kinetic energy*

This equation is routinely confirmed by experiments using high-energy particle accelerators.

At low speeds, where $u/c \ll 1$, Equation 9.20 should reduce to the classical expression $K = \frac{1}{2}mu^2$. We can check this by using the binomial expansion $(1 - x^2)^{-1/2} \approx 1 + \frac{1}{2}x^2 + \ \ldots$ for $x \ll 1$, where the higher-order powers of x are neglected in the expansion. In our case, $x = u/c$, so that

$$\frac{1}{\sqrt{1 - \dfrac{u^2}{c^2}}} = \left(1 - \frac{u^2}{c^2}\right)^{-1/2} \approx 1 + \frac{1}{2}\frac{u^2}{c^2} + \cdots$$

Substituting this into Equation 9.20 gives

$$K \approx mc^2\left(1 + \frac{1}{2}\frac{u^2}{c^2} + \cdots\right) - mc^2 = \frac{1}{2}mu^2$$

which agrees with the classical result. Figure 9.15 shows a comparison of the speed–kinetic energy relationships for a particle using the nonrelativistic expression for K (the blue curve) and the relativistic expression for K (the brown curve). The curves are in good agreement at low speeds but deviate at higher speeds. The nonrelativistic expression indicates a violation of physical law because it suggests that sufficient energy can be added to the particle to accelerate it to a speed larger than c. In the relativistic case, the particle speed never exceeds c, regardless of the kinetic energy. When an object's speed is less than one tenth the speed of light, the classical kinetic energy equation differs by less than 1% from the relativistic

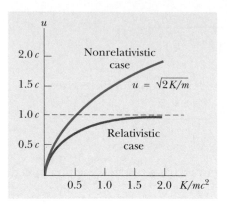

Figure 9.15 A graph comparing relativistic and nonrelativistic kinetic energy. The speeds are plotted versus energy. In the relativistic case, u is always less than c.

equation (which is experimentally verified at all speeds). Thus for practical calculations it is valid to use the classical equation when the object's speed is less than $0.1c$.

The constant term mc^2 in Equation 9.20, which is independent of the speed, is called the **rest energy** of the free particle E_R.

Rest energy •

$$E_R = mc^2 \qquad \text{[9.21]}$$

The term γmc^2 in Equation 9.20 depends on the particle speed and is the sum of the kinetic and rest energies. We define γmc^2 to be the **total energy** E—that is, total energy = kinetic energy + rest energy, or

$$E = \gamma mc^2 = K + mc^2 \qquad \text{[9.22]}$$

or, when γ is replaced by its equivalent,

Definition of total energy •

$$E = \frac{mc^2}{\sqrt{1 - \dfrac{u^2}{c^2}}} \qquad \text{[9.23]}$$

This, of course, is Einstein's famous mass–energy equivalence equation. The relation $E = \gamma mc^2 = \gamma E_R$ shows that **mass is one possible manifestation of energy.** Furthermore, this result shows that a small mass corresponds to an enormous amount of energy. This concept is fundamental to much of the field of nuclear physics.

In many situations, the momentum or energy of a particle is measured instead of its speed. It is therefore useful to have an expression relating the total energy E to the relativistic momentum p. This is accomplished by using the expressions $E = \gamma mc^2$ and $p = \gamma mu$. By squaring these equations and subtracting, we can eliminate u (Problem 35). The result, after some algebra, is

Energy–momentum • *relationship*

$$E^2 = p^2c^2 + (mc^2)^2 \qquad \text{[9.24]}$$

When the particle is at rest, $p = 0$, and so $E = E_R = mc^2$. That is, the total energy equals the rest energy. For the case of particles that have zero mass, such as photons (massless, chargeless particles of light), we set $m = 0$ in Equation 9.24, and we see that

$$E = pc \qquad [9.25]$$

• *Energy of a photon*

This equation is an exact expression relating energy and momentum for photons, which always travel at the speed of light.

Finally, note that because the mass m of a particle is independent of its motion, m must have the same value in all reference frames. For this reason, m is often called the *invariant mass*. However, the total energy and momentum of a particle depend on the reference frame in which they are measured, because they both depend on velocity. Because m is a constant, then according to Equation 9.24 the quantity $E^2 - p^2c^2$ must have the same value in all reference frames. That is, $E^2 - p^2c^2$ is invariant (its value remains the same) under a Lorentz transformation. These equations do not yet make provision for potential energy.

When dealing with subatomic particles, it is convenient to express their energy in electron volts (eV), because the particles are usually given this energy in particle accelerators. The conversion factor is

$$1 \text{ eV} = 1.60 \times 10^{-19} \text{ J}$$

For example, the mass of an electron is 9.11×10^{-31} kg. Hence, the rest energy of the electron is

$$m_e c^2 = (9.11 \times 10^{-31} \text{ kg})(3.00 \times 10^8 \text{ m/s})^2 = 8.20 \times 10^{-14} \text{ J}$$

Converting this to eV, we have

$$m_e c^2 = (8.20 \times 10^{-14} \text{ J})(1 \text{ eV}/1.60 \times 10^{-19} \text{ J}) = 0.511 \text{ MeV}$$

Thinking Physics 3

A common principle learned in chemistry is conservation of mass. In practice, if the mass of the reactants is measured before a reaction and the mass of the products is measured afterward, the results will be the same. In light of special relativity, should we stop teaching the principle of conservation of mass in chemistry classes?

Reasoning Consider a reaction that does not require energy input to occur. This type of reaction occurs because the products represent a lower overall rest energy than the reactants, the difference in rest energy carried away as kinetic energy of ejected particles or radiation. Because the rest energy of the reactants is smaller, according to relativity the mass of the reactants should be smaller than that of the products. Thus the law of conservation of mass is violated. The mass changes are so small, however, that, in practice the law of conservation of mass is still useful.

CONCEPTUAL PROBLEM 5

A photon has zero rest mass. If a photon is reflected from a surface, does it exert a force on the surface?

Example 9.8   The Energy of a Speedy Proton

The total energy of a proton is three times its rest energy.
(a) Find the proton's rest energy in electron volts.

Solution

$$E_R = m_p c^2 = (1.67 \times 10^{-27}\ \text{kg})(3.00 \times 10^8\ \text{m/s})^2$$
$$= (1.50 \times 10^{-10}\ \text{J})(1.00\ \text{eV}/1.60 \times 10^{-19}\ \text{J})$$
$$= \boxed{938\ \text{MeV}}$$

(b) With what speed is the proton moving?

Solution   Because the total energy E is three times the rest energy, $E = \gamma mc^2$ (Eq. 9.23) gives

$$E = 3m_p c^2 = \frac{m_p c^2}{\sqrt{1 - \dfrac{u^2}{c^2}}}$$

$$3 = \frac{1}{\sqrt{1 - \dfrac{u^2}{c^2}}}$$

Solving for u gives

$$\left(1 - \frac{u^2}{c^2}\right) = \frac{1}{9} \quad \text{or} \quad \frac{u^2}{c^2} = \frac{8}{9}$$

$$u = \frac{\sqrt{8}}{3}\,c = \boxed{2.83 \times 10^8\ \text{m/s}}$$

(c) Determine the kinetic energy of the proton in electron volts.

Solution

$$K = E - m_p c^2 = 3m_p c^2 - m_p c^2 = 2m_p c^2$$

Because $m_p c^2 = 938$ MeV, $K = \boxed{1876\ \text{MeV}}$

(d) What is the magnitude of the proton's momentum?

Solution   We can use Equation 9.24 to calculate the momentum with $E = 3m_p c^2$:

$$E^2 = p^2 c^2 + (m_p c^2)^2 = (3m_p c^2)^2$$
$$p^2 c^2 = 9(m_p c^2)^2 - (m_p c^2)^2 = 8(m_p c^2)^2$$
$$p = \sqrt{8}\,\frac{m_p c^2}{c} = \sqrt{8}\,\frac{(938\ \text{MeV})}{c} = \boxed{2650\ \frac{\text{MeV}}{c}}$$

The unit of momentum is written MeV/c for convenience.

EXERCISE 7   Find the speed at which the relativistic kinetic energy is two times the non-relativistic value.   Answer $0.786c$

EXERCISE 8   Find the speed of a particle whose total energy is twice its rest energy.   Answer $0.866c$

EXERCISE 9   An electron moves with a speed $u = 0.850c$. Find its (a) total energy and (b) kinetic energy in electron volts.   Answer   (a) 0.970 MeV   (b) 0.459 MeV

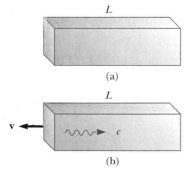

Figure 9.16   (a) A box of length L at rest. (b) When a light pulse is emitted at the left end of the box, the box recoils to the left until the pulse strikes the right end.

9.8 • MASS AS A MEASURE OF ENERGY

To understand the equivalence of mass and energy, consider the following thought experiment proposed by Einstein. Imagine a box of mass M and length L initially at rest, as in Figure 9.16a. Suppose that a pulse of light is emitted from the left side of the box, as in Figure 9.16b. From Equation 9.25, we know that the light of energy E carries momentum $p = E/c$. Hence, the box must recoil to the left with a speed v to conserve momentum. Assuming the box is very massive, the recoil speed is small compared with the speed of light. Conservation of momentum gives $Mv = E/c$, or

$$v = \frac{E}{Mc}$$

The time it takes the light to move the length of the box is approximately $\Delta t = L/c$ (where again, we assume that $v \ll c$). In this time interval, the box moves a small distance Δx to the left, where

$$\Delta x = v\,\Delta t = \frac{EL}{Mc^2}$$

The light then strikes the right end of the box, transfers its momentum to the box, causing the box to stop. With the box in its new position, it appears as if its center of mass has moved to the left. However, its center of mass cannot move because the box is an isolated system. Einstein resolved this perplexing situation by assuming that in addition to energy and momentum, light energy also has kinetic mass. If m_k is the kinetic mass carried by the pulse of light, and the center of mass of the box is to remain fixed, then

$$m_k L = M\,\Delta x$$

Solving for m_k, and using the previous expression for Δx, we get

$$m_k = \frac{M\,\Delta x}{L} = \frac{M}{L}\frac{EL}{Mc^2} = \frac{E}{c^2}$$

or

$$E = m_k c^2$$

Thus, Einstein reached the profound conclusion, "If a body gives off the energy E in the form of radiation, its mass diminishes by E/c^2 . . . The mass of a body is a measure of its energy content."

Although we derived the relationship $E = mc^2$ for light energy, the equivalence of mass and energy is universal. Equation 9.22, $E = \gamma mc^2$, which represents the total energy of any particle, suggests that even when a particle is at rest ($\gamma = 1$) it still possesses enormous energy through its mass. The clearest experimental proof of the equivalence of mass and energy occurs in nuclear and elementary particle interactions in which both the conversion of mass into energy and the conversion of energy into mass takes place. Because of this, we can no longer accept the separate classical laws of conservation of mass and conservation of energy; we must instead speak of a unified law of **conservation of mass–energy.** Simply put, this law requires that **the sum of the mass–energy of a system of particles before interaction must equal the sum of the mass–energy of the system after interaction,** where the mass–energy of the ith particle is defined as

• *Conservation of mass–energy*

$$E_i = \frac{m_i c^2}{\sqrt{1 - \dfrac{u_i^2}{c^2}}} \qquad \textbf{[9.26]}$$

The release of enormous energy, accompanied by the change in masses of particles after they have lost their excess energy as they are brought to rest, is the basis of atomic and hydrogen bombs. In a conventional nuclear reactor, the uranium nucleus undergoes fission, a reaction that results in several lighter fragments having considerable kinetic energy. In the case of ^{235}U (the parent nucleus), which undergoes fission, the fragments are two lighter nuclei and two neutrons. The total mass of the fragments is less than that of the parent nucleus by an amount Δm. The corresponding energy Δmc^2 associated with this mass difference is exactly equal to the total kinetic energy of the fragments. This kinetic energy is then used to produce hot water and steam for the generation of electrical power.

Next, consider the basic fusion reaction in which two deuterium atoms combine to form one helium atom. This reaction is of major importance in current research and development of controlled-fusion reactors. The decrease in mass that results from the creation of one helium atom from two deuterium atoms is $\Delta m = 4.25 \times 10^{-29}$ kg. Hence, the corresponding energy that results from one fusion reaction is $\Delta mc^2 = 3.83 \times 10^{-12}$ J $= 23.9$ MeV. To appreciate the magnitude of this result, if 1 g of deuterium is converted to helium, the energy released is about 10^{12} J! At the 1997 cost of electrical energy, this would be worth about \$65 000.

Example 9.9   Binding Energy of the Deuteron

The mass of the deuteron, which is the nucleus of "heavy hydrogen," is not equal to the sum of the masses of its constituents, which are the proton and neutron. Calculate this mass difference and determine its energy equivalence.

Solution   Using atomic mass units (u), we have

$$m_p = \text{mass of proton} = 1.007\ 276\ \text{u}$$

$$m_n = \text{mass of neutron} = 1.008\ 665\ \text{u}$$

$$m_p + m_n = 2.015\ 941\ \text{u}$$

Because the mass of the deuteron is 2.013 553 u (Appen-

dix A), we see that the mass difference Δm is 0.002 388 u. By definition, 1 u $= 1.66 \times 10^{-27}$ kg, and therefore

$$\Delta m = \boxed{0.002\ 388\ \text{u} = 3.96 \times 10^{-30}\ \text{kg}}$$

Using $E = \Delta mc^2$, we find that

$$E = \Delta mc^2 = (3.96 \times 10^{-30}\ \text{kg})(3.00 \times 10^8\ \text{m/s})^2$$

$$= 3.56 \times 10^{-13}\ \text{J} = \boxed{2.23\ \text{MeV}}$$

Therefore, the minimum energy required to separate the proton from the neutron of the deuterium nucleus (the binding energy) is 2.23 MeV.

OPTIONAL

9.9 • GENERAL RELATIVITY

Up to this point, we have sidestepped a curious puzzle. Mass has two seemingly different properties: a *gravitational attraction* for other masses and an *inertial* property that resists acceleration. To designate these two attributes, we use the subscripts *g* and *i* and write

$$\text{Gravitational property} \qquad F_g = m_g g$$

$$\text{Inertial property} \qquad F_i = m_i a$$

The value for the gravitational constant G was chosen to make the magnitudes of m_g and m_i numerically equal. Regardless of how G is chosen, however, the strict proportionality of m_g and m_i has been established experimentally to an extremely high degree: a few parts in 10^{12}. Thus, it appears that gravitational mass and inertial mass may indeed be exactly proportional.

But why? They seem to involve two entirely different concepts: a force of mutual gravitational attraction between two masses and the resistance of a single mass to being accelerated, regardless of what kind of force is producing the acceleration. This question, which puzzled Newton and many other physicists over the years, was answered when Einstein published his theory of gravitation, known as *general relativity*, in 1916. Because it is a mathematically complex theory, we merely offer a hint of its elegance and insight.

In Einstein's view, the remarkable coincidence that m_g and m_i seemed to be exactly proportional was evidence for a very intimate and basic connection between the two concepts. He pointed out that no mechanical experiment (such as dropping a mass) could distinguish between the two situations illustrated in Figures 9.17a and 9.17b. In each case, a mass released by the observer undergoes a downward acceleration of g relative to the floor.

Einstein carried this idea further and proposed that *no* experiment, mechanical or otherwise, could distinguish between the two cases. This extension to include all phenomena (not just mechanical ones) has interesting consequences. For example, suppose that a light pulse is sent horizontally across the box, as in Figure 9.17c. The trajectory of the light pulse bends downward as the box accelerates upward to meet it. Therefore, Einstein proposed that a beam of light should also be bent downward by a gravitational field. (No such bending is predicted in Newton's theory of gravitation.)

The two postulates of Einstein's **general relativity** are as follows:

- All the laws of nature have the same form for observers in any frame of reference, whether accelerated or not.
- In the vicinity of any given point, a gravitational field is equivalent to an accelerated frame of reference in the absence of gravitational effects. (This is the *principle of equivalence.*)

The second postulate implies that gravitational mass and inertial mass are completely equivalent, not just proportional. What were thought to be two different types of mass are actually identical.

One interesting effect predicted by general relativity is that time scales are altered by gravity. A clock in the presence of gravity runs more slowly than one in which gravity is negligible. As a consequence, the frequencies of radiation emitted by atoms in the presence of a strong gravitational field are shifted to lower fre-

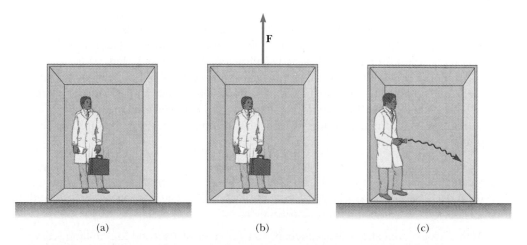

(a)   (b)   (c)

Figure 9.17   (a) The observer is at rest in a uniform gravitational field **g**. (b) The observer is in a region where gravity is negligible, but the frame of reference is accelerated by an external force **F** that produces an acceleration **g**. According to Einstein, the frames of reference in parts (a) and (b) are equivalent in every way. No local experiment could distinguish any difference between the two frames. (c) If parts (a) and (b) are truly equivalent, as Einstein proposed, then a ray of light would bend in a gravitational field.

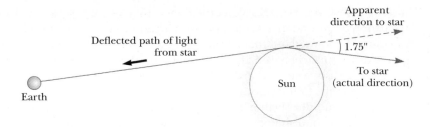

Figure 9.18   Deflection of starlight passing near the Sun. Because of this effect, the Sun and other remote objects can act as a *gravitational lens*. In his general theory of relativity, Einstein calculated that starlight just grazing the Sun's surface should be deflected by an angle of 1.75″.

quencies when compared with the same emissions in a weak field. This gravitational shift has been detected in spectral lines emitted by atoms in massive stars. It has also been verified on the Earth by comparing the frequencies of gamma rays (a high energy form of electromagnetic radiation) emitted from nuclei separated vertically by about 20 m.

The second postulate suggests that a gravitational field may be "transformed away" at any point if we choose an appropriate accelerated frame of reference—a freely falling one. Einstein developed an ingenious method of describing the acceleration necessary to make the gravitational field "disappear." He specified a certain quantity, the *curvature of space–time*, that describes the gravitational effect at every point. In fact, the curvature of space–time completely replaces Newton's gravitational theory. According to Einstein, there is no such thing as a gravitational force. Rather, the presence of a mass causes a curvature of space–time in the vicinity of the mass, and this curvature dictates the space–time path that all freely moving objects must follow. As one physicist said, "Mass one tells space–time how to curve; curved space–time tells mass two how to move." One important test of general relativity is the prediction that a light ray passing near the Sun should be deflected by some angle. This prediction was confirmed by astronomers as bending of starlight during a total solar eclipse shortly following World War I (Fig. 9.18).

If the concentration of mass becomes very great, as is believed to occur when a large star exhausts its nuclear fuel and collapses to a very small volume, a **black hole** may form. Here the curvature of space–time is so extreme that, within a certain distance from the center of the black hole, all matter and light become trapped.

Thinking Physics 4

Atomic clocks are extremely accurate; in fact, an error of 1 second in 3 million years is typical. This error can be described as about 1 part in 10^{14}. However, the atomic clock in Boulder, Colorado, is often 15 ns faster than the one in Washington after only 1 day. This is an error of about 1 part in 6×10^{12}, which is about 17 times larger than the previously expressed error. If atomic clocks are so accurate, why does a clock in Boulder not remain in synchronism with one in Washington? (*Hint:* Denver, near Boulder, is known as the mile-high city.)

Reasoning According to the general theory of relativity, the rate of passage of time depends on gravity—time runs more slowly in strong gravitational fields. Washington is at an elevation very close to sea level, whereas Boulder is about 1 mile higher in altitude. This will result in a weaker gravitational field at Boulder than at Washington. As a result, time runs more rapidly in Boulder than in Washington.

SUMMARY

The two basic postulates of special relativity are

- All the laws of physics are the same in all inertial reference frames.
- The speed of light in vacuum has the same value, $c = 3.00 \times 10^8$ m/s, in all inertial frames, regardless of the velocity of the observer or the velocity of the source emitting the light.

Three consequences of special relativity are

- Events that are simultaneous for one observer are not simultaneous for another observer who is in motion relative to the first.
- Clocks in motion relative to an observer appear to be slowed down by a factor γ. This is known as **time dilation.**
- Lengths of objects in motion appear to be contracted in the direction of motion.

To satisfy the postulates of special relativity, the Galilean transformations must be replaced by the **Lorentz transformations:**

$$
\begin{aligned}
x' &= \gamma(x - vt) \\
y' &= y \\
z' &= z
\end{aligned}
\qquad\qquad \text{[9.7]}
$$

$$
t' = \gamma\left(t - \frac{v}{c^2} x \right)
$$

where $\gamma = (1 - v^2/c^2)^{-1/2}$.

The relativistic form of the **velocity transformation** is

$$
u'_x = \frac{u_x - v}{1 - \dfrac{u_x v}{c^2}} \qquad\qquad \text{[9.13]}
$$

where u_x is the speed of an object as measured in the S frame and u'_x is its speed measured in the S' frame.

The relativistic expression for the **momentum** of a particle moving with a velocity **u** is

$$
\mathbf{p} \equiv \frac{m\mathbf{u}}{\sqrt{1 - \dfrac{u^2}{c^2}}} = \gamma m\mathbf{u} \qquad\qquad \text{[9.16]}
$$

The relativistic expression for the **kinetic energy** of a particle is

$$
K = \gamma mc^2 - mc^2 \qquad\qquad \text{[9.21]}
$$

where mc^2 is called the **rest energy** of the particle.

The total energy E of a particle is related to the **mass** through the **energy–mass** equivalence expression:

$$E = \gamma mc^2 = \frac{mc^2}{\sqrt{1 - \dfrac{u^2}{c^2}}}$$ [9.23]

The relativistic momentum is related to the total energy through the equation

$$E^2 = p^2 c^2 + (mc^2)^2$$ [9.24]

CONCEPTUAL QUESTIONS

1. What two speed measurements do two observers in relative motion *always* agree on?

2. The speed of light in water is 2.3×10^8 m/s. Suppose an electron is moving through water at 2.5×10^8 m/s. Does this violate the principles of relativity?

3. Two identical clocks are synchronized. One is put in orbit directed eastward around the Earth and the other remains on Earth. Which clock runs slower? When the moving clock returns to Earth, are the two still synchronized?

4. A train is approaching you as you stand next to the tracks. Just as an observer on the train passes, you both begin to play the same Beethoven symphony on portable compact disk players. (a) According to you, whose player finishes the symphony first? (b) According to the observer on the train, whose player finishes the symphony first? (c) Whose player *really* finishes the symphony first?

5. A spaceship in the shape of a sphere moves past an observer on Earth with a speed $0.5c$. What shape does the observer see as the spaceship moves past?

6. Explain why it is necessary, when defining length, to specify that the positions of the ends of a rod are to be measured simultaneously.

7. When we say that a moving clock runs slower than a stationary one, does this imply that there is something physically unusual about the moving clock?

8. Give a physical argument that shows that it is impossible to accelerate an object of mass m to the speed of light, even with a continuous force acting on it.

9. List some ways our day-to-day lives would change if the speed of light were only 50 m/s.

10. It is said that Einstein, in his teenage years, asked the question, "What would I see in a mirror if I carried it in my hands and ran at the speed of light?" How would you answer this question?

11. Some of the distant star-like objects, called quasars, are receding from us at half the speed of light (or greater). What is the speed of the light we receive from these quasars?

12. How is it possible that photons of light, which have zero rest mass, have momentum?

13. Relativistic quantities must make a smooth transition to their Newtonian counterparts as the speed of a system becomes small compared to the speed of light. Explain.

14. If the speed of a particle is doubled, what effect does it have on its momentum? (Assume that its initial speed is less than $c/2$.)

PROBLEMS

Section 9.1   The Principle of Newtonian Relativity

1. A 2000-kg car moving at 20.0 m/s collides with and sticks to a 1500-kg car at rest at a stop sign. Show that momentum is conserved in a reference frame moving at 10.0 m/s in the direction of the moving car.

2. A ball is thrown at 20.0 m/s inside a boxcar moving along the tracks at 40.0 m/s. What is the speed of the ball relative to the ground if the ball is thrown (a) forward? (b) backward? (c) out the side door?

3. In a laboratory frame of reference, an observer notes that Newton's second law is valid. Show that it is also valid for an observer moving at a constant speed, small compared to the speed of light, relative to the laboratory frame.

4. Show that Newton's second law is *not* valid in a reference frame moving past the laboratory frame of Problem 3 with a constant acceleration.

Section 9.4   Consequences of Special Relativity

5. How fast must a meter stick be moving if its length is observed to shrink to 0.500 m?

6. At what speed does a clock have to move in order to run at a rate that is one half the rate of a clock at rest?

7. An astronaut is traveling in a space vehicle that has a speed of $0.500c$ relative to the Earth. The astronaut measures his pulse rate at 75.0 per minute. Signals generated by the astronaut's pulse are radioed to Earth when the vehicle is moving perpendicularly to a line that connects the vehicle

with an Earth observer. What pulse rate does the Earth observer measure? What would be the pulse rate if the speed of the space vehicle were increased to 0.990c?

8. The proper length of one spaceship is three times that of another. The two spaceships are traveling in the same direction and, while both are passing overhead, an Earth observer measures the two spaceships to have the same length. If the slower spaceship is moving with a speed of 0.350c, determine the speed of the faster spaceship.

9. An atomic clock moves at 1000 km/h for 1 hour as measured by an identical clock on Earth. How many nanoseconds slow will the moving clock be at the end of the one-hour interval?

10. If astronauts could travel at $v = 0.950c$, we on Earth would say it takes $(4.20/0.950) = 4.42$ years to reach Alpha Centauri, 4.20 lightyears away. The astronauts disagree. (a) How much time passes on the astronauts' clocks? (b) What distance to Alpha Centauri do the astronauts measure?

11. A spaceship of proper length 300 m takes 0.750 μs to pass an Earth observer. Determine the speed of this spaceship as measured by the Earth observer.

12. A spaceship of proper length L_p takes time t to pass an Earth observer. Determine the speed of this spaceship as measured by the Earth observer.

13. A muon formed high in the Earth's atmosphere travels at speed $v = 0.990c$ for a distance of 4.60 km before it decays into an electron, a neutrino, and an antineutrino ($\mu^- \rightarrow e^- + \nu + \bar{\nu}$). (a) How long does the muon live, as measured in its reference frame? (b) How far does the muon travel, as measured in its frame?

Section 9.5 The Lorentz Transformation Equations

14. For what value of v does $\gamma = 1.01$? Observe that for speeds lower than this value, time dilation and length contraction will be less-than-one-percent effects.

15. A spaceship travels at 0.750c relative to Earth. If the spaceship fires a small rocket in the forward direction, how fast (relative to the ship) must it be fired for it to travel at 0.950c relative to Earth?

16. A certain quasar recedes from the Earth at $v = 0.870c$. A jet of material ejected from the quasar toward the Earth moves at 0.550c relative to the quasar. Find the speed of the ejected material relative to the Earth.

17. Two jets of material from the center of a radio galaxy fly away in opposite directions. Both jets move at 0.750c relative to the galaxy. Determine the speed of one jet relative to the other.

18. A friend travels by you at a high speed in a spaceship. He tells you that his ship is 20.0 m long and that the identically constructed ship you are sitting in is 19.0 m long. According to your observations, (a) how long is your ship, (b) how long is your friend's ship, and (c) what is the speed of your friend's ship?

19. Observer A measures the lengths of two rods, one stationary, the other moving with a speed of 0.955c. She finds that the rods have the same length. A second observer, B, travels along with the moving rod. What is the ratio of the length of A's rod to the length of B's rod, according to observer B?

20. A moving rod is 2.00 m long, and its length is oriented at an angle of 30.0° with respect to the direction of motion. The rod has a speed of 0.995c. (a) What is the proper length of the rod? (b) What is the orientation angle in the proper frame?

Section 9.6 Relativistic Momentum and the Relativistic Form of Newton's Laws

21. Calculate the momentum of an electron moving with a speed of (a) 0.0100c, (b) 0.500c, (c) 0.900c.

22. The *nonrelativistic* expression for the momentum of a particle, $p = mv$, can be used if $v \ll c$. For what speed does the use of this formula give an error in the momentum of (a) 1.00% and (b) 10.0%?

23. An electron has a momentum that is 90.0% greater than its classical momentum. (a) Find the speed of the electron. (b) How would your result change if the particle were a proton?

24. A golf ball travels with a speed of 90.0 m/s. By what fraction does its relativistic momentum, p, differ from its classical value, mv? That is, find the ratio $(p - mv)/mv$.

25. An unstable particle at rest breaks into two fragments of *unequal* mass. The rest mass of the lighter fragment is 2.50×10^{-28} kg, and that of the heavier fragment is 1.67×10^{-27} kg. If the lighter fragment has a speed of 0.893c after the breakup, what is the speed of the heavier fragment?

26. Show that the speed of an object having momentum p and mass m is

$$v = \frac{c}{\sqrt{1 + (mc/p)^2}}$$

Section 9.7 Relativistic Energy

27. A proton moves at 0.950c. Calculate its (a) rest energy, (b) total energy, and (c) kinetic energy (in eV).

28. Show that, for any object moving at less than one tenth the speed of light, the relativistic kinetic energy agrees with the result of the classical equation $K = \frac{1}{2}mv^2$ to within less than 1%. Thus for most purposes, the classical equation is good enough to describe these objects, whose motion we call *nonrelativistic*.

29. Determine the energy required to accelerate an electron from (a) 0.500c to 0.900c and (b) 0.900c to 0.990c.

30. A spaceship of mass 1.00×10^6 kg is to be accelerated to 0.600c. (a) How much energy does this require? (b) How

many kilograms of matter would it take to provide this much energy?

31. Make an order-of-magnitude estimate of the ratio of mass increase to the rest mass of a flag as you run it up a flagpole. In your solution explain what quantities you take as data and the values you estimate or measure for them.

32. When 1.00 g of hydrogen combines with 8.00 g of oxygen, 9.00 g of water is formed. During this chemical reaction, 2.86×10^5 J of energy is released. How much mass do the constituents of this reaction lose? Is the loss of mass likely to be detectable?

33. A cube of steel has a volume of 1.00 cm^3 and a mass of 8.00 g when at rest on the Earth. If this cube is now given a speed $v = 0.900c$, what is its density as measured by a stationary observer? Note that relativistic density is $m/V = E/c^2V$.

34. An electron has a kinetic energy five times greater than its rest energy. Find (a) its total energy and (b) its speed.

35. Show that the energy–momentum relationship $E^2 = p^2c^2 + (mc^2)^2$ follows from the expressions $E = \gamma mc^2$ and $p = \gamma mu$.

36. An unstable particle with a mass of 3.34×10^{-27} kg is initially at rest. The particle decays into two fragments that fly off with velocities of $0.987c$ and $-0.868c$. Find the rest masses of the fragments. (*Hint:* Conserve both mass–energy and momentum.)

Additional Problems

37. The net nuclear reaction inside the Sun is $4p \rightarrow He^4 + \Delta E$. If the rest mass of each proton is 938.2 MeV and the rest mass of the He^4 nucleus is 3727 MeV, calculate the percentage of the starting mass that is released as energy.

38. An electron has a speed of $0.750c$. Find the speed of a proton that has (a) the same kinetic energy as the electron, (b) the same momentum as the electron.

39. The cosmic rays of highest energy are protons, having kinetic energy on the order of 10^{13} MeV. (a) How long would it take a proton of this energy to travel across the Milky Way galaxy, of diameter 10^5 lightyears, as measured in the proton's frame? (b) From the point of view of the proton, how many kilometers across is the galaxy?

40. A spaceship moves away from the Earth at $0.500c$ and fires a shuttle craft in the forward direction at $0.500c$ relative to the ship. The pilot of the shuttle craft launches a probe at forward speed $0.500c$ relative to the shuttle craft. Determine (a) the speed of the shuttle craft relative to the Earth and (b) the speed of the probe relative to the Earth.

41. The average lifetime of a pi meson in its own frame of reference is 2.60×10^{-8} s. If the meson moves with a speed of $0.950c$, what are (a) its mean lifetime as measured by an observer on Earth and (b) the average distance it travels before decaying, as measured by an observer on Earth?

42. An astronaut wishes to visit the Andromeda galaxy (suppose it is 2.00 million lightyears away), making a one-way trip that will take 30.0 years in the spaceship's frame of reference. Assuming that her speed is constant, how fast must she travel relative to the Earth?

43. A physics professor on Earth gives an exam to her students who are on a rocket ship traveling at speed v relative to Earth. The moment the ship passes the professor, she signals the start of the exam. If she wishes her students to have time T_0 (rocket time) to complete the exam, show that she should wait a time (Earth time) of

$$T = T_0 \sqrt{\frac{1 - v/c}{1 + v/c}}$$

before sending a light signal telling them to stop. (*Hint:* Remember that it takes some time for the second light signal to travel from the professor to the students.)

44. Spaceship I, which contains students taking a physics exam, approaches Earth with a speed of $0.600c$ (relative to Earth), and spaceship II, which contains professors proctoring the exam, moves at $0.280c$ (relative to Earth) directly toward the students. If the professors stop the exam after 50.0 min have passed on their clock, how long does the exam last as measured by (a) the students? (b) an observer on Earth?

45. A supertrain (rest length = 100 m) travels at a speed of $0.950c$ as it passes through a tunnel (rest length = 50.0 m). As seen by a trackside observer, is the train ever completely within the tunnel? If so, with how much space to spare?

46. Energy reaches the upper atmosphere of the Earth from the Sun at the rate of 1.79×10^{17} W. If all of this energy were absorbed by the Earth and not re-emitted, how much would the mass of the Earth increase in 1 year?

47. A charged particle moves along a straight line in a uniform electric field **E** with a speed of v. The electric force on a charge q in an electric field is **E**. If the motion and the electric field are both in the x direction, (a) show that the acceleration of the charge q in the x direction is given by

$$a = \frac{dv}{dt} = \frac{qE}{m}\left(1 - \frac{v^2}{c^2}\right)^{3/2}$$

(b) Discuss the significance of the dependence of the acceleration on the speed. (c) If the particle starts from rest at $x = 0$ at $t = 0$, how would you proceed to find the speed of the particle and its position after a time t has elapsed?

48. Imagine that the entire Sun collapses to a sphere of radius R_g such that the work required to remove a small mass m from the surface would be equal to its rest energy mc^2. This radius is called the *gravitational radius* for the Sun. Find R_g. (It is believed that the ultimate fate of many stars is to collapse to their gravitational radii or smaller.)

Spreadsheet Problems

S1. Astronomers use the Doppler shift in the Balmer series of the hydrogen spectrum to determine the radial speed of a galaxy. The fractional change in the wavelength of the spectral line is given by

$$Z = \frac{\Delta\lambda}{\lambda_0} = \frac{\lambda - \lambda_0}{\lambda_0} = \sqrt{\frac{1 + v/c}{1 - v/c}} - 1$$

Once this quantity is measured for a particular receding galaxy, the speed of recession can be found by solving for v/c in terms of Z. Spreadsheet 9.1 calculates the speed Z values. (a) What is the speed for galaxies having $Z = 0.2, 0.5, 1.0$, and 2.0? (b) The largest Z values, $Z \approx 3.8$, have been measured for several quasars (quasi-stellar radio sources). How fast are these quasars moving away from us?

S2. Astronauts in a starship traveling at a speed v relative to a starbase are given instructions from mission control to call back in 1 h as measured by the starship clocks. Spreadsheet 9.2 calculates how long Mission Control has to wait for the call for different starship speeds. How long does Mission Control have to wait if the ship is traveling at $v = 0.1c, 0.4c, 0.6c, 0.8c, 0.9c, 0.995c, 0.9995c$?

S3. Most astronomers believe that the Universe began at some instant with an explosion called the Big Bang and that the observed recession of the galaxies is a direct result of this explosion. If the galaxies recede from each other at a constant rate, then we expect that the galaxies moving fastest are now farthest away from Earth. This result, called Hubble's law after Edwin Hubble, can be written $v = Hr$, where H can be determined from observation, v is the speed of recession of the galaxy, and r is its distance from Earth. Current estimates of H range from 15 to 30 km/s/Mly, where Mly is the distance light travels in one million years, and a conservative estimate is $H = 20$ km/s/Mly. Use Spreadsheet 9.1 to calculate the distance from Earth for each galaxy in Problem S1.

S4. Design a spreadsheet program to calculate and plot the relativistic kinetic energy (Eq. 9.20) and the classic kinetic energy ($\frac{1}{2}mu^2$) of a macroscopic object. Plot the relativistic and classical energies versus speed on the same graph. (a) For an object of mass $m = 3$ kg, at what speed does the classical kinetic energy underestimate the relativistic value by 1%? 5%? 50%? What is the relativistic kinetic energy at these speeds? Repeat part (a) for (b) an electron and (c) a proton.

ANSWERS TO CONCEPTUAL PROBLEMS

1. This scenario is not possible with light. Water waves move through a medium—the surface of the water—and therefore have an absolute velocity relative to the reference frame in which this medium is at rest. Thus, the relative velocity of the waves and your boat is determined by Galilean relativity. This makes it possible for you to move faster than the waves. Light waves, which do not require a medium, are described by the principles of special relativity. As you detect the light wave ahead of you and moving away from you (which would be a pretty good trick—think about it!), its velocity relative to you is c. No matter how fast you accelerate in that direction, trying to catch up to it, its velocity relative to you is still c. Thus, you will not be able to catch up to the light wave.

2. The answers to both of these questions is *no*. Both your clothing and your sleeping cabin are at rest in your reference frame, thus, they will have their proper length. There will be no change in measured lengths of objects within your spacecraft. Another observer, on a spacecraft traveling at a high speed relative to yours, will measure you as thinner (if your body is oriented in a direction perpendicular to your velocity vector relative to the other observer) or will claim that you are able to fit into a shorter sleeping cabin (if your body is oriented in a direction parallel to your velocity vector relative to the other observer).

3. Assuming that their on-duty time was kept on Earth, they will be pleasantly surprised with a large paycheck. Less time will have passed for the astronauts in their frame of reference than for their employer back on Earth.

4. The incoming rocket will not appear to have a longer length and a faster clock. Length contraction and time dilation depend only on the magnitude of the relative velocity, not on the direction.

5. A reflected photon does exert a force on a surface. Although a photon has zero rest mass, a photon does carry momentum. When it reflects from a surface, there is a change in the momentum, just like the change in momentum of a ball bouncing off the floor. According to the momentum interpretation of Newton's second law, a change in momentum results in a force on the surface. This concept is used in theoretical studies of *space sailing*. These studies propose building nonpowered spacecraft with huge reflective sails oriented perpendicularly to the rays from the Sun. The large number of photons from the Sun reflecting from the surface of the sail will exert a force which, although small, will provide a continuous acceleration. This would allow the spacecraft to travel to other planets without fuel.

10

Rotational Motion

In this chapter we investigate the dynamics of particles and systems of particles moving along a circular path and rigid objects rotating about a fixed axis. We shall encounter such terms as angular displacement, angular velocity, angular acceleration, torque, and angular momentum and show how these quantities are useful in describing rotational motion. Next we shall define a vector product, which is a convenient mathematical tool for expressing such physical quantities as torque and angular momentum.

One of the central points of this chapter is to develop the concept of the angular momentum of a system of particles. By analogy with the conservation of linear momentum (Chapter 8), we shall find that the angular momentum of any isolated system is always constant. The results we derive here will enable us to understand the rotational motions of a diverse range of objects in our environment, from an electron orbiting a nucleus to clusters of galaxies orbiting a common center. Although the focus of this chapter is the circular motion of particles, the exten-

Derek Swinson, professor of ▶ **physics at the University of New Mexico, demonstrating the "gyro-ski" technique. The skier initiates a turn by lifting the axle of the rotating bicycle wheel. The direction of the turn depends on whether the left or right hand is used to lift the axle from the horizontal. Ignoring friction and gravity, the angular momentum of the system (the skier and the bicycle wheel) remains constant.**
(Courtesy of Derek Swinson)

sion of this treatment to rigid bodies is reasonably straightforward and is treated in an optional section.

10.1 · ANGULAR VELOCITY AND ANGULAR ACCELERATION

We began our study of linear motion by defining the terms *displacement, velocity,* and *linear acceleration.* We will take the same basic approach now as we turn to the study of rotational motion. Let us begin by considering a circular disk rotating around a fixed axis that is perpendicular to the disk and goes through the point O (Fig. 10.1). A point P on the disk is at a fixed distance r from the origin and rotates about O in a circle of radius r. In fact, **every point on the disk undergoes circular motion about O.**

It is convenient to represent the position of the point P with its polar coordinates: (r, θ). The origin of the polar coordinates is chosen to coincide with the center of the circle. In this representation, the only coordinate that changes in time is the angle θ; r remains constant. As a point on the disk moves along the circle of radius r from the positive x axis ($\theta = 0$) to the point P, it moves through an arc of length s, which is related to the angular position θ through the relation $s = r\theta$, or

$$\theta = \frac{s}{r} \qquad [10.1]$$

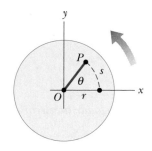

Figure 10.1 Rotation of a disk about a fixed axis through O perpendicular to the plane of the figure. (In other words, the axis of rotation is the z axis.) Note that a particle at P rotates in a circle of radius r centered at O.

It is important to note the units of θ as expressed by Equation 10.1. The angle θ is the ratio of an arc length and the radius of the circle, and hence is a pure number. However, we commonly refer to the unit of θ as a **radian** (rad). One radian is the angle subtended by an arc length equal to the radius of the arc. Because the circumference of a circle is $2\pi r$, it follows that 360° corresponds to an angle of $2\pi r/r$ rad or 2π rad (one revolution). Hence, 1 rad = $360°/2\pi \approx 57.3°$. To convert an angle in degrees to an angle in radians, we can use the fact that 2π radians = 360°, or π radians = 180°; hence,

$$\theta\,(\text{rad}) = \frac{2\pi}{360°}\,\theta\,(\text{deg}) = \frac{\pi}{180°}\,\theta\,(\text{deg})$$

For example, 60° equals $\pi/3$ rad, and 45° equals $\pi/4$ rad.

In Figure 10.2, as the particle travels from P to Q in a time Δt, the radius vector sweeps out an angle of $\Delta\theta = \theta_2 - \theta_1$, which equals the **angular displacement** during the time interval Δt. We define the **average angular speed,** $\overline{\omega}$ (omega), as the ratio of this angular displacement to the time interval Δt:

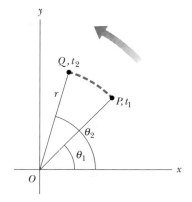

Figure 10.2 A particle on a rotating rigid object moves from P to Q along the arc of a circle. In the time interval $\Delta t = t_2 - t_1$, the radius vector sweeps out an angle $\Delta\theta = \theta_2 - \theta_1$.

$$\overline{\omega} \equiv \frac{\theta_2 - \theta_1}{t_2 - t_1} = \frac{\Delta\theta}{\Delta t} \qquad [10.2]$$

• *Average angular speed*

By analogy with linear speed, the **instantaneous angular speed,** ω, is defined as the limit of the ratio in Equation 10.2 as Δt approaches zero:

$$\omega \equiv \lim_{\Delta t \to 0} \frac{\Delta\theta}{\Delta t} = \frac{d\theta}{dt} \qquad [10.3]$$

• *Instantaneous angular speed*

Angular speed has units of rad/s (or s^{-1} because radians are not dimensional). Let us adopt the convention that the fixed axis of rotation for the disk is the z axis, as in Figure 10.1. We shall take ω to be positive when θ is increasing (counterclockwise motion) and negative when θ is decreasing (clockwise motion).

If the instantaneous angular speed of a particle changes from ω_1 to ω_2 in the time interval Δt, the particle has an angular acceleration. The **average angular acceleration,** $\overline{\alpha}$ (alpha), of a rotating particle is defined as the ratio of the change in the angular speed to the time interval, Δt:

Average angular acceleration •

$$\overline{\alpha} \equiv \frac{\omega_2 - \omega_1}{t_2 - t_1} = \frac{\Delta\omega}{\Delta t} \qquad [10.4]$$

By analogy with linear acceleration, the **instantaneous angular acceleration** is defined as the limit of the ratio $\Delta\omega/\Delta t$ as Δt approaches zero:

Instantaneous angular •
acceleration

$$\alpha \equiv \lim_{\Delta t \to 0} \frac{\Delta\omega}{\Delta t} = \frac{d\omega}{dt} \qquad [10.5]$$

Angular acceleration has units of rad/s^2 or s^{-2}.

Let us now generalize our argument from the circular disk to any rigid body. A rigid body is any object the elements of which remain fixed with respect to one another. **For rotation about a fixed axis, every particle on a rigid body, such as a circular disk, has the same angular velocity and the same angular acceleration.** That is, the vector quantities $\boldsymbol{\omega}$ and $\boldsymbol{\alpha}$ characterize the rotational motion of the *entire* rigid body. Using these quantities, we can greatly simplify the analysis of rigid-body rotation.

The angular displacement ($\boldsymbol{\theta}$), angular velocity ($\boldsymbol{\omega}$), and angular acceleration ($\boldsymbol{\alpha}$) are analogous to linear displacement ($\mathbf{x}$), linear velocity ($\mathbf{v}$), and linear acceleration ($\mathbf{a}$), respectively, for the corresponding motion discussed in Chapter 2. The variables θ, ω, and α differ dimensionally from the linear, variables, x, v, and a, only by a length factor.

We have indicated how the signs for $\boldsymbol{\omega}$ and $\boldsymbol{\alpha}$ are determined, but we have not specified any direction in space associated with these vector quantities.[1] For rotation about a fixed axis, the only direction in space that uniquely specifies the rotational motion is the direction along the axis of rotation. However, we must also decide on the sense of these quantities—that is, whether they point into or out of the plane of Figure 10.1.

The direction of $\boldsymbol{\omega}$ is along the axis of rotation, which is the z axis in Figure 10.1. By convention, we take the direction of $\boldsymbol{\omega}$ to be *out of* the plane of the diagram when the rotation is counterclockwise and *into* the plane of the diagram when the rotation is clockwise. To further illustrate this convention, it is convenient to use the **right-hand rule** illustrated by Figure 10.3a. The four fingers of the right hand are wrapped in the direction of the rotation. The extended right thumb points in

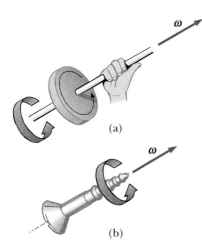

Figure 10.3 (a) The right-hand rule for determining the direction of the angular velocity vector. (b) The direction of $\boldsymbol{\omega}$ is in the direction of advance of a right-handed screw.

---

[1]Although we do not verify it here, the instantaneous angular velocity and instantaneous angular acceleration are vector quantities, but the corresponding average values are not. This is because angular displacement is not a vector quantity for finite rotations.

the direction of $\boldsymbol{\omega}$. Figure 10.3b illustrates that $\boldsymbol{\omega}$ is also in the direction of advance of a similarly rotating right-handed screw.

The sense of $\boldsymbol{\alpha}$ follows from its definition as $d\boldsymbol{\omega}/dt$. It is the same as $\boldsymbol{\omega}$ if the angular speed (the magnitude of $\boldsymbol{\omega}$) is increasing in time and antiparallel to $\boldsymbol{\omega}$ if the angular speed is decreasing in time.

10.2 · ROTATIONAL KINEMATICS

In our study of linear motion, we found that the simplest accelerated motion to analyze is motion under constant linear acceleration (Chapter 2). Likewise, for rotational motion about a fixed axis, the simplest accelerated motion to analyze is motion under constant angular acceleration. Therefore, we shall next develop kinematic relations for rotational motion under constant angular acceleration.

If we write Equation 10.5 in the form $d\omega = \alpha\, dt$ and let $\omega = \omega_0$ at $t_0 = 0$, we can integrate this expression directly:

$$\omega = \omega_0 + \alpha t \qquad (\alpha = \text{constant}) \tag{10.6}$$

Likewise, substituting Equation 10.6 into Equation 10.3 and integrating once more (with $\theta = \theta_0$ at $t_0 = 0$), we get

$$\theta = \theta_0 + \omega_0 t + \tfrac{1}{2}\alpha t^2 \tag{10.7}$$

• *Rotational kinematic equations (α = constant)*

If we eliminate t from Equations 10.6 and 10.7, we get

$$\omega^2 = \omega_0{}^2 + 2\alpha(\theta - \theta_0) \tag{10.8}$$

If we eliminate α, we obtain

$$\theta = \theta_0 + \tfrac{1}{2}(\omega_0 + \omega)t \tag{10.9}$$

Notice that these kinematic expressions for rotational motion under constant angular acceleration are of the *same form* as those for linear motion under constant linear acceleration, with the substitutions $x \to \theta$, $v \to \omega$, and $a \to \alpha$ (Table 10.1). Furthermore, the expressions are valid for both rigid-body rotation about a *fixed* axis and particle motion about a *fixed* axis.

TABLE 10.1 **A Comparison of Kinematic Equations for Rotational and Linear Motion Under Constant Acceleration**

Rotational Motion About a Fixed Axis with α = Constant	Linear Motion with a = Constant
Variables: θ and ω	Variables: x and v
$\omega = \omega_0 + \alpha t$	$v = v_0 + at$
$\theta = \theta_0 + \omega_0 t + \tfrac{1}{2}\alpha t^2$	$x = x_0 + v_0 t + \tfrac{1}{2}at^2$
$\theta = \theta_0 + \tfrac{1}{2}(\omega_0 + \omega)t$	$x = x_0 + \tfrac{1}{2}(v_0 + v)t$
$\omega^2 = \omega_0{}^2 + 2\alpha(\theta - \theta_0)$	$v^2 = v_0{}^2 + 2a(x - x_0)$

Example 10.1   Rotating Wheel

A wheel rotates with a constant angular acceleration of 3.50 rad/s^2. If the angular speed of the wheel is 2.00 rad/s at $t_0 = 0$, (a) what angle does the wheel rotate through in 2.00 s?

Solution

$$\theta - \theta_0 = \omega_0 t + \tfrac{1}{2}\alpha t^2 = \left(2.00 \, \frac{\text{rad}}{\text{s}}\right)(2.00 \text{ s})$$

$$+ \tfrac{1}{2}\left(3.50 \, \frac{\text{rad}}{\text{s}^2}\right)(2.00 \text{ s})^2$$

$$= 11.0 \text{ rad} = 630° = 1.75 \text{ rev}$$

(b) What is the angular speed at $t = 2.00$ s?

Solution

$$\omega = \omega_0 + \alpha t = 2.00 \text{ rad/s} + \left(3.50 \, \frac{\text{rad}}{\text{s}^2}\right)(2.00 \text{ s})$$

$$= 9.00 \text{ rad/s}$$

We could also obtain this result using Equation 10.8 and the results of part (a). Try it!

EXERCISE 1   Find the angle that the wheel rotates through between $t = 2.00$ s and $t = 3.00$ s.   Answer   10.8 rad

---

EXERCISE 2   A wheel starts from rest and rotates with constant angular acceleration to an angular speed of 12.0 rad/s in 3.00 s. Find (a) the magnitude of the angular acceleration of the wheel and (b) the angle in radians through which it rotates in this time.
Answer   (a) 4.00 rad/s^2 (b) 18.0 rad

10.3 · RELATIONS BETWEEN ANGULAR AND LINEAR QUANTITIES

In this section we shall derive some useful relations between the angular speed and angular acceleration of a rotating particle and its linear speed and linear acceleration. Keep in mind that, when a rigid body rotates about a fixed axis, *every* particle of the body moves in a circle whose center is the axis of rotation.

Consider a particle rotating in a circle of radius r about the z axis, as in Figure 10.4. Because the particle moves along a circular path, its linear velocity vector **v** is always tangent to the path; hence, we often call this quantity *tangential velocity*. The magnitude of the tangential velocity of the particle is, by definition, ds/dt, where s is the distance traveled by the particle along the circular path. Recalling from Equation 10.1 that $s = r\theta$, and noting that r is a constant, we get

$$v = \frac{ds}{dt} = r\frac{d\theta}{dt}$$

$$v = r\omega \qquad\qquad [10.10]$$

That is, the magnitude of the tangential velocity of the particle equals the distance of the particle from the axis of rotation multiplied by the particle's angular speed.

We can relate the angular acceleration of the particle to its tangential acceleration, a_t—which is the component of its acceleration tangent to the path of motion—by taking the time derivative of v:

$$a_t = \frac{dv}{dt} = r\frac{d\omega}{dt}$$

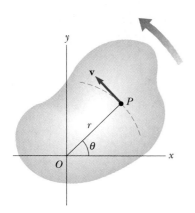

Figure 10.4   As a rigid object rotates about the fixed axis through O, the point P has a linear velocity **v** that is always tangent to the circular path of radius r.

Relationship between linear •
and angular speed

$$a_t = r\alpha \qquad\qquad [10.11]$$

That is, the tangential component of the linear acceleration of a particle undergoing circular motion equals the distance of the particle from the axis of rotation multiplied by the angular acceleration.

In Chapter 3 we found that a particle rotating in a circular path undergoes a centripetal, or radial, acceleration of magnitude v^2/r directed toward the center of rotation (Fig. 10.5). Because $v = r\omega$, we can express the centripetal acceleration of the particle as

$$a_r = \frac{v^2}{r} = r\omega^2 \qquad\qquad [10.12]$$

The *total linear acceleration* of the particle is $\mathbf{a} = \mathbf{a}_t + \mathbf{a}_r$. Therefore, the magnitude of the total linear acceleration of the particle is

$$a = \sqrt{a_t^2 + a_r^2} = \sqrt{r^2\alpha^2 + r^2\omega^4} = r\sqrt{\alpha^2 + \omega^4} \qquad\qquad [10.13]$$

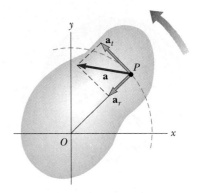

Figure 10.5 As a rigid object rotates about a fixed axis through O, the point P experiences a tangential component of linear acceleration, $\mathbf{a}_t$, and a centripetal component of linear acceleration, $\mathbf{a}_r$. The total linear acceleration of this point is $\mathbf{a} = \mathbf{a}_t + \mathbf{a}_r$.

Thinking Physics 1

A phonograph record (LP—for *long-playing*) is rotated at a constant *angular* speed. A compact disk (CD) is rotated so that the surface sweeps past the laser at a constant *tangential* speed. Consider two circular "grooves" of information on an LP—one near the outer edge and one near the inner edge. Suppose the outer groove "contains" 1.8 seconds of music. Does the inner groove also contain 1.8 seconds of music? How about the CD—do the inner and outer grooves contain the same time interval of music?

Reasoning On the LP, the inner and outer grooves must both rotate once in the same time period. Thus, each groove, regardless of where it is on the record, contains the same time interval of information. Of course, on the inner grooves, this same information must be compressed into a smaller circumference. On a CD, the constant tangential speed requires that there be no such compression—the digital pits representing the information are spaced uniformly everywhere on the surface. Thus, there is more information in an outer groove, due to its larger circumference, and, as a result, a longer time interval of music than the inner groove.

Thinking Physics 2

The launch area for the European Space Agency is not in Europe—it is in French Guiana, near the Equator in South America. Why?

Reasoning Placing a satellite in Earth orbit requires providing a large tangential speed to the satellite. This is the task of the rocket propulsion system. Anything that reduces the requirements on the rocket system is a welcome contribution. The surface of the Earth is already traveling toward the east at a high speed, due to the rotation of the Earth. Thus, if rockets are launched toward the east, the rotation of the Earth provides some initial tangential speed, reducing somewhat the requirements on the rocket. If rockets were launched from Europe, which is at a relatively large angle latitude, the contribution of the Earth's rotation is relatively small, because the distance between Europe and the rotation axis of the Earth is relatively small. The ideal place for launching is at the Equator, which is as far as one can be from the rotation axis of the Earth and still be on the surface of the Earth. This results in the largest possible tangential speed due to the Earth's rotation. The European Space Agency exploits this advantage by launching from French Guiana, which is only a few degrees north of the Equator.

CONCEPTUAL PROBLEM 1

When a tall smokestack falls over, it often breaks somewhere along its length before it hits the ground. Why does this happen?

CONCEPTUAL PROBLEM 2

When a wheel of radius R rotates about a fixed axis as in Figure 10.6, do all points on the wheel have the same angular speed? Do they all have the same linear speed? If the angular speed is constant and equal to ω, describe the linear speeds and linear accelerations of the points located at $r = 0$, $r = R/2$, and $r = R$, where the points are measured from the center of the wheel.

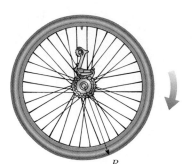

Figure 10.6   (Conceptual Problem 2) A wheel of radius R.

EXERCISE 3   A racing car travels on a circular track of radius 250 m. If the car moves with a constant linear speed of 45.0 m/s, find (a) its angular speed and (b) the magnitude and direction of its acceleration.   Answer   (a) 0.180 rad/s   (b) 8.10 m/s² toward the center of the track

EXERCISE 4   An automobile accelerates from zero to 30 m/s in 6.0 s. The wheels have diameters of 0.40 m. What is the angular acceleration of each wheel?
Answer   25 rad/s²

10.4 • ROTATIONAL KINETIC ENERGY

Let us consider a rigid body as a collection of particles, and let us assume that the body rotates about the fixed z axis with an angular speed of ω (Fig. 10.7). Each particle of the body has some kinetic energy, determined by its mass and speed. If the mass of the ith particle is m_i and its speed is v_i, the kinetic energy of this particle is

$$K_i = \tfrac{1}{2} m_i v_i^2$$

To proceed further, we must recall that, although every particle in the rigid body has the same angular speed ω, the individual linear speeds depend on the distance r_i from the axis of rotation, according to the expression $v_i = r_i \omega$ (Eq. 10.10). The *total* kinetic energy of the rotating rigid body is the sum of the kinetic energies of the individual particles:

$$K_R = \sum K_i = \sum \tfrac{1}{2} m_i v_i^2 = \tfrac{1}{2} \sum m_i r_i^2 \omega^2$$

$$K_R = \tfrac{1}{2} \left(\sum m_i r_i^2 \right) \omega^2$$

where we have factored ω^2 from the sum because it is common to every particle. The quantity in parentheses is called the **moment of inertia,** I:

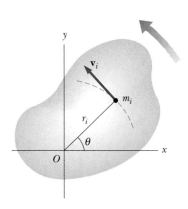

Figure 10.7   A rigid object rotating about the z axis with angular speed ω. The kinetic energy of the particle of mass m_i is $\tfrac{1}{2} m_i v_i^2$. The total energy of the object is $\tfrac{1}{2} I \omega^2$.

Moment of inertia for a •
system of particles

$$I = \sum m_i r_i^2 \qquad\qquad \text{[10.14]}$$

Therefore, we can express the kinetic energy of the rotating rigid body as

$$K_R = \tfrac{1}{2}I\omega^2 \qquad\qquad \text{[10.15]}$$

• *Kinetic energy of a rotating rigid body*

From the definition of moment of inertia, we see that it has dimensions of ML^2 ($kg \cdot m^2$ in SI units). The moment of inertia is a measure of a system's resistance to change in its angular velocity. It plays the role of mass in *all* rotational equations. However, the moment of inertia depends on the location and orientation of the axis, and is not a single property of an object the way its mass is. Although we shall commonly refer to the quantity $\tfrac{1}{2}I\omega^2$ as the **rotational kinetic energy,** it is not a new form of energy. It is ordinary kinetic energy, because it was derived from a sum over individual kinetic energies of the particles contained in the rigid body. However, the form of the kinetic energy given by Equation 10.15 is a convenient one for dealing with rotational motion, provided we know how to calculate *I*. It is important to recognize the analogy between kinetic energy associated with linear motion, $\tfrac{1}{2}mv^2$, and rotational kinetic energy, $\tfrac{1}{2}I\omega^2$. The quantities *I* and ω in rotational motion are analogous to *m* and *v* in linear motion, respectively.

Thinking Physics 3

The temperature of a gas is related to the average *translational* kinetic energy of the molecules of the gas. If we add energy to a gas, the average translational kinetic energy increases and, therefore, the temperature increases. Consider adding the same amount of energy to a monatomic gas, such as helium, and a diatomic gas, such as oxygen (O_2). The temperature of the monatomic gas will rise by more than that of the diatomic gas. Why?

Reasoning   In a monatomic gas, the only type of kinetic energy possible is translational kinetic energy. Thus, all of the energy added to the gas appears as translational kinetic energy, with a corresponding large rise in temperature. In the diatomic gas, some of the energy added to the gas can appear as rotational kinetic energy of the molecules. Thus, only part of the energy added appears as translational kinetic energy, and the temperature rises by a lesser amount than for the monatomic gas.

Example 10.2   The Oxygen Molecule

Consider the diatomic molecule oxygen, O_2, which is rotating in the *xy* plane about the *z* axis passing through its center, perpendicular to its length. The mass of each oxygen atom is 2.66×10^{-26} kg, and at room temperature, the average separation between the two oxygen atoms is $d = 1.21 \times 10^{-10}$ m (the atoms are treated as point masses). (a) Calculate the moment of inertia of the molecule about the *z* axis.

Solution   Because the distance of each atom from the *z* axis is $d/2$, the moment of inertia about the *z* axis is

$$I = \sum m_i r_i^2 = m\left(\frac{d}{2}\right)^2 + m\left(\frac{d}{2}\right)^2 = \frac{md^2}{2}$$

$$= \left(\frac{2.66 \times 10^{-26}}{2}\,\text{kg}\right)(1.21 \times 10^{-10}\text{ m})^2$$

$$= 1.95 \times 10^{-46}\text{ kg}\cdot\text{m}^2$$

(b) If the angular speed of the molecule about the *z* axis is 4.60×10^{12} rad/s, what is its rotational kinetic energy?

Solution

$$K_R = \tfrac{1}{2}I\omega^2$$

$$= \tfrac{1}{2}(1.95 \times 10^{-46}\text{ kg}\cdot\text{m}^2)\left(4.60 \times 10^{12}\,\frac{\text{rad}}{\text{s}}\right)^2$$

$$= 2.06 \times 10^{-22}\text{ J}$$

This is approximately equal to the average kinetic energy associated with the linear motion of the molecule at room temperature, which is about 6.2×10^{-21} J.

Example 10.3   Four Rotating Masses

Four point masses are fastened to the corners of a frame of negligible mass lying in the *xy* plane (Fig. 10.8). (a) If the rotation of the system occurs about the *y* axis with an angular speed ω, find the moment of inertia about the *y* axis and the rotational kinetic energy about this axis.

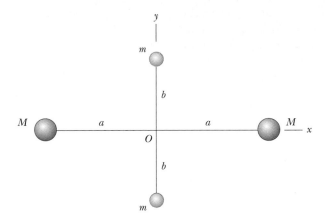

Figure 10.8   (Example 10.3) All four masses are at a fixed separation as shown. The moment of inertia depends on the axis about which it is evaluated.

Solution   First, note that the two masses *m* that lie on the *y* axis do not contribute to I_y (that is, $r_i = 0$ for these masses about this axis). Applying Equation 10.14, we get

$$I_y = \sum m_i r_i^2 = Ma^2 + Ma^2 = 2Ma^2$$

Therefore, the rotational kinetic energy about the *y* axis is

$$K_R = \tfrac{1}{2} I_y \omega^2 = \tfrac{1}{2}(2Ma^2)\,\omega^2 = \boxed{Ma^2\omega^2}$$

The fact that the masses *m* do not enter into this result makes sense, because they have no motion about the chosen axis of rotation; hence, they have no kinetic energy.

(b) Suppose the system rotates in the *xy* plane about an axis through *O* (the *z* axis). Calculate the moment of inertia about the *z* axis and the rotational energy about this axis.

Solution   Because r_i in Equation 10.14 is the *perpendicular* distance to the axis of rotation, we get

$$I_z = \sum m_i r_i^2 = Ma^2 + Ma^2 + mb^2 + mb^2 = \boxed{2Ma^2 + 2mb^2}$$

$$K_R = \tfrac{1}{2} I_z \omega^2 = \tfrac{1}{2}(2Ma^2 + 2mb^2)\,\omega^2 = \boxed{(Ma^2 + mb^2)\omega^2}$$

Comparing the results for (a) and (b), we conclude that the moment of inertia and, therefore, the rotational energy associated with a given angular speed depend on the axis of rotation. In (b), we expect the result to include all masses and distances because all four masses are in motion for rotation in the *xy* plane. Furthermore, the fact that the rotational energy in (a) is smaller than in (b) indicates that it would take less effort (work) to set the system into rotation about the *y* axis than about the *z* axis.

10.5 • TORQUE AND THE VECTOR PRODUCT

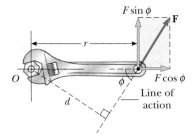

Figure 10.9   The force **F** has a greater rotating tendency about *O* as *F* increases and as the moment arm, *d*, increases. It is the component $F \sin \phi$ that tends to rotate the system about *O*.

When a force is exerted on a rigid body pivoted about some axis, the body tends to rotate about that axis. The tendency of a force to rotate a body about some axis is measured by a vector quantity called **torque** ($\boldsymbol{\tau}$). Consider the wrench pivoted about the axis through *O* in Figure 10.9. The applied force **F** generally can act at an angle of ϕ to the horizontal. We define the magnitude of the torque, $\boldsymbol{\tau}$ (Greek letter tau), resulting from the force **F** with the expression

$$\tau \equiv rF \sin \phi = Fd \qquad \text{[10.16]}$$

It is very important to recognize that **torque is defined only when a reference axis is specified.** The quantity $d = r \sin \phi$, called the **moment arm** (or *lever arm*) of the force **F**, represents the perpendicular distance from the rotation axis to the line of action of **F**. Note that the only component of **F** that tends to cause a rotation is $F \sin \phi$, the component perpendicular to *r*. The horizontal component, $F \cos \phi$, passes through *O* and has no tendency to produce a rotation.

If two or more forces are acting on a rigid body, as in Figure 10.10, then each has a tendency to produce a rotation about the pivot at *O*. For example, $\mathbf{F}_2$ tends to rotate the body clockwise, and $\mathbf{F}_1$ tends to rotate the body counterclockwise. We shall use the convention that the sign of the torque resulting from a force is positive if its turning tendency is counterclockwise and negative if its turning tendency is clockwise. For example, in Figure 10.10, the torque resulting from $\mathbf{F}_1$, which has a moment arm of d_1, is *positive* and equal to $+F_1d_1$; the torque from $\mathbf{F}_2$ is *negative* and equal to $-F_2d_2$. Hence, the *net* torque acting on the rigid body about O is

$$\tau_{net} = \tau_1 + \tau_2 = F_1d_1 - F_2d_2$$

From the definition of torque, we see that the rotating tendency increases as F increases and as d increases. For example, it is easier to close a door if we push at the doorknob rather than at a point close to the hinge. **Torque should not be confused with force.** Torque has units of force times length—N·m in SI units.

Now consider a force $\mathbf{F}$ acting on a particle located at the vector position $\mathbf{r}$ (Fig. 10.11). The origin O is assumed to be in an inertial frame, so Newton's second law is valid. The *magnitude* of the torque due to this force relative to the origin is, by definition, equal to $rF \sin \phi$, where ϕ is the angle between $\mathbf{r}$ and $\mathbf{F}$. The axis about which $\mathbf{F}$ would tend to produce rotation is perpendicular to the plane formed by $\mathbf{r}$ and $\mathbf{F}$. If the force lies in the xy plane, as in Figure 10.11, then the torque is represented by a vector parallel to the z axis. The force in Figure 10.11 creates a torque that tends to rotate the body counterclockwise looking down the z axis, and so the sense of τ is toward increasing z, and τ is in the positive z direction. If we reversed the direction of $\mathbf{F}$ in Figure 10.11, τ would then be in the negative z direction. The torque involves two vectors, $\mathbf{r}$ and $\mathbf{F}$, and is in fact defined to be equal to the **vector product,** or **cross product,** of $\mathbf{r}$ and $\mathbf{F}$:

$$\tau \equiv \mathbf{r} \times \mathbf{F} \qquad [10.17]$$

We now give a formal definition of the vector product. Given any two vectors $\mathbf{A}$ and $\mathbf{B}$, the vector product $\mathbf{A} \times \mathbf{B}$ is defined as a third vector, $\mathbf{C}$, the *magnitude* of which is $AB \sin \theta$, where θ is the angle included between $\mathbf{A}$ and $\mathbf{B}$:

$$\mathbf{C} = \mathbf{A} \times \mathbf{B} \qquad [10.18]$$

$$C \equiv |\mathbf{C}| = |AB \sin \theta| \qquad [10.19]$$

Note that the quantity $AB \sin \theta$ is equal to the area of the parallelogram formed by $\mathbf{A}$ and $\mathbf{B}$, as shown in Figure 10.12. The *direction* of $\mathbf{A} \times \mathbf{B}$ is perpendicular to the plane formed by $\mathbf{A}$ and $\mathbf{B}$, and its sense is determined by the advance of a right-handed screw when the screw is turned from $\mathbf{A}$ to $\mathbf{B}$ through the angle θ. A more convenient rule to use for the direction of $\mathbf{A} \times \mathbf{B}$ is the right-hand rule illustrated in Figure 10.12. The four fingers of the right hand are pointed along $\mathbf{A}$ and then

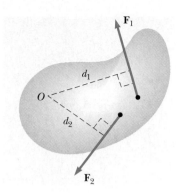

Figure 10.10 The force $\mathbf{F}_1$ tends to rotate the object counterclockwise about O, and $\mathbf{F}_2$ tends to rotate the object clockwise.

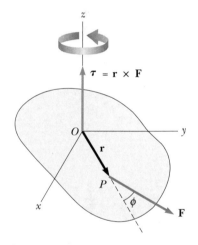

Figure 10.11 The torque vector τ lies in a direction perpendicular to the plane formed by the position vector $\mathbf{r}$ and the applied force $\mathbf{F}$.

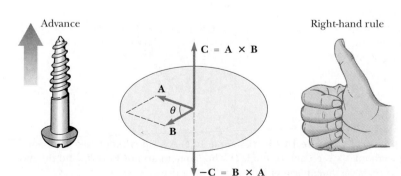

Figure 10.12 The vector product $\mathbf{A} \times \mathbf{B}$ is a third vector $\mathbf{C}$ having a magnitude $AB \sin \theta$ equal to the area of the parallelogram shown. The direction of $\mathbf{C}$ is perpendicular to the plane formed by $\mathbf{A}$ and $\mathbf{B}$, and its sense is determined by the right-hand rule.

"wrapped" into **B** through the angle θ. The direction of the erect right thumb is the direction of **A** × **B**. Because of the notation, **A** × **B** is often read "**A** cross **B**"; hence the term *cross product*.

Some properties of the vector product follow from its definition:

- Unlike the case of the scalar product, the order in which the two vectors are multiplied in a cross product is important; that is

$$\mathbf{A} \times \mathbf{B} = -(\mathbf{B} \times \mathbf{A}) \qquad [10.20]$$

Therefore, if you change the order of the cross product, you must change the sign. One could easily verify this relation with the right-hand rule (Fig. 10.12).

- If **A** is parallel to **B**($\theta = 0°$ or $180°$), then **A** × **B** = 0; therefore, it follows that **A** × **A** = 0.
- If **A** is perpendicular to **B**, then |**A** × **B**| = AB.

From Equations 10.18 and 10.19 and the definition of unit vectors, one finds that the cross products of the rectangular unit vectors **i**, **j**, and **k** obey the following expressions:

Cross products of unit vectors •

$$\mathbf{i} \times \mathbf{i} = \mathbf{j} \times \mathbf{j} = \mathbf{k} \times \mathbf{k} = 0$$
$$\mathbf{i} \times \mathbf{j} = -\mathbf{j} \times \mathbf{i} = \mathbf{k}$$
$$\mathbf{j} \times \mathbf{k} = -\mathbf{k} \times \mathbf{j} = \mathbf{i} \qquad [10.21]$$
$$\mathbf{k} \times \mathbf{i} = -\mathbf{i} \times \mathbf{k} = \mathbf{j}$$

Signs are interchangeable. For example, $\mathbf{i} \times (-\mathbf{j}) = -\mathbf{i} \times \mathbf{j} = -\mathbf{k}$.

The torque exerted on the nail by the hammer increases as the length of the handle (lever arm) increases. *(© Richard Megna, 1991, Fundamental Photographs)*

CONCEPTUAL PROBLEM 3

Both torque and work are products of force and distance. How are they different? Do they have the same units?

Example 10.4   The Net Torque on a Cylinder

A one-piece cylinder is shaped as in Figure 10.13, with a core section protruding from the larger drum. The cylinder is free to rotate around the central axis shown in the drawing. A rope wrapped around the drum, of radius R_1, exerts a force $\mathbf{F}_1$ to the right on the cylinder. A rope wrapped around the core, of radius R_2, exerts a force $\mathbf{F}_2$ downward on the cylinder. (a) What is the net torque acting on the cylinder about the rotation axis (which is the z axis in Fig. 10.13)?

Solution The torque due to $\mathbf{F}_1$ is $-R_1F_1$ and is negative because it tends to produce a clockwise rotation. The torque due to $\mathbf{F}_2$ is $+R_2F_2$ and is positive because it tends to produce a counterclockwise rotation. Therefore, the net torque about the rotation axis is

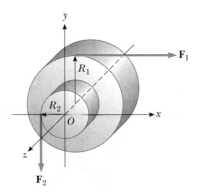

Figure 10.13   (Example 10.4) A solid cylinder pivoted about the z axis through O. The moment arm of $\mathbf{F}_1$ is R_1, and the moment arm of $\mathbf{F}_2$ is R_2.

$$\tau_{net} = \tau_1 + \tau_2 = \boxed{R_2F_2 - R_1F_1}$$

(b) Suppose $F_1 = 5.0$ N, $R_1 = 1.0$ m, $F_2 = 6.0$ N, and $R_2 = 0.50$ m. What is the net torque about the rotation axis and which way does the cylinder rotate?

Solution

$$\tau_{net} = (6.0 \text{ N})(0.50 \text{ m}) - (5.0 \text{ N})(1.0 \text{ m}) = \boxed{-2.0 \text{ N} \cdot \text{m}}$$

Because the net torque is negative, the cylinder rotates clockwise.

Example 10.5   The Cross Product

Two vectors lying in the xy plane are given by the equations $\mathbf{A} = 2\mathbf{i} + 3\mathbf{j}$ and $\mathbf{B} = -\mathbf{i} + 2\mathbf{j}$. Find $\mathbf{A} \times \mathbf{B}$, and verify explicitly that $\mathbf{A} \times \mathbf{B} = -\mathbf{B} \times \mathbf{A}$.

Solution   Using Equations 10.21 for the cross product of unit vectors gives

$$\mathbf{A} \times \mathbf{B} = (2\mathbf{i} + 3\mathbf{j}) \times (-\mathbf{i} + 2\mathbf{j})$$
$$= 2\mathbf{i} \times 2\mathbf{j} + 3\mathbf{j} \times (-\mathbf{i}) = 4\mathbf{k} + 3\mathbf{k} = \boxed{7\mathbf{k}}$$

(We have omitted the terms with $\mathbf{i} \times \mathbf{i}$ and $\mathbf{j} \times \mathbf{j}$, because they are zero.)

$$\mathbf{B} \times \mathbf{A} = (-\mathbf{i} + 2\mathbf{j}) \times (2\mathbf{i} + 3\mathbf{j})$$
$$= -\mathbf{i} \times 3\mathbf{j} + 2\mathbf{j} \times 2\mathbf{i} = -3\mathbf{k} - 4\mathbf{k} = -7\mathbf{k}$$

Therefore, $\mathbf{A} \times \mathbf{B} = -\mathbf{B} \times \mathbf{A}$.

EXERCISE 5   Use the results to this example and Equation 10.19 to find the angle between $\mathbf{A}$ and $\mathbf{B}$.   Answer   60.3°

EXERCISE 6   A particle is located at the vector position $\mathbf{r} = (\mathbf{i} + 3\mathbf{j})$ m, and the force acting on it is $\mathbf{F} = (3\mathbf{i} + 2\mathbf{j})$ N. What is the torque about (a) the origin and (b) the point having coordinates (0,6) m?   Answer   (a) $(-7\mathbf{k})$ N·m   (b) $(11\mathbf{k})$ N·m

EXERCISE 7   If $|\mathbf{A} \times \mathbf{B}| = \mathbf{A} \cdot \mathbf{B}$, what is the angle between $\mathbf{A}$ and $\mathbf{B}$?   Answer   45.0°

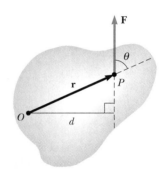

Figure 10.14   A single force $\mathbf{F}$ acts on a rigid object at the point P. The moment arm of $\mathbf{F}$ relative to O is the perpendicular distance d from O to the line of action of $\mathbf{F}$.

10.6 · EQUILIBRIUM OF A RIGID OBJECT

Consider a single force $\mathbf{F}$ acting on a rigid object as in Figure 10.14. The effect of the force depends on its point of application, P. If $\mathbf{r}$ is the position vector of this point relative to O, the torque associated with the force $\mathbf{F}$ about O is given by $\boldsymbol{\tau} = \mathbf{r} \times \mathbf{F}$.

As you can see from Figure 10.14, the tendency of $\mathbf{F}$ to make the object rotate about an axis through O depends on the moment arm d as well as on the magnitude of $\mathbf{F}$. By definition, the magnitude of $\boldsymbol{\tau}$ is Fd.

Now suppose either of two forces, $\mathbf{F}_1$ and $\mathbf{F}_2$, act on a rigid object. The two forces will have the same effect on the object only if they have the same magnitude, the same direction, and the same line of action. In other words, **two forces $\mathbf{F}_1$ and $\mathbf{F}_2$ have an equivalent effect on a rigid body if and only if $F_1 = F_2$ and if the two have the same torque about any given point**.

Two equal and opposite forces that are *not* equivalent are shown in Figure 10.15. The force directed to the right tends to rotate the object clockwise about an axis perpendicular to the diagram through O, whereas the force directed to the left tends to rotate it counterclockwise about that axis.

When pivoted about an axis through its center of mass, an object undergoes an angular acceleration about this axis if there is a nonzero torque acting. As an example, suppose an object is pivoted about an axis through its center of mass as in Figure 10.16. Two equal and opposite forces act in the directions shown, such

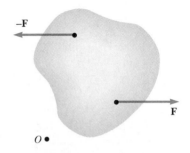

Figure 10.15   The two forces acting on the object are equal in magnitude and opposite in direction, yet the object is not in equilibrium.

that their lines of action do not pass through the center of mass. A pair of forces acting in this manner form what is called a **couple.** (The two forces shown in Figure 10.15 also form a couple.) Because each force produces the same torque, *Fd,* the net torque has a magnitude 2*Fd.* Clearly, the object rotates clockwise and undergoes an angular acceleration about the axis. This is a nonequilibrium situation as far as the rotational motion is concerned.

In general, an object is in rotational equilibrium only if its angular acceleration $\alpha = 0$. Because torque causes angular acceleration, **a necessary condition of equilibrium is that the net torque about any origin must be zero.** We now have **two necessary conditions for equilibrium of an object,** which can be stated as follows:

Two conditions for •
equilibrium of an object

The resultant external force must equal zero.	$\sum \mathbf{F} = 0$	**[10.22]**
The resultant external torque must be zero about *any* origin.	$\sum \boldsymbol{\tau} = 0$	**[10.23]**

The first condition is a statement of translational equilibrium; it tells us that the linear acceleration of the center of mass of the object must be zero when viewed from an inertial reference frame. The second condition is a statement of rotational equilibrium and tells us that the angular acceleration about any axis must be zero. In the special case of **static equilibrium,** the object is at rest so that it has no linear or angular speed (that is, $v_{\mathrm{CM}} = 0$ and $\omega = 0$). Note that if the resultant external force is zero, and the resultant external torque is zero about some origin, then the resultant external torque is zero about every other origin.

The two vector expressions given by Equations 10.22 and 10.23 are equivalent, in general, to six scalar equations, three from the first condition of equilibrium and three from the second (corresponding to *x, y,* and *z* components). Hence, in a complex system involving several forces acting in various directions, you would be faced with solving a set of equations with many unknowns. Here, we restrict our discussion to situations in which all the forces lie in the *xy* plane. (When a set of forces lie in the same plane, they are said to be *coplanar.*) With this restriction, we shall have to deal with only three scalar equations. Two of these come from balancing the forces in the *x* and *y* directions. The third comes from the torque equation—namely, that the net torque about *any* point in the *xy* plane must be zero. Hence, the two conditions of equilibrium provide the equations

$$\sum F_x = 0 \qquad \sum F_y = 0 \qquad \sum \tau_z = 0 \qquad \text{[10.24]}$$

where the axis of the torque equation is arbitrary, as we show later. In working static equilibrium problems, it is important to recognize external forces acting on the object. Failure to do so will result in an incorrect analysis. The following procedure is recommended when analyzing an object in equilibrium under the action of several external forces:

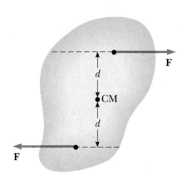

Figure 10.16 Two equal and opposite forces acting on the object form a couple. In this case, the object rotates clockwise. The magnitude of the net torque about the center of mass is 2*Fd.*

PROBLEM-SOLVING STRATEGY • Objects in Equilibrium

1. Make a sketch of the object under consideration.
2. Draw a free-body diagram and label all external forces acting on the object. Try to guess the correct direction for each force. If you select a direction that leads to a negative sign in your solution for a force, do not

be alarmed; this merely means that the direction of the force is the opposite of what you guessed.

3. Resolve all forces into rectangular components, choosing a convenient coordinate system. Then apply the first condition for equilibrium. Remember to keep track of the signs of the various force components.

4. Choose a convenient axis for calculating the net torque on the object. Remember that the choice of the origin for the torque equation is arbitrary; therefore, choose an origin that will simplify your calculation as much as possible. Usually the most convenient origin for calculating torques is the one at which most of the forces act.

5. The first and second conditions of equilibrium give a set of linear equations with several unknowns. All that is left is to solve the simultaneous equations for the unknowns in terms of the known quantities.

Example 10.6   Standing on a Horizontal Beam

A uniform horizontal beam of length 8.00 m and weight 200 N is attached to a wall by a pin connection. Its far end is supported by a cable that makes an angle of 53.0° with the horizontal (Fig. 10.17a). If a 600-N person stands 2.00 m from the wall, find the tension in the cable and the force exerted by the wall on the beam.

Solution   First we must identify all the external forces acting on the beam. These are the gravitational force, the force **T** exerted by the cable, the force **R** exerted by the wall at the pivot (the direction of this force is unknown), and the force exerted by the person on the beam. These are all indicated in the free-body diagram for the beam (Fig. 10.17b). If we resolve **T** and **R** into horizontal and vertical components and apply the first condition for equilibrium, we get

(1) $\quad \sum F_x = R \cos \theta - T \cos 53.0° = 0$

(2) $\quad \sum F_y = R \sin \theta + T \sin 53.0° - 600 \text{ N} - 200 \text{ N} = 0$

Because R, T, and θ are all unknown, we cannot obtain a solution from these expressions alone. (The number of simultaneous equations must equal the number of unknowns in order for us to be able to solve for the unknowns.)

Now let us invoke the condition for rotational equilibrium. A convenient axis to choose for our torque equation is the one that passes through the pivot at O. The feature that makes this point so convenient is that the force **R** and the horizontal component of **T** both have a lever arm of zero, and hence zero torque, about this pivot. Recalling our convention for the sign of the torque about an axis and noting that the lever

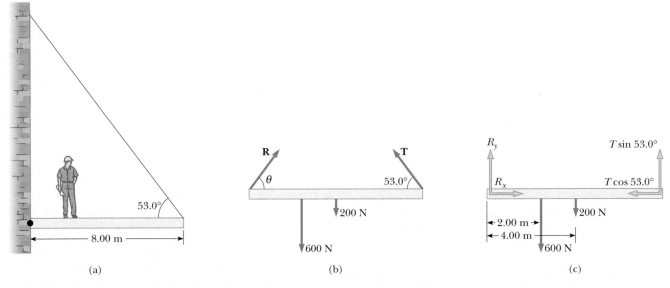

(a)   (b)   (c)

Figure 10.17   (Example 10.6) A uniform beam supported by a cable. (b) and (c) The free-body diagram for the beam.

arms of the 600-N, 200-N, and $T \sin 53°$ forces are 2.00 m, 4.00 m, and 8.00 m, respectively, we get

$$\Sigma \tau_O = (T \sin 53.0°)(8.00 \text{ m}) - (600 \text{ N})(2.00 \text{ m})$$
$$- (200 \text{ N})(4.00 \text{ m}) = 0$$

$$T = 313 \text{ N}$$

Thus the torque equation with this axis gives us one of the unknowns directly! This value is substituted into (1) and (2) to give

$$R \cos \theta = 188 \text{ N}$$

$$R \sin \theta = 550 \text{ N}$$

We divide these two equations and recall the trigonometric identity $\sin \theta / \cos \theta = \tan \theta$ to get

$$\tan \theta = \frac{550 \text{ N}}{188 \text{ N}} = 2.93$$

$$\theta = 71.1°$$

Finally,

$$R = \frac{188 \text{ N}}{\cos \theta} = \frac{188 \text{ N}}{\cos 71.1°} = 581 \text{ N}$$

If we had selected some other axis for the torque equation, the solution would have been the same. For example, if we had chosen to have the axis pass through the center of gravity of the beam, the torque equation would involve both T and R. However, this equation, coupled with (1) and (2), could still be solved for the unknowns. Try it!

When many forces are involved in a problem of this nature, it is convenient to set up a table of forces, lever arms, and torques. For instance, in the example just given, we would construct the following table. Setting the sum of the terms in the last column equal to zero represents the condition of rotational equilibrium.

Force Component	Lever Arm Relative to O (m)	Torque About O (N·m)
$T \sin 53.0°$	8.00	$(8.00)T \sin 53°$
$T \cos 53.0°$	0	0
200 N	4.00	$-(4.00)(200)$
600 N	2.00	$-(2.00)(600)$
$R \sin \theta$	0	0
$R \cos \theta$	0	0

Example 10.7   The Leaning Ladder

A uniform ladder of length ℓ and weight $w = 50$ N rests against a smooth, vertical wall (Fig. 10.18a). If the coefficient of static friction between ladder and ground is $\mu_s = 0.40$, find the minimum angle θ_{min} such that the ladder does not slip.

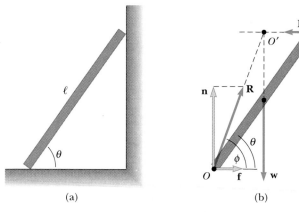

(a)   (b)

Figure 10.18 (Example 10.7) (a) A uniform ladder at rest, leaning against a frictionless wall. (b) The free-body diagram for the ladder. Note that the forces **R**, **w**, and **P** pass through a common point O'.

Solution The free-body diagram showing all the external forces acting on the ladder is illustrated in Figure 10.18b. The

reaction **R** exerted by the ground on the ladder is the vector sum of a normal force, **n**, and the force of friction, **f**. The reaction force **P** exerted by the wall on the ladder is horizontal, because the wall is smooth. From the first condition of equilibrium applied to the ladder, we have

$$\sum F_x = f - P = 0$$
$$\sum F_y = n - w = 0$$

Because $w = 50$ N, we see from the second equation here that $n = w = 50$ N. Furthermore, **when the ladder is on the verge of slipping, the force of friction must be a maximum,** given by $f_{max} = \mu_s n = 0.40(50 \text{ N}) = 20$ N. (Recall Eq. 5.9: $f_s \leq \mu_s n$.) Thus, at this angle, $P = 20$ N.

To find θ, we must use the second condition of equilibrium. When the torques are taken about the origin O at the bottom of the ladder, we get

$$\sum \tau_O = P\ell \sin \theta - w\frac{\ell}{2} \cos \theta - 0$$

But $P = 20$ N when the ladder is about to slip and $w = 50$ N, so that this expression gives

$$\tan \theta_{min} = \frac{w}{2P} = \frac{50 \text{ N}}{40 \text{ N}} = 1.25$$

$$\theta_{min} = 51°$$

It is interesting to note that the result does not depend on ℓ or w. The answer depends only on μ_s.

An alternative approach to analyzing this problem is to consider the intersection O' of the lines of action of forces **w** and **P**. Because the torque about any origin must be zero, the torque about O' must be zero. This requires that the line of action of **R** (the resultant of **n** and **f**) pass through O'. That is, because three forces act on this stationary object, the forces must be concurrent. With this condition, one could then obtain the angle ϕ that **R** makes with the horizontal (where ϕ is greater than θ), assuming the length of the ladder is known.

EXERCISE 8   For the angles labeled in Figure 10.18b, show that $\tan \phi = 2 \tan \theta$.

EXERCISE 9   Two children weighing 500 N and 350 N are on a uniform board weighing 40.0 N supported at its center (a see-saw). If the 500-N child is 1.50 m from the center, determine (a) the upward force exerted on the board by the support and (b) where the 350-N child must sit to balance the system.   Answer   (a) 890 N   (b) 2.14 m from the center

10.7 • RELATION BETWEEN TORQUE AND ANGULAR ACCELERATION

In this section we shall show that the angular acceleration of a particle rotating about a fixed axis is proportional to the net torque acting about that axis. The ideas embodied in this situation can easily be extended to the case of a rigid body rotating about a fixed axis.

Consider a particle of mass m rotating in a circle of radius r under the influence of both a tangential force, $\mathbf{F}_t$, and a radial force, $\mathbf{F}_r$, as in Figure 10.19. (The force $\mathbf{F}_r$, whose nature is not yet specified, *must* be present to keep the particle moving in its circular path.) According to Newton's second law, $\mathbf{F}_t = m\mathbf{a}_t$ and $\mathbf{F}_r = m\mathbf{a}_r$, where $\mathbf{a}_t$ is the tangential acceleration of the particle and $\mathbf{a}_r$ is its centripetal acceleration. The torque due to the force $\mathbf{F}_r$ about the origin is zero because $\mathbf{F}_r$ is antiparallel to $\mathbf{r}$; hence, $\mathbf{r} \times \mathbf{F}_r = 0$. The torque about the origin due to the tangential force is $\mathbf{r} \times \mathbf{F}_t$, but because $\mathbf{r}$ is perpendicular to $\mathbf{F}_t$, the magnitude of the torque is simply $F_t r$. Thus, the magnitude of the net torque on the particle is

$$\sum \tau = F_t r = (ma_t)r$$

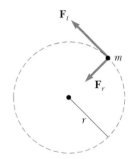

Figure 10.19   A particle rotating in a circle under the influence of a tangential force $\mathbf{F}_t$. A force $\mathbf{F}_r$ must also be present to maintain the circular motion.

Because the tangential acceleration is related to the angular acceleration through Equation 10.11, $a_t = r\alpha$, the torque can be expressed as

$$\sum \tau = (mr\alpha)r = (mr^2)\alpha$$

Recall from Equation 10.14 that the quantity mr^2 is the moment of inertia of the rotating mass about the z axis passing through the origin, so that

$$\sum \tau = I\alpha \qquad\qquad \textbf{[10.25]}$$

• *Relationship between net torque and angular acceleration*

That is, **the net torque acting on the particle is proportional to its angular acceleration,** and the proportionality constant is the moment of inertia. It is important to note that $\Sigma\tau = I\alpha$ is the rotational analogue of Newton's second law of motion, $\Sigma F = ma$.

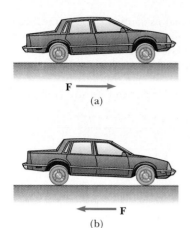

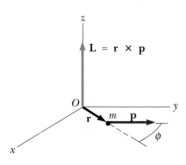

Figure TP10.4
(Thinking Physics 4)

Thinking Physics 4

When an automobile driver steps on the accelerator, the nose of the car moves upward. When the driver brakes, the nose moves downward. Why do these effects occur?

Reasoning When the driver steps on the accelerator, there is an increased force on the tires from the roadway. This force is parallel to the roadway and directed toward the front of the automobile, as suggested in Figure TP10.4a. This force provides a torque about the center of mass that tends to cause the car to rotate in the counterclockwise direction in the diagram. The result of this rotation is a "nosing up" of the car. When the driver steps on the brake, there is an increased force on the tires from the roadway, directed toward the rear of the automobile, as suggested in Figure TP10.4b. This force results in a torque that causes a clockwise rotation and the subsequent "nosing down" of the automobile.

10.8 • ANGULAR MOMENTUM

Again let us consider a particle of mass m, situated at the vector position $\mathbf{r}$ and moving with a momentum $\mathbf{p}$, as shown in Figure 10.20. The **instantaneous angular momentum, L,** of the particle relative to the origin O is defined by the cross product of its instantaneous vector position and the instantaneous linear momentum, $\mathbf{p}$:

• *Angular momentum of a particle*

$$\mathbf{L} = \mathbf{r} \times \mathbf{p} \qquad [10.26]$$

The SI units of angular momentum are $kg \cdot m^2/s$. It is important to note that both the magnitude and the direction of $\mathbf{L}$ depend on the choice of origin. The direction of $\mathbf{L}$ is perpendicular to the plane formed by $\mathbf{r}$ and $\mathbf{p}$, and the sense of $\mathbf{L}$ is governed by the right-hand rule. For example, in Figure 10.20, $\mathbf{r}$ and $\mathbf{p}$ are assumed to be in the xy plane, and $\mathbf{L}$ points in the z direction. Because $\mathbf{p} = m\mathbf{v}$, the magnitude of $\mathbf{L}$ is

$$L = mvr \sin \phi \qquad [10.27]$$

where ϕ is the angle between $\mathbf{r}$ and $\mathbf{p}$. It follows that $\mathbf{L}$ is zero when $\mathbf{r}$ is parallel to $\mathbf{p}$ ($\phi = 0°$ or $180°$). In other words, when the particle moves along a line that passes through the origin, it has zero angular momentum with respect to the origin. This is equivalent to stating that it has no tendency to rotate about the origin. However, if $\mathbf{r}$ is perpendicular to $\mathbf{p}$ ($\phi = 90°$), L is a maximum and equal to mvr. In this case, the particle has maximum tendency to rotate about the origin. In fact, at that instant the particle moves exactly as though it were on the rim of a wheel rotating about the origin in a plane defined by $\mathbf{r}$ and $\mathbf{p}$. A particle has nonzero angular momentum about some point if the position vector of the particle measured from that point rotates about the point.

For linear motion, we found that the resultant force on a particle equals the time rate of change of the particle's linear momentum (Eq. 8.3). We shall now show that Newton's second law implies that the resultant torque acting on a particle equals the time rate of change of the particle's angular momentum. Let us start by writing the torque on the particle in the form

Figure 10.20   The angular momentum $\mathbf{L}$ of a particle of mass m and momentum $\mathbf{p}$ located at the position $\mathbf{r}$ is a vector given by $\mathbf{L} = \mathbf{r} \times \mathbf{p}$. The value of $\mathbf{L}$ depends on the origin and is a vector perpendicular to both $\mathbf{r}$ and $\mathbf{p}$.

$$\boldsymbol{\tau} = \mathbf{r} \times \mathbf{F} = \mathbf{r} \times \frac{d\mathbf{p}}{dt} \qquad \text{[10.28]}$$

where we have used the fact that $\mathbf{F} = d\mathbf{p}/dt$. Now let us differentiate Equation 10.26 with respect to time, using the product rule.

$$\frac{d\mathbf{L}}{dt} = \frac{d}{dt}(\mathbf{r} \times \mathbf{p}) = \mathbf{r} \times \frac{d\mathbf{p}}{dt} + \frac{d\mathbf{r}}{dt} \times \mathbf{p}$$

It is important to adhere to the order of terms, because $\mathbf{A} \times \mathbf{B} = -\mathbf{B} \times \mathbf{A}.$

The last term on the right in the preceding equation is zero because $\mathbf{v} = d\mathbf{r}/dt$ is parallel to $\mathbf{p}$. Therefore,

$$\frac{d\mathbf{L}}{dt} = \mathbf{r} \times \frac{d\mathbf{p}}{dt} \qquad \text{[10.29]}$$

Comparing Equations 10.28 and 10.29, we see that

$$\boldsymbol{\tau} = \frac{d\mathbf{L}}{dt} \qquad \text{[10.30]}$$

• *Torque equals time rate of change of angular momentum*

This result, $\boldsymbol{\tau} = d\mathbf{L}/dt$, is the rotational analog of Newton's second law, $\mathbf{F} = d\mathbf{p}/dt$. Equation 10.30 says that the **torque** acting on a particle is equal to the time rate of change of the particle's angular momentum. It is important to note that Equation 10.30 is valid only if the origins of $\boldsymbol{\tau}$ and $\mathbf{L}$ are the *same*. Equation 10.30 is also valid when several forces are acting on the particle, in which case $\boldsymbol{\tau}$ is the *net* torque on the particle. **Furthermore, the expression is valid for any origin fixed in an inertial frame.** Of course, the same origin must be used in calculating all torques as well as the angular momentum.

A System of Particles

The total angular momentum, $\mathbf{L}$, of a system of particles about some point is defined as the vector sum of the angular momenta of the individual particles:

$$\mathbf{L} = \mathbf{L}_1 + \mathbf{L}_2 + \cdots + \mathbf{L}_n = \sum \mathbf{L}_i$$

where the vector sum is over all of the n particles in the system.

Because the individual momenta of the particles may change in time, the total angular momentum may also vary in time. In fact, from Equations 10.28 and 10.29, we find that the time rate of change of the total angular momentum equals the vector sum of *all* torques, including those associated with internal forces between particles and those associated with external forces. However, the net torque associated with internal forces is zero. To understand this, recall that Newton's third law tells us that the internal forces occur in equal and opposite pairs that lie along the line of separation of each pair of particles. Therefore, the torque due to each action–reaction force pair is zero. By summation, we see that *the net internal torque vanishes.* Finally, we conclude that the total angular momentum can vary with time *only* if there is a net *external* torque on the system, so that we have

$$\sum \boldsymbol{\tau}_{\text{ext}} = \sum \frac{d\mathbf{L}_i}{dt} = \frac{d}{dt} \sum \mathbf{L}_i = \frac{d\mathbf{L}}{dt} \qquad \text{[10.31]}$$

This is an aerial view of Hurricane Fran. The spiral-shaped, nonrigid mass of air undergoes rotation and has angular momentum.   *(Courtesy of NASA)*

That is, the time rate of change of the total angular momentum of the system about some origin in an inertial frame equals the net external torque acting on the system about that origin. Note that Equation 10.31 is the rotational analog of $\mathbf{F}_{ext} = d\mathbf{p}/dt$ (Eq. 8.35) for a system of particles.

Thinking Physics 5

Can a particle moving in a straight line have nonzero angular momentum?

Reasoning A particle has angular momentum about an origin when moving in a straight line as long as its line of motion does not pass through the origin. Its angular momentum is zero only if the line of motion passes through the origin, in which case $\mathbf{r}$ is parallel to $\mathbf{v}$, and $\mathbf{L} = \mathbf{r} \times \mathbf{p} = \mathbf{r} \times m\mathbf{v} = 0$.

Example 10.8   Circular Motion

A particle moves in the xy plane in a circular path of radius r, as in Figure 10.21. (a) Find the magnitude and direction of its angular momentum relative to O when its velocity is $\mathbf{v}$.

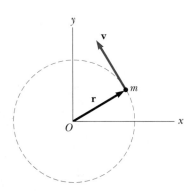

Figure 10.21   (Example 10.8) A particle moving in a circle of radius r has an angular momentum equal in magnitude to mvr relative to the center. The vector $\mathbf{L} = \mathbf{r} \times \mathbf{p}$ (not shown) points *out* of the diagram in this case.

Solution Because $\mathbf{r}$ is perpendicular to $\mathbf{v}$, $\phi = 90°$ and the magnitude of $\mathbf{L}$ is simply

$$L = mvr \sin 90° = mvr \quad \text{(for } \mathbf{r} \text{ perpendicular to } \mathbf{v})$$

The direction of $\mathbf{L}$ is perpendicular to the plane of the circle, and its sense depends on the direction of $\mathbf{v}$. If the sense of the rotation is counterclockwise, as in Figure 10.21, then by the right-hand rule, the direction of $\mathbf{L} = \mathbf{r} \times \mathbf{p}$ is *out* of the paper. Hence, we can write the vector expression for the angular momentum as $\mathbf{L} = (mvr)\mathbf{k}$. If the particle were to move clockwise, $\mathbf{L}$ would point into the paper.

(b) Find an alternative expression for L in terms of the angular velocity, ω.

Solution Because $v = r\omega$ for a particle rotating in a circle, we can express L as

$$L = mvr = m(r\omega)r = mr^2\omega = I\omega$$

where I is the moment of inertia of the particle about the z axis through O. In this case the angular momentum is in the *same* direction as the angular velocity vector, $\boldsymbol{\omega}$ (see Section 10.1), and so we can write $\mathbf{L} = I\boldsymbol{\omega} = I\omega\mathbf{k}$.

Example 10.9   Two Connected Masses

Two masses, m_1 and m_2, are connected by a light cord that passes over a pulley of radius R and moment of inertia I about its axle, as in Figure 10.22. The mass m_2 slides on a frictionless, horizontal surface. Determine the acceleration of the two masses using the concepts of angular momentum and torque.

Reasoning and Solution   First, calculate the angular momentum of the system, which consists of the two masses plus the

pulley. Then calculate the torque about an axis along the axle of the pulley through O. At the instant m_1 and m_2 have a speed v, the angular momentum of m_1 is m_1vR, and that of m_2 is m_2vR. At the same instant, the angular momentum of the pulley is $I\omega = Iv/R$. Therefore, the total angular momentum of the system is

$$(1) \quad L = m_1vR + m_2vR + I\frac{v}{R}$$

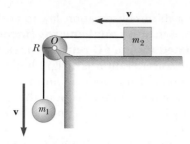

Figure 10.22   (Example 10.9)

Now let us evaluate the total external torque on the system about the axle. Because they have zero moment arms, the forces exerted by the axle on the pulley and by the weight of the pulley do not contribute to the torque. Furthermore, the normal force acting on m_2 is balanced by the force of gravity, $m_2\mathbf{g}$, and so these forces do not contribute to the torque. The external force $m_1\mathbf{g}$ produces a torque about the axle equal in magnitude to m_1gR, where R is the moment arm of the force

about the axle. This is the total external torque about O; that is, $\tau_{\text{ext}} = m_1gR$. Using this result, together with (1) and Equation 10.30 gives

$$\tau_{\text{ext}} = \frac{dL}{dt}$$

$$m_1gR = \frac{d}{dt}\left[(m_1 + m_2)Rv + I\frac{v}{R}\right]$$

$$(2) \quad m_1gR = (m_1 + m_2)R\frac{dv}{dt} + \frac{I}{R}\frac{dv}{dt}$$

Because $dv/dt = a$, we can solve this for a to get

$$a = \frac{m_1g}{(m_1 + m_2) + I/R^2}$$

You may wonder why we did not include the forces that the cord exerts on the objects in evaluating the net torque about the axle. The reason is that these forces are internal to the system under consideration. Only the external torques contribute to the change in angular momentum.

EXERCISE 10   A car of mass 1500 kg moves on a circular racetrack of radius 50 m with a speed of 40 m/s. What is the magnitude of its angular momentum relative to the center of the track?   Answer 3.0×10^6 kg·m²/s

10.9 · CONSERVATION OF ANGULAR MOMENTUM

In Chapter 8 we found that the total linear momentum of a system of particles remains constant when the net external force acting on the system is zero. In rotational motion, we have an analogous conservation law that states that **the total angular momentum of a system remains constant if the net external torque acting on the system is zero.**

Because the net torque acting on the system equals the time rate of change of the system's angular momentum, we see that if

$$\sum \boldsymbol{\tau}_{\text{ext}} = \frac{d\mathbf{L}}{dt} = 0 \qquad \qquad \textbf{[10.32]}$$

then

$$\mathbf{L} = \text{constant} \qquad \qquad \textbf{[10.33]}$$

For a system of particles, we can write this conservation law as $\sum\mathbf{L}_i = $ constant, where $\mathbf{L}_i$ is the angular momentum of the ith particle, given by $\mathbf{L}_i = \mathbf{r}_i \times \mathbf{p}_i$. Hence, we can express the conservation of angular momentum for a system of particles as

$$\sum \mathbf{L}_i = \sum \mathbf{r}_i \times \mathbf{p}_i = \text{constant} \qquad \qquad \textbf{[10.34]}$$

From Equation 10.34 we derive a third conservation law to add to our list of conserved quantities. We can now state that **the total energy, linear momentum, and angular momentum of an isolated system all remain constant.**

If the system of "particles" under consideration is a rigid body rotating about a fixed axis, then the angular momentum of the body about the axis has a magnitude given by $L = I\omega$, where I is the moment of inertia about the axis. In this case, if the net external torque on the body is zero, we can express the conservation of angular momentum as $I\omega$ = constant.

Although we do not prove it here, an important theorem concerns the angular momentum of a system of particles relative to the center of mass of the system. It can be stated as follows:

> The net torque acting on a system of particles about the center of mass equals the time rate of change of angular momentum, regardless of the motion of the center of mass.

This theorem applies even if the center of mass is accelerating, provided that both τ and **L** are evaluated relative to the center of mass.

Many examples can be used to demonstrate conservation of angular momentum; some of them should be familiar to you. You may have observed a figure skater

Figure 10.23 Nancy Kerrigan. In this spin, her angular momentum increases when she pulls her arms in close to her body, demonstrating that angular momentum is conserved. *(Reuters/Corbis-Bettmann)*

Figure 10.24 The Crab Nebula, in the constellation Taurus. This nebula is the remnant of a supernova explosion, which was seen on Earth in the year A.D. 1054. It is located some 6300 lightyears away and is approximately 6 lightyears in diameter, still expanding outward. *(David Malin, Anglo-Australian Observatory)*

spinning (Fig. 10.23). The angular speed of the skater increases as she pulls her hands and feet close to the trunk of her body. Neglecting friction between skater and ice, we see that there are no external torques on the skater. The moment of inertia of her body decreases as her hands and feet are brought in. The resulting change in angular speed is accounted for as follows. Because angular momentum must be conserved, the product $I\omega$ remains constant, and a decrease in I causes a corresponding increase in ω.

An interesting astrophysical example of conservation of angular momentum occurs when, at the end of its lifetime, a massive star uses up all its fuel and collapses under the influence of gravitational forces, causing a gigantic outburst of energy called a supernova explosion. The best-studied example of a remnant of a super-nova explosion is the Crab Nebula, a chaotic, expanding mass of gas (Fig. 10.24). Part of the star's mass is released into space, where it eventually condenses into new stars and planets. Most of what is left behind makes up a **neutron star,** an extremely dense sphere of matter with a diameter of about 10 km in comparison with the 10^6-km diameter of the original star. As the rotational inertia of the system de-creases, the star's rotational speed increases. More than 300 rapidly rotating neu-tron stars have been identified, with periods of rotation ranging from 1.6 ms to 4 s. The neutron star is a most dramatic system—an object with a mass greater than the Sun, rotating about its axis a few times each second!

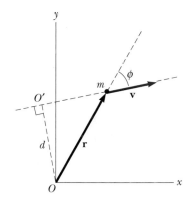

EXERCISE 11   A particle of mass m moves in the xy plane with a velocity **v** along a straight line, as in Figure E10.11. What are the magnitude and direction of its angular momentum (a) with respect to the origin O and (b) with respect to the origin O'?   Answer   (a) mvd into the diagram   (b) zero

Figure E10.11   (Exercise 11)

Example 10.10   A Revolving Ball on a Horizontal, Frictionless Surface

A ball of mass m on a horizontal, frictionless table is con-nected to a string that passes through a small hole in the table. The ball is set into circular motion of radius R, at which time its speed is v_0 (Fig. 10.25). If the string is pulled from

the bottom so that the radius of the circular path is decreased to r, what is the final speed v of the ball?

Solution   Let us take the torque about the center of rotation, O. Note that the force of gravity acting on the ball is balanced by the upward normal force, and so these forces cancel. The force, **F**, of the string on the ball acts toward the center of rotation, and the vector position **r** is directed away from O. Thus, we see that $\boldsymbol{\tau} = \mathbf{r} \times \mathbf{F} = 0$. Because $\boldsymbol{\tau} = d\mathbf{L}/dt = 0$, **L** is a constant of the motion. That is, $mv_0R = mvr$, or

$$v = \frac{v_0 R}{r}$$

From this result, we see that as r decreases, the speed v in-creases.

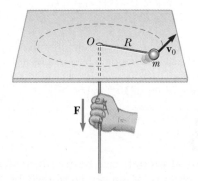

Figure 10.25   (Example 10.10)

10.10 · QUANTIZATION OF ANGULAR MOMENTUM

We have seen that the concept of angular momentum is very useful for describing the motions of macroscopic systems. The concept is also valid and equally useful on a submicroscopic scale, and it has been used extensively in the development of modern theories of atomic, molecular, and nuclear physics. In these developments, it was found that the angular momentum of a system is a *fundamental* quantity. The word "fundamental" in this context implies that angular momentum is an inherent property of atoms, molecules, and their constituents.

In order to explain the results of a variety of experiments on atomic and molecular systems, we must assign discrete values to—in modern terminology, **quantize**—the angular momentum. These discrete values are multiples of a fundamental unit of angular momentum that equals $h/2\pi$, where h is called **Planck's constant.**

$$\text{Fundamental unit of angular momentum} = \frac{h}{2\pi} = 1.054 \times 10^{-34} \text{ kg} \cdot \text{m}^2/\text{s}$$

Example 10.11   Estimating the Rotational Frequency of the Oxygen Molecule

Estimate the rotational frequency of the oxygen molecule using the facts that the "average" separation between the two oxygen atoms is 1.21×10^{-10} m and that the mass of an oxygen atom is 2.66×10^{-26} kg.

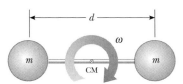

Figure 10.26   (Example 10.11) The rigid-rotor model of the diatomic molecule. The rotation occurs about the center of mass in the plane of the diagram.

Solution   Consider the O_2 molecule as a rigid rotor—that is, two oxygen atoms separated by a fixed distance d and rotating about their center of mass (Fig. 10.26). To find the rotational frequency, first we must evaluate the angular momentum of the molecule about an axis through its center of mass, perpendicular to the plane of rotation.

$$L = L_1 + L_2 = mvr + mvr = 2mvr$$

In this example, $r = d/2 = 0.605 \times 10^{-10}$ m. Also, it is useful to use the relation $v = r\omega$ (Eq. 10.10), giving

$$\begin{aligned} L &= 2mvr = 2mr^2\omega \\ &= 2(2.66 \times 10^{-26} \text{ kg})(0.605 \times 10^{-10} \text{ m})^2\omega \\ &= (1.95 \times 10^{-46} \text{ kg} \cdot \text{m}^2)\omega \end{aligned}$$

We can now estimate the lowest rotational frequency of the molecule by equating the rotational angular momentum with the fundamental unit of angular momentum, $h/2\pi$.

$$\begin{aligned} L &= (1.95 \times 10^{-46} \text{ kg} \cdot \text{m}^2)\omega = \frac{h}{2\pi} \\ &= 1.054 \times 10^{-34} \text{ kg} \cdot \text{m}^2/\text{s} \end{aligned}$$

Therefore,

$$\omega = \frac{1.054 \times 10^{-34} \text{ kg} \cdot \text{m}^2/\text{s}}{1.95 \times 10^{-46} \text{ kg} \cdot \text{m}^2} = 5.41 \times 10^{11} \text{ rad/s}$$

This result is in good agreement with measured rotational frequencies. The rotational frequencies are much lower than the vibrational frequencies of the molecule, which are typically of the order of 10^{13} rad/s.

Classical concepts and mechanical models can be useful in describing some features of atomic and molecular systems. However, as shown by the preceding example, a wide variety of submicroscopic phenomena can be explained only if one assumes discrete values of the angular momentum associated with a particular type of motion.

The Danish physicist Niels Bohr (1885–1962), in his theory of the hydrogen atom, suggested the radical idea of angular momentum quantization. Strictly classical models were unsuccessful in describing many properties of the hydrogen atom, such as the fact that the atom emits radiation at discrete frequencies. Bohr found that the electron could occupy only those orbits about the proton for which the orbital angular momentum was equal to $nh/2\pi$, where n is an integer, or quantum number. From this simple model, one can estimate the rotational frequencies of the electron in the allowed orbits.

Although Bohr's theory provided some insight concerning the behavior of matter at the atomic level, it is basically incorrect. Subsequent developments in quantum mechanics from 1924 to 1930 provided models and interpretations that are still accepted.

Later developments in atomic physics indicated that an electron possesses another kind of angular momentum, called **spin,** which is an inherent property of the electron itself, unrelated to its motion. The spin angular momentum is also restricted to discrete values. Later we shall return to this important property and discuss its great impact on modern physical science.

10.11 · ROTATION OF RIGID BODIES

OPTIONAL

In this section we shall investigate the rotational motion of rigid bodies, using as a basis much of what we have learned concerning the circular motion of particles. First we shall discuss the dynamics of a rigid body rotating about a fixed axis. Next we shall show how to determine the kinetic energy of a rotating rigid body and how its change is related to the work done by external forces. Finally we shall use the principle of conservation of energy to describe an object that rolls on a surface without slipping.

Rotational Dynamics

In Section 10.7 we learned that the magnitude of the net torque acting on a particle is given by Equation 10.25:

$$\tau_{\text{net}} = I\alpha$$

where I is the moment of inertia and α is the angular acceleration. It is important to note that this result also applies to a rigid body of arbitrary shape rotating about a fixed axis, because the body can be regarded as an infinite number of mass elements of infinitesimal size, each rotating in a circle about the rotation axis. The important and strikingly simple result given by $\tau_{\text{net}} = I\alpha$ is in complete agreement with experimental observations. Its simplicity is a result of how the motion is described.

> Although the points on a rigid body rotating about a fixed axis may not experience the same force, linear acceleration, or linear velocity, every point on the body has the same angular acceleration and angular velocity at any instant. Therefore, at any instant the rigid body as a whole is characterized by specific values for angular acceleration, net torque, and angular velocity.

TABLE 10.2   Moments of Inertia of Some Rigid Objects

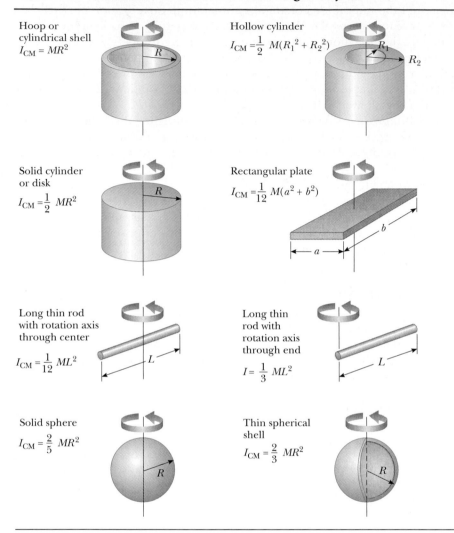

Hoop or cylindrical shell
$I_{CM} = MR^2$

Hollow cylinder
$I_{CM} = \frac{1}{2} M(R_1^2 + R_2^2)$

Solid cylinder or disk
$I_{CM} = \frac{1}{2} MR^2$

Rectangular plate
$I_{CM} = \frac{1}{12} M(a^2 + b^2)$

Long thin rod with rotation axis through center
$I_{CM} = \frac{1}{12} ML^2$

Long thin rod with rotation axis through end
$I = \frac{1}{3} ML^2$

Solid sphere
$I_{CM} = \frac{2}{5} MR^2$

Thin spherical shell
$I_{CM} = \frac{2}{3} MR^2$

The specific form of I depends on the axis of rotation and on the size and shape of the body. The moments of inertia of rigid bodies can be evaluated by methods of integral calculus. Table 10.2 gives the moments of inertia for a number of bodies about specific axes. In all cases, note that I is proportional to the mass of the object and to the square of a geometric factor.

Example 10.12   Angular Acceleration of a Wheel

A wheel of radius R, mass M, and moment of inertia I is mounted on a frictionless, horizontal axle as in Figure 10.27. A light cord wrapped around the wheel supports an object of mass m. Calculate the linear acceleration of the object, the angular acceleration of the wheel, and the tension in the cord.

Reasoning and Solution   The torque acting on the wheel about its axis of rotation is $\tau = TR$, where T is the force exerted by the cord on the rim of the wheel. (The weight of the wheel and the normal force exerted by the axle on the wheel pass through the axis of rotation and produce no torque.) Because $\tau = I\alpha$, we get

$$\tau = I\alpha = TR$$

$$(1) \quad \alpha = \frac{TR}{I}$$

Now let us apply Newton's second law to the motion of the suspended object, making use of the free-body diagram in Figure 10.27, taking the upward direction to be positive:

$$\sum F_y = T - mg = -ma$$

$$(2) \quad a = \frac{mg - T}{m}$$

The linear acceleration of the suspended object is equal to the tangential acceleration of a point on the rim of the wheel. Therefore, the angular acceleration of the wheel and this linear acceleration are related by $a = R\alpha$. Using this fact together with (1) and (2), gives

$$(3) \quad a = R\alpha = \frac{TR^2}{I} = \frac{mg - T}{m}$$

$$(4) \quad T = \frac{mg}{1 + mR^2/I}$$

Substituting (4) into (3), and solving for a and α gives

$$a = \frac{g}{1 + I/mR^2}$$

$$\alpha = \frac{a}{R} = \frac{g}{R + I/mR}$$

EXERCISE 12   The wheel in Figure 10.27 is a solid disk of $M = 2.00$ kg, $R = 30.0$ cm, and $I = 0.0900$ kg·m². The suspended object has a mass of $m = 0.500$ kg. Find the tension in the cord and the angular acceleration of the wheel. Answer   3.27 N; 10.9 rad/s²

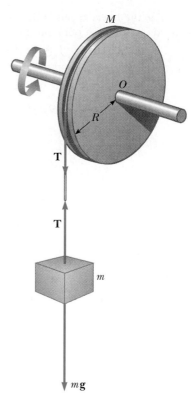

Figure 10.27   (Example 10.12) The cord attached to an object of mass m is wrapped around the wheel, and the tension in the cord produces a torque about the axle through O.

Example 10.13   Rotating Rod Revisited

A uniform rod of length L and mass M is free to rotate on a frictionless pin through one end (Fig. 10.28). The rod is released from rest in the horizontal position. (a) What is the angular speed of the rod at its lowest position?

Reasoning and Solution   The question can be answered by considering the mechanical energy of the system. When the rod is horizontal, it has no rotational energy. The potential energy relative to the lowest position of its center of mass (O') is $MgL/2$. When it reaches its lowest position, the energy is entirely rotational energy, $\frac{1}{2}I\omega^2$, where I is the moment of inertia about the pivot. Because $I = \frac{1}{3}ML^2$ (Table 10.2) and because mechanical energy is constant, we have

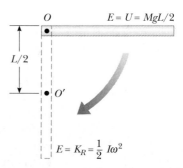

Figure 10.28   (Example 10.13) A uniform rigid rod pivoted at O rotates in a vertical plane under the action of the force of gravity.

$$\tfrac{1}{2}MgL = \tfrac{1}{2}I\omega^2 = \tfrac{1}{2}(\tfrac{1}{3}ML^2)\ \omega^2$$

$$\omega = \sqrt{\frac{3g}{L}}$$

(b) Determine the linear speed of the center of mass and the linear speed of the lowest point on the rod in the vertical position.

Solution

$$v_{\mathrm{CM}} = r\omega = \frac{L}{2}\ \omega = \tfrac{1}{2}\sqrt{3gL}$$

The lowest point on the rod has a linear speed equal to $2v_{\mathrm{CM}} = \sqrt{3gL}$.

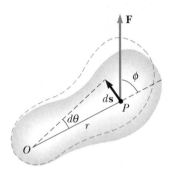

Figure 10.29 A rigid object rotates about an axis through *O* under the action of an external force **F** applied at *P*.

Work and Energy in Rotational Motion

Consider a rigid body pivoted at the point *O* in Figure 10.29. Suppose a single external force, **F**, is applied at the point *P* and *d***s** is the displacement of the point of application of the force. The work done by **F** as the body rotates through an infinitesimal distance $ds = r\,d\theta$ in a time *dt* is

$$dW = \mathbf{F}\cdot d\mathbf{s} = (F\sin\phi)r\,d\theta$$

where $F\sin\phi$ is the tangential component of **F** or the component of the force along the displacement. Note from Figure 10.29 that **the radial component of F does no work because it is perpendicular to the displacement.**

Because the magnitude of the torque due to **F** about the origin was defined as $rF\sin\phi$, we can write the work done for the infinitesimal rotation in the form

$$dW = \tau d\theta \qquad\qquad [10.35]$$

The rate at which work is being done by **F** for rotation about the fixed axis is obtained by formally dividing the left and right sides of Equation 10.35 by *dt*:

$$\frac{dW}{dt} = \tau\frac{d\theta}{dt} \qquad\qquad [10.36]$$

But the quantity *dW/dt* is, by definition, the instantaneous power, *P*, delivered by the force. Furthermore, because $d\theta/dt = \omega$, Equation 10.36 reduces to

Power delivered to a rigid • body

$$P = \frac{dW}{dt} = \tau\omega \qquad\qquad [10.37]$$

This expression is analogous to $P = Fv$ in the case of linear motion, and the expression $dW = \tau d\theta$ is analogous to $dW = F_x dx$.

In linear motion, we found the energy concept, and in particular the work-kinetic energy theorem, to be extremely useful in describing the motion of a system. The energy concept can be equally useful in simplifying the analysis of rotational motion. From what we learned of linear motion, we expect that for rotation of a symmetric object (such as a symmetric wheel) about a fixed axis, the work done by external forces will equal the change in the rotational kinetic energy. To show that this is in fact the case, let us begin with $\tau = I\alpha$. Using the chain rule from the calculus, we can express the torque as

$$\tau = I\alpha = I\frac{d\omega}{dt} = I\frac{d\omega}{d\theta}\frac{d\theta}{dt} = I\frac{d\omega}{d\theta}\omega$$

Figure 10.30 Light sources attached at the center and rim of a rolling cylinder illustrate the different paths these points take. The center moves in a straight line, as indicated by the green line, and a point on the rim moves in the path of a cycloid, as indicated by the red curve. (*Henry Leap and Jim Lehman*)

Rearranging this expression and noting that $\tau d\theta = dW$, we get

$$\tau d\theta = dW = I\omega\,d\omega$$

Integrating this expression, we get the total work done:

$$W = \int_{\theta_0}^{\theta} \tau\,d\theta = \int_{\omega_0}^{\omega} I\omega\,d\omega = \tfrac{1}{2}I\omega^2 - \tfrac{1}{2}I\omega_0{}^2 \qquad \textbf{[10.38]}$$

• *Work-kinetic energy relationship for rotational motion*

Rolling Motion of a Rigid Body

Suppose a cylinder is rolling on a straight path as in Figure 10.30. The center of mass moves in a straight line, and a point on the rim moves in a more complex path corresponding to the shape of a cycloid. Let us further assume that the cylinder of radius R is uniform and rolls on a surface with friction. As the cylinder rotates through an angle of θ, its center of mass moves a distance of $s = r\theta$. Therefore, the speed and acceleration of the center of mass for **pure rolling motion** are

$$v_{CM} = \frac{ds}{dt} = R\frac{d\theta}{dt} = R\omega \qquad \textbf{[10.39]}$$

$$a_{CM} = \frac{dv_{CM}}{dt} = R\frac{d\omega}{dt} = R\alpha \qquad \textbf{[10.40]}$$

The linear velocities of various points on the rolling cylinder are illustrated in Figure 10.31. Note that the linear velocity of any point is in a direction perpendicular to the line from that point to the contact point. At any instant, the point P is at rest relative to the surface, because sliding does not occur.

A general point on the cylinder, such as Q, has both horizontal and vertical components of velocity. However, the points P and P' and the point at the center of mass are unique and of special interest. Relative to the surface on which the cylinder is moving, the center of mass moves with a speed of $v_{CM} = R\omega$, whereas the contact point P has zero velocity. The point P' has a speed equal to $2v_{CM} = 2R\omega$, because all points on the cylinder have the same angular speed.

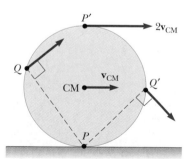

Figure 10.31 All points on a rolling body move in a direction perpendicular to an axis through the contact point P. The center of the body moves with a velocity $\mathbf{v}_{CM}$, and the point P' moves with a velocity $2\mathbf{v}_{CM}$.

We can express the total kinetic energy of the rolling cylinder as

$$K = \tfrac{1}{2}I_p\omega^2 \tag{10.41}$$

where I_p is the moment of inertia about the axis through P.

A useful theorem called the **parallel axis theorem** enables us to calculate the moment of inertia, I_p, through any axis parallel to the axis through the center of mass of a body. This theorem states that

$$I_p = I_{CM} + MR^2 \tag{10.42}$$

where R is the distance from the center-of-mass axis to the parallel axis, and M is the total mass of the body. Substitution of this expression into Equation 10.41 gives

$$K = \tfrac{1}{2}I_{CM}\omega^2 + \tfrac{1}{2}MR^2\omega^2$$

$$K = \tfrac{1}{2}I_{CM}\omega^2 + \tfrac{1}{2}Mv_{CM}{}^2 \tag{10.43}$$

Total kinetic energy of a •
rolling body

where we have used the fact that $v_{CM} = R\omega$.

We can think of Equation 10.43 as follows: The first term on the right, $\tfrac{1}{2}I_{CM}\omega^2$, represents the rotational kinetic energy about the center of mass, and the term $\tfrac{1}{2}Mv_{CM}{}^2$ represents the kinetic energy the cylinder would have if it were just translating through space without rotating. Thus, we can say that **the total kinetic energy of an object undergoing rolling motion is the sum of a rotational kinetic energy about the center of mass and the translational kinetic energy of the center of mass.**

We can use energy methods to treat a class of problems concerning the rolling motion of a rigid body down a rough incline. We shall assume that the rigid body in Figure 10.32 does not slip and is released from rest at the top of the incline. Note that rolling motion is possible only if a frictional force is present between the object and the incline to produce a net torque about the center of mass. Despite the presence of friction, no loss of mechanical energy takes place because the contact point is at rest relative to the surface at any instant. However, if the rigid body were to slide, mechanical energy would decrease as motion progressed.

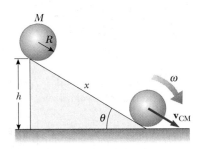

Figure 10.32   A round object rolling down an incline. Mechanical energy remains constant if no slipping occurs and there is no rolling friction.

Thinking Physics 6

A common classroom demonstration is to spin on a rotating stool with heavy weights in your outstretched hands and bring the weights close to your body. The result is an increase in the angular velocity, in agreement with the principle of conservation of angular momentum (see Fig. 10.33). Let us imagine we perform such a demonstration, and moving the weights inward results in halving the moment of inertia, and, therefore, doubling the angular velocity. If we consider the rotational kinetic energy, we see that the energy is *doubled* in this situation. Thus, angular momentum is conserved, but kinetic energy is not. Where does this extra energy come from?

Reasoning As you spin with the weights in your hands, there is a tension force in your arms that is causing the weights to move in a circular path. As you bring the weights in, this force from your arms is doing work on the weights, because it is moving them through a radial displacement. As the angular velocity increases in response to the weights moving inward, you must apply more and more force to keep the weights moving in the smaller radius circular path. Thus, the work that you do in bringing the weights in increases as they are brought in. It is this work that you do with your arms

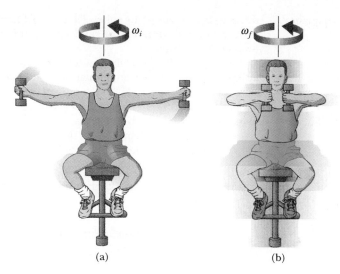

(a)   (b)

Figure 10.33   (Thinking Physics 6) (a) The student is given an initial angular speed while holding two masses as shown. (b) When the masses are pulled in close to the body, the angular speed of the system increases.

that appears as the increase in rotational kinetic energy of the system. You can perform this work due to the energy stored in your body from previous meals, so you are transforming potential energy stored in your body into rotational kinetic energy.

CONCEPTUAL PROBLEM 4

Suppose a pencil is balanced on a perfectly frictionless table. If it falls over, what is the path followed by the center of mass of the pencil?

CONCEPTUAL PROBLEM 5

In some motorcycle races, the riders drive over small hills, and the motorcycle becomes airborne for a short time. If the motorcycle racer keeps the throttle open while leaving the hill and going into the air, the motorcycle tends to nose upward. Why does this happen?

CONCEPTUAL PROBLEM 6

Suppose you are designing a car for a coasting race—the cars in this race have no engines, they simply coast down a hill. Do you want large wheels or small wheels? Do you want solid, disk-like wheels or hoop-like wheels? Should the wheels be heavy or light?

(Conceptual Problem 5)
(Tom Raymond/Tony Stone Images)

Example 10.14   Sphere Rolling Down an Incline

If the object in Figure 10.32 is a solid sphere, calculate the speed of its center of mass at the bottom and determine the magnitude of the linear acceleration of the center of mass.

Solution   For a uniform solid sphere, $I_{CM} = \frac{2}{5}MR^2$, and therefore, conservation of mechanical energy gives

$$v_{CM} = \left(\frac{2gh}{1 + \frac{\frac{2}{5}MR^2}{MR^2}} \right)^{1/2} = \left(\tfrac{10}{7}gh \right)^{1/2}$$

The vertical displacement is related to the displacement x along the incline through the relationship $h = x \sin \theta$.

Hence, after squaring both sides, we can express the equation above as

$$v_{CM}^2 = \tfrac{10}{7} gx \sin \theta$$

Comparing this with the familiar expression from kinematics, $v_{CM}^2 = 2a_{CM}x$, we see that the acceleration of the center of mass is

$$a_{CM} = \tfrac{5}{7} g \sin \theta$$

These results are quite interesting in that both the speed and the acceleration of the center of mass are *independent* of the mass and radius of the sphere! That is, **all homogeneous solid spheres experience the same speed and acceleration on a given incline.**

If we repeated the calculations for a hollow sphere, a solid cylinder, or a hoop, we would obtain similar results. The constant factors that appear in the expressions for v_{CM} and a_{CM} depend only on the moment of inertia about the center of mass for the specific body. In all cases, the acceleration of the center of mass is *less* than $g \sin \theta$, the value it would have if the plane were frictionless and no rolling occurred.

EXERCISE 13 The center of mass of a pitched baseball (radius = 3.8 cm) moves at 38 m/s. The ball spins about an axis through its center of mass with an angular speed of 125 rad/s. Calculate the ratio of the rotational energy to the translational kinetic energy. Treat the ball as a uniform sphere.   Answer   1/160

EXERCISE 14 An automobile tire, considered as a solid disk, has a radius of 25 cm and a mass of 6.0 kg. Find its rotational kinetic energy when it is rotating about an axis through its center at an angular speed of 2.0 rev/s.   Answer   15 J

SUMMARY

The **instantaneous angular speed** of a particle rotating in a circle or of a rigid body rotating about a fixed axis is

$$\omega = \frac{d\theta}{dt} \qquad \text{[10.3]}$$

where ω is in rad/s or in s^{-1}.

The **instantaneous angular acceleration** of a rotating body is

$$\alpha = \frac{d\omega}{dt} \qquad \text{[10.5]}$$

and has units of rad/s^2 or s^{-2}.

When a rigid body rotates about a fixed axis, every part of the body has the same angular velocity and the same angular acceleration. However, different parts of the body, in general, have different linear velocities and different linear accelerations.

If a particle (or body) undergoes rotational motion about a fixed axis under constant angular acceleration, α, one can apply equations of kinematics by analogy with kinematic equations for linear motion with constant linear acceleration:

$$\omega = \omega_0 + \alpha t \qquad \text{[10.6]}$$

$$\theta = \theta_0 + \omega_0 t + \tfrac{1}{2}\alpha t^2 \qquad \text{[10.7]}$$

$$\omega^2 = \omega_0^2 + 2\alpha(\theta - \theta_0) \qquad \text{[10.8]}$$

$$\theta = \theta_0 + \tfrac{1}{2}(\omega_0 + \omega)t \qquad \text{[10.9]}$$

When a particle rotates about a fixed axis, the angular speed and angular acceleration are related to the linear speed and tangential linear acceleration through the relationships

$$v = r\omega \qquad\qquad [10.10]$$

$$a_t = r\alpha \qquad\qquad [10.11]$$

The **moment of inertia of a system of particles** is

$$I = \sum m_i r_i^2 \qquad\qquad [10.14]$$

If a rigid body rotates about a fixed axis with angular speed ω, its **kinetic energy** can be written

$$K = \tfrac{1}{2} I \omega^2 \qquad\qquad [10.15]$$

where I is the moment of inertia about the axis of rotation.

The **torque, τ,** due to a force $\mathbf{F}$ about an origin in an inertial frame is defined to be

$$\tau \equiv \mathbf{r} \times \mathbf{F} \qquad\qquad [10.17]$$

Given two vectors, $\mathbf{A}$ and $\mathbf{B}$, their **cross product, $\mathbf{A} \times \mathbf{B}$,** is a vector $\mathbf{C}$ having the magnitude

$$C \equiv |AB \sin \theta| \qquad\qquad [10.19]$$

where θ is the angle between $\mathbf{A}$ and $\mathbf{B}$. The direction of $\mathbf{C}$ is perpendicular to the plane formed by $\mathbf{A}$ and $\mathbf{B}$, and its sense is determined by the right-hand rule. Some properties of the cross product include the facts that $\mathbf{A} \times \mathbf{B} = -\mathbf{B} \times \mathbf{A}$ and $\mathbf{A} \times \mathbf{A} = 0$.

The net torque acting on a particle is proportional to the angular acceleration of the particle, and the proportionality constant is the moment of inertia, I:

$$\sum \tau = I\alpha \qquad\qquad [10.22]$$

The **angular momentum, L,** of a particle of linear momentum $\mathbf{p} = m\mathbf{v}$ is

$$\mathbf{L} \equiv \mathbf{r} \times \mathbf{p} \qquad\qquad [10.23]$$

where $\mathbf{r}$ is the vector position of the particle relative to an origin in an inertial frame. If ϕ is the angle between $\mathbf{r}$ and $\mathbf{p}$, the magnitude of $\mathbf{L}$ is

$$L = mvr \sin \phi = I\omega \qquad\qquad [10.24]$$

The **net torque** acting on a particle is equal to the time rate of change of its angular momentum:

$$\sum \tau = \frac{d\mathbf{L}}{dt} \qquad\qquad [10.27]$$

The law of conservation of angular momentum states that the total angular momentum of a system remains constant if the net external torque acting on the system is zero:

$$\text{When } \sum \tau_{\text{ext}} = \frac{d\mathbf{L}}{dt} = 0, \text{ then} \qquad\qquad [10.29]$$

$$\mathbf{L} = \text{constant} \qquad\qquad [10.30]$$

Angular momentum is an inherent property of atoms, molecules, and their constituents. Such systems have discrete (quantized) values of angular momenta that are integer multiples of $h/2\pi$, where h is Planck's constant.

The **total kinetic energy** of a rigid body, such as a cylinder, that is rolling on a rough surface without slipping equals the rotational kinetic energy about the body's center of mass $\tfrac{1}{2} I_{\text{CM}} \omega^2$ plus the translational kinetic energy of the center of mass $\tfrac{1}{2} M v_{\text{CM}}^2$:

$$K = \tfrac{1}{2} I_{\text{CM}} \omega^2 + \tfrac{1}{2} M v_{\text{CM}}^2 \qquad\qquad [10.43]$$

In this expression, v_{CM} is the speed of the center of mass and $v_{\text{CM}} = R\omega$ for pure rolling motion.

CONCEPTUAL QUESTIONS

1. In a tape recorder, the tape is pulled past the read-and-write heads at a constant speed by the drive mechanism. Consider the reel from which the tape is pulled—as the tape is pulled off it, the radius of the roll of remaining tape decreases. How does the torque on the reel change with time? How does the angular velocity of the reel change with time? If the tape mechanism is suddenly turned so that the tape is quickly pulled with a large force, is the tape more likely to break when pulled from a nearly full reel or a nearly empty reel?

2. If a car's wheels are replaced with wheels of a larger diameter, will the reading of the speedometer change? Explain.

3. What is the magnitude of the angular velocity, ω, of the second hand of a clock? What is the direction of ω as you view a clock hanging vertically? What is the magnitude of the angular acceleration, α, of the second hand?

4. If you see an object rotating, is there necessarily a net torque acting on it?

5. In order for a helicopter to be stable as it flies, it must have two propellers. Why?

6. The moment of inertia of an object depends on the choice of rotation axis, as suggested by the parallel axis theorem. Argue that an axis passing through the center of mass of an object must be the axis with the smallest moment of inertia.

7. Suppose you remove two eggs from the refrigerator, one hard-boiled and the other uncooked. You wish to determine which is the hard-boiled egg without breaking the eggs. This can be done by spinning the two eggs on the floor and comparing the rotational motions. Which egg spins faster? Which rotates more uniformly? Explain.

8. A ladder rests inclined against a wall. Would you feel safer climbing up the ladder if you were told that the floor is frictionless but the wall is rough, or that the wall is frictionless but the floor is rough? Justify your answer.

9. Often when a high diver wants to flip in midair, she will draw her legs up against her chest. Why does this make her rotate faster? What should she do when she wants to come out of her flip?

10. If the net force acting on a system is zero, then is it necessarily true that the net torque on it is also zero?

11. Why do tightrope walkers carry a long pole to help balance themselves?

12. Two solid spheres are rolled down a hill—a large, massive sphere and a small sphere with low mass. Which one reaches the bottom of the hill first? Next, we roll a large, low density sphere, and a small high density sphere, and both spheres have the same mass. Which one wins in this case?

13. Consider an object in the shape of a hoop lying in the xy plane with all of its mass concentrated on its rim. In two separate experiments, the hoop is rotated by an external agent from rest to an angular speed ω. In one experiment, the rotation occurs about the z axis through the center of the hoop. In the other experiment, the rotation occurs about an axis parallel to z passing through a point P on the rim of the hoop. Which rotation requires more work?

14. Vector **A** is in the negative y direction and vector **B** is in the negative x direction. What are the directions of (a) **A** $\times$ **B** and (b) **B** $\times$ **A**?

15. If global warming occurs over the next century, it is likely that the polar ice caps of the Earth will melt and the water will be distributed closer to the Equator. How would this change the moment of inertia of the Earth? Would the length of the day (one revolution) increase or decrease?

PROBLEMS

Section 10.2   Rotational Kinematics

1. What is the angular speed in radians per second of (a) the Earth in its orbit about the Sun and (b) the Moon in its orbit about the Earth?

2. An airliner arrives at the terminal and the engines are shut off. The rotor of one of the engines has an initial clockwise angular speed of 2000 rad/s. The engine's rotation slows with an angular acceleration of magnitude 80.0 rad/s^2. (a) Determine the angular speed after 10.0 s. (b) How long does it take the rotor to come to rest?

3. An electric motor rotating a grinding wheel at 100 rev/min is switched off. Assuming constant negative acceleration of magnitude 2.00 rad/s^2, (a) how long does it take the wheel to stop? (b) Through how many radians does it turn during the time found in (a)?

4. A centrifuge in a medical laboratory rotates at an angular speed of 3600 rev/min. When switched off, it rotates 50.0 times before coming to rest. Find the constant angular acceleration of the centrifuge.

5. A dentist's drill starts from rest. After 3.20 s of constant angular acceleration, the drill rotates at a rate of 2.51×10^4 rev/min. (a) Find the drill's angular acceleration. (b) Determine the angle (in radians) through which the drill rotates during this period.

6. The angular position of a swinging door is described by $\theta = 5.00 + 10.0t + 2.00t^2$ rad. Determine the angular po-

sition, angular speed, and angular acceleration of the door at $t = 0$ and $t = 3.00$ s.

7. The tub of a washer goes into its spin cycle, starting from rest and reaching an angular speed of 5.00 rev/s in 8.00 s. At this point the person doing the laundry opens the lid, and a safety switch turns off the washer. The tub slows to rest in 12.0 s. Through how many revolutions does the tub turn? Assume constant angular acceleration while the machine is starting and stopping.

8. A rotating wheel requires 3.00 s to rotate 37.0 revolutions. Its angular velocity at the end of the 3.00 s interval is 98.0 rad/s. What is the constant angular acceleration of the wheel?

Section 10.3   Relations Between Angular and Linear Quantities

9. A discus thrower accelerates a discus from rest to a speed of 25.0 m/s by whirling it through 1.25 rev. Assume the discus moves on the arc of a circle 1.00 m in radius. (a) Calculate the final angular speed of the discus. (b) Determine the magnitude of the angular acceleration of the discus, assuming it to be constant. (c) Calculate the acceleration time.

10. A car accelerates uniformly from rest and reaches a speed of 22.0 m/s in 9.00 s. If the diameter of a tire is 58.0 cm, find (a) the number of revolutions the tire makes during this motion, assuming no slipping. (b) What is the final rotational speed of a tire in revolutions per second?

11. A disk 8.00 cm in radius rotates at a constant rate of 1200 rev/min about its central axis. Determine (a) its angular speed, (b) the linear speed at a point 3.00 cm from its center, (c) the radial acceleration of a point on the rim, and (d) the total distance a point on the rim moves in 2.00 s.

12. A 6.00-kg block is released from A on a frictionless track shown in Figure P10.12. Determine the radial and tangential components of acceleration for the block at P.

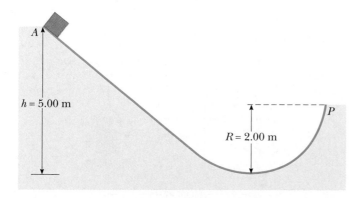

Figure P10.12

13. A car traveling on a flat (unbanked) circular track accelerates uniformly from rest with a tangential acceleration of 1.70 m/s². The car makes it one quarter of the way around the circle before it skids off the track. Determine the coefficient of static friction between the car and track from these data.

Section 10.4   Rotational Kinetic Energy

14. Three particles are connected by rigid rods of negligible mass lying along the y axis (Fig. P10.14). If the system rotates about the x axis with an angular speed of 2.00 rad/s, find (a) the moment of inertia about the x axis and the total rotational energy evaluated from $\frac{1}{2}I\omega^2$ and (b) the linear speed of each particle and the total energy evaluated from $\Sigma \frac{1}{2}m_i v_i^2$.

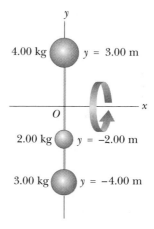

Figure P10.14

15. The four particles in Figure P10.15 are connected by rigid rods of negligible mass. The origin is at the center of the rectangle. If the system rotates in the xy plane about the z axis with an angular speed of 6.00 rad/s, calculate (a) the moment of inertia of the system about the z axis and (b) the rotational energy of the system.

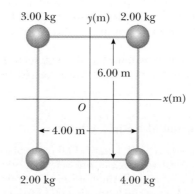

Figure P10.15

Section 10.5   Torque and the Vector Product

16. The fishing pole in Figure P10.16 makes an angle of 20.0°
with the horizontal. What is the torque exerted by the fish
about an axis perpendicular to the page and passing
through the fisher's hand?

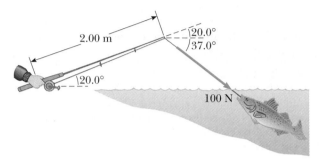

Figure P10.16

17. In Figure P10.17, find the net torque on the wheel about
the axle through *O* if *a* = 10.0 cm and *b* = 25.0 cm.

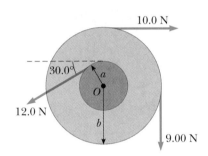

Figure P10.17

18. Given **M** = 6**i** + 2**j** − **k** and **N** = 2**i** − **j** − 3**k**, calculate
the vector product **M** × **N**.

19. Two vectors are given by **A** = −3**i** + 4**j** and **B** = 2**i** + 3**j**.
Find (a) **A** × **B** and (b) the angle between **A** and **B**.

20. A force of **F** = 2.00**i** + 3.00**j** N is applied to an object that
is pivoted about a fixed axis aligned along the *z* coordinate
axis. If the force is applied at the point whose vector position
is **r** = (4.00**i** + 5.00**j** + 0**k**) m, find (a) the magnitude of
the net torque about the *z* axis and (b) the direction of the
torque vector, **τ**.

21. A student claims that she has found a vector **A** such that
(2**i** − 3**j** + 4**k**) × **A** = (4**i** + 3**j** − **k**). Do you believe this
claim? Explain.

Section 10.6   Equilibrium of a Rigid Object

22. A ladder of weight 400 N and length 10.0 m is placed
against a smooth vertical wall. A person weighing 800 N
stands on the ladder 2.00 m from the bottom as measured
along the ladder. The foot of the ladder is 8.00 m from the

bottom of the wall. Draw a free-body diagram of the ladder.
Calculate the force exerted by the wall and the normal force
exerted by the floor on the ladder.

23. A 1500 kg automobile has a wheel base (the distance between the axles) of 3.00 m. The center of mass of the automobile is on the center line at a point 1.20 m behind the
front axle. Find the force exerted by the ground on each
wheel.

24. A uniform plank of length 6.00 m and mass 30.0 kg rests
horizontally on a scaffold, with 1.50 m of the plank hanging
over one end of the scaffold. Draw a free-body diagram of
the plank. How far can a painter of mass 70.0 kg walk on
the overhanging part of the plank before it tips?

25. A 1200-N uniform boom is supported by a cable as in Figure
P10.25. The boom is pivoted at the bottom, and a 2000-N
object hangs from its top. Find the tension in the cable and
the components of the reaction force on the boom by the
floor.

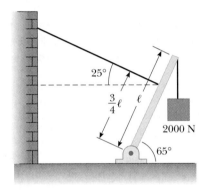

Figure P10.25

26. A crane of mass 3000 kg supports a load of 10 000 kg as in
Figure P10.26. The crane is pivoted with a smooth pin at *A*
and rests against a smooth support at *B*. Find the reaction
forces at *A* and *B*.

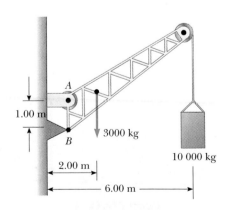

Figure P10.26

Section 10.7 Relation Between Torque and Angular Acceleration

27. A model airplane the mass of which is 0.750 kg is tethered by a wire so that it flies in a circle 30.0 m in radius. The airplane engine provides a net thrust of 0.800 N perpendicular to the tethering wire. (a) Find the torque that the net thrust produces about the center of the circle. (b) Find the angular acceleration of the airplane when it is in level flight. (c) Find the linear acceleration of the airplane tangent to its flight path.

28. The combination of an applied force and a frictional force produces a constant total torque of 36.0 N·m on a wheel rotating about a fixed axis. The applied force acts for 6.00 s, during which time the angular speed of the wheel increases from 0 to 10.0 rad/s. The applied force is then removed, and the wheel comes to rest in 60.0 s. Find (a) the moment of inertia of the wheel, (b) the magnitude of the frictional torque, and (c) the total number of revolutions of the wheel.

Section 10.8 Angular Momentum

29. A light rigid rod 1.00 m in length rotates in the xy plane about a pivot through the rod's center. Two particles of masses of 4.00 kg and 3.00 kg are connected to its ends (Fig. P10.29). Determine the angular momentum of the system about the origin at the instant the speed of each particle is 5.00 m/s.

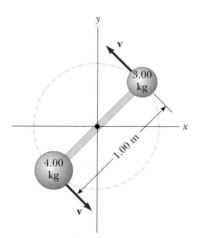

Figure P10.29

30. A 1.50-kg particle moves in the xy plane with a velocity of $\mathbf{v} = (4.20\mathbf{i} - 3.60\mathbf{j})$ m/s. Determine the angular momentum of the particle when its position vector is given by $\mathbf{r} = (1.50\mathbf{i} + 2.20\mathbf{j})$ m.

31. The position vector of a particle of mass 2.00 kg is given as a function of time by $\mathbf{r} = (6.00\,\mathbf{i} + 5.00t\mathbf{j})$ m. Determine the angular momentum of the particle as a function of time.

32. An airplane of mass 12 000 kg flies level to the ground at an altitude of 10.0 km with a constant speed of 175 m/s relative to the Earth. (a) What is the magnitude of the airplane's angular momentum relative to a ground observer directly below the airplane? (b) Does this value change as the airplane continues its motion along a straight line?

Section 10.9 Conservation of Angular Momentum

Section 10.10 Quantization of Angular Momentum (Optional)

33. A ball having mass m is fastened at the end of a flagpole that is connected to the side of a tall building at point P shown in Figure P10.33. The length of the flagpole is ℓ and it makes an angle θ with the horizontal. If the ball become loose and starts to fall, determine the angular momentum (as a function of time) of the ball about point P. Neglect air resistance.

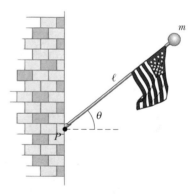

Figure P10.33

34. A student sits on a rotating stool holding two weights, each of mass 3.00 kg. When his arms are extended horizontally, the weights are 1.00 m from the axis of rotation and he rotates with an angular speed of 0.750 rad/s. The moment of inertia of the student plus stool is 3.00 kg·m² and is assumed to be constant. The student pulls the weights horizontally to 0.300 m from the rotation axis. (a) Find the new angular speed of the student. (b) Find the kinetic energy of the student before and after the weights are pulled in.

35. A 60.0-kg woman stands at the rim of a horizontal turntable having a moment of inertia of 500 kg·m² and a radius of

2.00 m. The turntable is initially at rest and is free to rotate about a frictionless, vertical axle through its center. The woman then starts walking around the rim clockwise (as viewed from above the system) at a constant speed of 1.50 m/s relative to the Earth. (a) In what direction and with what angular speed does the turntable rotate? (b) How much work does the woman do to set herself and the turntable into motion?

36. The ball in Figure 10.25 has a mass of 0.120 kg. The distance of the ball from the center of rotation is originally 40.0 cm, and the ball is moving with a speed of 80.0 cm/s. The string is pulled downward 15.0 cm through the hole in the frictionless table. Determine the work done on the ball. (*Hint:* Consider the change of kinetic energy.)

37. A puck of mass 80.0 g and radius 4.00 cm slides along an air table at a speed of 1.50 m/s as shown in Figure P10.37a. It makes a glancing collision with a second puck of radius 6.00 cm and mass 120 g (initially at rest) such that their rims just touch. The pucks stick together and spin after the collision (Fig. P10.37b). (a) What is the angular momentum of the system relative to the center of mass? (b) What is the angular velocity about the center of mass?

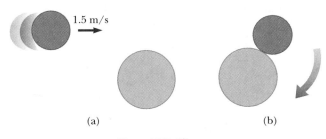

(a)   (b)

Figure P10.37

38. In the Bohr model of the hydrogen atom, the electron moves in a circular orbit of radius 0.529×10^{-10} m around the proton. Assuming the orbital angular momentum of the electron is equal to $h/2\pi$, calculate (a) the orbital speed of the electron, (b) the kinetic energy of the electron, and (c) the angular frequency of the electron's motion.

Section 10.11   Rotation of Rigid Bodies (Optional)

39. Attention! About face! Compute an order-of-magnitude estimate for the moment of inertia of your body as you stand tall and turn around an axis through the top of your head and the point half way between your ankles. In your solution state the quantities you measure or estimate and their values.

40. A horizontal 800-N merry-go-round is a solid disk of radius 1.50 m, started from rest by a constant horizontal force of 50.0 N applied tangentially at its edge. Find the kinetic energy of the disk after 3.00 s.

41. A cylinder of mass 10.0 kg rolls without slipping on a horizontal surface. At the instant its center of mass has a speed of 10.0 m/s, determine (a) the translational kinetic energy of its center of mass, (b) the rotational energy about its center of mass, and (c) its total energy.

42. Consider two masses connected by a string that passes over a pulley having a moment of inertia of I about its axis of rotation, as in Figure P10.42. The string does not slip on the pulley, and the system is released from rest. Use the principle of conservation of energy to find the linear speeds of the masses after m_2 descends through a distance h and the angular speed of the pulley at this time.

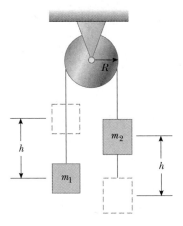

Figure P10.42

43. A uniform rod of length L and mass M is free to rotate about a frictionless pivot at one end, as in Figure P10.43. The rod is released from rest in the horizontal position. What are the *initial* angular acceleration of the rod and the *initial* linear acceleration of the right end of the rod?

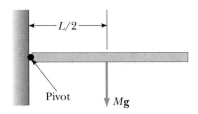

Figure P10.43

44. A bicycle wheel has a diameter of 64.0 cm and a mass of 1.80 kg. The bicycle is placed on a stationary stand on rollers and a resistive force of 120 N is applied to the rim of the tire. (Assume that the wheel is a hoop with all the mass concentrated on the outside radius.) What force must be

applied by a chain passing over a 9.00-cm diameter sprocket in order to give the wheel an acceleration of 4.50 rad/s²? What if you shifted to a 5.60-cm diameter sprocket?

45. (a) A uniform solid disk of radius R and mass M is free to rotate on a frictionless pivot through a point on its rim (Fig. P10.45). If the disk is released from rest in the position shown by the blue circle, what is the speed of its center of mass when the disk reaches the position indicated by the dashed circle? (b) What is the speed of the lowest point on the disk in the dashed position? (c) Repeat part (a) using a uniform hoop.

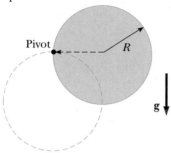

Figure P10.45

46. A block of mass $m_1 = 2.00$ kg and a block of mass $m_2 = 6.00$ kg are connected by a massless string over a pulley in the shape of a disk having radius $R = 0.25$ m, mass $M = 10.0$ kg. In addition, they are allowed to move on a fixed block-wedge of angle $\theta = 30.0°$ as in Figure P10.46. The coefficient of kinetic friction is 0.360 for both blocks. Draw free-body diagrams of both blocks and of the pulley. Determine (a) the acceleration of the two blocks, and (b) the tensions in the string on both sides of the pulley.

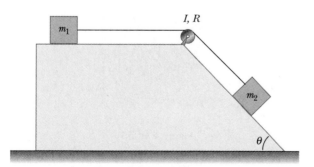

Figure P10.46

47. A can of condensed mushroom soup has mass 215 g, height 10.8 cm, and outer diameter 6.38 cm. It is placed at rest on the top of a 3.00-m long incline, which is at 25.0° to the

horizontal. Assuming energy conservation, calculate the moment of inertia of the can if it takes 1.50 s to reach the bottom of the incline.

Additional Problems

48. An electric motor can accelerate a Ferris wheel of moment of inertia $I = 20\,000$ kg·m² from rest to 10.0 rev/min in 12.0 s. When the motor is turned off, friction causes the wheel to slow down from 10.0 to 8.00 rev/min in 10.0 s. Determine (a) the torque generated by the motor to bring the wheel to 10.0 rev/min and (b) the power needed to maintain this rotational speed.

49. Two blocks, as shown in Figure P10.49, are connected by a string of negligible mass passing over a pulley of radius 0.250 m and moment of inertia I. The block on the frictionless incline is moving up with a constant acceleration of 2.00 m/s². (a) Determine T_1 and T_2, the tensions in the two parts of the string. (b) Find the moment of inertia of the pulley.

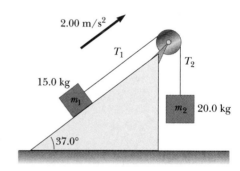

Figure P10.49

50. As a result of friction, the angular speed of a wheel changes with time according to

$$\frac{d\theta}{dt} = \omega_0 e^{-\sigma t}$$

where ω_0 and σ are constants. The angular speed changes from 3.50 rad/s at $t = 0$ to 2.00 rad/s at $t = 9.30$ s. Use this information to determine σ and ω_0. Then determine (a) the magnitude of the angular acceleration at $t = 3.00$ s, (b) the number of revolutions the wheel makes in the first 2.50 s, and (c) the number of revolutions it makes before coming to rest.

51. The pulley shown in Figure P10.51 has radius R and moment of inertia I. One end of the mass m is connected to a spring of force constant k, and the other end is fastened to a cord wrapped around the pulley. The pulley axle and the incline are frictionless. If the pulley is wound counterclock-

wise in order to stretch the spring a distance d from its un-stretched position and then released from rest, find (a) the angular speed of the pulley when the spring is again unstretched and (b) a numerical value for the angular speed at this point if $I = 1.00$ kg · m^2, $R = 0.300$ m, $k = 50.0$ N/m, $m = 0.500$ kg, $d = 0.200$ m, and $\theta = 37.0°$.

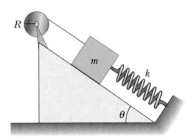

Figure P10.51

52. A solid sphere of mass m and radius r rolls without slipping along the track shown in Figure P10.52. It starts from rest with the lowest point of the sphere at height h above the bottom of the loop of radius R. (a) What is the minimum value of h (in terms of r and R) such that the sphere completes the loop? (b) What are the force components on the sphere at the point P if $h = 3R$?

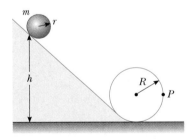

Figure P10.52

53. Two astronauts (Fig. P10.53), each having a mass of 75.0 kg, are connected by a 10.0-m rope of negligible mass. They are isolated in space, orbiting their center of mass at speeds of 5.00 m/s. (a) Calculate the magnitude of the angular momentum of the system by treating the astronauts as particles and (b) the rotational energy of the system. By pulling on the rope, the astronauts shorten the distance between them to 5.00 m. (c) What is the new angular momentum of the system? (d) What are their new speeds? (e) What is the new rotational energy of the system? (f) How much work is done by the astronauts in shortening the rope?

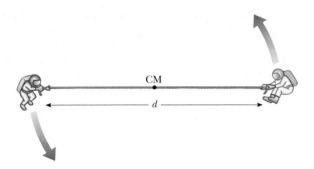

Figure P10.53

54. Two astronauts (Fig. P10.53), each having a mass M, are connected by a rope of length d having negligible mass. They are isolated in space, orbiting their center of mass at speeds v. Calculate (a) the magnitude of the angular momentum of the system by treating the astronauts as particles and (b) the rotational energy of the system. By pulling on the rope, the astronauts shorten the distance between them to $d/2$. (c) What is the new angular momentum of the system? (d) What are their new speeds? (e) What is the new rotational energy of the system? (f) How much work is done by the astronauts in shortening the rope?

55. Figure P10.55 shows a vertical force applied tangentially to a uniform cylinder of weight w. The coefficient of static friction between the cylinder and all surfaces is 0.500. Find, in terms of w, the maximum force **F** that can be applied without causing the cylinder to rotate. (*Hint:* When the cylinder is on the verge of slipping, both friction forces are at their maximum values. Why?)

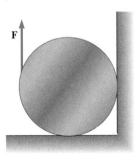

Figure P10.55

56. A common physics demonstration (Fig. P10.56) consists of a ball resting at the end of a board of length ℓ that is elevated at angle θ with the horizontal. A light cup is attached to the board at r_c so that it will catch the ball when the support stick is suddenly removed. (a) Show that the ball will lag behind the falling board when θ is less than 35.3°, and (b) the ball will fall into the cup when the board is supported at this limiting angle and the cup is placed at

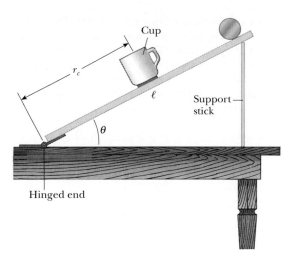

Figure P10.56

$$r_c = \frac{2\ell}{3 \cos \theta}$$

(c) If a ball is at the end of a 1.00-m stick at this critical angle, show that the cup must be 18.4 cm from the moving end.

📊 Spreadsheet Problem

S1. A disk having a moment of inertia of 100 kg·m² is free to rotate about a fixed axis through its center as in Figure 10.1. A tangential force whose magnitude ranges from $F = 0$ to $F = 50.0$ N can be applied at any distance ranging from $R = 0$ to $R = 3.00$ m measured from the axis of rotation. Using Spreadsheet 10.1, find values of F and R that will cause the disk to complete two revolutions in 10.0 s. Are there unique values for F and R?

ANSWERS TO CONCEPTUAL PROBLEMS

1. As the smokestack rotates around its base, each higher portion of the smokestack falls with an increasing tangential acceleration. The tangential acceleration of a given point on the smokestack is proportional to the distance of that portion from the base. As the acceleration increases, eventually higher portions of the smokestack will experience an acceleration higher than that which could result from gravity alone. This can only happen if these portions are being pulled downward by a force in addition to the gravitational force. The force that does this is the shear force from lower portions of the smokestack. Eventually, the shear force to provide this acceleration is larger than the smokestack can withstand, and it breaks.

2. Yes, all points on the wheel have the same angular speed. This is why we use angular quantities to describe rotational motion. Not all points on the wheel have the same linear speed. The point at $r = 0$ has zero linear speed and zero linear acceleration; a point at $r = R/2$ has a linear speed $v = R\omega/2$ and a linear acceleration equal to the centripetal acceleration $v^2/(R/2) = R\omega^2/2$ (the tangential acceleration is zero at all points because ω is constant). A point on the rim at $r = R$ has a linear speed $v = R\omega$ and a linear acceleration $R\omega^2$.

3. There are two major differences between torque and work. The primary difference is that the displacement in the expression for work is directed *along* the force, while the important distance in the torque expression is *perpendicular* to the force. The second difference involves whether there is motion or not—in the case of work, there is work done only if the force succeeds in causing a displacement of the point of application of the force. On the other hand, a force applied at a perpendicular distance from a rotation axis results in a torque whether there is motion or not.

As far as units are concerned, the mathematical expressions for both work and torque result in the product of newtons and meters, but this product is called a joule in the case of work and remains as a newton-meter in the case of torque.

4. On a frictionless table, the only forces on the pencil are the gravitational force, at the center of mass, and the normal force from the table, at the tip. These forces result in a torque on the pencil, causing it to rotate. Both of these forces, however, are vertical—there are no horizontal forces on the pencil. As a result, the center of mass of the pencil must fall straight downward. As the pencil falls, the tip slides to the side, so that the center of mass of the pencil follows a straight line downward to the table.

5. As the motorcycle tire leaves the ground, the friction between the tire and the ground suddenly disappears. If the motorcycle driver keeps the throttle open while leaving the ground, the drive tire will increase its angular velocity, since it no longer experiences the friction force from the ground. The airborne motorcycle is now an isolated system, and its angular momentum must be conserved. The increase in angular momentum of the tire (directed to the left of the motorcycle) must be compensated by an increase in angular momentum of the entire motorcycle (to the right). This ro-

tation results in the nose of the motorcycle rising and the tail dropping.

6. In general, you want the rotational kinetic energy of the system to be as small a fraction as possible of the total energy—you want translation, not rotation! You want the wheels to have as little moment of inertia as possible, so that they represent the lowest resistance to changes in rotational motion. Disk-like wheels would have lower moment of inertia than hoop-like wheels, so disks are preferable. The lower the mass of the wheels, the less is the moment of inertia, so light wheels are preferable. The smaller the radius of the wheels, the less is the moment of inertia, so smaller wheels are preferable, within limits—you want the wheels to be large enough to be able to travel relatively smoothly over irregularities in the road.

11

Orbital Motions and the Hydrogen Atom

We began our study of mechanics with translational motion and the forces that cause it. If we know the forces acting on a system and its initial conditions, we can predict the future of the system. However, describing motion in this manner is often tedious and time consuming. Fortunately, we can often follow the simpler approach of using conservation principles. We easily solved many interesting problems involving motion by recognizing that certain fundamental quantities—such as energy and momentum—were conserved.

In this chapter we return to Newton's universal law of gravitation—one of the fundamental force laws in nature—and show how it, together with Newton's laws of motion, enables us to understand a variety of familiar orbital motions, such as the motions of planets and Earth satellites. Newton's theory of gravitation evolved out of his studies of the motions of the

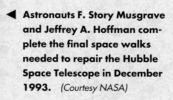

◀ **Astronauts F. Story Musgrave and Jeffrey A. Hoffman complete the final space walks needed to repair the Hubble Space Telescope in December 1993.** *(Courtesy NASA)*

Moon and planets, for which he used the groundwork provided by Copernicus, Brahe, Kepler, and other astronomers. In examining this evolution, we shall once again make use of conservation of energy and angular momentum.

We conclude this chapter with a discussion of Niels Bohr's famous model of the hydrogen atom, which represents an interesting mixture of classical physics (Newton's laws of motion and the electromagnetic force) and quantum physics (quantization of angular momentum). Although Bohr's theory contains ideas that are contrary to classical physics, his model successfully predicts the observed spectral lines of hydrogen.

11.1 • NEWTON'S UNIVERSAL LAW OF GRAVITY REVISITED

Prior to 1686, many data had been collected on the motions of the Moon and the planets, but a clear understanding of the forces that caused those motions was not yet attainable. In that year, Isaac Newton provided the key that unlocked the secrets of the heavens. He knew, from the first law, that a net force had to be acting on the Moon. If not, the Moon would move in a straight-line path rather than in its almost circular orbit. Newton reasoned that this force between Moon and Earth was an attractive force. He also concluded that there could be nothing special about the Earth–Moon system or the Sun and its planets that would cause gravitational forces to act on them alone. He wrote,

> I deduced that the forces which keep the planets in their orbs must be reciprocally as the squares of their distances from the centers about which they revolve; and thereby compared the force requisite to keep the Moon in her orb with force of gravity at the surface of the Earth; and found them answer pretty nearly.

As you should recall from Chapter 5, every particle in the Universe attracts every other particle with a force that is directly proportional to the product of their masses and inversely proportional to the square of the distance between them. If two particles have masses m_1 and m_2 and are separated by a distance r, the magnitude of the gravitational force between them is

Universal law of gravity •

$$F_g = G \frac{m_1 m_2}{r^2} \qquad [11.1]$$

where G is the *gravitational constant* the value of which in SI units is

$$G = 6.672 \times 10^{-11} \frac{\text{N} \cdot \text{m}^2}{\text{kg}^2} \qquad [11.2]$$

The force law given by Equation 11.1 is often referred to as an **inverse-square law** because the magnitude of the force varies as the inverse square of the separation of the particles. We can express this attractive force in vector form by defining a unit vector, $\hat{\mathbf{r}}$, directed from m_1 to m_2, as shown in Figure 11.1. The force on m_1 due to m_2 is

$$\mathbf{F}_{12} = G \frac{m_1 m_2}{r^2} \hat{\mathbf{r}} \qquad [11.3]$$

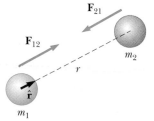

Figure 11.1 The gravitational force between two particles is attractive. The unit vector $\hat{\mathbf{r}}$ is directed from m_1 to m_2. Note that $\mathbf{F}_{21} = -\mathbf{F}_{12}$.

Likewise, by Newton's third law, the force on m_2 due to m_1, designated $\mathbf{F}_{21}$, is equal in magnitude to $\mathbf{F}_{12}$ and in the opposite direction. That is, these forces form an action–reaction pair, and $\mathbf{F}_{21} = -\mathbf{F}_{12}$.

The gravitational force exerted by a finite-sized, spherically symmetric mass distribution on a particle outside the sphere is the same as if the entire mass of the sphere were concentrated at its center. For example, the force on a particle of mass m at the Earth's surface has the magnitude

$$F_g = G \frac{M_E m}{R_E{}^2}$$

where M_E is the Earth's mass and R_E is the Earth's radius. This force is directed toward the center of the Earth.

Measurement of the Gravitational Constant

The gravitational constant, G, was first measured in an important experiment by Sir Henry Cavendish in 1798. The apparatus he used consisted of two small spheres, each of mass m, fixed to the ends of a light horizontal rod suspended by a thin wire, as in Figure 11.2a. Two large spheres, each of mass M, are then placed near the smaller spheres. The attractive force between the smaller and larger spheres causes the rod to rotate and twist the wire. If the system is oriented as shown in Figure 11.2a, the rod rotates clockwise when viewed from the top. The angle through which it rotates is measured by the deflection of a light beam that is reflected from a mirror attached to the wire. The experiment is carefully repeated with different masses at various separations. In addition to providing a value for G, the results show that the force is attractive, proportional to the product mM, and inversely proportional to the square of the distance r.

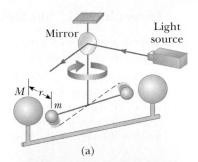

(a)

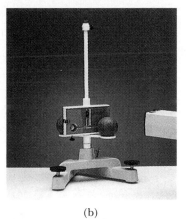

(b)

Figure 11.2 (a) Schematic diagram of the Cavendish apparatus for measuring G. The smaller spheres of mass m are attracted to the large spheres of mass M, and the rod rotates through a small angle. A light beam reflected from a mirror on the rotating apparatus measures the angle of rotation. The dashed line represents the original position of the rod. (b) Photograph of a student Cavendish apparatus. *(Courtesy of PASCO Scientific)*

Thinking Physics 1

A student drops a ball and considers why the ball falls to the ground as opposed to the situation in which the ball stays stationary and the Earth moves up to meet it. The student comes up with the following explanation: "The Earth is much more massive than the ball, so the Earth pulls much harder on the ball than the ball pulls on the Earth. Thus, the ball falls while the Earth remains stationary." What do you think about this explanation?

Reasoning According to Newton's universal law of gravity, the force between the ball and the Earth depends on the product of their masses, so both forces—that of the ball on the Earth and that of the Earth on the ball—are equal in magnitude. This follows also, of course, from Newton's third law. The ball has large motion compared to the Earth because, according to Newton's second law, the force of equal magnitude gives a much greater acceleration to the small mass of the ball.

Example 11.1   The Mass of the Earth

Use the gravitational force law to find an approximate value for the mass of the Earth.

Solution   Figure 11.3, obviously not to scale, shows a baseball falling toward the Earth at a location where the acceleration due to gravity is *g*. We know from Chapter 5 that the magnitude of the gravitational force exerted on the baseball by the Earth is the same as the weight of the ball, $w = m_b g$. That is,

$$m_b g = G \frac{M_E m_b}{R_E^2}$$

We can divide each side of this equation by m_b and solve for the mass of the Earth, M_E:

$$M_E = \frac{g R_E^2}{G}$$

The falling baseball is close enough to the Earth so that the distance of separation between the center of the ball and the center of the Earth can be taken as the radius of the Earth, 6.38×10^6 m. Thus, the mass of the Earth is

$$M_E = \frac{(9.80 \text{ m/s}^2)(6.38 \times 10^6 \text{ m})^2}{6.67 \times 10^{-11} \text{ N} \cdot \text{m}^2/\text{kg}^2} = 5.98 \times 10^{24} \text{ kg}$$

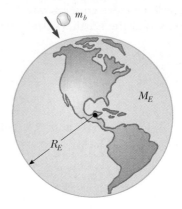

Figure 11.3   (Example 11.2) A baseball falling toward the Earth (not drawn to scale).

Example 11.2   Gravity and Altitude

Derive an expression that shows, for any point above the Earth's surface, how the acceleration due to gravity varies with distance from the center of the Earth.

Solution   The falling baseball of Example 11.1 can be used here also. Now, however, assume that the ball is situated some arbitrary distance, $r > R_E$, from the Earth's center. The first equation in Example 11.1, with *r* replacing R_E and m_b removed from both sides, becomes

$$g = G \frac{M_E}{r^2}$$

This indicates that the acceleration due to gravity at a point above the Earth's surface decreases as the inverse square of the distance between the point and the center of the Earth. Our assumption of Chapter 2 that objects fall with a constant acceleration is obviously incorrect in light of this example. For short falls, however, the change in *g* is so small that neglecting the variation does not introduce a significant effect in our results.

Because the weight of an object is *mg*, a change in the value of *g* produces a change in weight. For example, if you weigh 800 N at the surface of the Earth, you will weigh only 200 N at a height above the Earth equal to the radius of the Earth.

Also, if the distance of an object from the Earth becomes infinitely large, the weight approaches zero. Values of *g* at various altitudes are given in Table 11.1.

TABLE 11.1   **Free-Fall Acceleration, *g*, at Various Altitudes Above the Earth's Surface**

Altitude *h* (km)	*g*(m/s²)
1000	7.33
2000	5.68
3000	4.53
4000	3.70
5000	3.08
6000	2.60
7000	2.23
8000	1.93
9000	1.69
10 000	1.49
50 000	0.13
∞	0

EXERCISE 1   If an object weighs 270 N at the Earth's surface, what will it weigh at an altitude equal to twice the radius of the Earth?   Answer   30 N

EXERCISE 2   Determine the magnitude of the acceleration due to gravity at an altitude of 500 km. By what percentage is the weight of a body reduced at this altitude?
Answer   8.43 m/s^2; 14%

11.2 · KEPLER'S LAWS

The movements of the planets, stars, and other celestial bodies have been observed by people for thousands of years. Early in history, scientists regarded the Earth as the center of the Universe. This so-called geocentric model was elaborated and formalized by the Greek astronomer Claudius Ptolemy in the second century A.D. and was accepted for the next 1400 years. In 1543 the Polish astronomer Nicolaus Copernicus (1473–1543) suggested that the Earth and the other planets revolve in circular orbits about the Sun (the heliocentric hypothesis).

The Danish astronomer Tycho Brahe (1546–1601) made accurate astronomical measurements over a period of 20 years and provided the basis for the currently accepted model of the Solar System. It is interesting to note that these precise observations, made on the planets and 777 stars, were carried out with nothing more elaborate than a large sextant and compass; the telescope had not yet been invented.

The German astronomer Johannes Kepler, who was Brahe's assistant, acquired Brahe's astronomical data and spent about 16 years trying to deduce a mathematical model for the motions of the planets. After many laborious calculations, he found that Brahe's precise data on the revolution of Mars about the Sun provided the answer. Kepler's analysis first showed that the concept of circular orbits about the Sun had to be abandoned. He eventually discovered that the orbit of Mars could be accurately described by an ellipse with the Sun at one focus. He then generalized this analysis to include the motions of all planets. The complete analysis is summarized in three statements, known as **Kepler's laws:**

Johannes Kepler (1571–1630)

A German astronomer, Kepler is best known for developing the laws of planetary motion based on the careful observations of Tycho Brahe. After spending several years trying to work out a "regular-solid theory" of the planets, he concluded that the Copernican view of circular planetary orbits had to be abandoned for the view that the planetary orbits are ellipses with the Sun always at one of the foci. (Art Resource)

1. Every planet moves in an elliptical orbit with the Sun at one of the focal points.
2. The radius vector drawn from the Sun to any planet sweeps out equal areas in equal time intervals.
3. The square of the orbital period of any planet is proportional to the cube of the semimajor axis of the elliptical orbit.

• *Kepler's laws*

Newton demonstrated that these laws were consequences of a simple force that exists between any two masses. Newton's universal law of gravity, together with his laws of motion, provides the basis for a full mathematical solution to the motion of planets and satellites. More important, Newton's universal law of gravity correctly describes the gravitational attractive force between *any* two masses.

11.3 · THE UNIVERSAL LAW OF GRAVITY AND THE MOTIONS OF PLANETS

In formulating his universal law of gravity, Newton built on an observation that suggests that the gravitational force between two bodies is proportional to the inverse square of the separation. Let us compare the acceleration of the Moon in its orbit with the acceleration of an object falling near the Earth's surface, such as an apple (Fig. 11.4). Assume that the two accelerations have the same cause—namely, the gravitational attraction of the Earth. From the inverse-square law, Newton found that the acceleration of the Moon toward the Earth should be proportional to $1/r_M^2$, where r_M is the separation between centers of the Earth and Moon. Furthermore, the acceleration of the apple toward the Earth should be proportional to $1/R_E^2$, where R_E is the radius of the Earth. When the values $r_M = 3.84 \times 10^8$ m and $R_E = 6.37 \times 10^6$ m are used, the ratio of the Moon's acceleration, a_M, to the apple's acceleration, g, is predicted to be

$$\frac{a_M}{g} = \frac{(1/r_M)^2}{(1/R_E)^2} = \left(\frac{R_E}{r_M}\right)^2 = \left(\frac{6.37 \times 10^6 \text{ m}}{3.84 \times 10^8 \text{ m}}\right)^2 = 2.75 \times 10^{-4}$$

Therefore,

$$a_M = (2.75 \times 10^{-4})(9.80 \text{ m/s}^2) = 2.70 \times 10^{-3} \text{ m/s}^2$$

The centripetal acceleration of the Moon can also be calculated kinematically from a knowledge of the Moon's orbital period, T, where $T = 27.32$ days $= 2.36 \times 10^6$ s and its mean distance from the Earth, r_M. In the time T, the Moon travels a distance $2\pi r_M$, which equals the circumference of its orbit. Therefore, its orbital speed is $2\pi r_M/T$, and its centripetal acceleration is

$$a_M = \frac{v^2}{r_M} = \frac{(2\pi r_M/T)^2}{r_M} = \frac{4\pi^2 r_M}{T^2} = \frac{4\pi^2(3.84 \times 10^8 \text{ m})}{(2.36 \times 10^6 \text{ s})^2} = 2.72 \times 10^{-3} \text{ m/s}^2$$

This agreement provides strong evidence that the inverse-square law of force is correct.

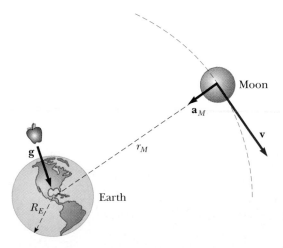

Figure 11.4 As it revolves about the Earth, the Moon experiences a centripetal acceleration $\mathbf{a}_M$ directed toward the Earth. An object near the Earth's surface, such as the apple shown here, experiences an acceleration $\mathbf{g}$. (Dimensions are not to scale.)

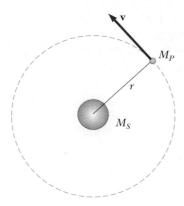

Figure 11.5 A planet of mass M_P moving in a circular orbit about the Sun. The orbits of all planets except Mars, Mercury, and Pluto are nearly circular.

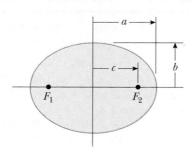

Figure 11.6 Plot of an ellipse. The semimajor axis has a length a, and the semiminor axis has a length b. The focal points are located at a distance c from the center, where $a^2 = b^2 + c^2$, and the eccentricity is defined as $e = c/a$.

Kepler's Third Law

Kepler's third law can be predicted from the inverse-square law for circular orbits.[1] Consider a planet of mass M_P which is assumed to be moving about the Sun (mass M_S) in a circular orbit, as in Figure 11.5. Because the gravitational force on the planet provides the centripetal acceleration as it moves in a circle, we can use Equation 5.3 and write

$$\frac{GM_S M_P}{r^2} = \frac{M_P v^2}{r}$$

But the orbital speed of the planet is simply $2\pi r/T$, where T is the period; therefore, the preceding expression becomes

$$\frac{GM_S}{r^2} = \frac{(2\pi r/T)^2}{r}$$

$$T^2 = \left(\frac{4\pi^2}{GM_S}\right) r^3 = K_S r^3 \qquad\qquad \textbf{[11.4]} \qquad \bullet \ \textit{Kepler's third law}$$

where K_S is a constant given by

$$K_S = \frac{4\pi^2}{GM_S} = 2.97 \times 10^{-19} \ \text{s}^2/\text{m}^3$$

Equation 11.4 is Kepler's third law. It is also valid for elliptical orbits if we replace r with the length of the semimajor axis, a (Fig. 11.6). Note that the constant of proportionality, K_S, is independent of the mass of the planet. Equation 11.4 is therefore valid for *any* planet. If we were to consider the orbit of a satellite about the Earth, such as the Moon, then the constant would have a different value, with

---

[1]The orbits of all planets except Mars, Mercury, and Pluto are very close to being circular. For example, the ratio of the semiminor axis to the semimajor axis for the Earth is $b/a = 0.999\ 86$. The closeness of the ratio b/a to 1 indicates that the difference between perihelion and aphelion is small as expected.

TABLE 11.2   Useful Planetary Data

Body	Mass (kg)	Mean Radius (m)	Period (s)	Distance from Sun (m)	$\dfrac{T^2}{r^3}\left(\dfrac{s^2}{m^3}\right)$
Mercury	3.18×10^{23}	2.43×10^6	7.60×10^6	5.79×10^{10}	2.97×10^{-19}
Venus	4.88×10^{24}	6.06×10^6	1.94×10^7	1.08×10^{11}	2.99×10^{-19}
Earth	5.98×10^{24}	6.37×10^6	3.156×10^7	1.496×10^{11}	2.97×10^{-19}
Mars	6.42×10^{23}	3.37×10^6	5.94×10^7	2.28×10^{11}	2.98×10^{-19}
Jupiter	1.90×10^{27}	6.99×10^7	3.74×10^8	7.78×10^{11}	2.97×10^{-19}
Saturn	5.68×10^{26}	5.85×10^7	9.35×10^8	1.43×10^{12}	2.99×10^{-19}
Uranus	8.68×10^{25}	2.33×10^7	2.64×10^9	2.87×10^{12}	2.95×10^{-19}
Neptune	1.03×10^{26}	2.21×10^7	5.22×10^9	4.50×10^{12}	2.99×10^{-19}
Pluto	$\approx 1.4 \times 10^{22}$	$\approx 1.5 \times 10^6$	7.82×10^9	5.91×10^{12}	2.96×10^{-19}
Moon	7.36×10^{22}	1.74×10^6	—	—	—
Sun	1.991×10^{30}	6.96×10^8	—	—	—

the Sun's mass replaced by the Earth's mass. In this case, the proportionality constant would equal $4\pi^2/GM_E$.

Table 11.2 is a collection of useful planetary data. The last column verifies that the ratio T^2/r^3 is constant.

Example 11.3   An Earth Satellite

A satellite of mass m moves in a circular orbit about the Earth with a constant speed of v, at a height of $h = 1000$ km above the Earth's surface, as in Figure 11.7. (For clarity, this figure is not drawn to scale.) Find the orbital speed of the satellite. The radius of the Earth is 6.37×10^6 m, and its mass is 5.98×10^{24} kg.

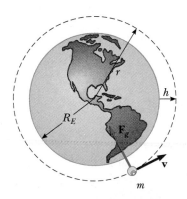

Figure 11.7   (Example 11.3) A satellite of mass m moving around the Earth in a circular orbit of radius r and with constant speed v. The only force acting on the satellite is the gravitational force $\mathbf{F}_g$. (Not drawn to scale.)

Solution   The only external force on the satellite is the gravitational attraction exerted by the Earth. This force is directed toward the center of the satellite's circular path and is the force causing the centripetal acceleration of the satellite. Because the magnitude of the force of gravity is $GM_E m/r^2$, we find that

$$F_g = G\frac{M_E m}{r^2} = m\frac{v^2}{r}$$

$$v^2 = \frac{GM_E}{r}$$

In this expression, the distance r is the Earth's radius plus the height of the satellite; that is, $r = R_E + h = 7.37 \times 10^6$ m, so that

$$v^2 = \frac{(6.67 \times 10^{-11}\ \text{N} \cdot \text{m}^2/\text{kg}^2)(5.98 \times 10^{24}\ \text{kg})}{7.37 \times 10^6\ \text{m}}$$

$$- 5.41 \times 10^7\ \text{m}^2/\text{s}^2$$

Therefore,

$$v = 7.36 \times 10^3\ \text{m/s} \approx \boxed{16\ 400\ \text{mi/h}}$$

Note that v *is independent of the mass of the satellite!*

EXERCISE 3   Calculate the period of revolution, T, of the satellite.   Answer   105 min

Kepler's Second Law and Conservation of Angular Momentum

Consider a planet of mass M_P moving about the Sun in an elliptical orbit (Fig. 11.8). The gravitational force acting on the planet is a central force, always along the radius vector, directed toward the Sun. The torque on the planet due to this central force is clearly zero, because $\mathbf{F}$ is parallel to $\mathbf{r}$. That is

$$\boldsymbol{\tau} = \mathbf{r} \times \mathbf{F} = \mathbf{r} \times F(r)\hat{\mathbf{r}} = 0$$

But recall that the torque equals the time rate of change of angular momentum; that is, $\boldsymbol{\tau} = d\mathbf{L}/dt$. Therefore,

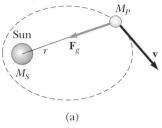

(a)

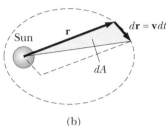

(b)

> because $\boldsymbol{\tau} = 0$, the angular momentum, $\mathbf{L}$, of the planet is a constant of the motion:

$$\mathbf{L} = \mathbf{r} \times \mathbf{p} = m\mathbf{r} \times \mathbf{v} = \text{constant}$$

Because $\mathbf{L}$ is a constant of the motion, the planet's motion at any instant is restricted to the plane formed by $\mathbf{r}$ and $\mathbf{v}$.

We can relate this result to the following geometric consideration. In a time of dt, the radius vector $\mathbf{r}$ in Figure 11.9b sweeps out the area dA, which equals one half the area $|\mathbf{r} \times d\mathbf{r}|$ of the parallelogram formed by the vectors $\mathbf{r}$ and $d\mathbf{r}$. Because the displacement of the planet in the time dt is given by $d\mathbf{r} = \mathbf{v}\, dt$, we get

$$dA = \tfrac{1}{2}|\mathbf{r} \times d\mathbf{r}| = \tfrac{1}{2}|\mathbf{r} \times \mathbf{v}\, dt| = \frac{L}{2m}\, dt$$

$$\frac{dA}{dt} = \frac{L}{2m} = \text{constant} \qquad\qquad [11.5]$$

Figure 11.8 (a) The gravitational force acting on a planet acts toward the Sun, along the radius vector. (b) As a planet orbits the Sun, the area swept out by the radius vector in a time dt is equal to one half the area of the parallelogram formed by the vectors $\mathbf{r}$ and $d\mathbf{r} = \mathbf{v}\, dt$.

• *Kepler's second law*

(Left) This image of Saturn, taken by the *Voyager I* spacecraft on October 18, 1980, was color-enhanced to increase the visibility of large, bright features in Saturn's North Temperate Belt. The distinct color difference between the North Equatorial Belt and other belts may be due to a thicker haze layer covering the northern belt. *(Right)* Jupiter and its four planet-size moons were photographed by *Voyager I* and assembled into this collage. The moons are not to scale but are in their relative positions. Nine other smaller moons circle Jupiter. Not visible is Jupiter's faint ring of particles, seen for the first time by *Voyager I*. *(NASA photos)*

where L and m are both constants of the motion. Thus, we conclude that

> the radius vector from the Sun to any planet sweeps out equal areas in equal times.

It is important to recognize that this result, which is Kepler's second law, is a consequence of the fact that the force of gravity is a central force, which in turn implies that angular momentum of the planet is constant. Therefore, the law applies to *any* situation that involves a central force, whether inverse square or not.

The inverse-square nature of the force of gravity is not revealed by Kepler's second law. Although we do not prove it here, Kepler's first law (as well as his third law) is a direct consequence of the fact that the gravitational force varies as $1/r^2$. That is, under an inverse-square force law, the orbits of the planets can be shown to be ellipses with the Sun at one focus.

Thinking Physics 2

The Earth is closer to the Sun when it is winter in the Northern Hemisphere than when it is summer. July and January both have 31 days. In which month, if either, does the Earth move through a longer distance in its orbit?

Reasoning The Earth is in a slightly elliptical orbit around the Sun. In order to conserve angular momentum, the Earth must move more rapidly when it is close to the Sun and more slowly when it is farther away. Thus, since it is closer to the Sun in January, it is moving faster, and will cover more distance in its orbit than it will in July.

CONCEPTUAL PROBLEM 1

Kepler's first law indicates that planets travel in elliptical orbits. This is intimately related to the fact that the gravitational force varies as the inverse square of the separation between the Sun and the planet. Suppose the gravitational force varied as the inverse *cube*, rather than the inverse square. This opens up some new types of orbits. What about the second and third laws? Will they change if the force depends on the inverse cube of the separation?

CONCEPTUAL PROBLEM 2

A satellite in orbit is not truly traveling through a vacuum—it is moving through very thin air. Does the resulting air friction cause the satellite to slow down?

CONCEPTUAL PROBLEM 3

How would you explain the fact that Saturn and Jupiter have periods much greater than one year?

EXERCISE 4   At its aphelion the planet Mercury is 6.99×10^{10} m from the Sun, and at its perihelion it is 4.60×10^{10} m from the Sun. If its orbital speed is 3.88×10^4 m/s at the aphelion, what is its orbital speed at the perihelion?   Answer 5.90×10^4 m/s

11.4 • ENERGY CONSIDERATIONS IN PLANETARY AND SATELLITE MOTION

Consider a body of mass m moving with a speed of v in the vicinity of a massive body of mass $M \gg m$. The system might be a planet moving around the Sun or a satellite in orbit around the Earth. If we assume that M is at rest in an inertial reference frame,[2] then the total mechanical energy, E, of the two-body system when the bodies are separated by a distance r is the sum of the kinetic energy of the mass m and the potential energy of the system:

$$E = K + U$$

Recall from Chapter 7, Equation 7.22, that the gravitational potential energy, U_g, associated with *any pair* of particles of masses m_1 and m_2 separated by a distance r is given by

$$U_g = -\frac{Gm_1 m_2}{r}$$

Therefore, in our case,

$$E = \tfrac{1}{2}mv^2 - \frac{GMm}{r} \qquad \textbf{[11.6]}$$

Furthermore, the total mechanical energy is constant if we assume that the system is isolated. Therefore, as the mass m moves from P to Q in Figure 11.9, the total energy remains constant and Equation 11.6 gives

$$E = \tfrac{1}{2}mv_i^2 - \frac{GMm}{r_i} = \tfrac{1}{2}mv_f^2 - \frac{GMm}{r_f} \qquad \textbf{[11.7]}$$

Equation 11.6 shows that E may be positive, negative, or zero, depending on the value of v. However, for a bound system, such as the Earth and Sun, E is necessarily *less than zero* if we use the arbitrary convention $U \rightarrow 0$ as $r \rightarrow \infty$. We can easily establish that $E < 0$ for the system consisting of a mass m moving in a circular orbit about a body of mass M. Newton's second law applied to the body of mass m gives

$$\frac{GMm}{r^2} = \frac{mv^2}{r}$$

Multiplying both sides by r and dividing by 2,

$$\tfrac{1}{2}mv^2 = \frac{GMm}{2r} \qquad \textbf{[11.8]}$$

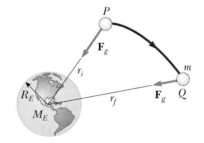

Figure 11.9   As a particle of mass m moves from P to Q above the Earth's surface, the total mechanical energy of the system remains constant.

[2]To see that this is reasonable, consider an object of mass m falling toward the Earth. Because the center of mass of the object–Earth system is stationary, it follows that $mv = M_E v_E$. Thus, the Earth acquires a kinetic energy equal to

$$\tfrac{1}{2}M_E v_E^2 = \tfrac{1}{2}\frac{m^2}{M_E}v^2 = \frac{m}{M_E}K$$

where K is the kinetic energy of the object. Because $M_E \gg m$, the kinetic energy of the Earth is negligible.

Substituting this into Equation 11.6, we obtain

$$E = \frac{GMm}{2r} - \frac{GMm}{r}$$

Total mechanical energy for •
circular orbits

$$E = -\frac{GMm}{2r} \qquad \text{[11.9]}$$

This clearly shows that **the total mechanical energy must be negative in the case of circular orbits.** Furthermore, **the kinetic energy is positive and equal to one half the magnitude of the potential energy** (when the potential energy is chosen to be zero at infinity). The absolute value of E is also equal to the binding energy of the system.

The total mechanical energy is also negative in the case of elliptical orbits. The expression for E for elliptical orbits is the same as Equation 11.9, with r replaced by the semimajor axis length, a. The total energy, the total angular momentum, and the total linear momentum of a planet–Sun system are constants of the motion.

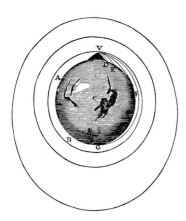

"... the greater the velocity ... with which (a stone) is projected, the farther it goes before it falls to the Earth. We may therefore suppose the velocity to be so increased, that it would describe an arc of 1, 2, 5, 10, 100, 1000 miles before it arrived at the Earth, till at last, exceeding the limits of the Earth, it should pass into space without touching."—Newton, *System of the World.*

Thinking Physics 3

Icebound, by Dean Koontz (Ballantine Books, New York, 1995), is a story of a group of scientists trapped on a floating iceberg near the North Pole. One of the devices that the scientists have with them on their task is a transmitter with which they can fix their position with "the aid of a *geosynchronous polar satellite.*" Can a satellite in a *polar* orbit be *geosynchronous*?

Reasoning A geosynchronous satellite is one that stays over one location on the Earth's surface at all times. Thus, an antenna receiving signals from the satellite, such as a television dish, can stay pointed in a fixed direction toward the sky. The satellite must be in an orbit with the correct radius such that its orbital period is the same as that of the Earth's rotation. This would result in the satellite appearing to have no east–west motion relative to the observer at the chosen location. Another requirement is that a geosynchronous satellite *must be in orbit over the Equator*. Otherwise it would appear to undergo a north–south oscillation during one orbit. Thus, it would be impossible to have a geosynchronous satellite in a *polar* orbit. Even if such a satellite were at the proper distance from the Earth, it would be moving rapidly in the north–south direction, resulting in the necessity of accurate tracking equipment. What's more, it would be below the horizon for long periods of time, making it useless for determining one's position.

Thinking Physics 4

A satellite is in a circular orbit in space. Is it necessary that there be a massive object at the center of the orbit?

Reasoning This is the usual situation—satellites are normally in orbit around the Earth, a massive object. Imagine, however, tying a rock somewhere along a length of string and twirling the rock in a circle as shown in Figure 11.10. In this case, there is no entity at the center of the circular path of the rock, nor along its radius, that is exerting a force on the rock. The net force holding the rock in the circular orbit is a

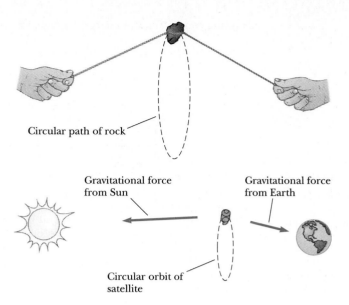

Circular path of rock

Gravitational force from Sun

Gravitational force from Earth

Circular orbit of satellite

Figure 11.10 (Thinking Physics 4)

combination of the radial components of two forces—one from the string on each side of the rock.

The question posed was not about rocks on strings; it was about satellites. Is there a gravitational counterpart to this situation? There is. Imagine moving from the Earth to the Sun along the line between them. At some point along that line, the gravitational force of attraction to the Earth will balance that toward the Sun. This is similar to the balance along the string of the tension on either side of the rock—the rock is in equilibrium between two balanced forces. Now imagine pulling the satellite perpendicularly off the Earth–Sun line a small amount—as long as the distance is relatively small, the components of the forces from the Earth and Sun along the Sun–Earth centerline will cancel. The components of the two forces perpendicular to the centerline will add, to provide a net force pulling the satellite back toward the centerline. Now, suppose we set the satellite into circular motion with just the right velocity. The satellite will be in orbit around a center point with no mass located there.

The International Sun–Earth Explorer 3 satellite (ISEE-3) was launched in 1978 and placed in just such an orbit in order to study the solar wind. The advantage over a Sun-centered orbit is that the satellite always resides on the side of the Sun toward the Earth, so that it is constantly in communication with Earth. In the absence of a specific mission to Comet Halley, the ISEE-3 satellite was removed from this orbit in 1982 and placed in orbit around the Moon, in anticipation of a flyby of Comet Halley. On the way to this flyby, it passed through the tail of Comet Giacobini-Zinner and successfully gathered important data on both comets.

Example 11.4 A Satellite in an Elliptical Orbit

A satellite moves in an elliptical orbit about the Earth, as in Figure 11.11. The minimum and maximum distances from the surface of the Earth are 400 km and 3000 km. Find the speeds of the satellite at apogee and perigee.

Solution Because the mass of the satellite is negligible compared with the Earth's mass, we take the center of mass of the Earth to be at rest. Gravity is a *central* force, and so the angular momentum of the satellite about the Earth's center of

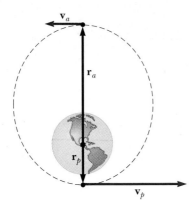

Figure 11.11   (Example 11.4) A satellite in an elliptical orbit about the Earth.

mass remains constant in time. With subscripts a and p for the apogee and perigee positions, conservation of angular momentum gives $L_p = L_a$, or

$$mv_p r_p = mv_a r_a$$

$$v_p r_p = v_a r_a \qquad\qquad \text{[1]}$$

Applying conservation of energy, we obtain $E_p = E_a$, or

$$U_p + K_p = U_a + K_a$$

$$-G\frac{M_E m}{r_p} + \tfrac{1}{2}mv_p^2 = -G\frac{M_E m}{r_a} + \tfrac{1}{2}mv_a^2$$

$$2GM_E\left(\frac{1}{r_a} - \frac{1}{r_p}\right) = (v_a^2 - v_p^2) \qquad \text{[2]}$$

Assuming the Earth's radius is 6.37×10^6 m and using the given data, we get $r_a = 9.37 \times 10^6$ m and $r_p = 6.77 \times 10^6$ m. Because we know the numerical values of G, M_E, r_p, and r_a, we can use Equations (1) and (2) to determine the two unknowns, (v_p and v_a). Solving the equations simultaneously, we obtain

$$v_p = \boxed{8.27 \text{ km/s}} \qquad v_a = \boxed{5.98 \text{ km/s}}$$

Escape Speed

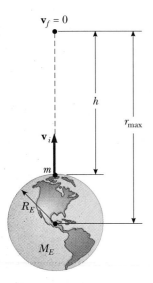

Figure 11.12 An object of mass m projected upward from the Earth's surface with an initial velocity $\mathbf{v}_i$ reaches a maximum altitude h.

Suppose an object of mass m is projected vertically from the Earth's surface with an initial speed v_i, as in Figure 11.12. We can use energy considerations to find the minimum value of the initial speed such that the object will escape the Earth's gravitational field. Equation 11.6 gives the total energy of the object at any point when its speed and distance from the center of the Earth are known. At the surface of the Earth, where $v_i = v$, $r_i = R_E$. When the object reaches its maximum altitude, $v_f = 0$ and $r_f = r_{\max}$. Because the total energy of the system is constant, substitution of these conditions into Equation 11.7 gives

$$\tfrac{1}{2}mv_i^2 - \frac{GM_E m}{R_E} = -\frac{GM_E m}{r_{\max}}$$

Solving for v_i^2 gives

$$v_i^2 = 2GM_E\left(\frac{1}{R_E} - \frac{1}{r_{\max}}\right) \qquad \text{[11.10]}$$

Therefore, if the initial speed is known, this expression can be used to calculate the maximum altitude, h, because we know that $h = r_{\max} - R_E$.

We are now in a position to calculate the minimum speed the object must have at the Earth's surface in order to escape from the influence of the Earth's gravitational field. This corresponds to the situation in which the object will continue to move away forever, with the speed asymptotically approaching *zero*. Setting $r_{\max} = \infty$ in Equation 11.10 and taking $v_i = v_{\text{esc}}$ (the escape speed), we get

$$v_{esc} = \sqrt{\frac{2GM_E}{R_E}} \qquad\qquad [11.11]$$

• *Escape speed*

Note that this expression for v_{esc} is independent of the mass of the object projected from the Earth. For example, a spacecraft has the same escape speed as a molecule. Furthermore, the result is independent of the *direction* of the velocity, provided the object is not thrown into the ground.

If the object is given an initial speed equal to v_{esc}, its *total* energy is equal to zero. This can be seen by noting that as $r \to \infty$, the object's kinetic energy and its potential energy are both zero. If v_i is greater than v_{esc}, the *total* energy is greater than zero and the object is still moving as $r \to \infty$.

Finally, you should note that Equations 11.10 and 11.11 can be applied to objects projected from *any* planet. That is, in general, the escape speed from any planet of mass M and radius R is

$$v_{esc} = \sqrt{\frac{2GM}{R}} \qquad\qquad [11.12]$$

A list of escape speeds for the planets, the Moon, and the Sun is given in Table 11.3. Note that the values vary from 1.1 km/s for Pluto to about 618 km/s for the Sun. These results, together with some ideas from the kinetic theory of gases (Chapter 16), explain why some planets have atmospheres and others do not. As we shall see later, a gas molecule has an average kinetic energy that depends on its temperature. Lighter atoms, such as hydrogen and helium, have higher average speeds than do the heavier species. When the average speed of the lighter atoms is not much less than the escape speed, a significant fraction of the molecules have a chance to escape from the planet. This mechanism also explains why the Earth does not retain hydrogen and helium molecules in its atmosphere but does retain much heavier molecules, such as oxygen and nitrogen. By contrast, Jupiter has a very large escape speed (60 km/s), which enables it to retain hydrogen, the primary component of its atmosphere.

TABLE 11.3 Escape Speeds from the Surfaces of the Planets, the Moon, and the Sun

Planet	v_{esc}(km/s)
Mercury	4.3
Venus	10.3
Earth	11.2
Mars	5.0
Jupiter	60.0
Saturn	36.0
Uranus	22.0
Neptune	24.0
Pluto	1.1
Moon	2.3
Sun	618.0

Black Holes

In Chapter 10 we briefly described a rare event called a supernova—the catastrophic explosion of a very massive star. The material that remains in the central core of such an object continues to collapse, but the core's ultimate fate depends on its mass. If the core has a mass less than 1.4 times the mass of our Sun, it gradually cools down and ends its life as a white dwarf star. However, if the core's mass is greater than this, the star may collapse further due to gravitational forces. What remains is a neutron star, a region compressed to a radius of about 10 km. (On Earth, a teaspoon of this material would weigh about 5 billion tons!)

An even more unusual star death may occur when the core has a mass greater than about three solar masses. No known forces in nature are strong enough to prevent the collapse of such a star. The collapse may continue until the star becomes a small object in space, commonly referred to as a **black hole.** In effect, black holes are remains of stars that have collapsed under their own weight. Once an object such as a spaceship comes near a black hole, it experiences an extremely strong gravitational force and is trapped forever (Fig. 11.13).

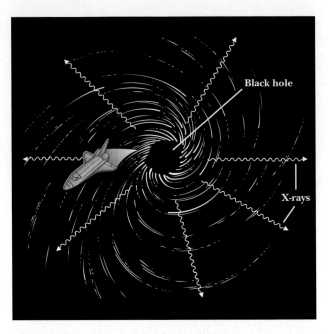

Figure 11.13 The gravitational field in the vicinity of a black hole is so strong that nothing can escape. Any object moving close to a black hole can emit x-rays.

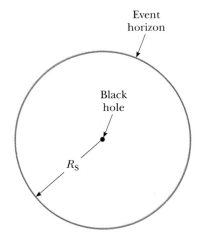

Figure 11.14 A black hole. The distance R_S equals the Schwarzschild radius. Any event occurring within the boundary of radius R_S, called the event horizon, is invisible to an outside observer.

The escape speed from any spherical body can be calculated from Equation 11.12. If the escape speed exceeds the speed of light, c, radiation within the body (such as visible light) cannot escape and the body appears to be black; hence the origin of the terminology "black hole." The critical radius, R_S, for which this occurs is called the **Schwarzschild radius** (Fig. 11.14). Taking $v_{esc} = c$ in Equation 11.12 and solving for R_S, we obtain $R_S = 2GM/c^2$. For example, the value for R_S for a black hole with a mass equal to that of the Sun is calculated to be 3.0 km (about 2 mi); a black hole the mass of which equals that of the Earth has a radius of about 9 mm (about the size of a dime).

Although light from a black hole cannot escape, light from events taking place near the black hole should be visible. A companion star captured by the strong gravitational field of a black hole should emit x-rays (Fig. 11.13). Based on this reasoning, several candidates for black holes have been detected, the most famous being Cygnus X-1, the first x-ray source detected in the constellation Cygnus. There is also evidence that supermassive black holes exist at the centers of galaxies.

EXERCISE 5 The escape speed from the surface of the Earth is 11.2 km/s. Estimate the escape speed for a spacecraft from the surface of the Moon. The Moon has a mass $\frac{1}{81}$ that of Earth and a radius $\frac{1}{4}$ that of Earth.   Answer   2.49 km/s

EXERCISE 6   (a) Calculate the minimum energy required to send a 3000-kg spacecraft from the Earth to a distant point in space where Earth's gravity is negligible. (b) If the journey is to take three weeks, what *average* power will the engines have to supply?
Answer   (a) 1.88×10^{11} J   (b) 103 kW

EXERCISE 7 A satellite moves in an elliptical orbit about the Earth such that, at perigee and apogee positions, the distances from the Earth's center are, respectively, D and $4D$. Find the ratio of the speeds at the two positions, v_p/v_a.   Answer   4

11.5 · ATOMIC SPECTRA AND THE BOHR THEORY OF HYDROGEN

As you may have already learned in chemistry, the hydrogen atom is the simplest known atomic system and an especially important one to understand. Much of what is learned about the hydrogen atom (which consists of one proton and one electron) can be extended to other single-electron ions such as He^+ and Li^{2+}. Furthermore, a thorough understanding of the physics underlying the hydrogen atom can then be used to describe more complex atoms and the periodic table of the elements.

Suppose an evacuated glass tube is filled with hydrogen (or some other gas). If a voltage applied between metal electrodes in the tube is great enough to produce an electric current in the gas, the tube emits light the color of which is characteristic of the gas (this is how a neon sign works). When the emitted light is analyzed with a device called a spectroscope, a series of discrete lines is observed, each line corresponding to a different wavelength, or color, of light. Such a series of spectral lines is commonly referred to as an **emission spectrum.** The wavelengths contained in a given line spectrum are characteristic of the element emitting the light (Fig. 11.15). Because no two elements emit the same line spectrum, this phenomenon represents a marvelous and reliable technique for identifying elements in a substance.

As you shall learn in more detail in Chapter 13, a mechanical wave is a disturbance that transports energy as it moves through a system without transporting matter. A common form of periodic wave is the sinusoidal wave, the shape of which

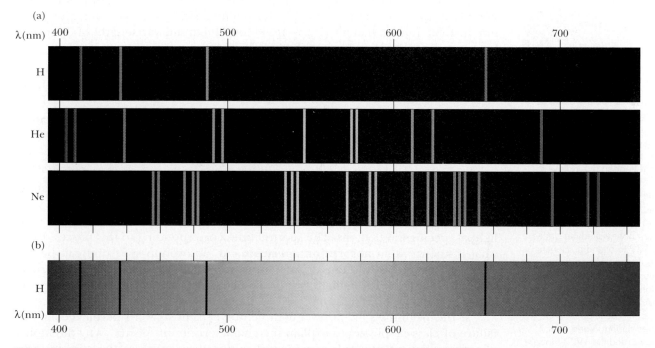

Figure 11.15 Visible spectra. (a) Line spectra produced by emission in the visible range for the elements hydrogen, helium, and neon. (b) The absorption spectrum for hydrogen. The dark absorption lines occur at the same wavelengths as the emission line for hydrogen shown in (a).

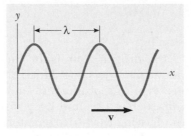

Figure 11.16 A sinusoidal wave traveling to the right. Any point on the wave moves a distance of one wavelength, λ, in a time equal to the period of the wave.

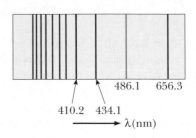

Figure 11.17 A series of spectral lines for atomic hydrogen. The prominent lines labeled are part of the Balmer series.

Niels Bohr (1885–1962)

A Danish physicist, Bohr was an active participant in the early development of quantum mechanics and provided much of its philosophical framework. During the 1920s and 1930s, Bohr headed the Institute for Advanced Studies in Copenhagen. This institute was a magnet for many of the world's best physicists and provided a forum for the exchange of ideas. When Bohr visited the United States in 1939 to attend a scientific conference, he brought news that the fission of uranium had been discovered by Hahn and Strassman in Berlin. The results were the foundations of the atomic bomb developed in the United States during World War II. Bohr was awarded the 1922 Nobel prize for his investigation of the structure of atoms and of the radiation emanating from them.

is depicted in Figure 11.16. The distance between two consecutive crests of the wave is called the **wavelength,** λ. As the wave travels to the right with speed v, any point on the wave travels a distance of one wavelength in a time interval of one period, T (the time for one cycle), so the wave speed is $v = \lambda / T$. The inverse of the period, $1/T$, is called the **frequency,** f, of the wave; it represents the number of cycles per second. Thus, the speed of the wave is often written as $v = \lambda f$. In this section, because we shall deal with electromagnetic waves—which travel at the speed of light, c—the appropriate relation is

$$c = \lambda f \qquad \qquad \textbf{[11.13]}$$

The emission spectrum of hydrogen shown in Figure 11.17 includes four prominent lines that occur at wavelengths of 656.3 nm, 486.1 nm, 434.1 nm, and 410.2 nm. In 1885, Johann Balmer (1825–1898) found that the wavelengths of these and less prominent lines can be described by this simple empirical equation:

$$\frac{1}{\lambda} = R_H \left(\frac{1}{2^2} - \frac{1}{n^2} \right) \qquad \text{Balmer series} \qquad \textbf{[11.14]}$$

where n may have integral values of 3, 4, 5, . . . , and R_H is a constant, now called the **Rydberg constant.** If the wavelength is in meters, R_H has the value

$$R_H = 1.0973732 \times 10^7 \ \text{m}^{-1}$$

The first line in the Balmer series, at 656.3 nm, corresponds to $n = 3$ in Equation 11.14; the line of 486.1 nm corresponds to $n = 4$; and so on.

In addition to emitting light at specific wavelengths, an element can also absorb light at specific wavelengths. The spectral lines corresponding to this process form what is known as an **absorption spectrum.** An absorption spectrum can be obtained by passing a continuous radiation spectrum (one containing all wavelengths) through a vapor of the element being analyzed. The absorption spectrum consists of a series of dark lines superimposed on the otherwise continuous spectrum.

At the beginning of the twentieth century, scientists were perplexed by the failure of classical physics to explain the characteristics of spectra. Why did atoms of a given element emit only certain wavelengths of radiation, so that the emission spectrum displayed discrete lines? Furthermore, why did the atoms absorb only those wavelengths that they emitted? In 1913 Bohr provided an explanation of atomic spectra that included some features of the currently accepted theory. Using the simplest atom, hydrogen, Bohr described a model of what he thought must be the atom's structure. His model of the hydrogen atom contains some classical features as well as some revolutionary postulates that could not be justified within the framework of classical physics.[3] The basic assumptions of the Bohr theory as it applies to the hydrogen atom are as follows:

1. The electron moves in circular orbits about the proton under the influence of the Coulomb force of attraction, as in Figure 11.18.
2. Only certain electron orbits are stable. These are orbits in which the hydrogen atom does not emit energy in the form of radiation. Hence, the total energy of the atom remains constant, and classical mechanics can be used to describe the electron's motion.
3. Radiation is emitted by the hydrogen atom when the electron "jumps" from a more energetic initial state to a lower state. The jump cannot be visualized or treated classically. In particular, the frequency, f, of the radiation emitted in the jump is related to the change in the atom's energy and **is independent of the frequency of the electron's orbital motion.** The frequency of the emitted radiation is found from

$$E_i - E_f = hf \qquad \textbf{[11.15]}$$

where E_i is the energy of the initial state, E_f is the energy of the final state, h is Planck's constant (see Section 10.9), and $E_i > E_f$.

4. The size of the allowed electron orbits is determined by a condition imposed on the electron's orbital angular momentum: the allowed orbits are those for which the electron's orbital angular momentum about the nucleus is an integral multiple of $\hbar = h/2\pi$.

$$m_e v r = n\hbar \qquad n = 1, 2, 3, \ldots \qquad \textbf{[11.16]}$$

• *Assumptions of the Bohr theory*

Using these four assumptions, we can calculate the allowed energy levels and emission wavelengths of the hydrogen atom. The electrical potential energy of the system shown in Figure 11.18 is given by $U_e = -k_e e^2/r$, where k_e is the Coulomb constant, e is the charge on the electron, and r is the electron–proton separation. Thus, the total energy of the atom, which contains both kinetic and potential energy terms, is

$$E = K + U_e = \tfrac{1}{2} m_e v^2 - k_e \frac{e^2}{r} \qquad \textbf{[11.17]}$$

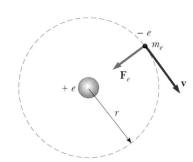

Figure 11.18 Diagram representing Bohr's model of the hydrogen atom in which the orbiting electron is allowed to be only in specific orbits of discrete radii.

Applying Newton's second law to this system, we see that the Coulomb attractive force on the electron, $k_e e^2/r^2$ (Eq. 5.15), must equal the mass times the centripetal acceleration $(a = v^2/r)$ of the electron:

$$\frac{k_e e^2}{r^2} = \frac{m_e v^2}{r}$$

---

[3]The Bohr model can be applied successfully to such hydrogen-like ions as singly ionized helium and doubly ionized lithium, but the theory does not properly describe the visible spectra of more complex atoms and ions.

From this expression, we find the kinetic energy to be

$$K = \tfrac{1}{2} m_e v^2 = \frac{k_e e^2}{2r}$$ [11.18]

Substituting this value of K into Equation 11.17, we find that the total energy of the atom is

Total energy of the hydrogen •
atom

$$E = -\frac{k_e e^2}{2r}$$ [11.19]

Note that the total energy is negative, indicating a bound electron–proton system. This means that energy in the amount of $k_e e^2/2r$ must be added to the atom just to remove the electron and make the total energy zero. An expression for r, the radius of the allowed orbits, can be obtained by eliminating v by substitution between Equations 11.16 and 11.18:

Radii of Bohr orbits in •
hydrogen

$$r_n = \frac{n^2 \hbar^2}{m_e k_e e^2} \qquad n = 1, 2, 3, \ldots$$ [11.20]

This result shows that the radii have discrete values, or are *quantized*.

The orbit for which $n = 1$ has the smallest radius; it is called the **Bohr radius,** a_0, and has the value

The Bohr radius •

$$a_0 = \frac{\hbar^2}{m_e k_e e^2} = 0.0529 \text{ nm}$$ [11.21]

The first three Bohr orbits are shown to scale in Figure 11.19.

The quantization of the orbit radii immediately leads to energy quantization. This can be seen by substituting $r_n = n^2 a_0$ into Equation 11.19. The allowed energy levels are found to be

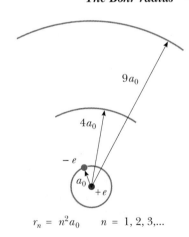

$$E_n = -\frac{k_e e^2}{2a_0} \left(\frac{1}{n^2} \right) \qquad n = 1, 2, 3, \ldots$$ [11.22]

Insertion of numerical values into Equation 11.22 gives

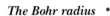

$$E_n = -\frac{13.6}{n^2} \text{ eV} \qquad n = 1, 2, 3, \ldots$$ [11.23]

$r_n = n^2 a_0 \qquad n = 1, 2, 3,...$

Figure 11.19 The first three Bohr orbits for hydrogen.

(Recall from Section 9.7 that 1 eV = 1.6×10^{-19} J.) The lowest stationary state corresponding to $n = 1$, called the **ground state,** has an energy of $E_1 = -13.6$ eV. The next state, the **first excited state,** has $n = 2$ and an energy of $E_2 = E_1/2^2 = -3.4$ eV. Figure 11.20 is an energy-level diagram showing the energies of these discrete energy states and the corresponding quantum numbers. The uppermost level, corresponding to $n = \infty$ (or $r = \infty$) and $E = 0$, represents the state for which the electron is removed from the atom. The minimum energy required to ionize the atom (that is, to completely remove an electron in the ground state from the proton's influence) is called the **ionization energy.** As can be seen from Figure 11.20, the ionization energy for hydrogen, based on Bohr's calculation, is 13.6 eV. This constituted another major achievement for the Bohr theory, because the ionization energy for hydrogen had already been measured to be precisely 13.6 eV.

Figure 11.20 also shows other spectral series (the Lyman series and the Paschen series) that were found after Balmer's discovery. These spectra obey other empirical formulas, which are reconciled with the Bohr model.

Equation 11.22, together with Bohr's third postulate, can be used to calculate the frequency of the radiation that is emitted when the electron jumps from an outer orbit to an inner orbit:

$$f = \frac{E_i - E_f}{h} = \frac{k_e e^2}{2a_0 h}\left(\frac{1}{n_f^2} - \frac{1}{n_i^2}\right) \qquad [11.24]$$

Because the quantity being measured is wavelength, it is convenient to convert frequency to wavelength, using $c = f\lambda$, to get

$$\frac{1}{\lambda} = \frac{f}{c} = \frac{k_e e^2}{2a_0 hc}\left(\frac{1}{n_f^2} - \frac{1}{n_i^2}\right) \qquad [11.25]$$

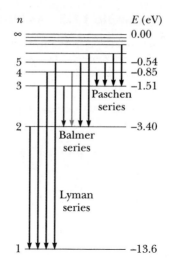

Figure 11.20 An energy level diagram for hydrogen. The discrete allowed energies are plotted on the vertical axis. Nothing is plotted on the horizontal axis, but the horizontal extent of the diagram is made large enough to show allowed transitions. Quantum numbers are given on the left.

The remarkable fact is that the *theoretical* expression, Equation 11.25, is identical to a generalized form of the empirical relations discovered by Balmer and others (see Eq. 11.14),

$$\frac{1}{\lambda} = R_H\left(\frac{1}{n_f^2} - \frac{1}{n_i^2}\right) \qquad [11.26]$$

provided that the combination of constants $k_e e^2/2a_0 hc$ is equal to the experimentally determined Rydberg constant. After Bohr demonstrated the agreement of these two quantities to a precision of about 1%, it was soon recognized as the crowning achievement of his new theory of quantum mechanics. Furthermore, Bohr showed that all of the spectral series for hydrogen have a natural interpretation in his theory. Figure 11.20 shows these spectral series as transitions between energy levels.

Bohr immediately extended his model for hydrogen to other elements in which all but one electron had been removed. Ionized elements such as He^+, Li^{2+}, and Be^{3+} were suspected to exist in hot stellar atmospheres, where frequent atomic collisions occur with enough energy to completely remove one or more atomic electrons. Bohr showed that many mysterious lines observed in the Sun and several stars could not be due to hydrogen but were correctly predicted by his theory if attributed to singly ionized helium.

Although the Bohr model had some success in predicting the spectra of single valence electron atoms, it has some serious drawbacks. For example, it cannot account for the visible spectra of more complex atoms, and it is unable to predict many subtle spectral details of hydrogen and other simple atoms.

CONCEPTUAL PROBLEM 4

Under normal experimental conditions, hydrogen atoms show more lines in emission than in absorption. Why?

CONCEPTUAL PROBLEM 5

Suppose the energy of the electron in a hydrogen atom is $-E$. What is the kinetic energy? What is the potential energy?

Example 11.5   An Electronic Transition in Hydrogen

The electron in the hydrogen atom makes a transition from the $n = 2$ energy state to the ground state (corresponding to $n = 1$). Find the wavelength and frequency of the emitted radiation.

Solution   We can use Equation 11.26 directly to obtain λ, with $n_i = 2$ and $n_f = 1$:

$$\frac{1}{\lambda} = R_H \left(\frac{1}{n_f^2} - \frac{1}{n_i^2} \right)$$

$$\frac{1}{\lambda} = R_H \left(\frac{1}{1^2} - \frac{1}{2^2} \right) = \frac{3R_H}{4}$$

$$\lambda = \frac{4}{3R_H} = \frac{4}{3(1.097 \times 10^{-7} \text{ m}^{-1})}$$

$$= 1.215 \times 10^{-7} \text{ m} = \boxed{121.5 \text{ nm}} \quad \text{(ultraviolet)}$$

Because $c = f\lambda$, the frequency of the radiation is

$$f = \frac{c}{\lambda} = \frac{3.00 \times 10^8 \text{ m/s}}{1.215 \times 10^{-7} \text{ m}} = \boxed{2.47 \times 10^{15} \text{ s}^{-1}}$$

EXERCISE 8   What is the wavelength of the radiation emitted by hydrogen when the electron makes a transition from the $n = 3$ state to the $n = 1$ state?

Answer $\dfrac{9}{8R_H} = 102.6$ nm

Bohr's Correspondence Principle

In our study of relativity in Chapter 9, we found that newtonian mechanics cannot be used to describe phenomena that occur at speeds approaching the speed of light. Newtonian mechanics is a special case of relativistic mechanics and is usable only when v is much less than c. In a similar way, **quantum mechanics is in agreement with classical physics where the energy differences between quantized levels vanish.** This principle, first set forth by Bohr, is called the **correspondence principle.**

For example, consider an electron orbiting the hydrogen atom with $n > 10\,000$. For such large values of n, the energy differences between adjacent levels approach zero and the levels are nearly continuous. As a consequence, the classical model is reasonably accurate in describing the system for large values of n. According to the classical picture, the frequency of the light emitted by the atom is equal to the frequency of revolution of the electron in its orbit about the nucleus. Calculations show that for $n > 10\,000$, this frequency differs from that predicted by quantum mechanics by less than 0.015%.

SUMMARY

Newton's universal law of gravity states that the gravitational force of attraction between any two particles of masses m_1 and m_2 separated by a distance r has the magnitude

$$F_g = G \frac{m_1 m_2}{r^2} \qquad \text{[11.1]}$$

where G is the gravitational constant, 6.672×10^{-11} N·m²/kg².
Kepler's laws of planetary motion state the following:

1. Every planet moves in an elliptical orbit with the Sun at one of the focal points.
2. The radius vector drawn from the Sun to any planet sweeps out equal areas in equal time intervals.

3. The square of the orbital period of any planet is proportional to the cube of the semimajor axis for the elliptical orbit.

Kepler's second law is a consequence of the fact that the force of gravity is a *central force*. This implies that the angular momentum of the planet–Sun system is a constant of the motion.

Kepler's first and third laws are a consequence of the inverse-square nature of the universal law of gravity. Newton's second law, together with the force law given by Equation 11.1, verifies that the period, T, and radius, r, of the orbit of a planet about the Sun are related by

$$T^2 = \left(\frac{4\pi^2}{GM_S}\right) r^3 \qquad \textbf{[11.4]}$$

where M_S is the mass of the Sun. Most planets have nearly circular orbits about the Sun. For elliptical orbits, Equation 11.4 is valid if r is replaced by the semimajor axis, a.

If an isolated system consists of a particle of mass m moving with a speed of v in the vicinity of a massive body of mass M, the *total energy* of the system is constant and is

$$E = \tfrac{1}{2}mv^2 - \frac{GMm}{r} \qquad \textbf{[11.6]}$$

If m moves in a circular orbit of radius r about M, where $M \gg m$, the total energy of the system is

$$E = -\frac{GMm}{2r} \qquad \textbf{[11.9]}$$

The total energy is negative for any bound system—that is, one in which the orbit is closed, such as a circular or an elliptical orbit.

The minimum speed an object must have to escape the gravitational field of a uniform sphere of mass M and radius R is

$$v_{\text{esc}} = \sqrt{\frac{2GM}{R}} \qquad \textbf{[11.12]}$$

The Bohr model of the atom is successful in describing the spectra of atomic hydrogen and hydrogen-like ions. One of the basic assumptions of the model is that the electron can exist only in discrete orbits such that the angular momentum mvr is an integral multiple of $h/2\pi = \hbar$. Assuming circular orbits and a simple coulombic attraction between the electron and proton, the energies of the quantum states for hydrogen are calculated to be

$$E_n = -\frac{k_e e^2}{2a_0}\left(\frac{1}{n^2}\right) \qquad \textbf{[11.22]}$$

where k_e is the Coulomb constant, e is electronic charge, $n = 1, 2, 3, \cdots$ is a positive integer called the **quantum number** of the state, and $a_0 = 0.0529$ nm is the **Bohr radius.**

If the electron in the hydrogen atom makes a transition from an orbit with a quantum number of n_i to one with a quantum number of n_f, where $n_f < n_i$, the frequency of the radiation emitted by the atom is

$$f = \frac{k_e e^2}{2a_0 h}\left(\frac{1}{n_f^2} - \frac{1}{n_i^2}\right) \qquad \textbf{[11.24]}$$

Using $E = hf = hc/\lambda$, one can calculate the wavelengths of the radiation for various transitions in which there is a change in quantum number, $n_i \rightarrow n_f$. The calculated wavelengths are in excellent agreement with observed line spectra.

CONCEPTUAL QUESTIONS

1. If the gravitational force on an object is directly proportional to its mass, why don't large masses fall with greater acceleration than small ones?

2. The gravitational force that the Sun exerts on the Moon is about twice as great as the gravitational force that the Earth exerts on the Moon. Why doesn't the Sun pull the Moon away from the Earth during a total eclipse of the Sun?

3. Explain why it takes more fuel for a spacecraft to travel from the Earth to the Moon than for the return trip. Estimate the difference.

4. Explain why there is no work done on a planet as it moves in a circular orbit around the Sun, even though a gravitational force is acting on the planet. What is the *net* work done on a planet during each revolution as it moves around the Sun in an elliptical orbit?

5. At what position in its elliptical orbit is the speed of a planet a maximum? At what position is the speed a minimum?

6. Why don't we put a communications satellite in orbit around the 45th parallel? Wouldn't this be more useful in the United States than one in orbit around the Equator?

7. In his 1798 experiment, Cavendish was said to have "weighed the Earth." Explain this statement.

8. If a hole could be dug to the center of the Earth, do you think that the force on a mass *m* would still obey Equation 11.1 there? What do you think the force on *m* would be at the center of the Earth?

9. The *Voyager* spacecraft was accelerated toward escape speed from the Sun by the gravitational force exerted on the spacecraft by Jupiter. How is this possible?

10. The *Apollo 13* spaceship developed trouble in the oxygen system about halfway to the Moon. Why did the mission continue on around the Moon and then return home rather than immediately turn back to Earth?

11. Discuss the similarities and differences between the classical description of planetary motion and the Bohr model of the hydrogen atom.

12. The Bohr theory of the hydrogen atom is based on several assumptions. Discuss those assumptions and their significance. Do any of them contradict classical physics?

13. Explain the significance behind the fact that the total energy of the atom in the Bohr model is negative.

14. Suppose that the electron in the hydrogen atom obeyed classical mechanics rather than quantum mechanics. Why should such a "hypothetical" atom emit a continuous spectrum rather than the observed line spectrum?

PROBLEMS

Section 11.1   Newton's Universal Law of Gravity Revisited

1. Which exerts a greater force of gravitational attraction on objects on the Earth: the Moon or the Sun? Calculate these forces on a 1.00-kg mass.

2. Two ocean liners, each with a mass of 40 000 metric tons, are moving on parallel courses, 100 m apart. What is the magnitude of the acceleration of one of the liners toward the other due to the mutual gravitational attraction? (Treat the ships as spheres.)

3. The gravitational field on the surface of the Moon is about one sixth that on the surface of the Earth. If the radius of the Moon is about one quarter that of the Earth, find the ratio of the average mass density of the Moon to the average mass density of the Earth.

4. A student proposes to measure the gravitational constant *G* by suspending two spherical masses from the ceiling of a tall cathedral and measuring the deflection of the cables from the vertical. Draw a free-body diagram for one of the masses. If two 100.0 kg masses are suspended at the end of 45.00 m long cables and the cables are attached to the ceiling 1.000 m apart, what is the separation of the masses?

5. On the way to the Moon the Apollo astronauts reach a point at which the Moon's gravitational pull becomes equal to that of Earth. Determine the distance of this point from the center of the Earth.

Section 11.2   Kepler's Laws

Section 11.3   The Universal Law of Gravity and the Motions of Planets

6. A satellite is in a circular orbit just above the surface of the Moon. (The radius of the Moon is 1738 km.) (a) What is the acceleration of the satellite? (b) What is the speed of the satellite? (c) What is the period of the satellite orbit?

7. The *Explorer VIII* satellite, placed into orbit November 3, 1960, to investigate the ionosphere, had the following orbit parameters: perigee 459 km and apogee 2289 km (both distances above the Earth's surface); period, 112.7 min. Find the ratio v_p/v_a.

8. Halley's Comet approaches the Sun to within 0.570 A.U., and its orbital period is 75.6 years. (A.U. is the abbreviation for astronomical unit, where 1 A.U. = 1.50×10^{11} m is the

mean Earth–Sun distance.) How far from the Sun will Halley's Comet travel before it starts its return journey? (See Fig. P11.8.)

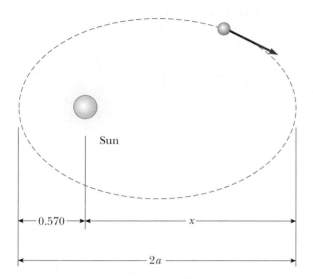

Figure P11.8

9. Io, a satellite of Jupiter, has an orbital period of 1.77 days and an orbital radius of 4.22×10^5 km. From these data, determine the mass of Jupiter.

10. Two planets X and Y travel counterclockwise in circular orbits about a star, as in Figure P11.10. The radii of their orbits is in the ratio $3:1$. At some time, they are aligned as in Figure P11.10a, making a straight line with the star. Five years later, Planet X has rotated through $90°$ as in Figure P11.10b. Where is planet Y at this time?

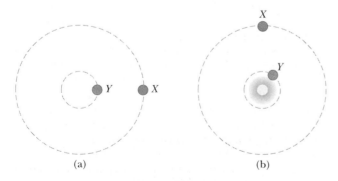

Figure P11.10

11. A synchronous satellite, which always remains above the same point on a planet's equator, is put in orbit around Jupiter to study the famous red spot. Jupiter rotates once every 9.84 h. Use the data of Table 11.2 to find the altitude of the satellite.

Section 11.4 Energy Considerations in Planetary and Satellite Motion

12. A satellite of mass 200 kg is placed in Earth orbit at a height of 200 km above the surface. (a) Assuming a circular orbit, how long does the satellite take to complete one orbit? (b) What is the satellite's speed? (c) What is the minimum energy necessary to place this satellite in orbit (assuming no air friction)?

13. A spaceship is fired from the Earth's surface with an initial speed 2.00×10^4 m/s. What will its speed be when it is very far from the Earth? (Neglect friction.)

14. A satellite of the Earth has a mass of 100 kg and is at an altitude of 2.00×10^6 m. (a) What is the potential energy of the satellite–Earth system? (b) What is the magnitude of the gravitational force exerted by the Earth on the satellite?

15. A 1000-kg satellite orbits the Earth at an altitude of 100 km. It is desired to increase the altitude of the circular orbit to 200 km. How much energy must be added to the system to effect this change in altitude?

16. A satellite of mass m orbits the Earth at an altitude h_1. It is desired to increase the altitude of the circular orbit to h_2. How much energy must be added to the system to effect this change in altitude?

17. A satellite moves in a circular orbit just above the surface of a planet. Show that the orbital speed v and escape speed of the satellite are related by the expression $v_{esc} = \sqrt{2}v$.

18. The planet Uranus has a mass about 14.0 times the Earth's mass, and its radius is equal to about 3.70 Earth radii. (a) By setting up ratios with the corresponding Earth values, find the acceleration due to gravity at the cloud tops of Uranus. (b) Ignoring the rotation of the planet, find the minimum escape speed from Uranus.

19. Determine the escape velocity for a rocket on the far side of Ganymede, the largest of Jupiter's moons. Ganymede's radius is 2.64×10^6 m, and its mass is 1.495×10^{23} kg. The mass of Jupiter is 1.90×10^{27} kg, and the distance between Jupiter and Ganymede is 1.071×10^9 m. Be sure to include the gravitational effect due to Jupiter, but you may ignore the motion of Jupiter and Ganymede as they revolve about their center of mass.

20. How much work is done by the Moon's gravitational field as a 1000 kg meteor comes in from outer space and impacts on the Moon's surface?

21. In Robert Heinlein's *The Moon Is a Harsh Mistress*, the colonial inhabitants of the Moon threaten to launch rocks down onto the Earth if they are not given independence (or at least representation). Assume that a rail gun can launch a rock at twice the speed that would be required to escape from an isolated stationary Moon. Calculate the speed of this rock as it enters the Earth's atmosphere.

22. A spacecraft in the shape of a long cylinder has a length of 100 m and its mass with occupants is 1000 kg. It has strayed in too close to a 1.0-m radius black hole having a mass 100 times that of the Sun (Fig. P11.22). If the nose of the space-

craft points toward the center of the black hole, and if distance between the nose of the spaceship and the black hole's center is 10.0 km, (a) determine the average acceleration of the spaceship. (b) What is the difference in the acceleration experienced by the occupants in the nose of the ship and those in the rear of the ship farthest from the black hole?

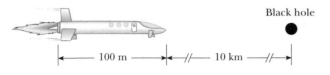

Figure P11.22

Section 11.5   Atomic Spectra and the Bohr Theory of Hydrogen

23. For a hydrogen atom in its ground state, use the Bohr model to compute (a) the orbital speed of the electron, (b) the kinetic energy of the electron, and (c) the electrical potential energy of the atom.

24. Radiation is emitted from a hydrogen atom, which undergoes a transition from the $n = 3$ state to the $n = 2$ state. Calculate (a) the energy, (b) the wavelength, and (c) the frequency of the emitted light.

25. A hydrogen atom is in its first excited state ($n = 2$). Using the Bohr theory of the atom, calculate (a) the radius of the orbit, (b) the linear momentum of the electron, (c) the angular momentum of the electron, and its (d) kinetic energy, (e) potential energy, and (f) total energy.

26. How much energy is required to ionize hydrogen (a) when it is in the ground state? (b) when it is in the state for which $n = 3$?

27. Show that the speed of the electron in the nth Bohr orbit in hydrogen is given by

$$v_n = \frac{k_e e^2}{n\hbar}$$

28. (a) Calculate the angular momentum of the Moon due to its orbital motion about the Earth. In your calculation, use 3.84×10^8 m as the average Earth–Moon distance and 2.36×10^6 s as the period of the Moon in its orbit. (b) Determine the corresponding quantum number. (c) By what fraction would the Earth–Moon distance have to be increased to increase the quantum number by 1?

Additional Problems

29. *Voyagers 1* and *2* surveyed the surface of Jupiter's moon Io and photographed active volcanoes spewing liquid sulfur to heights of 70 km above the surface of this moon. Estimate the speed with which the liquid sulfur left the volcano. Io's mass is 8.9×10^{22} kg and its radius is 1820 km.

30. You are an astronaut, and you observe a small planet to be spherical. After landing on the planet, you set off, walk straight ahead, and find yourself returning to your spacecraft from the opposite side after completing a lap of 25.0 km. You hold a hammer and a falcon feather at a height of 1.40 m, release them, and observe them to fall together to the surface in 29.2 s. Determine the mass of the planet.

31. A cylindrical habitat in space 6.00 km in diameter and 30 km long has been proposed (by G. K. O'Neill, 1974). Such a habitat would have cities, land, and lakes on the inside surface and air and clouds in the center. This would all be held in place by rotation of the cylinder about its long axis. How fast would the cylinder have to rotate to imitate the Earth's gravitational field at the walls of the cylinder?

32. A "treetop satellite" is a satellite that orbits just above the surface of a spherical object, assumed to offer no air resistance. (a) Find the period of a treetop satellite of Earth. (b) Prove that its speed is given by $v = R\sqrt{4\pi G\rho/3}$, where ρ is the density of the planet.

33. In introductory physics laboratories, a typical Cavendish balance for measuring the gravitational constant G uses lead spheres of masses 1.50 kg and 15.0 g the centers of which are separated by about 4.50 cm. Calculate the gravitational force between these spheres, treating each as a point mass located at the center of the sphere.

34. Show that the escape speed from the surface of a planet of uniform density is directly proportional to the radius of the planet.

35. Two hypothetical planets of masses m_1 and m_2 and radii r_1 and r_2, respectively, are at rest when they are an infinite distance apart. Because of their gravitational attraction, they head toward each other on a collision course. (a) When their center-to-center separation is d, find the speed of each planet and their *relative* velocity. (b) Find the kinetic energy of each planet *just* before they collide if $m_1 = 2.00 \times 10^{24}$ kg, $m_2 = 8.00 \times 10^{24}$ kg, $r_1 = 3.00 \times 10^6$ m, and $r_2 = 5.00 \times 10^6$ m. (*Hint:* Both energy and momentum are conserved.)

36. The maximum distance from the Earth to the Sun (at our aphelion) is 1.521×10^{11} m and the distance of closest approach (at perihelion) is 1.471×10^{11} m. If the Earth's orbital speed at perihelion is 3.027×10^4 m/s, determine (a) the Earth's orbital speed at aphelion, (b) the kinetic and potential energy at perihelion, and (c) the kinetic and potential energy at aphelion. Is the total energy constant? (Neglect the effect of the Moon and other planets.)

37. (a) Determine the amount of work (in joules) that must be done on a 100-kg payload to elevate it to a height of 1000 km above the Earth's surface. (b) Determine the amount of additional work that is required to put the payload into circular orbit at this elevation.

38. After a supernova explosion, a star may undergo a gravitational collapse to an extremely dense state known as a neutron star, in which all the electrons and protons are

squeezed together to form neutrons. A neutron star having a mass about equal to that of the Sun would have a radius of about 10.0 km. Find (a) the free-fall acceleration at its surface, (b) the weight of a 70.0-kg person at its surface, and (c) the energy required to remove a neutron of mass 1.67×10^{-27} kg from its surface to infinity.

39. When it orbited the Moon, the *Apollo 11* spacecraft's mass was 9.979×10^3 kg, its period was 119 min, and its mean distance from the Moon's center was 1.849×10^6 m. Assuming its orbit to be circular and the Moon to be a uniform sphere, find (a) the mass of the Moon, (b) the orbital speed of the spacecraft, and (c) the minimum energy required for the craft to leave the orbit and escape the Moon's gravitational field.

40. Three point objects having masses m, $2m$, and $3m$, are fixed at the corners of a square of side length a such that the lighter object is at the upper left-hand corner, the heavier object is at the lower left-hand corner, and the remaining object is at the upper right-hand corner. Determine the magnitude and direction of the resulting gravitational acceleration at the center of the square.

41. X-ray pulses from Cygnus X-1, a celestial x-ray source, have been recorded during high-altitude rocket flights. The signals can be interpreted as originating when a blob of ionized matter orbits a black hole with a period of 5.00 ms. If the blob were in a circular orbit about a black hole the mass of which is $20.0 M_{Sun}$, what is the orbital radius?

42. Studies of the relationship of the Sun to its galaxy—the Milky Way—have revealed that the Sun is located near the outer edge of the galactic disc, about 30 000 lightyears from the center. Furthermore, it has been found that the Sun has an orbital speed of approximately 250 km/s around the galactic center. (a) What is the period of the Sun's galactic motion? (b) What is the order of magnitude of the mass of the Milky Way galaxy? Supposing that the galaxy is made mostly of stars of which the Sun is typical, estimate the number of stars in the Milky Way.

43. Four possible transitions for a hydrogen atom are as follows:

$$(A) \ n_i = 2; \ n_f = 5 \qquad (B) \ n_i = 5; \ n_f = 3$$

$$(C) \ n_i = 7; \ n_f = 4 \qquad (D) \ n_i = 4; \ n_f = 7$$

(a) Which transition emits the shortest wavelength photon? (b) In which transition does the atom gain the most energy? (c) In which transition(s) does the atom lose energy?

44. *Vanguard I*, launched March 3, 1958, is the oldest human-made satellite still in orbit. Its initial orbit had an apogee of 3970 km and a perigee of 650 km. Its maximum speed was 8.23 km/s and it had a mass of 1.60 kg. (a) Determine the period of the orbit. (Use the semimajor axis.) (b) Determine the speeds at apogee and perigee. (c) Find the total energy of the satellite.

45. Two stars of masses M and m, separated by a distance d, revolve in circular orbits about their center of mass (Fig. P11.45). Show that each star has a period given by

$$T^2 = \frac{4\pi^2}{G(M + m)} d^3$$

(*Hint:* Apply Newton's second law to each star, and note that the center-of-mass condition requires that $Mr_2 = mr_1$, where $r_1 + r_2 = d$.)

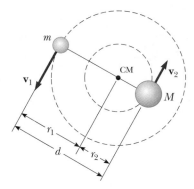

Figure P11.45

46. In 1978, astronomers at the U.S. Naval Observatory discovered that Pluto has a moon, called Charon, that eclipses the planet every 6.39 days. Given that the center-to-center separation between Pluto and Charon is 19 700 km, find the total mass $(M + m)$ of the two bodies. (*Hint:* See Problem 45.)

47. Two hydrogen atoms collide head-on and end up with zero kinetic energy. Each then emits a 121.6-nm photon ($n = 2$ to $n = 1$ transition). At what speed were the atoms moving before the collision?

48. A rocket is given an initial speed of $v_0 = 2\sqrt{Rg}$ vertically upward at the surface of the Earth, which has radius R and surface free-fall acceleration g. The rocket motors are then cut off, and thereafter the rocket coasts under the action of gravitational forces only. (Ignore atmospheric friction and the Earth's rotation.) Derive an expression for the subsequent speed, v, as a function of the distance, r, from the center of the Earth in terms of g, R, and r.

📊 Spreadsheet Problems

S1. Four point masses are fixed as shown in Figure PS11.1. The gravitational potential energy for a test particle of mass m moving along the x axis in the gravitational field of the fixed masses can be written as

$$U(x) = -\frac{GM_1 m}{r_1} - \frac{GM_2 m}{r_2} - \frac{GM_3 m}{r_3} - \frac{GM_4 m}{r_4}$$

where $r_2 = r_4 = [b^2 + (x + a)^2]^{1/2}$ and $r_2 = r_3 = [b^2 + (x - a)^2]^{1/2}$. Spreadsheet 11.1 calculates $U(x)$ and the force $F(x) = -dU(x)/dx$ exerted on the test particle. The quantities M_1, M_2, M_3, M_4, a, and b are input parameters.

The mass of the test particle is 1.00 kg. (This is the case of two particles on the y axis at $y = +1.00$ m and -1.00 m.) (a) Plot $U(x)$ and $F(x)$ versus x for $x = -4.00$ m to $+4.00$ m. (b) If the test particle has an energy of -7.50×10^{-11} J, describe the particle's motion. (c) How does the force on the particle vary as the particle moves along the x axis?

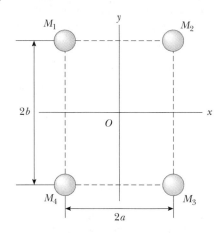

Figure PS11.1

S2. In Spreadsheet 11.1, let $a = 1.00$ m, $b = 0.50$ m, and $M_1 = M_2 = M_3 = M_4 = 1.00$ kg. The mass of the particle is 1.00 kg. (a) Plot $U(x)$ and $F(x)$ versus x. (b) If the test particle has an energy of -2.00×10^{-10} J, what motion is possible?

S3. If a projectile is fired straight up from the surface of the Earth with an initial speed v_0 that is less than the escape speed $v_{esc} = \sqrt{2gR_E} = 11.2$ km/s, the projectile will rise to a maximum height h above the Earth's surface and return. Neglecting air resistance and using conservation of energy, show that

$$h = \frac{R_E v_0^{\,2}}{2gR_E - v_0}$$

where R_E is the Earth's radius. Write a spreadsheet to calculate h for $v_0 = 1, 2, 3, \ldots, 11$ km/s. Plot h/R_E versus v_0. What happens as v_0 approaches the escape speed v_{esc}?

S4. The acceleration of an object moving in the gravitational field of the Earth is

$$\mathbf{a} = -GM_E \frac{\mathbf{r}}{r^3}$$

where $\mathbf{r}$ is the position vector directed from the center of the Earth to the object. Choosing the origin at the center of the Earth and assuming the object is moving in the xy plane, the acceleration has rectangular components

$$a_x = -\frac{GM_E x}{(x^2 + y^2)^{3/2}} \qquad a_y = -\frac{GM_E y}{(x^2 + y^2)^{3/2}}$$

Write a spreadsheet (or computer program) to find the position of the object as a function of time. (Use the techniques developed in Spreadsheet 11.1 as a model for your own program. Assume that the initial position of the object is at $x = 0$ and $y = 2R_E$, where R_E is the radius of the Earth, and give the object an initial velocity of 5 km/s in the x direction. The time increment should be made as small as practical. Try 5 s. Plot the x and y coordinates of the object as functions of time. Does the object hit the Earth? Vary the initial velocity until a circular orbit is found.

S5. Modify your spreadsheet (or program) from Problem S4 to calculate the kinetic, potential, and total energies as functions of time. How do they vary with time for the various cases?

S6. Modify your spreadsheet (or program) from Problem S4 to calculate the angular momentum of the object. In rectangular coordinates, $L = m(xv_y - yv_x)$. How does the angular momentum vary with time? Kepler's second law of planetary motion implies that L remains constant. Is your calculated L constant?

ANSWERS TO CONCEPTUAL PROBLEMS

1. Kepler's second law is a consequence of conservation of angular momentum. Angular momentum is conserved as long as there is no torque, which is true for any central force, regardless of the particular dependence on separation. Thus, Kepler's second law would be unchanged. The particular mathematical form of Kepler's third law depends on the inverse square character of the force. If the force were an inverse cube, there would still be a proportional relationship between a power of the period and a power of the semimajor axis, but the powers would be different than in the existing law.

2. Air friction causes a decrease in the mechanical energy of the satellite–Earth system. This reduces the major axis of the orbit, bringing the satellite closer to the surface of the Earth. A satellite in a smaller orbit, however, must travel faster. Thus, the effect of the air friction is to speed up the satellite!

3. Kepler's third law (Eq. 11.4), which applies to all the planets, tells us that the period of a planet is proportional to $r^{3/2}$. Because Saturn and Jupiter are farther from the Sun than Earth, they have longer periods. The Sun's gravitational field (whose magnitude is GM_S/r^2) is much weaker at a distant Jovian planet. Thus, an outer planet experiences much

smaller centripetal acceleration than Earth, and a correspondingly longer period.

4. In an emission spectrum, a gas is initially excited to higher quantum states. The emission spectrum is provided by subsequent downward transitions which can occur to all lower states. If the transition is not to the lowest state, the atom can make a subsequent transition from the intermediate state to the lowest state. For example, a transition can be made from $n = 10$ to $n = 5$, followed by a transition from $n = 5$ to $n = 1$. Thus, there is a rich mixture of spectral lines in the spectrum, corresponding to this large number of possible transitions. An absorption spectrum is normally performed with the gas at room temperature, at which almost all of the atoms are in the ground state. Thus, when radiation is absorbed, upward transitions involving *only the ground state* are possible. Thus, there are fewer transitions than for the emission spectrum and fewer lines.

5. In the discussion of the hydrogen atom, it can be seen that the kinetic energy is equal to half of the magnitude of the potential energy. Thus, we must have $K = E$ and $U_e = -2E$.

B.C.   by John Hart

By permission of John Hart and Field Enterprises, Inc.

12

Oscillatory Motion

I f a force acting on a body varies in time, the acceleration of the body also changes with time. If this force always acts toward the equilibrium position of the body, a repetitive back-and-forth motion about that position results. The motion is an example of what is called *periodic* or *oscillatory* motion.

You are most likely familiar with several examples of periodic motion, such as the oscillations of a mass on a spring, the motion of a pendulum, and the vibrations of a stringed musical instrument. Numerous systems exhibit oscillatory motion. For example, the molecules in a solid oscillate about their equilibrium positions; electromagnetic waves, such as light waves, radar, and radio waves, are characterized by oscillating electric and

Time exposure of a pendulum, ▶ which consists of a small sphere suspended by a light string. The time it takes the pendulum to undergo one complete oscillation is called the period of its motion, and the maximum displacement of the pendulum from the vertical position is called the *amplitude*. As we shall see in this chapter, the period of its motion for small amplitudes depends only on the length of the pendulum and the value of the free-fall acceleration. *(James Stevenson/ SPL/Photo Researchers)*

magnetic field vectors; in alternating-current circuits, voltage, current, and electrical charge vary periodically with time.

Much of the material in this chapter deals with **simple harmonic motion.** This special kind of motion occurs when a restoring force acts on a body. In simple harmonic motion, an object oscillates between two spatial positions for an indefinite period of time. In real mechanical systems, retarding (frictional) forces are always present. Such forces reduce the mechanical energy of the system with time, and the oscillations are said to be *damped.* If an external driving force is applied, we call the motion a *forced oscillation.*

12.1 · SIMPLE HARMONIC MOTION

A particle moving along the *x* axis is said to exhibit **simple harmonic motion** when *x*, its displacement from equilibrium, varies in time according to the relationship

$$x = A\cos(\omega t + \phi)$$ [12.1]

• *Displacement versus time for simple harmonic motion*

where A, ω, and ϕ are constants of the motion. In order to give physical significance to these constants, it is convenient to plot *x* as a function of *t*, as in Figure 12.1. First, we note that A, called the **amplitude** of the motion, is simply **the maximum displacement of the particle in either the positive or negative *x* direction.** The constant ω is called the **angular frequency** (defined in Eq. 12.4). The constant angle ϕ is called the **phase constant** (or phase angle) and, along with the amplitude A, is determined uniquely by the initial displacement and velocity of the particle. The constants ϕ and A tell us what the displacement and velocity were at time $t = 0$. The quantity $(\omega t + \phi)$ is called the **phase** of the motion and is useful for comparing the motions of two particles. Note that the function *x* is periodic and repeats itself each time ωt increases by 2π radians.

The **period,** T, of the motion is the time required for the particle to go through one full cycle of its motion. That is, the value of *x* at time *t* equals the value of *x* at time $t + T$. We can show that the period is given by $T = 2\pi/\omega$ by using the fact that the phase increases by 2π radians in a time of T:

$$\omega t + \phi + 2\pi = \omega(t + T) + \phi$$

Simplifying this expression, we see that $\omega T = 2\pi$, or

$$T = \frac{2\pi}{\omega}$$ [12.2]

• *Period*

The inverse of the period is called the **frequency** of the motion, *f*. The frequency represents the **number of oscillations the particle makes per unit time:**

$$f = \frac{1}{T} = \frac{\omega}{2\pi}$$ [12.3]

• *Frequency*

The units of *f* are cycles per second, or hertz (Hz). Rearranging Equation 12.3 gives

$$\omega = 2\pi f = \frac{2\pi}{T}$$ [12.4]

• *Angular frequency*

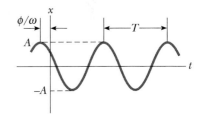

Figure 12.1   Displacement versus time for a particle undergoing simple harmonic motion. The amplitude of the motion is A and the period is T.

The angular frequency ω has the units radians per second. We shall discuss the geometric significance of ω in Section 12.4.

We can obtain the velocity of a particle undergoing simple harmonic motion by differentiating Equation 12.1 with respect to time:

Velocity in simple harmonic •
motion

$$v = \frac{dx}{dt} = -\omega A \sin(\omega t + \phi) \qquad [12.5]$$

The acceleration of the particle is dv/dt:

Acceleration in simple •
harmonic motion

$$a = \frac{dv}{dt} = -\omega^2 A \cos(\omega t + \phi) \qquad [12.6]$$

Because $x = A \cos(\omega t + \phi)$, we can express Equation 12.6 in the form

$$a = -\omega^2 x \qquad [12.7]$$

From Equation 12.5 we see that, because the sine and cosine functions oscillate between ± 1, the extreme values of v are $\pm \omega A$. Likewise, Equation 12.6 tells us that the extreme values of the acceleration are $\pm \omega^2 A$. Therefore, the *maximum* values of the velocity and acceleration are

Maximum values of velocity •
and acceleration in simple
harmonic motion

$$v_{\text{max}} = \omega A \qquad [12.8]$$

$$a_{\text{max}} = \omega^2 A \qquad [12.9]$$

Figure 12.2a plots displacement versus time for an arbitrary value of the phase constant. The velocity and acceleration-versus-time curves are illustrated in Figures 12.2b and 12.2c. They show that the phase of the velocity differs from the phase of the displacement by $\pi/2$ rad, or 90°. That is, when x is a maximum or a minimum,

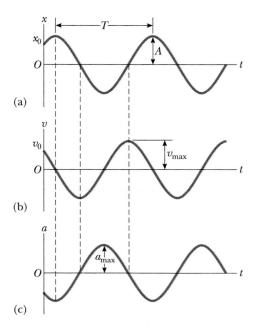

(a)

(b)

(c)

Figure 12.2 Graphical representation of simple harmonic motion: (a) displacement versus time, (b) velocity versus time, and (c) acceleration versus time. Note that at any specified time the velocity is 90° out of phase with the displacement and the acceleration is 180° out of phase with the displacement.

the velocity is zero. Likewise, when x is zero, the speed is a maximum. Furthermore, note that the phase of the acceleration differs from the phase of the displacement by π radians, or 180°. That is, when x is a maximum, a is a maximum in the opposite direction.

As we stated earlier, $x = A \cos(\omega t + \phi)$ is a general expression for the displacement of the particle from equilibrium, where the phase constant ϕ and the amplitude A must be chosen to meet the initial conditions of the motion. Suppose that the initial position, x_0, and initial velocity, v_0, of a single oscillator of known angular frequency are given; that is, at $t = 0$, $x = x_0$ and $v = v_0$. Under these conditions, Equations 12.1 and 12.5 give

$$x_0 = A \cos \phi \qquad \text{and} \qquad v_0 = -\omega A \sin \phi \qquad \text{[12.10]}$$

Dividing these two equations eliminates A, giving $\dfrac{v_0}{x_0} = -\omega \tan \phi$, or

$$\tan \phi = -\frac{v_0}{\omega x_0} \qquad \text{[12.11]}$$

Furthermore, if we take the sum $x_0{}^2 + (v_0/\omega)^2 = A^2 \cos^2 \phi + A^2 \sin^2 \phi$ (where we have used Eq. 12.10) and solve for A, we find that

$$A = \sqrt{x_0{}^2 + \left(\frac{v_0}{\omega}\right)^2} \qquad \text{[12.12]}$$

• *The phase angle ϕ and amplitude A can be obtained from the initial conditions.*

Thus, we see that ϕ and A are known if x_0, ω, and v_0 are specified. We shall treat a few specific cases in the next section.

We conclude this section by pointing out the following important properties of a particle moving in simple harmonic motion:

• The displacement, velocity, and acceleration all vary sinusoidally with time but are not in phase, as shown in Figure 12.2.
• The acceleration is proportional to the displacement, but in the opposite direction.
• The frequency and the period of motion are independent of the amplitude.

• *Properties of simple harmonic motion*

Example 12.1   An Oscillating Body

A body oscillates with simple harmonic motion along the x axis. Its displacement varies with time according to the equation

$$x = (4.00 \text{ m}) \cos\left(\pi t + \frac{\pi}{4}\right)$$

where t is in seconds and the angles in the parentheses are in radians.

(a) Determine the amplitude, frequency, and period of the motion.

Solution   By comparing this equation with the general equa-

tion for simple harmonic motion, $x = A \cos(\omega t + \phi)$, we see that $A = 4.00$ m and $\omega = \pi$ rad/s; therefore we find $f = \omega/2\pi = \pi/2\pi = 0.500 \text{ s}^{-1}$ and $T = 1/f = 2.00$ s.

(b) Calculate the velocity and acceleration of the body at any time t.

Solution

$$v = \frac{dx}{dt} = -4.00 \sin\left(\pi t + \frac{\pi}{4}\right) \frac{d}{dt}(\pi t)$$

$$= -(4.00\pi \text{ m/s}) \sin\left(\pi t + \frac{\pi}{4}\right)$$

$$a = \frac{dv}{dt} = -4.00\pi \cos\left(\pi t + \frac{\pi}{4}\right)\frac{d}{dt}(\pi t)$$

$$= -(4.00\pi^2 \text{ m/s}^2)\cos\left(\pi t + \frac{\pi}{4}\right)$$

EXERCISE 1   Using the results to part (b), determine the position, velocity, and acceleration of the body at $t = 1.00$ s.
Answer $x = -2.83$ m; $v = 8.89$ m/s; $a = 27.9$ m/s^2

EXERCISE 2   Determine the maximum speed and the maximum acceleration of the body.   Answer   From the general expressions for v and a found in part (b), we see that the maximum values of the sine and cosine functions are unity. Therefore, v varies between $\pm 4.00\pi$ m/s, and a varies between $\pm 4.00\pi^2$ m/s^2. Thus, $v_{\text{max}} = 4.00\pi$ m/s and $a_{\text{max}} = 4.00\pi^2$ m/s^2.

EXERCISE 3   A particle moving with simple harmonic motion travels a total distance of 20.0 cm in each cycle of its motion, and its maximum acceleration is 50.0 m/s^2. Find (a) the angular frequency of the motion and (b) the maximum speed of the particle.
Answer   (a) 31.6 rad/s   (b) 1.58 m/s

12.2 • MOTION OF A MASS ATTACHED TO A SPRING

Consider a physical system consisting of a mass, m, attached to the end of a spring where the mass is free to move on a horizontal track (Fig. 12.3). We know from experience that such a system oscillates back and forth if disturbed from the equilibrium position $x = 0$, where the spring is unstretched. If the surface is frictionless, the mass can exhibit simple harmonic motion. One experimental arrangement that clearly demonstrates that such a system exhibits simple harmonic motion is illustrated in Figure 12.4. A mass oscillating vertically on a spring has a marking pen attached to it. While the mass is in motion, a sheet of paper is moved horizontally as shown, and the marking pen traces out a sinusoidal pattern. We can understand this qualitatively by first recalling that when the mass is displaced a small distance, x, from equilibrium, the spring exerts a force on it, given by **Hooke's law,**

$$F_s = -kx \qquad [12.13]$$

where k is the force constant of the spring. We call this a **linear restoring force** because it is linearly proportional to the displacement and always directed toward the equilibrium position *opposite* the displacement. That is, when the mass is displaced to the right in Figure 12.3, x is positive and the restoring force is to the left. When the mass is displaced to the left of $x = 0$, then x is negative and the restoring force is to the right.

If we apply Newton's second law to the motion of the mass in the x direction, we get

$$F_s = -kx = ma$$

$$a = -\frac{k}{m}x \qquad [12.14]$$

That is, just as we learned in Section 12.1, **the acceleration is proportional to the displacement of the mass from equilibrium and is in the opposite direction.** If the mass is displaced a maximum distance, $x = A$, at some initial time and released from rest, its *initial* acceleration is $-kA/m$ (that is, the acceleration has its extreme negative value). When the mass passes through the equilibrium position,

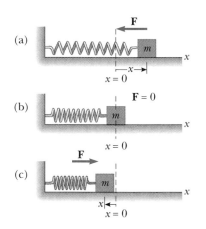

Figure 12.3   A mass attached to a spring on a frictionless track moves in simple harmonic motion.
(a) When the mass is displaced to the right of equilibrium, the displacement is positive and the acceleration is negative. (b) At the equilibrium position, $x = 0$, the acceleration is zero but the speed is a maximum. (c) When the displacement is negative, the acceleration is positive.

$x = 0$ and its acceleration is zero. At this instant, its speed is a maximum. It then continues to travel to the left of equilibrium and finally reaches $x = -A$, at which time its acceleration is kA/m (maximum positive) and its speed is again zero. Thus, we see that the mass oscillates between the turning points $x = \pm A$. In one full cycle of its motion it travels a distance of $4A$.

Let us now describe this motion in a quantitative fashion. Recall that, by definition, $a = dv/dt = d^2x/dt^2$, and so we can express Equation 12.14 as

$$\frac{d^2x}{dt^2} = -\frac{k}{m}x \qquad\qquad [12.15]$$

If we denote the ratio k/m with the symbol ω^2,

$$\omega^2 = \frac{k}{m} \qquad\qquad [12.16]$$

then Equation 12.15 can be written in the form

$$\frac{d^2x}{dt^2} = -\omega^2 x \qquad\qquad [12.17]$$

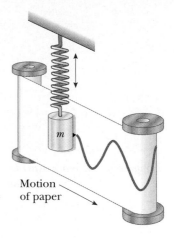

Figure 12.4 An experimental apparatus for demonstrating simple harmonic motion. A pen attached to the oscillating mass traces out a sine wave on the moving chart paper.

What we now require is a solution to Equation 12.17—that is, a function $x(t)$ that satisfies this second-order differential equation. Because Equations 12.17 and 12.7 are equivalent, we see that the solution must be that of simple harmonic motion:

$$x(t) = A\cos(\omega t + \phi)$$

To see this explicitly, note that if $x = A\cos(\omega t + \phi)$, then

$$\frac{dx}{dt} = A\frac{d}{dt}\cos(\omega t + \phi) = -\omega A\sin(\omega t + \phi)$$

$$\frac{d^2x}{dt^2} = -\omega A\frac{d}{dt}\sin(\omega t + \phi) = -\omega^2 A\cos(\omega t + \phi)$$

Comparing the expressions for x and d^2x/dt^2, we see that $d^2x/dt^2 = -\omega^2 x$ and Equation 12.17 is satisfied.

The following general statement can be made based on the foregoing discussion:

> Whenever the force acting on a particle is linearly proportional to the displacement and in the opposite direction, the particle exhibits simple harmonic motion.

We shall give additional physical examples in subsequent sections.

Because the period of simple harmonic motion is $T = 2\pi/\omega$ and the frequency is the inverse of the period, we can express the period and frequency of the motion for this mass–spring system as

$$T = \frac{2\pi}{\omega} = 2\pi\sqrt{\frac{m}{k}} \qquad\qquad [12.18]$$

• *Period of a mass–spring system*

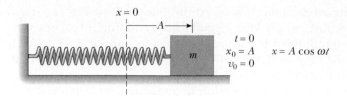

Figure 12.5   A mass–spring system that starts from rest at $x_0 = A$. In this case, $\phi = 0$, and so $x = A \cos \omega t$.

Frequency of a mass–spring •
system

$$f = \frac{1}{T} = \frac{1}{2\pi}\sqrt{\frac{k}{m}} = \frac{\omega}{2\pi} \qquad [12.19]$$

That is, the period and frequency depend *only* on the mass and on the force constant of the spring. As we might expect, the frequency is larger for a stiffer spring (larger value of k) and decreases with increasing mass.

Special Case I   In order to better understand the physical significance of our solution of the equation of motion, let us consider the following special case. Suppose we pull the mass from equilibrium by a distance of A and release it from rest in this stretched position, as in Figure 12.5. We must then require that our solution for $x(t)$ obey the initial conditions that at $t = 0$, $x_0 = A$ and $v_0 = 0$. These conditions are met if we choose $\phi = 0$, giving $x = A \cos \omega t$ as our solution. To check this solution, we note that it satisfies the condition that $x_0 = A$ at $t = 0$, because $\cos 0 = 1$. Thus, we see that A and ϕ contain the information on initial conditions.

Now let us investigate the behavior of the velocity and acceleration in this special case. Because $x = A \cos \omega t$, we have

$$v = \frac{dx}{dt} = -\omega A \sin \omega t \qquad \text{and} \qquad a = \frac{dv}{dt} = -\omega^2 A \cos \omega t$$

From the preceding velocity expression we see that at $t = 0$, $v_0 = 0$, as we require. The expression for the acceleration tells us that at $t = 0$, $a = -\omega^2 A$. Physically this makes sense, because the force on the mass is to the left when the displacement is positive. In fact, at this position $F_s = -kA$ (to the left), and the initial acceleration is $-kA/m$.

We could also use a more formal approach to show that $x = A \cos \omega t$ is the correct solution by using the relation $\tan \phi = -v_0/\omega x_0$ (Eq. 12.10). Because $v_0 = 0$ at $t = 0$, $\tan \phi = 0$ and so $\phi = 0$.

The displacement, velocity, and acceleration versus time are plotted in Figure 12.6 for this special case. Note that the acceleration reaches extreme values of $\pm \omega^2 A$ when the displacement has extreme values of $\mp A$. Furthermore, the velocity has extreme values of $\pm \omega A$, which both occur at $x = 0$. Hence, the quantitative solution agrees with our qualitative description of this system.

Special Case II   Now suppose that the mass is given an initial velocity of $\mathbf{v}_0$ to the *right* at the unstretched position of the spring, so that at $t = 0$, $x_0 = 0$ and $v = v_0$ (Fig. 12.7). Our particular solution must now satisfy these initial conditions.

Applying Equation 12.11, $\tan \phi = -v_0/\omega x_0$, and the initial condition that $x_0 = 0$ at $t = 0$ gives $\tan \phi = -\infty$ or $\phi = -\pi/2$. Hence, the solution is given by $x = A \cos(\omega t - \pi/2)$, which can be written $x = A \sin \omega t$. Furthermore, from Equation 12.11 we see that $A = v_0/\omega$; therefore, we can express our solution as

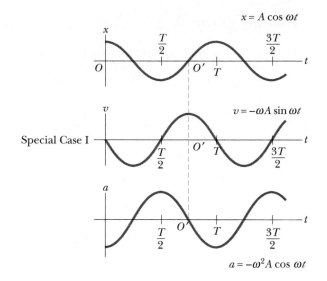

Figure 12.6 Displacement, velocity, and acceleration versus time for a particle undergoing simple harmonic motion under the initial conditions that at $t = 0$, $x = A$, and $v = 0$.

$$x = \frac{v_0}{\omega} \sin \omega t$$

The velocity and acceleration in this case are

$$v = \frac{dx}{dt} = v_0 \cos \omega t \quad \text{and} \quad a = \frac{dv}{dt} = -\omega v_0 \sin \omega t$$

This is consistent with the fact that the mass always has its maximum speed at $x = 0$, and the force and acceleration are zero at that position. The graphs of these functions versus time in Figure 12.6 correspond to the origin at O'. What would be the solution for x if the mass were initially moving to the left in Figure 12.7?

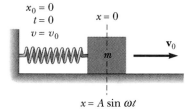

Figure 12.7 The mass–spring system starts its motion at the equilibrium position, $x = 0$ at $t = 0$. If its initial velocity is v_0 to the right, its x coordinate varies as $x = \dfrac{v_0}{\omega} \sin \omega t$.

Thinking Physics 1

We know that the period of oscillation of a mass hung on a spring is proportional to the square root of the mass. Thus, if we perform an experiment in which we hang a range of masses on the end of a spring and measure the period of oscillation, a graph of the square of the period against the mass will result in a straight line, as suggested in Figure TP12.1. But we find that the line does not go through the origin. Why not?

Reasoning The reason that the line does not go through the origin is that the spring itself has mass. Thus, the resistance to changes in motion of the system is a combination of the mass hung on the end of the spring and the mass of the oscillating coils of the spring. The entire mass of the spring is not oscillating, however. The bottom-most coil is oscillating over the same amplitude as the mass, and the top-most coil is not oscillating at all. For a cylindrical spring, energy arguments can be used to show that the effective additional mass representing the oscillations of the spring is one third of the mass of the spring. The square of the period is proportional to the total oscillating mass, but the graph in the diagram shows the square of the period against only the mass hung on the spring. A graph of period squared against total mass (mass hung on the spring plus the effective oscillating mass of the spring) would pass through the origin.

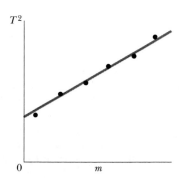

Figure TP12.1 (Thinking Physics 1)

Thinking Physics 2

If you go bungee jumping, you will bounce up and down at the end of the cord after your daring dive off a bridge. Suppose you perform this dive and measure the frequency of your bouncing. You then move to another bridge. You discover that the bungee cord is too long for dives off this bridge—you will hit the ground. Thus, you fold the bungee cord in half and make the dive from the doubled bungee cord. How does the frequency of your bouncing at the end of this dive compare to the frequency after the dive from the first bridge?

Reasoning Let us model the bungee cord as a spring. The force exerted by a spring is proportional to the separation of the coils as the spring is extended. Imagine that we extend a spring by a given distance and measure the distance between coils. We then cut the spring in half. If one of the half-springs is now extended by the same distance, the coils will be twice as far apart as they were for the complete spring. Thus, it takes twice as much force to stretch the half-spring, from which we conclude that the half-spring has a spring constant that is twice that of the complete spring. Now consider the folded bungee cord that we model as two half-springs in parallel. Each half has a spring constant that is twice the original spring constant of the bungee cord. In addition, a given mass hanging on the folded bungee cord will experience two forces—one from each half-spring. As a result, the required force for a given extension will be four times as much as for the original bungee cord. The effective spring constant of the folded bungee cord is, therefore, four times as large as the original spring constant. Because the frequency of oscillation is proportional to the square root of the spring constant, your bouncing frequency on the folded cord will be twice that of the original cord.

(Thinking Physics 2) Bungee jumping. *(Gamma)*

CONCEPTUAL PROBLEM 1

Is a bouncing ball an example of simple harmonic motion? Is the daily movement of a student from home to school and back simple harmonic motion?

CONCEPTUAL PROBLEM 2

A mass is hung on a spring and the frequency of oscillation of the system, f, is measured. The mass, a second identical mass, and the spring are carried in the Space Shuttle to space. The two masses are attached to the ends of the spring and the system is taken out into space on a space walk. The spring is extended and the system is released to oscillate while floating in space. What is the frequency of oscillation for this system, in terms of f?

Example 12.2   A Mass–Spring System

A mass of 200 g is connected to a light spring of force constant 5.00 N/m and is free to oscillate on a horizontal, frictionless surface. If the mass is displaced 5.00 cm from equilibrium and released from rest, as in Figure 12.5, (a) find the period of its motion.

Solution This situation corresponds to Special Case I, in which $x = A \cos \omega t$ and $A = 5.00 \times 10^{-2}$ m. Therefore,

$$\omega = \sqrt{\frac{k}{m}} = \sqrt{\frac{5.00 \text{ N/m}}{200 \times 10^{-3} \text{ kg}}} = 5.00 \text{ rad/s}$$

and

$$T = \frac{2\pi}{\omega} = \frac{2\pi}{5.00} = 1.26 \text{ s}$$

(b) Determine the maximum speed and maximum acceleration of the mass.

Solution

$$v_{max} = \omega A = (5.00 \text{ rad/s})(5.00 \times 10^{-2} \text{ m}) = 0.250 \text{ m/s}$$

$$a_{max} = \omega^2 A = (5.00 \text{ rad/s})^2(5.00 \times 10^{-2} \text{ m}) = 1.25 \text{ m/s}^2$$

(c) Express the displacement, speed, and acceleration as functions of time.

Solution The expression $x = A \cos \omega t$ is our solution for Special Case I, and so we can use the results from (a), (b), and (c) to get

$x = A \cos \omega t = $ (0.0500 m) $\cos 5.00t$

$v = -\omega A \sin \omega t = $ $-$(0.250 m/s) $\sin 5.00t$

$a = -\omega^2 A \cos \omega t = $ $-$(1.25 m/s^2) $\cos 5.00t$

EXERCISE 4 (a) A 400-g mass is suspended from a spring hanging vertically, and the spring is found to stretch 8.0 cm. Find the spring constant. (b) How much will the spring stretch if the suspended mass is 575 g?   Answer   (a) 49 N/m   (b) 12 cm

EXERCISE 5 A 3.0-kg mass is attached to a spring and pulled out horizontally to a maximum displacement from equilibrium of 0.50 m. What spring constant must the spring have if the mass is to achieve an acceleration equal to the free-fall acceleration?   Answer   59 N/m

12.3 • ENERGY OF THE SIMPLE HARMONIC OSCILLATOR

Let us examine the mechanical energy of the mass–spring described in Figure 12.5. Because the surface is frictionless, we expect that the total mechanical energy is constant. We can use Equation 12.5 to express the kinetic energy as

$$K = \tfrac{1}{2}mv^2 = \tfrac{1}{2}m\omega^2 A^2 \sin^2(\omega t + \phi) \qquad [12.20]$$

• *Kinetic energy of a simple harmonic oscillator*

The elastic potential energy stored in the spring for any elongation x is $\tfrac{1}{2}kx^2$. Using Equation 12.1, we get

$$U = \tfrac{1}{2}kx^2 = \tfrac{1}{2}kA^2 \cos^2(\omega t + \phi) \qquad [12.21]$$

• *Potential energy of a simple harmonic oscillator*

We see that K and U are always positive quantities. Because $\omega^2 = k/m$, we can express the *total energy* of the simple harmonic oscillator as

$$E = K + U = \tfrac{1}{2}kA^2[\sin^2(\omega t + \phi) + \cos^2(\omega t + \phi)]$$

But $\sin^2 \theta + \cos^2 \theta = 1$; therefore, this equation reduces to

$$E = \tfrac{1}{2}kA^2 \qquad [12.22]$$

• *Total energy of a simple harmonic oscillator*

That is, the energy of a simple harmonic oscillator is a constant of the motion and proportional to the square of the amplitude. In fact, the total mechanical energy is just equal to the maximum potential energy stored in the spring when $x = \pm A$. At these points, $v = 0$ and there is no kinetic energy. At the equilibrium position, $x = 0$ and $U = 0$, so that the total energy is all in the form of kinetic energy. That is, at $x = 0$, $E = \tfrac{1}{2}mv_{\max}^2 = \tfrac{1}{2}m\omega^2 A^2$.

Plots of the kinetic and potential energies versus time are shown in Figure 12.8a, where $\phi = 0$. In this situation, both K and U are always positive, and their sum at all times is a constant equal to $\tfrac{1}{2}kA^2$, the total energy of the system. The variations of K and U with displacement are plotted in Figure 12.8b. Energy is continuously being transferred between potential energy stored in the spring and the kinetic energy of the mass. Figure 12.9 illustrates the position, velocity, acceleration, kinetic energy, and potential energy of the mass–spring system for one full period of the motion. Most of the ideas discussed so far are incorporated in this important figure. We suggest that you study it carefully.

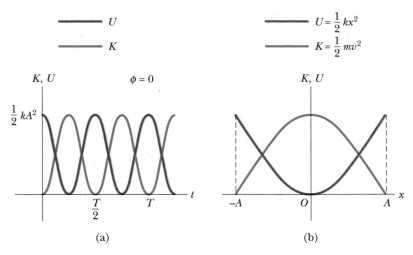

Figure 12.8 (a) Kinetic energy and potential energy versus time for a simple harmonic oscillator with $\phi = 0$. (b) Kinetic energy and potential energy versus displacement for a simple harmonic oscillator. In either plot, note that $K + U = $ constant.

Finally, we can use energy to obtain the velocity for an arbitrary displacement, x, expressing the total energy at some arbitrary position as

$$E = K + U = \tfrac{1}{2} mv^2 + \tfrac{1}{2} kx^2 = \tfrac{1}{2} kA^2$$

Velocity as a function of position for a simple harmonic oscillator

$$v = \pm \sqrt{\frac{k}{m}(A^2 - x^2)} = \pm \omega \sqrt{A^2 - x^2} \qquad \text{[12.23]}$$

Again, this expression substantiates the fact that the speed is a maximum at $x = 0$ and zero at the turning points, $x = \pm A$.

Thinking Physics 3

A mass oscillating on the end of a horizontal spring slides back and forth over a frictionless surface. During one oscillation, you set an identical mass with Velcro (a sticky fabric) attached at the maximum displacement point. Just as the oscillating mass reaches its largest displacement and is momentarily at rest, it adheres to the new mass by means of the Velcro, and the two masses continue the oscillation together. Does the period of the oscillation change? Does the amplitude of oscillation change? Does the energy of the oscillation change?

Reasoning The period of oscillation *does* change, because the period depends on the mass that is oscillating. The amplitude does not change. Because the new mass was added while the original mass was at rest, the combined masses are at rest at this point, also, defining the amplitude as the same as in the original oscillation. The energy does not change, either. At the maximum displacement point, the energy is all potential energy stored in the spring, which depends only on the spring constant and the amplitude, not the mass. The increased mass will pass through the equilibrium point with less velocity than in the original oscillation, but with the same kinetic energy. Another approach is to think about how energy could be transferred into the oscillating system—no work was done (nor was there any other form of energy transfer), so the energy in the system cannot change.

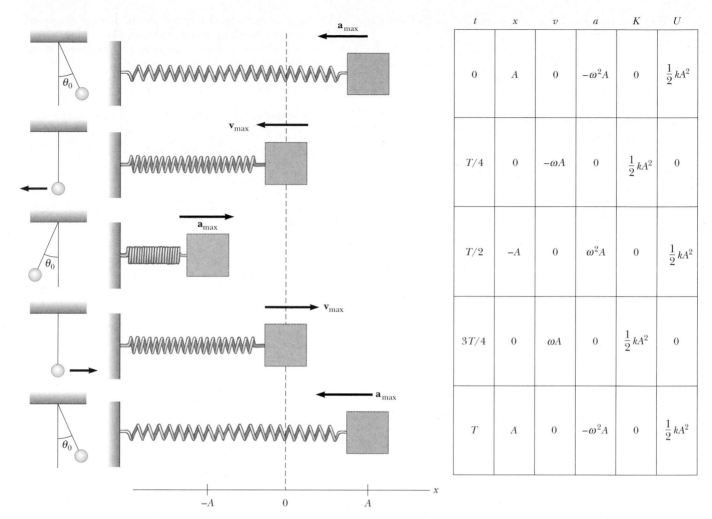

Figure 12.9   Simple harmonic motion for a mass–spring system and its analogy to the motion of a simple pendulum. The parameters in the table at the right refer to the mass–spring system, assuming that at $t = 0$, $x = A$ so that $x = A \cos \omega t$ (Special Case I).

CONCEPTUAL PROBLEM 3

A mass–spring system undergoes simple harmonic motion with an amplitude A. Does the total energy change if the mass is doubled but the amplitude is not changed? Are the kinetic and potential energies at a given point in its motion affected by the change in mass? Explain.

Example 12.3   Oscillations on a Horizontal Surface

A 0.50-kg mass connected to a light spring of force constant 20.0 N/m oscillates on a horizontal, frictionless track. (a) Calculate the total energy of the system and the maximum speed of the mass if the amplitude of the motion is 3.0 cm.

Solution   Using Equation 12.22, we get

$$E = \tfrac{1}{2}kA^2 = \tfrac{1}{2}\left(20.0\,\frac{N}{m}\right)(3.0 \times 10^{-2}\,m)^2 = 9.0 \times 10^{-3}\,J$$

When the mass is at $x = 0$, $U = 0$ and $E = \frac{1}{2}mv_{max}^2$; therefore

$$\frac{1}{2}mv_{max}^2 = 9.0 \times 10^{-3}\,J$$

$$v_{max} = \sqrt{\frac{18.0 \times 10^{-3}\,J}{0.50\,kg}} = 0.19\,m/s$$

(b) What is the velocity of the mass when the displacement is equal to 2.0 cm?

Solution   We can apply Equation 12.23 directly:

$$v = \pm\sqrt{\frac{k}{m}(A^2 - x^2)}$$

$$= \pm\sqrt{\frac{20.0}{0.50}(3.0^2 - 2.0^2) \times 10^{-4}}$$

$$= \pm 0.14\,m/s$$

The positive and negative signs indicate that the mass could be moving to the right or left at this instant.

(c) Compute the kinetic and potential energies of the system when the displacement equals 2.0 cm.

Solution   Using the result to part (b), we get

$$K = \frac{1}{2}mv^2 = \frac{1}{2}(0.50\,kg)(0.14\,m/s)^2 = 5.0 \times 10^{-3}\,J$$

$$U = \frac{1}{2}kx^2 = \frac{1}{2}\left(20.0\,\frac{N}{m}\right)(2.0 \times 10^{-2}\,m)^2$$

$$= 4.0 \times 10^{-3}\,J$$

Note that the sum $K + U$ equals the total mechanical energy, E.

EXERCISE 6   For what values of x does the speed of the mass equal 0.10 m/s?   Answer ± 2.6 cm

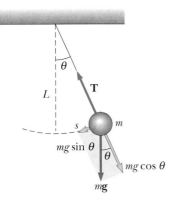

Figure 12.10   When θ is small, the simple pendulum oscillates in simple harmonic motion about the equilibrium position ($\theta = 0$). The restoring force is $mg\sin\theta$, the component of the force of gravity tangent to the circle.

12.4 • MOTION OF A PENDULUM

The **simple pendulum** is another mechanical system that exhibits periodic motion. It consists of a point mass, m, suspended by a light string of length L, where the upper end of the string is fixed as in Figure 12.10. The motion occurs in a vertical plane and is driven by the force of gravity. We shall show that the motion is that of a simple harmonic oscillator, provided the angle, θ, that the pendulum makes with the vertical is small.

The forces acting on the mass are the tension force, **T**, acting along the string, and the force of gravity, $m\mathbf{g}$. The tangential component of the force of gravity, $mg\sin\theta$, always acts toward $\theta = 0$, opposite the displacement. Therefore, the tangential force is a restoring force, and we can use Newton's second law to write the equation of motion in the tangential direction as

$$F_t = -mg\sin\theta = m\frac{d^2s}{dt^2}$$

where s is the displacement measured along the arc in Figure 12.10 and the minus sign indicates that F_t acts toward the equilibrium position. Because $s = L\theta$ and L is constant, this equation reduces to

$$\frac{d^2\theta}{dt^2} = -\frac{g}{L}\sin\theta$$

The right side is proportional to $\sin\theta$ rather than to θ; hence, we conclude that the motion is not simple harmonic motion, because it is not of the form of Equation 12.17. However, if we assume that θ is *small*, we can use the approximation $\sin \theta \approx \theta$, where θ is measured in radians,[1] and the equation of motion becomes

$$\frac{d^2\theta}{dt^2} = -\frac{g}{L}\theta \qquad [12.24]$$

• *Equation of motion for the simple pendulum (small θ)*

Now we have an expression with exactly the same form as Equation 12.17, and so we conclude that the motion is approximately simple harmonic motion for small amplitudes. Therefore, θ can be written as $\theta = \theta_0 \cos(\omega t + \phi)$, where θ_0 is the **maximum angular displacement** and the angular frequency ω is

$$\omega = \sqrt{\frac{g}{L}} \qquad [12.25]$$

• *Angular frequency of motion for a simple pendulum*

The period of the motion is

$$T = \frac{2\pi}{\omega} = 2\pi\sqrt{\frac{L}{g}} \qquad [12.26]$$

• *Period of motion for a simple pendulum*

In other words, **the period and frequency of a simple pendulum oscillating at small angles depend only on the length of the string and the free-fall acceleration.** Because the period is *independent* of the mass, we conclude that *all* simple pendula of equal length at the same location oscillate with equal periods. Experiment shows that this is correct. The analogy between the motion of a simple pendulum and the mass–spring system is illustrated in Figure 12.9.

The simple pendulum can be used as a timekeeper. It is also a convenient device for making precise measurements of the free-fall acceleration. Such measurements are important, because variations in local values of **g** can provide information on the locations of oil and other valuable underground resources.

CONCEPTUAL PROBLEM 4

A simple pendulum is suspended from the ceiling of a stationary elevator, and the period is determined. Describe the changes, if any, in the period when the elevator (a) accelerates upward, (b) accelerates downward, and (c) moves with constant velocity.

CONCEPTUAL PROBLEM 5

Imagine that a pendulum is hanging from the ceiling of a car. As the car coasts freely down a hill, is the equilibrium position of the pendulum vertical? Does the period of oscillation change from that in a stationary car?

CONCEPTUAL PROBLEM 6

A grandfather clock depends on the period of a pendulum to keep correct time. Suppose a grandfather clock is calibrated correctly and then the temperature of the room in which it resides increases. Does the grandfather clock run slow, fast, or correctly? *Hint:* A metal expands when its temperature is raised.

[1]See Appendix A for more details on the small-angle approximation.

Example 12.4   A Measure of Height

A man enters a tall tower, needing to know its height. He notes that a long pendulum extends from the ceiling almost to the floor and that its period is 12.0 s. How tall is the tower?

Solution   If we use $T = 2\pi\sqrt{L/g}$ and solve for L, we get

$$L = \frac{gT^2}{4\pi^2} = \frac{(9.80 \text{ m/s}^2)(12.0 \text{ s})^2}{4\pi^2} = \boxed{35.7 \text{ m}}$$

EXERCISE 7   If the pendulum described in this example is taken to the moon, where the free-fall acceleration is 1.67 m/s², what is the period there?   Answer   29.1 s

O P T I O N A L

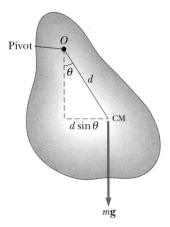

Figure 12.11   The physical pendulum consists of a rigid body pivoted at the point O, which is not at the center of mass. At equilibrium, the weight vector passes through O, corresponding to $\theta = 0$. The restoring torque about O when the system is displaced through an angle θ is $mgd \sin \theta$.

The Physical Pendulum

If a hanging object oscillates about a fixed axis that does not pass through its center of mass, and the object cannot be accurately approximated as a point mass, then it must be treated as a physical, or compound, pendulum. Consider a rigid body pivoted at a point O that is a distance of d from the center of mass (Fig. 12.11). The torque about O is provided by the force of gravity, and its magnitude is $mgd \sin \theta$. Using the fact that $\tau = I\alpha$, where I is the moment of inertia about the axis through O, we get

$$-mgd \sin \theta = I\frac{d^2\theta}{dt^2}$$

The minus sign on the left indicates that the torque about O tends to decrease θ. That is, the force of gravity produces a restoring torque.

If we again assume that θ is small, then the approximation $\sin \theta \approx \theta$ is valid and the equation of motion reduces to

$$\frac{d^2\theta}{dt^2} = -\left(\frac{mgd}{I}\right)\theta = -\omega^2\theta \qquad [12.27]$$

Note that the equation has the same form as Equation 12.17, and so the motion is approximately simple harmonic motion for small amplitudes. That is, the solution of Equation 12.27 is $\theta = \theta_0 \cos(\omega t + \phi)$, where θ_0 is the maximum angular displacement and

$$\omega = \sqrt{\frac{mgd}{I}}$$

The period is

Period of motion for a •
physical pendulum

$$T = \frac{2\pi}{\omega} = 2\pi\sqrt{\frac{I}{mgd}} \qquad [12.28]$$

One can use this result to measure the moment of inertia of a planar rigid body. If the location of the center of mass and, hence, the distance d are known, the moment of inertia can be obtained through a measurement of the period. Finally, note that Equation 12.28 reduces to the period of a simple pendulum (Eq. 12.26) when $I = md^2$—that is, when all the mass is concentrated at the center of mass.

CONCEPTUAL PROBLEM 7

Two students are watching the swinging of a Foucault pendulum with a large bob in a museum. One student says, "I'm going to sneak past the fence and stick some chewing gum on the top of the pendulum bob, to change its frequency of oscillation." The other student says, "That won't change the frequency—the frequency of a pendulum is independent of mass." Which student is correct?

Example 12.5   A Swinging Rod

A uniform rod of mass M and length L is pivoted about one end and oscillates in a vertical plane (Fig. 12.12). Find the period of oscillation if the amplitude of the motion is small.

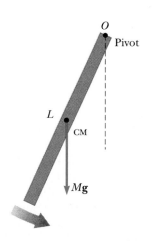

Figure 12.12   (Example 12.5) A rigid rod oscillating about a pivot through one end is a physical pendulum with $d = L/2$ and from Table 10.2, $I_0 = \frac{1}{3}ML^2$.

Solution   The moment of inertia of a uniform rod about an axis through one end is $\frac{1}{3}ML^2$. The distance d from the pivot to the center of mass is $L/2$. Substituting these quantities into Equation 12.28 gives

$$T = 2\pi \sqrt{\frac{\frac{1}{3}ML^2}{Mg\frac{L}{2}}} = 2\pi \sqrt{\frac{2L}{3g}}$$

Comment   In one of the Moon landings, an astronaut walking on the Moon's surface had a belt hanging from his space suit, and the belt oscillated as a compound pendulum. A scientist on Earth observed this motion on TV and from it was able to estimate the free-fall acceleration on the Moon. How do you suppose this calculation was done?

EXERCISE 8   Calculate the period of a meter stick pivoted about one end and oscillating in a vertical plane as in Figure 12.12.   Answer   1.64 s

EXERCISE 9   A simple pendulum has a length of 3.00 m. Determine the change in its period if it is taken from a point where $g = 9.80$ m/s² to an elevation where the free-fall acceleration decreases to 9.79 m/s².   Answer 1.78×10^{-3} s

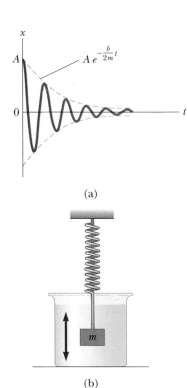

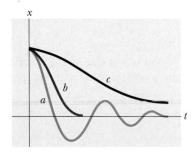

Figure 12.13 (a) Graph of the displacement versus time for a damped oscillator. Note the decrease in amplitude with time. (b) One example of a damped oscillator is a mass on a spring submersed in a liquid.

12.5 • DAMPED OSCILLATIONS

The oscillatory motions we have considered so far have occurred in the context of an ideal system — that is, one that oscillates indefinitely under the action of a linear restoring force. In realistic systems, dissipative forces, such as friction, are present and retard the motion of the system. As a consequence, the mechanical energy of the system diminishes in time, and the motion is said to be *damped.*

One common type of drag force, which we discussed in Chapter 5, is proportional to the velocity and acts in the direction opposite the motion. This type of drag is often observed when an object is oscillating in air, for instance. Because the drag force can be expressed as **R** $= -b\mathbf{v}$, where b is a constant, and the restoring force exerted on the system is $-kx$, we can write Newton's second law as

$$\sum F_x = -kx - bv = ma_x \qquad \text{[12.29]}$$

$$-kx - b\frac{dx}{dt} = m\frac{d^2x}{dt^2}$$

The solution of this equation requires mathematics that may not yet be familiar to you, and so it will simply be stated without proof. When the parameters of the system are such that $b < \sqrt{4mk}$, the solution to Equation 12.29 is

$$x = Ae^{-(b/2m)t}\cos(\omega t + \phi) \qquad \text{[12.30]}$$

where the angular frequency of motion is

$$\omega = \sqrt{\frac{k}{m} - \left(\frac{b}{2m}\right)^2} \qquad \text{[12.31]}$$

This result can be verified by substituting Equation 12.30 into Equation 12.29. Figure 12.13a shows the displacement as a function of time in this case. We see that **when the drag force is relatively small, the oscillatory character of the motion is preserved but the amplitude of vibration decreases in time** and the motion ultimately ceases. This is known as an **underdamped oscillator.** The dashed blue lines in Figure 12.13a, which outline the *envelope* of the oscillatory curve, represent the exponential factor that appears in Equation 12.30. The exponential factor shows that *the amplitude decays exponentially with time.* For motion with a given spring constant and particle mass, the oscillations dampen more rapidly as the maximum value of the drag force approaches the maximum value of the restoring force. One example of a damped harmonic oscillator is a mass immersed in a fluid, as in Figure 12.13b.

It is convenient to express the angular frequency of vibration of a damped system (Eq. 12.31) in the form

$$\omega = \sqrt{\omega_0{}^2 - \left(\frac{b}{2m}\right)^2}$$

where $\omega_0 = \sqrt{k/m}$ represents the angular frequency of oscillation in the absence of a drag force (the undamped oscillator). In other words, when $b = 0$, the drag force is zero and the system oscillates with its natural frequency, ω_0. As the magnitude of the drag force increases, the oscillations dampen more rapidly. When b reaches a critical value, b_c, so that $b_c/2m = \omega_0$, the system does not oscillate and is said to be

Figure 12.14 Plots of displacement versus time for (a) an underdamped oscillator, (b) a critically damped oscillator, and (c) an overdamped oscillator.

critically damped. In this case it returns to equilibrium in an exponential manner with time, as in Figure 12.14.

If the medium is highly viscous and the parameters meet the condition that $b/2m > \omega_0$, the system is **overdamped.** Again, the displaced system does not oscillate but simply returns to its equilibrium position. As the damping increases, the time it takes the displacement to reach equilibrium also increases, as indicated in Figure 12.14. In any case, when a drag force is present, the energy of the oscillator eventually falls to zero. The lost mechanical energy dissipates into thermal energy in the resistive medium.

12.6 · FORCED OSCILLATIONS

O P T I O N A L

We have seen that the energy of a damped oscillator decreases in time as a result of the drag force. It is possible to compensate for this energy loss by applying an external force that does positive work on the system. At any instant, energy can be put into the system by an applied force that acts in the direction of motion of the oscillator. For example, a child on a swing can be kept in motion by appropriately timed "pushes." The amplitude of motion remains constant if the energy input per cycle of motion exactly equals the energy lost as a result of air drag.

A common example of a forced oscillator is a damped oscillator driven by an external force that varies periodically, such as $F = F_0 \cos \omega t$, where ω is the angular frequency of the force and F_0 is a constant. Adding this driving force to the left side of Equation 12.29 gives

$$F_0 \cos \omega t - b\frac{dx}{dt} - kx = m\frac{d^2x}{dt^2} \qquad \text{[12.32]}$$

Again, the solution of this equation is rather lengthy and will not be presented. However, after a sufficiently long period of time, when the energy input per cycle equals the energy lost per cycle, a steady-state condition is reached in which the oscillations proceed with constant amplitude. At this time, Equation 12.32 has the solution

$$x = A \cos(\omega t + \phi) \qquad \text{[12.33]}$$

where

$$A = \frac{F_0/m}{\sqrt{(\omega^2 - \omega_0^2)^2 + \left(\dfrac{b\omega}{m}\right)^2}} \qquad \text{[12.34]}$$

and where $\omega_0 = \sqrt{k/m}$ is the angular frequency of the undamped oscillator $(b = 0)$. From a physical point of view, one can argue that in a steady state the oscillator must have the same angular frequency as the driving force, and so the solution given by Equation 12.33 is expected.

Equation 12.34 shows that the amplitude of the forced oscillator is constant, because it is being driven by an external force. That is, the external agent provides the energy necessary to overcome the losses due to the drag force. Note that the mass oscillates at the angular frequency of the driving force, ω. For small damping, the amplitude becomes large when the frequency of the driving force is near the

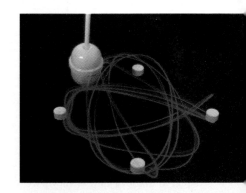

This is a computer-generated simulation of the chaotic pattern of a driven pendulum. The nonlinearity in the motion is provided by the gravitational torque, which is proportional to the sine of the angle. An important research problem is to predict the conditions under which different physical systems exhibit chaotic behavior. *(© Yoav Levy/Phototake)*

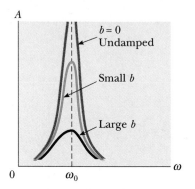

Figure 12.15 Graph of amplitude versus frequency for a damped oscillator when a periodic driving force is present. When the frequency of the driving force equals the natural frequency, ω_0, resonance occurs. Note that the shape of the resonance curve depends on the size of the damping coefficient, b.

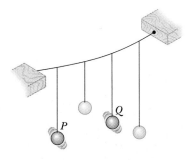

Figure 12.16 If pendulum P is set into oscillation, pendulum Q will eventually oscillate with the greatest amplitude because its length is equal to that of P and so they have the same natural frequency of vibration. The pendula are oscillating in a direction perpendicular to the plane formed by the stationary strings.

natural frequency of oscillation or when $\omega \approx \omega_0$. The dramatic increase in amplitude near the natural frequency is called **resonance,** and the angular frequency, ω_0, is called the **resonance frequency** of the system.

The reason for large-amplitude oscillations at the resonance frequency is that energy is being transferred to the system under the most favorable conditions. This can be better understood by taking the first time derivative of x, which gives an expression for the velocity of the oscillator. In doing so, one finds that v is proportional to $\sin(\omega t + \phi)$. When the applied force is in phase with **v**, the rate at which work is done on the oscillator by the force **F** (in other words, the power) equals Fv. Because the quantity Fv is always positive when **F** and **v** are in phase, we conclude that **at resonance the applied force is in phase with the velocity, and the power transferred to the oscillator is a maximum.**

Figure 12.15 is a graph of amplitude as a function of frequency for the forced oscillator, with and without a resistive force. Note that the amplitude increases with decreasing damping ($b \rightarrow 0$) and that the resonance curve broadens as the damping increases. Under steady-state conditions, and at any driving frequency, the energy transferred into the system equals the energy lost because of the damping force; hence, the average total energy of the oscillator remains constant. In the absence of a damping force ($b = 0$), we see from Equation 12.34 that the steady-state amplitude approaches infinity as $\omega \rightarrow \omega_0$. In other words, if there are no losses in the system and we continue to drive an initially motionless oscillator with a sinusoidal force that is in phase with the velocity, the amplitude of motion will build up without limit (Fig. 12.15). This does not occur in practice because some damping is always present. That is, at resonance the amplitude is large but finite for small damping.

One experiment that demonstrates a resonance phenomenon is illustrated in Figure 12.16. Several pendula of different lengths are suspended from a stretched string. If one of them, such as P, is set swinging in the plane perpendicular to the plane of the string, the others will begin to oscillate, because they are coupled by the stretched string. Of those that are forced into oscillation by this coupling, pendulum Q, which is the same length as P (hence, the two pendula have the same natural frequency), will oscillate with the greatest amplitude.

Resonance appears in other areas of physics. For example, certain electrical circuits have natural (resonant) frequencies. A structure such as a bridge has natural frequencies and can be set into resonance by an appropriate driving force. A striking example of such a structural resonance occurred in 1940, when the Tacoma Narrows Bridge in Washington was destroyed by resonant vibrations (Fig. 12.17). The winds were not particularly strong on that occasion, but the bridge still collapsed because vortices (turbulences) generated by the wind blowing through the bridge occurred at a frequency that matched the natural frequency of the bridge.

Many other examples of resonant vibrations can be cited. A resonant vibration you may have experienced is the "singing" of telephone wires in the wind. Machines often break if one vibrating part is at resonance with some other moving part.

Thinking Physics 4

In an earthquake region, should damping in a building be large or small?

Reasoning The damping in a building should be large, so that the resonance curve is broad and flat. The building should respond to a wide range of earthquake frequen-

(a)   (b)

Figure 12.17   (a) High winds set up vibrations in the bridge, causing it to oscillate at a frequency near one of the natural frequencies of the bridge structure. (b) Once established, this resonance condition led to the bridge's collapse.   *(UPI/Bettmann Newsphotos)*

cies, with a relatively small response. If the damping were small, so that the resonance curve were tall and skinny, the building would reject many earthquake frequencies but would have a disastrously large response to a narrow range of frequencies.

CONCEPTUAL PROBLEM 8

You stand on the end of a diving board and bounce to set it into oscillation. You find a maximum response, in terms of the amplitude of oscillation of the end of the board, when you bounce at frequency *f*. You now move to the middle of the board and repeat the experiment. Is the resonance frequency for forced oscillations at this point higher than *f*, lower than *f*, or the same as *f*? Why?

CONCEPTUAL PROBLEM 9

Some parachutes have holes in them to allow air to move smoothly through the chute. Without the holes, the air gathered under the chute as the parachutist falls is sometimes released from under the edges of the chute alternately and periodically from one side and then the other. Why might this periodic release of air cause a problem?

SUMMARY

The position of a simple harmonic oscillator varies periodically in time according to the relation

$$x = A \cos(\omega t + \phi) \qquad \text{[12.1]}$$

where A is the amplitude of the motion, ω is the angular frequency, and ϕ is the phase constant. The values of A and ϕ depend on the initial position and velocity of the oscillator.

The time for one complete oscillation is called the **period** of the motion, defined by

$$T = \frac{2\pi}{\omega} \qquad \text{[12.2]}$$

The inverse of the period is the **frequency** of the motion, which equals the number of oscillations per second.

The **velocity** and **acceleration** of a simple harmonic oscillator are

$$v = \frac{dx}{dt} = -\omega A \sin(\omega t + \phi) \qquad \text{[12.5]}$$

$$a = \frac{dv}{dt} = -\omega^2 A \cos(\omega t + \phi) \qquad \text{[12.6]}$$

Thus, the maximum velocity is ωA, and the maximum acceleration is $\omega^2 A$. The velocity is zero when the oscillator is at its turning points, $x = \pm A$, and the speed is a maximum at the equilibrium position, $x = 0$. The magnitude of the acceleration is a maximum at the turning points and is zero at the equilibrium position.

A mass–spring system moving on a frictionless track exhibits simple harmonic motion with the period

$$T = \frac{2\pi}{\omega} = 2\pi \sqrt{\frac{m}{k}} \qquad \text{[12.18]}$$

where k is the force constant of the spring and m is the mass attached to the spring.

The kinetic energy and potential energy of a simple harmonic oscillator vary with time and are

$$K = \tfrac{1}{2}mv^2 = \tfrac{1}{2}m\omega^2 A^2 \sin^2(\omega t + \phi) \qquad \text{[12.20]}$$

$$U = \tfrac{1}{2}kx^2 = \tfrac{1}{2}kA^2 \cos^2(\omega t + \phi) \qquad \text{[12.21]}$$

The **total energy** of a simple harmonic oscillator is a constant of the motion and is

$$E = \tfrac{1}{2}kA^2 \qquad \text{[12.22]}$$

The potential energy of a simple harmonic oscillator is a maximum when the particle is at its turning points (maximum displacement from equilibrium) and is zero at the equilibrium position. The kinetic energy is zero at the turning points and is a maximum at the equilibrium position.

A **simple pendulum** of length L exhibits simple harmonic motion for small angular displacements from the vertical, with a period of

$$T = 2\pi \sqrt{\frac{L}{g}} \qquad \text{[12.26]}$$

That is, the period is independent of the suspended mass.

A **physical pendulum** exhibits simple harmonic motion about a pivot that does not go through the center of mass. The period of this motion is

$$T = 2\pi \sqrt{\frac{I}{mgd}} \qquad \text{[12.28]}$$

where I is the moment of inertia about an axis through the pivot and d is the distance from the pivot to the center of mass.

Damped oscillations occur in a system in which a drag force opposes the linear restoring force. If such a system is set in motion and then left to itself, its mechanical energy decreases in time because of the presence of the nonconservative drag force. It is possible to compensate for this loss in energy by driving the system with an external periodic force that is in phase with the motion of the system. When the frequency of the driving force matches the natural frequency of the *undamped* oscillator that starts its motion from rest, energy is continuously transferred to the oscillator and its amplitude increases without limit.

CONCEPTUAL QUESTIONS

1. Does the acceleration of a simple harmonic oscillator remain constant during its motion? Is the acceleration ever zero? Explain.
2. Explain why the kinetic and potential energies of a mass–spring system can never be negative.
3. What is the total distance traveled by a body executing simple harmonic motion in a time equal to its period if its amplitude is A?
4. If a mass–spring system is hung vertically and set into oscillation, why does the motion eventually stop?
5. If a pendulum clock keeps perfect time at the base of a mountain, will it also keep perfect time when moved to the top of the mountain? Explain.
6. If a grandfather clock were running slow, how could we adjust the length of the pendulum to correct the time?
7. A pendulum bob is made from a ball filled with water. What happens to the frequency of vibration of this pendulum if the ball has a hole that allows water to slowly leak out?
8. A platoon of soldiers marches in step along a road. Why are they ordered to break step when crossing a bridge?
9. If the length of a simple pendulum is quadrupled, what happens to (a) its frequency and (b) its period?
10. The amplitude of a system moving in simple harmonic motion is doubled. Determine the change in (a) the total energy, (b) the maximum speed, (c) the maximum acceleration, and (d) the period.

11. A simple harmonic oscillator has a total energy of E. (a) Determine the kinetic and potential energies when the displacement equals one-half the amplitude. (b) For what value of the displacement does the kinetic energy equal the potential energy?
12. If the coordinate of a particle varies as $x = -A \cos \omega t$, what is the phase constant in Equation 12.1? At what position does the particle begin its motion?
13. Does the displacement of an oscillating particle between $t = 0$ and a later time t necessarily equal the position of the particle at time t? Explain.
14. Determine whether or not the following quantities can be in the same direction for a simple harmonic oscillator: (a) displacement and velocity, (b) velocity and acceleration, (c) displacement and acceleration.
15. Can the amplitude A and phase constant ϕ be determined for an oscillator if only the position is specified at $t = 0$? Explain.
16. A simple pendulum undergoes simple harmonic motion when θ is small. Is the motion periodic when θ is large? How does the period of motion change as θ increases?
17. Will damped oscillations occur for any values of b and k? Explain.
18. Is it possible to have damped oscillations when a system is at resonance? Explain.

PROBLEMS

Section 12.1   Simple Harmonic Motion

1. The displacement of a particle is given by the expression $x = (4.00 \text{ m}) \cos(3.00 \pi t + \pi)$, where x is in meters and t is in seconds. Determine (a) the frequency and period of the motion, (b) the amplitude of the motion, (c) the phase constant, and (d) the displacement of the particle at $t = 0.250$ s.
2. A particle oscillates with simple harmonic motion so that its displacement varies according to the expression $x = (5.00 \text{ cm}) \cos(2t + \pi/6)$, where x is in centimeters and t is in seconds. At $t = 0$, find (a) the displacement of the particle, (b) its velocity, and (c) its acceleration. (d) Find the period and amplitude of the motion.
3. A piston in an automobile engine is in simple harmonic motion. If its amplitude of oscillation from the centerline is ± 5.00 cm and its mass is 2.00 kg, find the maximum velocity and acceleration of the piston when the auto engine is running at the rate of 3600 rev/min.
4. A 20.0-g particle moves in simple harmonic motion with a frequency of 3.00 oscillations/s and an amplitude of 5.00 cm. (a) Through what total distance does the particle move during one cycle of its motion? (b) What is its maximum speed? Where does this occur? (c) Find the maximum acceleration of the particle. Where in the motion does the maximum acceleration occur?

5. A particle moving along the x axis in simple harmonic motion starts from the origin at $t = 0$ and moves to the right. If the amplitude of its motion is 2.00 cm and the frequency is 1.50 Hz, (a) show that its displacement is given by the expression $x = (2.00 \text{ cm}) \sin(3.00 \pi t)$. Determine (b) the maximum speed and the earliest time ($t > 0$) at which the particle has this speed, (c) the maximum acceleration and the earliest time ($t > 0$) at which the particle has this acceleration to the right, and (d) the total distance traveled between $t = 0$ and $t = 1.00$ s.
6. If the initial position, velocity, and acceleration of an object moving in simple harmonic motion are x_0, v_0, and a_0, and

if the angular frequency of oscillation is ω, (a) show that the position and velocity of the object for all time can be written as

$$x(t) = x_0 \cos \omega t + \left(\frac{v_0}{\omega}\right) \sin \omega t$$

$$v(t) = -x_0 a \sin \omega t + v_0 \cos \omega t$$

(b) If the amplitude of the motion is A, show that

$$v^2 - ax = v_0{}^2 - a_0 x_0 = A^2 \omega^2$$

Section 12.2   Motion of a Mass Attached to a Spring

(*Note:* Neglect the mass of the spring in all of these problems.)

7. An archer pulls her bow string back 0.400 m by exerting a force that increases uniformly from zero to 230 N. (a) What is the equivalent spring constant of the bow? (b) How much work is done in pulling the bow?

8. A 1.00-kg mass attached to a spring of force constant 25.0 N/m oscillates on a horizontal, frictionless track. At $t = 0$, the mass is released from rest at $x = -3.00$ cm. (That is, the spring is compressed by 3.00 cm.) Find (a) the period of its motion, (b) the maximum values of its speed and acceleration, and (c) the displacement, velocity, and acceleration as functions of time.

9. A 0.500-kg mass attached to a spring of force constant 8.00 N/m vibrates in simple harmonic motion with an amplitude of 10.0 cm. Calculate (a) the maximum value of its speed and acceleration, (b) the speed and acceleration when the mass is 6.00 cm from the equilibrium position, and (c) the time it takes the mass to move from $x = 0$ to $x = 8.00$ cm.

10. A block of unknown mass is attached to a spring of force constant 6.50 N/m and undergoes simple harmonic motion with an amplitude of 10.0 cm. When the mass is halfway between its equilibrium position and the endpoint, its speed is measured to be +30.0 cm/s. Calculate (a) the mass of the block, (b) the period of the motion, and (c) the maximum acceleration of the block.

11. A 1.00-kg mass is attached to a horizontal spring. The spring is initially stretched by 0.100 m and the mass is released from rest there. After 0.500 s, the speed of the mass is zero. What is the maximum speed of the mass?

12. A particle that hangs from a spring oscillates with an angular frequency of 2.00 rad/s. The spring is suspended from the ceiling of an elevator car and hangs motionless (relative to the elevator car) as the car descends at a constant speed of 1.50 m/s. The car then stops suddenly. (a) With what amplitude does the particle oscillate? (b) What is the equation of motion for the particle? (Choose the upward direction to be positive.)

Section 12.3   Energy of the Simple Harmonic Oscillator

(Neglect spring masses.)

13. An automobile having a mass of 1000 kg is driven into a brick wall in a safety test. The bumper behaves like a spring of constant 5.00×10^6 N/m and compresses 3.16 cm as the car is brought to rest. What was the speed of the car before impact, assuming no energy is lost during impact with the wall?

14. An ore car of mass 4000 kg starts from rest and rolls downhill from the mine on tracks. At the end of the tracks, 10.0 m lower in elevation, is a spring with force constant $k = 400\,000$ N/m. How much is the spring compressed in stopping the ore car? Ignore friction.

15. A 200-g mass is attached to a spring and executes simple harmonic motion with a period of 0.250 s. If the total energy of the system is 2.00 J, find (a) the force constant of the spring and (b) the amplitude of the motion.

16. A mass–spring system oscillates with an amplitude of 3.50 cm. If the force constant of the spring is 250 N/m and the mass is 0.500 kg, determine (a) the mechanical energy of the system, (b) the maximum speed of the mass, and (c) the maximum acceleration.

17. A 50.0-g mass connected to a spring of force constant 35.0 N/m oscillates on a horizontal, frictionless surface with an amplitude of 4.00 cm. Find (a) the total energy of the system and (b) the speed of the mass when the displacement is 1.00 cm. When the displacement is 3.00 cm, find (c) the kinetic energy and (d) the potential energy.

18. A 2.00-kg mass is attached to a spring and placed on a horizontal, smooth surface. A horizontal force of 20.0 N is required to hold the mass at rest when it is pulled 0.200 m from its equilibrium position (the origin of the x axis). The mass is now released from rest with an initial displacement of $x_0 = 0.200$ m, and it subsequently undergoes simple harmonic oscillations. Find (a) the force constant of the spring, (b) the frequency of the oscillations, and (c) the maximum speed of the mass. Where does this maximum speed occur? (d) Find the maximum acceleration of the mass. Where does it occur? (e) Find the total energy of the oscillating system. When the displacement equals one third the maximum value, find (f) the speed and (g) the acceleration.

Section 12.4   Motion of a Pendulum

19. A simple pendulum has a period of 2.50 s. (a) What is its length? (b) What would its period be on the Moon, where $g_M = 1.67$ m/s^2?

20. A "seconds" pendulum is one that moves through its equilibrium position once each second. (The period of the pendulum is 2.000 s.) The length of a seconds pendulum is 0.9927 m at Tokyo and 0.9942 m at Cambridge, England.

What is the ratio of the free-fall accelerations at these two locations?

21. A simple pendulum has a mass of 0.250 kg and a length of 1.00 m. It is displaced through an angle of 15.0° and then released. What are (a) the maximum speed? (b) the maximum angular acceleration? (c) the maximum restoring force?

22. The angular displacement of a pendulum is represented by the equation $\theta = 0.320 \cos \omega t$, where θ is in radians and $\omega = 4.43$ rad/s. Determine the period and length of the pendulum.

23. A particle of mass m slides inside a frictionless hemispherical bowl of radius R. Show that if it starts from rest with a small displacement from equilibrium, the particle moves in simple harmonic motion with an angular frequency equal to that of a simple pendulum of length R. That is, $\omega = \sqrt{g/R}$.

24. A simple pendulum is 5.00 m long. (a) What is the period of simple harmonic motion for this pendulum if it is located in an elevator accelerating upward at 5.00 m/s²? (b) What is its period if the elevator is accelerating downward at 5.00 m/s²? (c) What is the period of simple harmonic motion for this pendulum if it is placed in a truck that is accelerating horizontally at 5.00 m/s²?

25. A very light rod extends rigidly 0.500 m out from one end of a meter stick. The stick is suspended from a pivot at the far end of the rod and set into oscillation. (a) Determine the period of oscillation. (b) By what percentage does this differ from a 1.00–m-long simple pendulum?

26. Consider the physical pendulum of Figure 12.11. (a) If its moment of inertia about an axis passing through its center of mass and parallel to the axis passing through its pivot point is I_{CM}, show that its period is

$$T = 2\pi \sqrt{\frac{I_{CM} + md^2}{mgd}}$$

where d is the distance between the pivot point and center of mass. (b) Show that the period has a minimum value when d satisfies $md^2 = I_{CM}$.

Section 12.5   Damped Oscillations (Optional)

27. A pendulum of length 1.00 m is released from an initial angle of 15.0°. After 1000 s, its amplitude is reduced by friction to 5.50°. What is the value of $b/2m$?

28. Show that Equation 12.30 is a solution of Equation 12.29 provided that $b^2 < 4mk$.

29. Show that the time rate of change of mechanical energy for a damped, undriven oscillator is given by $dE/dt = -bv^2$ and hence is always negative. (*Hint:* Differentiate the expression for the mechanical energy of an oscillator, $E = \frac{1}{2}mv^2 + \frac{1}{2}kx^2$, and use Eq. 12.29.)

Section 12.6   Forced Oscillations (Optional)

30. Calculate the resonant frequencies of (a) a 3.00-kg mass attached to a spring of force constant 240 N/m and (b) a simple pendulum 1.50 m in length.

31. A 2.00-kg mass attached to a spring is driven by an external force $F = (3.00 \text{ N}) \cos(2\pi t)$. If the force constant of the spring is 20.0 N/m, determine (a) the period and (b) the amplitude of the motion. (*Hint:* Assume there is no damping—that is, $b = 0$, and use Eq. 12.34.)

32. Consider an *undamped*, forced oscillator ($b = 0$), and show that Equation 12.33 is a solution of Equation 12.32, with an amplitude given by Equation 12.34.

Additional Problems

33. A mass m is oscillating freely on a vertical spring (Fig. P12.33). When $m = 0.810$ kg, the period is 0.910 s. An unknown mass on the same spring has a period of 1.16 s. Determine (a) the spring constant k and (b) the unknown mass.

Figure P12.33

34. A mass m is oscillating freely on a vertical spring with a period T (Fig. P12.33). An unknown mass m' on the same spring oscillates with a period T'. Determine (a) the force constant k and (b) the unknown mass m'.

35. A pendulum of length L and mass M has a spring of force constant k connected to it at a distance h below its point of suspension (Fig. P12.35). Find the frequency of vibration of the system for small values of the amplitude (small θ). (Assume the vertical suspension of length L is rigid but neglect its mass.)

36. A horizontal plank of mass m and length L is pivoted at one end, and the opposite end is attached to a spring of force constant k (Fig. P12.36). The moment of inertia of the plank about the pivot is $\frac{1}{3}mL^2$. When the plank is displaced a small angle θ from its horizontal equilibrium position and released, show that it moves with simple harmonic motion with an angular frequency $\omega = \sqrt{3k/m}$.

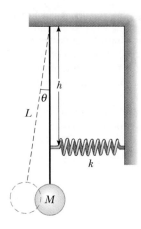

Figure P12.35

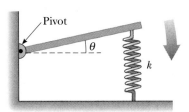

Figure P12.36

37. A large block *P* executes horizontal simple harmonic motion by sliding across a frictionless surface with a frequency *f* = 1.50 Hz. Block *B* rests on it, as shown in Figure P12.37, and the coefficient of static friction between the two is μ_s = 0.600. What maximum amplitude of oscillation can the system have if the block is not to slip?

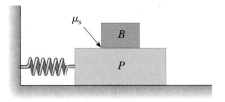

Figure P12.37

38. A 2.00-kg block hangs without vibrating at the end of a spring (*k* = 500 N/m) that is attached to the ceiling of an elevator car. The car is rising with an upward acceleration of *g*/3 when the acceleration suddenly ceases (at *t* = 0). (a) What is the angular frequency of oscillation of the block after the acceleration ceases? (b) By what amount is the spring stretched during the time that the elevator car is ac-

celerating? (c) What are the amplitude of the oscillation and the initial phase angle observed by a rider in the car? Take the upward direction to be positive.

39. A simple pendulum having a length of 2.23 m and a mass of 6.74 kg is given an initial speed of 2.06 m/s at its equilibrium position. Assume it undergoes simple harmonic motion and determine its (a) period, (b) total energy, and (c) maximum angular displacement.

40. People who ride motorcycles and bicycles learn to look out for bumps in the road—and especially for washboarding, a condition of many equally spaced ridges or bumps a certain distance apart. What is so bad about washboarding? A motorcycle has several springs and shock absorbers in its suspension, but you can model it as a single spring supporting a mass. You can estimate the spring constant by thinking about how far the spring compresses when a big biker sits down on the seat. Compute, on the basis of reasonable estimates, the distance between the washboard bumps of which a motorcyclist must be particularly careful when he or she is traveling at highway speed.

41. A mass *M* is connected to a spring of mass *m* and oscillates in simple harmonic motion on a horizontal, frictionless track (Fig. P12.41). The force constant of the spring is *k* and the equilibrium length is ℓ. Find (a) the kinetic energy of the system when the mass has a speed of *v* and (b) the period of oscillation. [*Hint:* Assume that all portions of the spring oscillate in phase and that the velocity of a segment *dx* is proportional to the distance from the fixed end; that is, $v_x = (x/\ell)v$. Also, note that the mass of a segment of the spring is $dm = (m/\ell)dx$.]

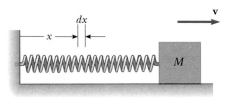

Figure P12.41

42. When a mass *M*, connected to the end of a spring of mass m_s = 7.40 g and force constant *k* is set into simple harmonic motion, the period of its motion is

$$T = 2\pi \sqrt{\frac{M + (m_s/3)}{k}}$$

A two-part experiment is conducted using various masses suspended vertically from the spring, as in Figure P12.33. (a) Displacements of 17.0, 29.3, 35.3, 41.3, 47.1, and 49.3 cm are measured for *M* values of 20.0, 40.0, 50.0, 60.0, 70.0, and 80.0 g, respectively. Construct a graph of *Mg* versus *x*, and perform a linear least-squares fit to the data. From the slope of your graph, determine a value for *k* for this spring. (b) The system is now set into simple harmonic motion, and periods are measured with a stopwatch. With $M = 80.0$ g, the total time for 10 oscillations is measured to be 13.41 s. The experiment is repeated with M values of 70.0, 60.0, 50.0, 40.0, and 20.0 g, with corresponding times for 10 oscillations of 12.52, 11.67, 10.67, 9.62, and 7.03 s. Obtain experimental values for T for each of these M values. Plot a graph of T^2 versus M and determine a value for k from the slope of the linear least-squares fit through the data points. Compare this value of k with that obtained in part (a). (c) Obtain a value for m_s from your graph and compare it with the given value of 7.40 g.

43. A mass m is connected to two rubber bands of length L, each under tension F, as in Figure P12.43. The mass is displaced vertically by a small distance, y. Assuming the tension does not change, show that (a) the restoring force is $-(2F/L)y$ and (b) the system exhibits simple harmonic motion with an angular frequency $\omega = \sqrt{2F/mL}$.

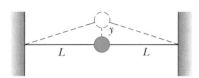

Figure P12.43

44. Consider a damped oscillator as illustrated in Figure 12.13. Assume the mass is 375 g, the spring constant is 100 N/m, and $b = 0.100$ kg/s. (a) How long does it take for the amplitude to drop to half its initial value? (b) How long does it take for the mechanical energy to drop to half its initial value? (c) Show that, in general, the fractional rate at which the amplitude decreases in a damped harmonic oscillator is one half the fractional rate at which the mechanical energy decreases.

45. A small, thin disk of radius r and mass m is attached rigidly to the face of a second thin disk of radius R and mass M as shown in Figure P12.45. The center of the small disk is located at the edge of the large disk. The large disk is mounted at its center on a frictionless axle. The assembly is rotated through a small angle θ from its equilibrium position and released. (a) Show that the speed of the center of the small disk as it passes through the equilibrium position is

$$v = 2 \left[\frac{Rg(1 - \cos\theta)}{(M/m) + (r/R)^2 + 2} \right]^{1/2}$$

(b) Show that the period of the motion is

$$T = 2\pi \left[\frac{(M + 2m)R^2 + mr^2}{2mgR} \right]^{1/2}$$

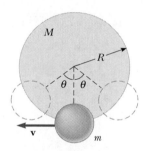

Figure P12.45

46. A mass m is connected to two springs of force constants k_1 and k_2, as in Figures P12.46a and P12.46b. In each case, the mass moves on a frictionless table and is displaced from equilibrium and released. Show that in each case the mass exhibits simple harmonic motion with periods

(a) $\quad T = 2\pi \sqrt{\dfrac{m(k_1 + k_2)}{k_1 k_2}}$

(b) $\quad T = 2\pi \sqrt{\dfrac{m}{k_1 + k_2}}$

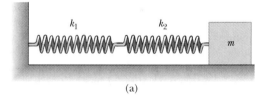

(a)

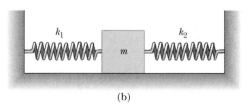

(b)

Figure P12.46

47. Imagine that a hole is drilled through the center of the Earth to its other side. It can be shown that an object of mass m at a distance r from the center of the Earth is pulled toward the center of the Earth only by the mass within the sphere of radius r (the reddish region in Fig. P12.47). Write Newton's law of gravitation for an object at the distance r from the center of the Earth, and show that the force on it is of Hooke's law form, $F = -kr$, where the effective force constant is $k = (4/3)\pi\rho Gm$, where ρ is the density of the Earth (assumed uniform) and G is the gravitational constant. Show that a sack of mail dropped into the hole will execute simple harmonic motion. When will it arrive at the other side of the Earth?

Earth

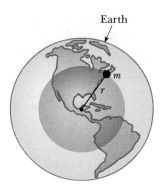

Figure P12.47

Spreadsheet Problems

S1. Use a spreadsheet to plot the position x of an object undergoing simple harmonic motion as a function of time. (a) Use $A = 2.00$ m, $\omega = 5.00$ rad/s, and $\phi = 0$. On the same graph, plot the function for $\phi = \pi/4$, $\pi/2$, and π. Explain how the different phase constants affect the position of the object. (b) Repeat part (a) with $\omega = 5.00$ rad/s. How does the change in ω affect the plots?

S2. The potential and kinetic energies of a particle of mass m oscillating on a spring are given by Equations 12.20 and 12.21. Use a spreadsheet to plot $U(t)$, $K(t)$, and their sum as functions of time on the same graph. The input variables should be k, A, m, and ϕ. What happens to the sum of $U(t)$ and $K(t)$ as the input variables are changed?

S3. Consider a simple pendulum of length L that is displaced from the vertical by an angle θ_0 and released. In this problem, you are to determine the angular displacement for any value of the initial angular displacement. Prepare a spreadsheet to integrate the equation of motion of a simple pendulum:

$$\frac{d^2\theta}{dt^2} = -\frac{g}{L}\sin\theta$$

Take the initial condition to be $\theta = \theta_0$ and $d\theta/dt = 0$ at $t = 0$. Choose values for θ_0 and L, and find the angular displacement θ as a function of time. Using the same values of θ_0, compare your results for θ with those obtained from $\theta(t) = \theta_0 \cos \omega_0 t$ where $\omega_0 = \sqrt{g/L}$. Repeat for different values of θ_0. Be sure to include large initial angular displace-

ments. Does the period change when the initial angular displacement is changed? How do the periods for large values of θ_0 compare to those for small values of θ_0? (*Note:* Using the modified Euler's method to solve this differential equation you may find that the amplitude tends to increase with time. The fourth-order Runge-Kutta method discussed in the *Spreadsheet Investigations* Supplement would be a better choice to solve the differential equation. However, if you choose dt small enough, the solution using the modified Euler's method will still be good.)

S4. In the real world friction is always present. Modify your spreadsheet for Problem S3 to include the force of air resistance: $F_{\text{air}} = -bv$. In this situation, the equation of motion is

$$\frac{d^2\theta}{dt^2} = -\frac{g}{L}\sin\theta - \frac{b}{m}\frac{d\theta}{dt}$$

where m is the mass of the pendulum and b is the drag coefficient. Integrate this equation and calculate the sum of the kinetic and potential energies as functions of time using your calculated values of θ and $d\theta/dt$. Because energy is not constant, when in the pendulum's cycle is energy dissipated the fastest?

S5. The differential equation for the position x of a driven, damped harmonic oscillator of mass m is

$$m\frac{d^2x}{dt^2} + b\frac{dx}{dt} + m\omega_0^2 x = F_0 \cos \omega t$$

where ω_0 and b are constants, F_0 is the amplitude of the driving force, and ω is the driving angular frequency. For the case of a mass on a spring, $\omega_0 = \sqrt{k/m}$, where k is the force constant. Write a computer program or spreadsheet to integrate this equation of motion. Your input variables are the damping coefficient, b, the mass m, ω (the natural frequency of the system), and F_0. Choose the values for the initial velocity and position and calculate the position x as a function of time. Vary the driving angular frequency ω. Pick some values near ω_0. Does resonance occur? (*Note:* Your program may crash or blow up near resonance. The amplitudes of the motion may have large values when $\omega = \omega_0$. Also, the Euler method usually tends to overestimate the solution. Other methods such as the fourth-order Runge-Kutta method discussed in the *Spreadsheet Investigations* Supplement give a more accurate solution.)

ANSWERS TO CONCEPTUAL PROBLEMS

1. The bouncing ball is not an example of simple harmonic motion. The ball does not follow a sinusoidal function for its position as a function of time. The daily movement of a student is also not simple harmonic motion, since the student stays at a fixed position—school—for a long period of time. If this motion were sinusoidal, the student would move more

and more slowly as she approached her desk, and, as soon as she sat down at the desk, she would start to move back toward home again!

2. When the spring with two masses is set into oscillation in space, the coil in the exact center of the spring does not move. Thus, we can imagine clamping the center coil in place without affecting the motion. If we do this, we have two separate oscillating systems, one on each side of the clamp. The half-spring on each side of the clamp has twice the spring constant of the full spring, as shown by the following argument. The force exerted by a spring is proportional to the separation of the coils as the spring is extended. Imagine that we extend a spring by a given distance and measure the distance between coils. We then cut the spring in half. If one of the half-springs is now extended by the same distance, the coils will be twice as far apart as they were for the complete spring. Thus, it takes twice as much force to stretch the half-spring, from which we conclude that the half-spring has a spring constant which is twice that of the complete spring. Thus, our clamped system of masses on two half-springs will vibrate with a frequency that is higher than f by a factor of $\sqrt{2}$.

3. No. Because $E = \frac{1}{2}kA^2$, changing the mass while keeping A constant has no effect on the total energy. The potential energy $U = \frac{1}{2}kx^2$ depends only on position, not mass. Because $K = E - U$, the kinetic energy will also depend on position and not mass.

4. If it accelerates upward, the effective "g" is greater than the free-fall acceleration, so the period decreases. If it accelerates downward, the effective "g" is less than the free-fall acceleration, so the period increases. If it moves with constant velocity, the period does not change. (If the pendulum is in free-fall, it does not oscillate.)

5. Since the car is accelerating downhill, the principle of equivalence tells us that there is an effective gravitational field component uphill. Combining this with the actual gravitational field, the net effective gravitational field is no longer vertical. Thus, the equilibrium position of the pendulum will not be vertical. In addition, the net effective gravitational field has changed in value from the actual field, so the period of the pendulum will change.

6. As the temperature increases, the length of the pendulum will increase, due to thermal expansion. With a longer length, the period of the pendulum will increase. Thus, it will take longer to execute each swing, so that each second according to the clock will take longer than an actual second. Thus, the clock will run *slow*.

7. The first student is correct. While changing the mass of a simple pendulum does not change the frequency, the Foucault pendulum is a *physical* pendulum—the bob has a size. Assuming that the cord is very light, when the gum is placed on top of the bob, the center of mass of the physical pendulum is shifted very slightly upward. This shortens the effective length of the pendulum, which alters the frequency.

8. The diving board is acting as a physical pendulum. With your mass at the end of the board, you and the board have the maximum possible moment of inertia. When you move to the center of the board, the moment of inertia of the system decreases, and the natural frequency of oscillation increases above f.

9. If the period of the alternating release of air matched the swinging period of the parachutist, there could be a resonance response. The swinging of the parachutist could become so energetic that the parachute could become fouled and its braking action could disappear.

13

Wave Motion

Most of us experienced waves as children, when we dropped pebbles into a pond. The disturbance created by a pebble manifests itself as ripple waves that move outward until they finally reach the shore. If you were to carefully examine the motion of a leaf floating near the point where the pebble entered the water, you would see that the leaf moves up and down and sideways about its original position, but does not undergo any net displacement away from or toward the source of the disturbance. The water wave (the disturbance) moves from one place to another, *yet the water is not carried with it*.

The world is full of other kinds of waves, including sound waves, waves on strings, earthquake waves, radio waves, and x-rays. Most waves can be placed in one of two categories. **Mechanical waves** are waves that disturb and propagate through a medium; sound waves, for which air is the medium, and earthquake waves, for which the Earth's crust is the medium, are two examples of mechanical waves. **Electromagnetic waves** are a special class of waves that do not require a medium in order to propagate; light waves and radio waves are two familiar examples. In this chapter we shall confine our attention to the study of mechanical waves, deferring our study of electromagnetic waves to Chapter 24.

The wave concept is abstract. When we observe a water wave, what we see is a

Large waves such as this travel ▶ great distances over the surface of the ocean, yet the water does not flow with the wave. The crests and troughs of the wave often form repetitive patterns. *(Superstock)*

rearrangement of the water's surface. Without the water, there would be no wave. In the case of this or any other mechanical wave, what we interpret as a wave corresponds to the disturbance of a medium. Therefore, we can consider a mechanical wave to be **the motion of a disturbance.** This motion is not to be confused with the motion of the particles that make up the medium. In general, we describe mechanical wave motion by specifying the positions of all particles of the disturbed medium as a function of time.

13.1 • THREE WAVE CHARACTERISTICS

All mechanical waves require (1) some source of disturbance, (2) a medium that can be disturbed, and (3) some physical mechanism through which particles of the medium can influence each other. All waves carry energy and momentum, but the amount of energy transmitted through a medium and the mechanism responsible for the energy transport differ from case to case. For instance, the power of ocean waves during a storm is much greater than the power of sound waves generated by a musical instrument.

Three physical characteristics are important in describing waves: wavelength, frequency, and wave speed. **One wavelength is the minimum distance between any two points on a wave that behave identically**—for example, adjacent crests or adjacent troughs. Figure 13.1a shows displacement versus position for a sinusoidal wave at a specific time. The symbol λ (Greek lambda) is used to denote wavelength.

Most waves are periodic. **The frequency of such waves is the rate at which the disturbance repeats itself.** The period of the wave is the minimum time it takes the disturbance to repeat itself and is equal to the inverse of the frequency. Figure 13.1b shows displacement versus time for a sinusoidal wave at a fixed position, where the period is the time between identical displacements of a point on the wave.

Waves travel, or *propagate,* with a specific speed, which depends on the properties of the medium being disturbed. For instance, sound waves travel through air at 20°C with a speed of about 343 m/s, whereas the speed of sound in most solids is higher than 343 m/s.

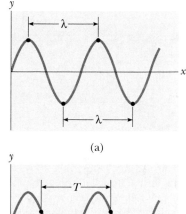

(a)

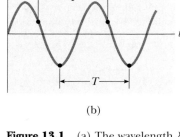

(b)

Figure 13.1   (a) The wavelength λ of a wave is the distance between adjacent crests or adjacent troughs. (b) The period T of a wave is the time it takes the wave to travel one wavelength.

Thinking Physics 1

Why is it important to be quiet in avalanche country? In the movie *On Her Majesty's Secret Service* (United Artists, 1969), the bad guys try to stop James Bond, who is escaping on skis, by firing a gun and causing an avalanche. Why did this happen?

Reasoning The essence of a wave is the propagation of a *disturbance* through a medium. An impulsive sound, like the gunshot in the James Bond movie, can cause an acoustical disturbance that propagates through the air and can impact a ledge of snow that is just ready to break free to begin an avalanche. Such a disastrous event occurred in 1916 during World War I, when Austrian soldiers in the Alps were smothered by an avalanche caused by cannon fire.

13.2 • TYPES OF WAVES

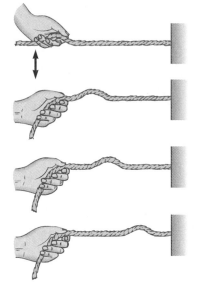

Figure 13.2 A wave pulse traveling down a stretched rope. The shape of the pulse is approximately unchanged as it travels along the rope.

One way to demonstrate wave motion is to flip the free end of a long rope that is under tension and has its opposite end fixed, as in Figure 13.2. In this manner, a single wave bump (or pulse) is formed and travels (to the right in Fig. 13.2) with a definite speed. This type of disturbance is called a **traveling wave,** and the rope is the medium through which it travels. Figure 13.2 represents four consecutive "snapshots" of the traveling wave. The shape of the wave pulse changes very little as it travels along the rope.[1]

Note that **as the wave pulse travels, each rope segment that is disturbed moves in a direction perpendicular to the wave motion.** Figure 13.3 illustrates this point for a particular segment, labeled *P*. Note that there is no motion of any part of the rope in the direction of the wave. A traveling wave such as this, in which the particles of the disturbed medium move perpendicularly to the wave velocity, is called a **transverse wave.**

In another class of waves, called **longitudinal waves,** the particles of the medium undergo displacements *parallel* to the direction of wave motion. Sound waves in air, for instance, are longitudinal. Their disturbance corresponds to a series of high- and low-pressure regions that may travel through air or through any material medium with a certain speed. A longitudinal pulse can be easily produced in a stretched spring, as in Figure 13.4. The free end is pumped back and forth along the length of the spring. This action produces compressed and stretched regions of the coil that travel along the spring, parallel to the wave motion.[2]

Some waves are neither transverse nor longitudinal but a combination of the two. Surface water waves are a good example. Figure 13.5 shows the motion of water particles at the surface as a wave moves to the right. Each particle moves in a circular path, and hence the disturbance has both transverse and longitudinal components. As the wave passes, water particles at the crests are moving in the direction of the wave, and particles at the troughs move in the opposite direction. Hence, no *net* displacement of water particles takes place. A cork bobbing on a pond surface as a wave passes exhibits this circular motion.

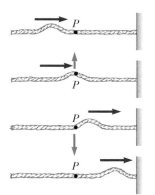

Figure 13.3 A pulse traveling on a stretched rope is a transverse wave. That is, any element *P* on the rope moves (blue arrows) in a direction perpendicular to the wave motion (red arrows).

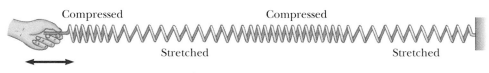

Longitudinal wave

Figure 13.4 A longitudinal pulse along a stretched spring. The displacement of the coils is in the direction of the wave motion. For the starting motion described in the text, the compressed region is followed by a stretched region.

---

[1]Strictly speaking, the pulse changes its shape and gradually spreads out during the motion. This effect is called *dispersion* and is common to many mechanical waves.

[2]In the case of longitudinal waves in a gas, each compressed area is a region of higher-than-average pressure and density, and each stretched region is a region of lower-than-average pressure and density.

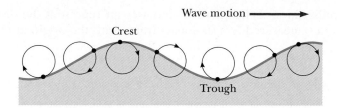

Figure 13.5 Wave motion on the surface of water. The molecules at the water's surface move in nearly circular paths. Each molecule is displaced horizontally and vertically from its equilibrium position, represented by circles.

CONCEPTUAL PROBLEM 1

In a long line of people waiting to buy tickets at a movie theater, when the first person leaves, a pulse of motion occurs, as people step forward to fill in the gap. The gap moves through the line of people. What determines the speed of this pulse? Is it transverse or longitudinal? How about "the wave" at a baseball game, where people in the stands stand up and shout as the wave arrives at their location, and this pulse moves around the stadium—what determines the speed of this pulse? Is it transverse or longitudinal?

13.3 • ONE-DIMENSIONAL TRANSVERSE TRAVELING WAVES

So far we have provided only verbal and graphical descriptions of a traveling wave. Let us now develop a mathematical description of a one-dimensional traveling wave. Consider a wave pulse traveling to the right with constant speed, v, along a long, stretched string, as in Figure 13.6. The pulse moves along the x axis (the axis of the string), and the transverse (up-and-down) displacement of the string is measured with the coordinate y.

Figure 13.6a represents the shape and position of the pulse at time $t = 0$. At this time, the shape of the pulse, whatever it may be, can be represented as $y = f(x)$. That is, y is some definite function of x. The *maximum displacement* of the string, y_m, is called the **amplitude** of the wave. Because the speed of the wave pulse is v, the pulse travels to the right a distance of vt in the time t (Fig. 13.6b). If the shape

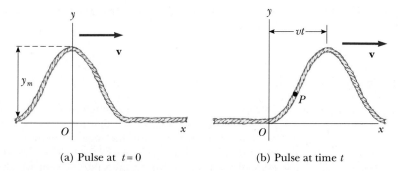

(a) Pulse at $t = 0$ (b) Pulse at time t

Figure 13.6 A one-dimensional wave pulse traveling to the right with a speed v. (a) At $t = 0$, the shape of the pulse is given by $y = f(x)$. (b) At some later time t, the shape remains unchanged and the vertical displacement of any point P of the medium is given by $y = f(x - vt)$.

of the wave pulse doesn't change with time, we can represent the displacement y for all later times, measured in a stationary frame with the origin at O, as

Pulse traveling to the right •

$$y = f(x - vt) \qquad\qquad [13.1]$$

In the same way, if the wave pulse travels to the left, the displacement of the string is

Pulse traveling to the left •

$$y = f(x + vt) \qquad\qquad [13.2]$$

The displacement y, sometimes called the **wave function,** depends on the two variables x and t. For this reason, it is often written $y(x, t)$, which is read "y as a function of x and t." It is important to understand the meaning of y.

Consider a point, P, on the string, identified by a particular value of its coordinates. As the wave pulse passes through P, the y coordinate of this point increases, reaches a maximum, and then decreases to zero. **The wave function $y(x, t)$ represents the y coordinate of any point P at any time t.** Furthermore, if t is fixed, then the wave function y as a function of x, sometimes called the *waveform,* defines a curve representing the actual shape of the pulse at that time. This is equivalent to a snapshot of the pulse at time t.

For a pulse that moves without changing its shape, the velocity of the pulse is the same as the velocity of any point along the pulse, such as the crest. To find the velocity of the pulse, we can calculate how far the crest moves horizontally in a short time and then divide that distance by the time interval. In order to follow the motion of the crest, some particular value, say x_0, must be substituted for $x - vt$ in Equation 13.1. (The value x_0 is called the *argument* of the function y.) Regardless of how x and t change individually, we must require that $x - vt = x_0$ in order to stay with the crest. This equation, therefore, represents the motion of the crest. At $t = 0$, the crest is at $x = x_0$; after an interval of dt, the crest is at $x = x_0 + v\,dt$. Therefore, in the time dt the crest has moved a distance of $dx = (x_0 + v\,dt) - x_0 = v\,dt$. Clearly, the wave speed, often called the **phase speed,** is

Phase speed •

$$v = \frac{dx}{dt} \qquad\qquad [13.3]$$

The *wave velocity,* or *phase velocity,* must not be confused with the *transverse velocity* of a particle in the medium (which is always perpendicular to the wave velocity). The phase velocity is the same at any point along the pulse.

The following example illustrates how a specific wave function is used to describe the motion of a traveling wave pulse.

Example 13.1   A Pulse Moving to the Right

A wave pulse moving to the right along the x axis is represented by the wave function

$$y(x, t) = \frac{2}{(x - 3.0t)^2 + 1}$$

where x and y are measured in centimeters and t is in seconds. Let us plot the waveform at $t = 0$, $t = 1.0$ s, and $t = 2.0$ s.

Solution First, note that this function is of the form $y = f(x - vt)$. By inspection, we see that the speed of the wave is $v = 3.0$ cm/s. Furthermore, the wave amplitude (the maximum value of y) is given by $A = 2.0$ cm. The location of the peak of the pulse occurs at the value of x for which the denominator is a minimum, that is, where $(x - 3.0t) = 0$. At times $t = 0$, $t = 1.0$ s, and $t = 2.0$ s, the wave function expressions are

$$y(x, 0) = \frac{2}{x^2 + 1} \qquad \text{at } t = 0$$

$$y(x, 1.0) = \frac{2}{(x - 3.0)^2 + 1} \qquad \text{at } t = 1.0 \text{ s}$$

$$y(x, 2.0) = \frac{2}{(x - 6.0)^2 + 1} \qquad \text{at } t = 2.0 \text{ s}$$

We can now use these expressions to plot the wave function versus x at these times. For example, let us evaluate $y(x, 0)$ at $x = 0.50$ cm:

$$y(0.50, 0) = \frac{2}{(0.50)^2 + 1} = 1.6 \text{ cm}$$

Likewise, $y(1.0, 0) = 1.0$ cm, $y(2.0, 0) = 0.40$ cm, and so on. A continuation of this procedure for other values of x yields the waveform shown in Figure 13.7a. In a similar manner, one obtains the graphs of $y(x, 1.0)$ and $y(x, 2.0)$, shown in Figures 13.7b and 13.7c, respectively. These snapshots show that the wave pulse moves to the right without changing its shape and has a constant speed of 3.0 cm/s.

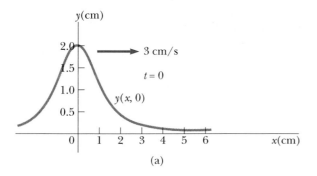

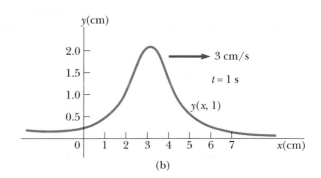

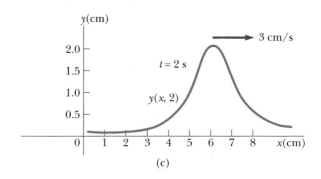

Figure 13.7 (Example 13.1) Graphs of the function $y(x, t) = 2/[(x - 3t)^2 + 1]$. (a) $t = 0$, (b) $t = 1$ s, and (c) $t = 2$ s.

13.4 · SINUSOIDAL TRAVELING WAVES

In this section we introduce an important periodic waveform known as a **sinusoidal wave** (Fig. 13.8). The red curve represents a snapshot of a sinusoidal traveling wave at $t = 0$, and the blue curve represents a snapshot of the wave at some later t. At $t = 0$, the displacement of the curve can be written

$$y = A \sin \left(\frac{2\pi}{\lambda} x \right) \qquad [13.4]$$

where the amplitude, A, as usual, represents the maximum value of the displacement, and λ is the wavelength as defined in Figure 13.1a. Thus, we see that the displacement repeats itself when x is increased by an integral multiple of λ. If the wave moves to the right with a phase speed of v, the wave function at some later t is

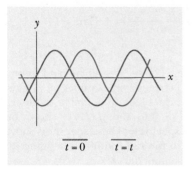

Figure 13.8 A one-dimensional sinusoidal wave traveling to the right with a speed v. The red curve represents a snapshot of the wave at $t = 0$, and the blue curve represents a snapshot at some later time t.

$$y = A \sin \left[\frac{2\pi}{\lambda} (x - vt) \right] \qquad \text{[13.5]}$$

That is, the sinusoidal wave moves to the right a distance of vt in the time t, as in Figure 13.8. Note that the wave function has the form $f(x - vt)$ and represents a wave traveling to the right. If the wave were traveling to the left, the quantity $x - vt$ would be replaced by $x + vt$, just as in the case of the traveling pulse described by Equations 13.1 and 13.2.

Because the period T is the time it takes the wave to travel a distance of one wavelength, the phase speed, wavelength, and period are related by

$$v = \frac{\lambda}{T} \qquad \text{[13.6]}$$

or

$$\lambda = vT \qquad \text{[13.7]}$$

Substituting Equation 13.6 into Equation 13.5, we find that

$$y = A \sin \left[2\pi \left(\frac{x}{\lambda} - \frac{t}{T} \right) \right] \qquad \text{[13.8]}$$

This form of the wave function clearly shows the periodic nature of y. That is, at any given time t (a snapshot of the wave), y has the same value at the positions x, $x + \lambda$, $x + 2\lambda$, and so on. Furthermore, at any given position x, the value of y at times t, $t + T$, $t + 2T$, and so on, is the same.

We can express the sinusoidal wave function in a convenient form by defining two other quantities: **angular wave number,** k, and **angular frequency,** ω:

Angular wave number •

$$k \equiv \frac{2\pi}{\lambda} \qquad \text{[13.9]}$$

Angular frequency •

$$\omega \equiv \frac{2\pi}{T} = 2\pi f \qquad \text{[13.10]}$$

Note that in Equation 13.10, we used the definition of frequency, $f = 1/T$. Using these definitions, we see that Equation 13.8 can be written in the more compact form

$$y = A \sin(kx - \omega t) \qquad \text{[13.11]}$$

We shall use this form most frequently.

Using Equations 13.9 and 13.10, we can express the phase speed, v, in the alternative forms

$$v = \frac{\omega}{k} \qquad \text{[13.12]}$$

Speed of a traveling •
sinusoidal wave

$$v = f\lambda \qquad \text{[13.13]}$$

The wave function given by Equation 13.11 assumes that the displacement, y, is zero at $x = 0$ and $t = 0$. This need not be the case. If the transverse displacement is not zero at $x = 0$ and $t = 0$, we generally express the wave function in the form

$$y = A \sin(kx - \omega t - \phi) \qquad \text{[13.14]}$$

• *Wave function for a traveling sinusoidal wave*

where ϕ is called the **phase constant** and can be determined from the initial conditions.

Thinking Physics 2

The quantity that we call the angular wave number k is sometimes called *spatial frequency*. Why?

Reasoning The angular frequency ω is 2π divided by the *time* interval for one cycle of oscillation for the wave, the period, T. The angular wave number k is 2π divided by the *space* interval for one cycle of oscillation for the wave, the wavelength, λ (the spatial interval of one cycle of the wave). Due to the similarity in these parameters, it is reasonable to consider ω the *temporal* frequency and k the *spatial* frequency.

CONCEPTUAL PROBLEM 2

Sound waves from a musical performance are encoded onto radio waves for transmission from the radio studio to your home receiver. A radio wave with a wavelength of 3 meters carries a sound wave that has a wavelength of 3 meters in air. Which has the higher frequency, the radio wave or the sound wave?

Example 13.2   A Traveling Sinusoidal Wave

A sinusoidal wave traveling in the positive x direction has an amplitude of 15.0 cm, a wavelength of 40.0 cm, and a frequency of 8.00 Hz. The vertical displacement of the medium at $t = 0$ and $x = 0$ is also 15.0 cm, as shown in Figure 13.9. (a) Find the angular wave number, period, angular frequency, and speed of the wave.

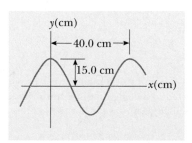

Figure 13.9   (Example 13.2) A sinusoidal wave of wavelength $\lambda = 40.0$ cm and amplitude $A = 15.0$ cm. The wave function can be written in the form $y = A\cos(kx - \omega t)$.

Solution   Using Equations 13.9, 13.10, and 13.13, we find the following:

$$k = \frac{2\pi}{\lambda} = \frac{2\pi \text{ rad}}{40.0 \text{ cm}} = \boxed{0.157 \text{ rad/cm}}$$

$$T = \frac{1}{f} = \frac{1}{8.00 \text{ s}^{-1}} = \boxed{0.125 \text{ s}}$$

$$\omega = 2\pi f = 2\pi(8.00 \text{ s}^{-1}) = \boxed{50.3 \text{ rad/s}}$$

$$v = f\lambda = (8.00 \text{ s}^{-1})(40.0 \text{ cm}) = \boxed{320 \text{ cm/s}}$$

(b) Determine the phase constant ϕ, and write a general expression for the wave function.

Solution   Because $A = 15.0$ cm and because it is given that $y = 15.0$ cm at $x = 0$ and $t = 0$, substitution into Equation 13.14 gives

$$15 = 15 \sin(-\phi) \qquad \text{or} \qquad \sin(-\phi) = 1$$

Because $\sin(-\phi) = -\sin \phi$, we see that $\phi = -\pi/2$ rad (or $-90°$). Hence, the wave function is of the form

$$y = A \sin\left(kx - \omega t + \frac{\pi}{2} \right) = A \cos(kx - \omega t)$$

That the wave function must have this form can be seen by inspection, noting that the cosine argument is displaced by $90°$ from the sine function. Substituting the values for A, k, and ω into this expression gives

$$y = (15.0 \text{ cm})\cos(0.157x - 50.3t)$$

Sinusoidal Waves on Strings

One method of producing a traveling sinusoidal wave on a very long string is shown in Figure 13.10. One end of the string is connected to a blade that is set in vibration. As the blade oscillates vertically with simple harmonic motion, a traveling wave moving to the right is set up on the string. Figure 13.10 represents snapshots of the wave at intervals of one quarter of a period. Note that **each particle of the string, such as P, oscillates vertically in the y direction with simple harmonic motion.** This must be the case because each particle follows the simple harmonic motion of the blade. Therefore, every segment of the string can be treated as a simple harmonic oscillator vibrating with a frequency equal to the frequency of vibration of the blade that drives the string.[3] Note that although each segment oscillates in the y direction, the wave (or disturbance) travels in the x direction with speed v.

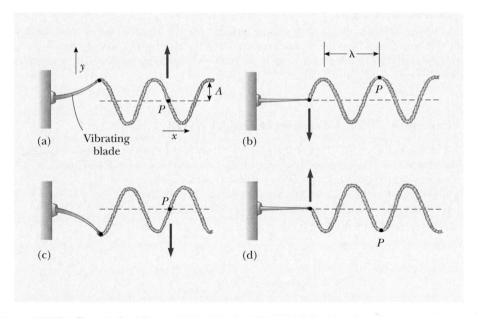

Figure 13.10 One method for producing a train of sinusoidal wave pulses on a continuous string. The left end of the string is connected to a blade that is set into vibration. Every segment of the string, such as the point *P*, oscillates with simple harmonic motion in the vertical direction.

---

[3]In this arrangement, we are assuming that the segment always oscillates in a vertical line. The tension in the string would vary if the segment were allowed to move sideways. Such a motion would make the analysis very complex.

Of course, this is the definition of a transverse wave. In this case, the energy carried by the traveling wave is supplied by the vibrating blade.

If the waveform at $t = 0$ is as described in Figure 13.10b, then the wave function can be written

$$y = A \sin(kx - \omega t)$$

We can use this expression to describe the motion of any point on the string. The point P (or any other point on the string) moves vertically, and so its x coordinate *remains constant*. Therefore, the **transverse speed,** v_y, of the point P (not to be confused with the wave speed, v) and its **transverse acceleration,** a_y, are

$$v_y = \frac{dy}{dt}\bigg]_{x=\text{constant}} = \frac{\partial y}{\partial t} = -\omega A \cos(kx - \omega t) \qquad \textbf{[13.15]}$$

$$a_y = \frac{dv_y}{dt}\bigg]_{x=\text{constant}} = \frac{\partial v_y}{\partial t} = -\omega^2 A \sin(kx - \omega t) \qquad \textbf{[13.16]}$$

The maximum values of these quantities are simply the absolute values of the co-efficients of the cosine and sine functions:

$$(v_y)_{\text{max}} = \omega A \qquad \textbf{[13.17]}$$

$$(a_y)_{\text{max}} = \omega^2 A \qquad \textbf{[13.18]}$$

You should recognize that the transverse speed and transverse acceleration of any point on the string do not reach their maximum values simultaneously. In fact, the transverse speed reaches its maximum value (ωA) when the displacement $y = 0$, whereas the transverse acceleration reaches its maximum value ($\omega^2 A$) when $y = -A$. Finally, Equations 13.17 and 13.18 are identical to the corresponding equations for simple harmonic motion.

Example 13.3   A Sinusoidally Driven String

The string shown in Figure 13.10 is driven at a frequency of 5.00 Hz. The amplitude of the motion is 12.0 cm, and the wave speed is 20.0 m/s. Furthermore, the wave is such that $y = 0$ at $x = 0$ and $t = 0$. Determine the angular frequency and wave number for this wave, and write an expression for the wave function.

Solution   Using Equations 13.10 and 13.12 gives

$$\omega = \frac{2\pi}{T} = 2\pi f = 2\pi(5.00 \text{ Hz}) = \boxed{31.4 \text{ rad/s}}$$

$$k = \frac{\omega}{v} = \frac{31.4 \text{ rad/s}}{20.0 \text{ m/s}} = \boxed{1.57 \text{ rad/m}}$$

Because $A = 12.0$ cm $= 0.120$ m, we have

$$y = A \sin(kx - \omega t) = (0.120 \text{ m})\sin(1.57x - 31.4t)$$

EXERCISE 1   Calculate the maximum values for the transverse speed and transverse acceleration of any point on the string.   Answer   3.77 m/s; 118 m/s²

EXERCISE 2   When a particular wire is vibrating with a frequency of 4.00 Hz, a transverse wave of wavelength 60.0 cm is produced. Determine the speed of wave pulses along the wire. Answer   2.40 m/s

EXERCISE 3   For a certain transverse wave, the distance between two successive maxima is 1.2 m and eight maxima pass a given point along the direction of travel every 12 s. Calculate the wave speed.   Answer   0.80 m/s

13.5 • SUPERPOSITION AND INTERFERENCE OF WAVES

Many interesting wave phenomena in nature cannot be described by a single moving pulse. Instead, one must analyze complex waveforms in terms of a combination of many traveling waves. To analyze such wave combinations, one can make use of the **superposition principle,** which states that

> if two or more traveling waves are moving through a medium, the resultant wave function at any point is the sum of the wave functions of the individual waves.

Linear waves obey the • superposition principle.

This rather striking property is exhibited by many waves in nature. Waves that obey this principle are called *linear waves,* and they are generally characterized by small amplitudes. Waves that violate the superposition principle are called *nonlinear waves* and are often characterized by large amplitudes. In this book, we shall deal only with linear waves.

One consequence of the superposition principle is that **two traveling waves can pass through each other without being destroyed or even altered.** For instance, when two pebbles are thrown into a pond, the expanding circular surface waves do not destroy each other. In fact, the ripples pass through each other. The complex pattern that is observed can be viewed as two independent sets of expanding circles. Likewise, when sound waves from two sources move through air, they

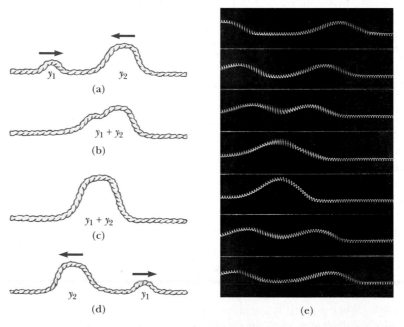

Figure 13.11   *(Left)* Two wave pulses traveling on a stretched string in opposite directions pass through each other. When the pulses overlap, as in (b) and (c), the net displacement of the string equals the sum of the displacements produced by each pulse. Because each pulse produces positive displacements of the string, we refer to their superposition as *constructive interference.*   *(Right)* Photograph of superposition of two equal and symmetric pulses traveling in opposite directions on a stretched spring.   *(Photo, Education Development Center, Newton, Mass.)*

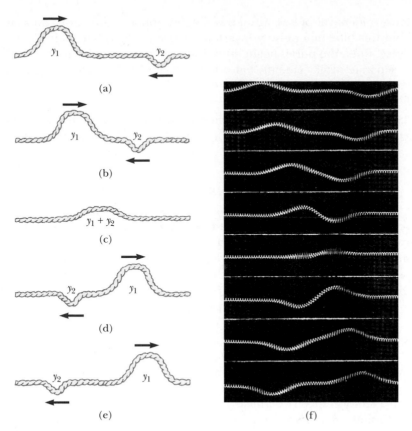

(a)

(b)

(c)

(d)

(e)

(f)

Figure 13.12 *(Left)* Two wave pulses traveling in opposite directions with displacements that are inverted relative to each other. When the two overlap as in (c), their displacements subtract from each other. *(Right)* Photograph of superposition of two symmetric pulses traveling in opposite directions, where one is inverted relative to the other. *(Photo, Education Development Center, Newton, Mass.)*

also can pass through each other. The sound one hears at a given point is the resultant of both disturbances.

A simple pictorial representation of the superposition principle is obtained by considering two pulses traveling in opposite directions on a stretched string, as in Figure 13.11. The wave function for the pulse moving to the right is y_1, and the wave function for the pulse moving to the left is y_2. The pulses have the same speed but different shapes. Each pulse is assumed to be symmetric (although this is not a necessary condition), and in both cases the vertical displacements of the string are taken to be positive. When the waves begin to overlap (Fig. 13.11b), the resulting complex waveform is given by $y_1 + y_2$. When the crests of the pulses exactly coincide (Fig. 13.11c), the resulting waveform, $y_1 + y_2$, is symmetric. The two pulses finally separate and continue moving in their original directions (Fig. 13.11d). Note that the final waveforms remain unchanged, as if the two pulses had never met! The combination of separate waves in the same region of space to produce a resultant wave is called **interference.**

For the two pulses shown in Figure 13.11, the vertical displacements caused by the pulses are in the same direction, and so the resultant waveform (when the pulses overlap) exhibits a displacement greater than those of the individual pulses. Now

Interference patterns produced by outward spreading waves from several drops of water falling into a pond. *(Martin Dohrn/SPL/Photo Researchers)*

consider two identical pulses, again traveling in opposite directions on a stretched string, but this time one pulse is inverted relative to the other, as in Figure 13.12. In this case, when the pulses begin to overlap, the resultant waveform is the sum of the two separate displacements, but one displacement is negative. Again, the two pulses pass through each other.

EXERCISE 4 Two waves are traveling in the same direction along a stretched string. Each has an amplitude of 4.0 cm and they are 90° out of phase. Find the amplitude of the resultant wave.   Answer   5.7 cm

13.6 • THE SPEED OF TRANSVERSE WAVES ON STRINGS

For linear mechanical waves, **the speed depends only on the properties of the medium through which the disturbance travels.** In this section we shall focus our attention on determining the speed of a transverse pulse traveling on a stretched string. We shall show that if the tension in the string is F and its mass per unit length is μ, then the wave speed, v, is

Speed of a wave on a •
stretched string

$$v = \sqrt{\frac{F}{\mu}}$$ [13.19]

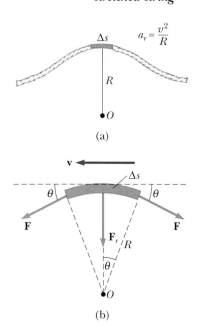

Now let us use a mechanical analysis to derive the preceding expression for the speed of a pulse traveling on a stretched string. Consider a pulse moving to the right with a uniform speed of v, measured relative to a stationary frame of reference. Instead of using a stationary frame, however, it is more convenient to choose as our reference frame one that moves along with the pulse at the same speed, so that the pulse appears to be at rest in the frame, as in Figure 13.13a. This is permitted because Newton's laws are valid in either a stationary frame or one that moves with constant velocity. A small segment of the string, of length Δs, forms the approximate arc of a circle of radius R, as shown in Figure 13.13a and magnified in Figure 13.13b. In the pulse's frame of reference, the segment moves to the left with the speed v. This segment has a centripetal acceleration of v^2/R, which is supplied by the tension, F, in the string. The force **F** acts on each side of the segment, tangent to the arc, as in Figure 13.13b. The horizontal components of **F** cancel, and each vertical component $F \sin \theta$ acts radially inward toward the center of the arc. Hence, the total radial force is $2F \sin \theta$. Because the segment is small, θ is small, and we can use the small-angle approximation $\sin \theta \approx \theta$. Therefore, the total radial force can be expressed as

$$F_r = 2F \sin \theta \approx 2F\theta$$

Figure 13.13 (a) To obtain the speed v of a wave on a stretched string, it is convenient to describe the motion of a small segment of the string in a moving frame of reference. (b) The net force on a small segment of length Δs is in the radial direction. The horizontal components of the tension force cancel.

The segment has the mass $m = \mu \, \Delta s$, where μ is the mass per unit length of the string. Because the segment forms part of a circle and subtends an angle of 2θ at the center, $\Delta s = R(2\theta)$, and hence

$$m = \mu \, \Delta s = 2\mu R\theta$$

If we apply Newton's second law to this segment, the radial component of motion gives

$$F_r = \frac{mv^2}{R} \qquad \text{or} \qquad 2F\theta = \frac{2\mu R\theta v^2}{R}$$

where F_r is the force that supplies the centripetal acceleration of the segment and maintains the curvature of this segment.

Solving for v gives

$$v = \sqrt{\frac{F}{\mu}}$$

Notice that this derivation is based on the assumption that the pulse height is small relative to the length of the string. Using this assumption, we were able to use the approximation $\sin \theta \approx \theta$. Furthermore, the model assumes that the tension, F, is not affected by the presence of the pulse, so that F is the same at all points on the string. Finally, this proof does *not* assume any particular shape for the pulse. Therefore, we conclude that a pulse of *any shape* will travel on the string with speed $v = \sqrt{F/\mu}$, without changing its shape.

Thinking Physics 3

A secret agent is trapped in a building on top of the elevator car at a lower floor. She attempts to signal a fellow agent on the roof by tapping a message on the elevator cable. As the pulses from this tapping move up the cable to the accomplice, does the speed with which they move stay the same, increase, or decrease? If the pulses are sent one second apart, are they received one second apart by her partner?

Reasoning The elevator cable can be modeled as a heavy string. The speed of waves on the cable is a function of the tension in the cable. As the waves move higher on the cable, they encounter increased tension, because each higher point on the cable must support the weight of all the cable below it (and the elevator). Thus, the speed of the pulses increases as they move higher on the cable. The frequency of the pulses will not be affected, because each pulse takes the same total time to reach the top—they will still arrive at the top of the cable at intervals of one second.

CONCEPTUAL PROBLEM 3

In mechanics, massless strings are often assumed. Why is this not a good assumption when discussing waves on strings?

Example 13.4   The Speed of a Pulse on a Cord

A uniform cord has a mass of 0.300 kg and a total length of 6.00 m. Tension is maintained in the cord by suspending a 2.00-kg mass from one end (Fig. 13.14). Find the speed of a pulse on this cord. Assume the tension is not affected by the mass of the cord.

Solution   The tension F in the cord is equal to the weight of the suspended 2.00-kg mass:

$$F = mg = (2.00 \text{ kg})(9.80 \text{ m/s}^2) = 19.6 \text{ N}$$

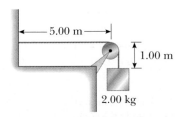

Figure 13.14   (Example 13.4) The tension F in the cord is maintained by the suspended mass. The wave speed is given by the expression $v = \sqrt{F/\mu}$.

(This calculation of the tension neglects the small mass of the cord. Strictly speaking, the cord can never be exactly horizontal, and therefore the tension is not uniform.)

The mass per unit length μ is

$$\mu = \frac{m}{\ell} = \frac{0.300 \text{ kg}}{6.00 \text{ m}} = 0.0500 \text{ kg/m}$$

Therefore, the wave speed is

$$v = \sqrt{\frac{F}{\mu}} = \sqrt{\frac{19.6 \text{ N}}{0.0500 \text{ kg/m}}} = 19.8 \text{ m/s}$$

EXERCISE 5   Show that Equation 13.19 is dimensionally correct using the known dimensions of F and μ.

EXERCISE 6   Transverse waves travel with a speed of 20.0 m/s in a string under a tension of 6.00 N. What tension is required for a wave speed of 30.0 m/s in the same string?
Answer   13.5 N

13.7 • REFLECTION AND TRANSMISSION OF WAVES

Whenever a traveling wave pulse reaches a boundary, part or all of the pulse is *reflected*. Any part not reflected is said to be *transmitted* through the boundary. Consider a pulse traveling on a string that is fixed at one end (Fig. 13.15). When the pulse reaches the fixed boundary, it is reflected. Because the support attaching the string to the wall is rigid, none of the pulse is transmitted through the wall.

Note that the reflected pulse has exactly the same amplitude as the incoming pulse but is inverted. The inversion can be explained as follows. When the pulse meets the end of the string that is fixed at the support, the string produces an upward force on the support. By Newton's third law, the support must exert an equal and opposite reaction force on the string. This downward force causes the pulse to invert on reflection.

Now consider another situation in which there is total reflection and zero transmission. In this case the pulse arrives at the end of a string that is free to move vertically, as in Figure 13.16. The tension at the free end is maintained by tying the string to a ring of negligible mass that is free to slide vertically on a frictionless post. Again, the pulse is reflected, but this time it is not inverted. As the pulse reaches the post, it exerts a force on the free end, causing the ring to accelerate upward. In the process, the ring has upward momentum as it reaches the top of its motion and is then returned to its original position by the downward component of the tension force. This produces a reflected pulse that is not inverted, the amplitude of which is the same as that of the incoming pulse.

Finally, we may have a situation in which the boundary is intermediate between these two extreme cases; that is, it is neither completely rigid nor completely free. In this case, part of the wave is transmitted and part is reflected. For instance, suppose a light string is attached to a heavier string, as in Figure 13.17. When a pulse traveling on the light string reaches the boundary between the two strings, part of the pulse is reflected and inverted and part is transmitted to the heavier string. As one would expect, both the reflected pulse and the transmitted one have a smaller amplitude than the incident pulse. The inversion in the reflected wave is similar to the behavior of a pulse meeting a rigid boundary.

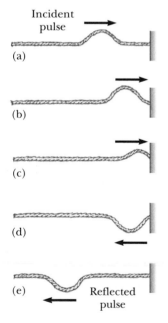

Incident pulse

(a)

(b)

(c)

(d)

(e)   Reflected pulse

Figure 13.15   The reflection of a traveling wave pulse at the fixed end of a stretched string. The reflected pulse is inverted, but its shape remains the same.

When a pulse traveling on a heavy string strikes the boundary of a lighter string, as in Figure 13.18, again part is reflected and part transmitted. This time, however, the reflected pulse is not inverted. In both the lighter-to-heavier case and the heavier-to-lighter case, the relative heights of the reflected and transmitted pulses depend on the relative densities of the two strings. In an extreme case, if the second string were infinitely more dense, it would behave like a rigid wall.

In the preceding section, we found that the speed of a wave on a string increases as the mass per unit length of the string decreases. In other words, a pulse travels more slowly on a heavy string than on a light string if both are under the same tension. The following general rules apply to reflected waves:

When a wave pulse travels from medium A to medium B and $v_A > v_B$ (that is, when B is more massive than A), the reflected part of the pulse is inverted on reflection. When a wave pulse travels from medium A to medium B and $v_A < v_B$ (A is more massive than B), the reflected pulse is not inverted.

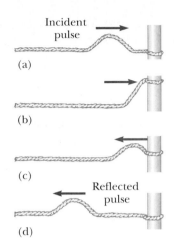

Figure 13.16 The reflection of a traveling wave pulse at the free end of a stretched string. In this case, the reflected pulse is not inverted.

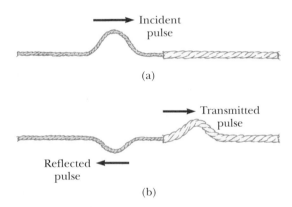

Figure 13.17 (a) A pulse traveling to the right on a light string attached to a heavier string. (b) Part of the incident pulse is reflected (and inverted), and part is transmitted to the heavier string. (Note that the change in pulse width is not shown.)

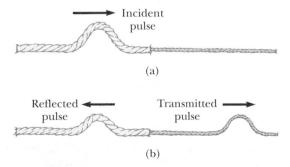

Figure 13.18 (a) A pulse traveling to the right on a heavy string attached to a lighter string. (b) The incident pulse is partially reflected and partially transmitted. In this case, the reflected pulse is not inverted. (Note that the change in pulse width is not shown.)

13.8 · ENERGY TRANSMITTED BY SINUSOIDAL WAVES ON STRINGS

As waves propagate through a medium, they transport energy. This is easily demonstrated by hanging a mass on a stretched string and then sending a pulse down the string, as in Figure 13.19. When the pulse meets the suspended mass, the mass is momentarily displaced, as in Figure 13.19b. In the process, energy is transferred to the mass, because work must be done in moving it upward.

This section examines the rate at which energy is transported along a string. We shall assume that this one-dimensional wave is sinusoidal when we calculate the power transferred. Later we shall extend the same ideas to three-dimensional waves.

Consider a sinusoidal wave traveling on a string (Fig. 13.20). The source of the energy is some external agent at the left end of the string, which does work in producing the oscillations. Let us focus our attention on an element of the string of length Δx and mass Δm. Each such segment moves vertically with simple harmonic motion. Furthermore, each segment has the same angular frequency, ω, and the same amplitude, A. As we found in Chapter 12, the total energy, E, associated with a particle moving with simple harmonic motion is $E = \frac{1}{2}kA^2 = \frac{1}{2}m\omega^2A^2$ (Eq. 12.22), where k is the effective force constant of the restoring force. If we apply this equation to the element of length Δx, we see that the total energy of this element is

$$\Delta E = \tfrac{1}{2}(\Delta m)\omega^2A^2$$

If μ is the mass per unit length of the string, then the element of length Δx has a mass, Δm, that is equal to $\mu \Delta x$. Hence, we can express the energy as

$$\Delta E = \tfrac{1}{2}(\mu \Delta x)\omega^2A^2 \qquad [13.20]$$

If the wave travels from left to right, as in Figure 13.20, the energy ΔE arises from the work done on the element Δm by the string element to the left of Δm. Similarly, the element Δm does work on the element to its right, so we see that energy is transmitted to the right. The rate at which energy is transmitted along the string, or the power (P), is dE/dt. If we let Δx approach 0, Equation 13.20 gives

$$P = \frac{dE}{dt} = \tfrac{1}{2}\left(\mu \frac{dx}{dt}\right)\omega^2A^2$$

Figure 13.19   (a) A pulse traveling to the right on a stretched string on which a mass has been suspended. (b) Energy is transmitted to the suspended mass when the pulse arrives.

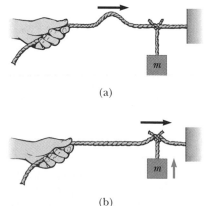

(a)

(b)

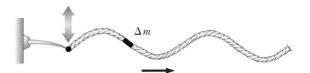

Figure 13.20   A sinusoidal wave traveling along the *x* axis on a stretched string. Every segment moves vertically, and each segment has the same total energy. The power transmitted by the wave equals the energy contained in one wavelength divided by the period of the wave.

Because dx/dt is equal to the wave speed, v, we have

$$P = \tfrac{1}{2}\mu\omega^2 A^2 v \qquad\qquad \textbf{[13.21]}$$ • ***Power***

This shows that the power transmitted by a sinusoidal wave on a string is proportional to (1) the wave speed, (2) the square of the frequency, and (3) the square of the amplitude. In fact, *all* sinusoidal waves have the following general property: **The power transmitted by any sinusoidal wave is proportional to the square of the angular frequency and to the square of the amplitude.**

Thus, we see that a wave traveling through a medium corresponds to energy transport through the medium, with no net transfer of matter. An oscillating source provides the energy and produces a harmonic disturbance of the medium. The disturbance is able to propagate through the medium as the result of the interaction between adjacent particles. In order to verify Equation 13.21 by direct experiment, one would have to design some device at the far end of the string to extract the energy of the wave without producing any reflections.

Thinking Physics 4

We have shown that waves carry energy—not only waves on strings, but all waves. Devise a demonstration that shows that waves transfer momentum as well as energy.

Reasoning Imagine that we cover the ends of a cardboard tube with rubber sheets, secured with rubber bands, to form a crude drum with two heads. We now lay the tube horizontally on the table and suspend a Ping-Pong ball from a string so that the Ping-Pong ball is just touching one of the rubber sheets. If we now thump the other rubber sheet with a finger, the sound will travel down the tube, strike the first rubber sheet, and the Ping-Pong ball will bounce away from the rubber sheet. This demonstrates that the sound wave is carrying momentum to the Ping-Pong ball.

Example 13.5   Power Supplied to a Vibrating String

A string of mass per unit length $\mu = 5.00 \times 10^{-2}$ kg/m is under a tension of 80.0 N. How much power must be supplied to the rope to generate sinusoidal waves at a frequency of 60.0 Hz and an amplitude of 6.00 cm?

Solution   The wave speed on the string is

$$v = \sqrt{\frac{F}{\mu}} = \left(\frac{80.0 \text{ N}}{5.00 \times 10^{-2} \text{ kg/m}}\right)^{1/2} = 40.0 \text{ m/s}$$

Because $f = 60.0$ Hz, the angular frequency ω of the sinusoidal waves on the string has the value

$$\omega = 2\pi f = 2\pi(60.0 \text{ Hz}) = 377 \text{ s}^{-1}$$

Using these values in Equation 13.21 for the power, with $A = 6.00 \times 10^{-2}$ m, gives

$$\text{Power} = \tfrac{1}{2}\mu\omega^2 A^2 v$$

$$= \tfrac{1}{2}(5.00 \times 10^{-2} \text{ kg/m})(377 \text{ s}^{-1})^2$$

$$\times (6.00 \times 10^{-2} \text{ m})^2(40.0 \text{ m/s})$$

$$= 512 \text{ W}$$

13.9 • SOUND WAVES

Let us turn our attention from transverse waves to longitudinal ones. As stated in Section 13.2, longitudinal waves are waves in which the particles of the medium undergo displacements parallel to the direction of wave motion. Sound waves in air are the most important example of longitudinal waves. They can travel through any material medium, their speed depending on the properties of that medium.

The displacements accompanying a sound wave in air are longitudinal displacements of individual molecules from their equilibrium positions. Such displacements result if the source of the waves, such as the diaphragm of a loudspeaker, oscillates with simple harmonic motion. For instance, you can produce a one-dimensional harmonic sound wave in a long, narrow tube containing a gas by means of a vibrating piston at one end, as in Figure 13.21. The darker regions in the figure represent regions in which the gas is compressed and, consequently, the density and pressure are *above* their equilibrium values. Such a compressed layer of gas, called a **condensation,** is formed when the piston is being pushed into the tube. The condensation moves down the tube as a pulse, continuously compressing the layers in front of it. When the piston is withdrawn from the tube, the gas in front of it expands and consequently the pressure and density in this region fall below their equilibrium values. These low-pressure regions, called **rarefactions,** are represented by the lighter areas in Figure 13.21. The rarefactions also propagate along the tube, following the condensations. Both regions move with a speed equal to the speed of sound in that medium. The speed of sound waves in air at 20°C is about 343 m/s.

As the piston oscillates back and forth in a sinusoidal fashion, regions of condensation and rarefaction are continuously set up. The distance between two successive condensations (or two successive rarefactions) equals the wavelength, λ. As these regions travel down the tube, any small volume of the medium moves with simple harmonic motion parallel to the direction of the wave (in other words, longitudinally). If $s(x, t)$ is the displacement of a small volume element measured from its equilibrium position, we can express this displacement function as

$$s(x, t) = s_{max} \cos(kx - \omega t) \qquad \text{[13.22]}$$

where s_{max} is the **maximum displacement from equilibrium** (the displacement amplitude), k is the angular wave number, and ω is the angular frequency of the piston. Note that the displacement of the volume element is along x, the direction of motion of the sound wave, which of course means we are describing a longitudinal wave. The variation in the pressure of the gas, ΔP, measured from its equilibrium value is also sinusoidal; it is given by

$$\Delta P = \Delta P_{max} \sin(kx - \omega t) \qquad \text{[13.23]}$$

The **pressure amplitude,** ΔP_{max}, is the **maximum change in pressure from the equilibrium value.** It is proportional to the displacement amplitude, s_{max}:

Pressure amplitude •

$$\Delta P_{max} = \rho v \omega s_{max} \qquad \text{[13.24]}$$

where ρ is the density of the medium, v is the wave speed, and ωs_{max} is the maximum longitudinal speed of the medium in front of the piston.

Thus, we see that a sound wave may be considered as either a displacement wave or a pressure wave. A comparison of Equations 13.22 and 13.23 shows that **the**

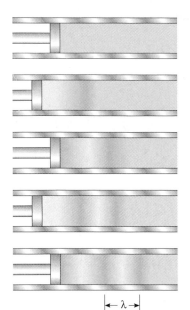

Figure 13.21 A sinusoidal longitudinal wave propagating down a tube filled with a compressible gas. The source of the wave is a vibrating piston at the left. The high- and low-pressure regions are dark and light, respectively.

$|\leftarrow \lambda \rightarrow|$

pressure wave is 90° out of phase with the displacement wave. Graphs of these functions are shown in Figure 13.22. Note that the pressure variation is a maximum when the displacement is zero, whereas the displacement is a maximum when the pressure variation is zero. Because the pressure is proportional to the density of the medium, the expression that describes how that density varies from its equilibrium value is similar to Equation 13.23.

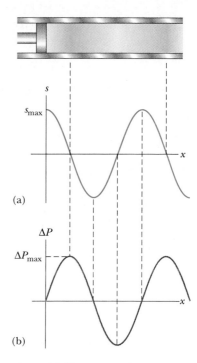

Thinking Physics 5

Garth Brooks had a hit song with "The Thunder Rolls" (Caged Panther Music, Inc., 1990). Why *does* thunder produce an extended "rolling" sound? And how does lightning produce thunder in the first place?

Reasoning Let us assume we are at ground level and neglect ground reflections. When lightning strikes, a channel of ionized air carries a very large electric current from a cloud to the ground. This results in a very rapid temperature increase of this channel of air as the current flows through it. The temperature increase causes a sudden expansion of the air. This expansion is so sudden and so intense that a tremendous disturbance is produced in the air—thunder. The thunder rolls due to the fact that the lightning channel is a long extended source—the entire length of the channel produces the sound at essentially the same instant of time. Sound produced at the bottom of the channel reaches you first, if you are on the ground, because that is the point closest to you, and then sounds from progressively higher portions of the channel reach you. If the lightning channel were a perfectly straight line, the resulting sound might be a steady roar, but the zig-zagged shape of the path results in the rolling variation in loudness, with sound from some portions of the channel arriving at your location simultaneously. This results in periods of loud sounds, interspersed with other instants when the net sound reaching you is low in intensity.

Figure 13.22 (a) Displacement amplitude versus position and (b) pressure amplitude versus position for a sinusoidal longitudinal wave. The displacement wave is 90° out of phase with the pressure wave.

13.10 • THE DOPPLER EFFECT

When a vehicle sounds its horn as it travels along a highway, the frequency of the sound you hear is higher as the vehicle approaches you than it is as the vehicle moves away from you. This is one example of the **Doppler effect,** named after the Austrian physicist Christian Johann Doppler (1803–1853).

> In general, a Doppler effect is experienced whenever there is relative motion between the source of sound and the observer. When the source and observer are moving toward each other, the observer hears a frequency that is higher than the true frequency of the source. When the source and observer are moving away from each other, the observer hears a frequency that is lower than the true frequency of the source.[4]

[4]Although the Doppler effect is most commonly experienced with sound waves, it is a phenomenon common to all harmonic waves. For example, a shift in frequencies of light waves (electromagnetic waves) is produced by the relative motion of source and observer. The relativistic Doppler effect demands that there can be no distinction between observer moving toward source and source moving toward observer. In 1842 Doppler first reported the frequency shift in connection with light emitted by stars revolving about each other in double-star systems.

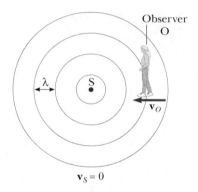

Observer
O

λ

S

v_O

$\mathbf{v}_S = 0$

Figure 13.23 An observer O moving with a speed v_O toward a stationary point source S hears a frequency f' that is greater than the source frequency.

"I love hearing that lonesome wail of the train whistle as the magnitude of the frequency of the wave changes due to the Doppler effect."

The Doppler effect for electromagnetic waves is used in police radar systems to measure the speeds of motor vehicles. Likewise, astronomers use the effect to determine the relative motions of stars, galaxies, and other celestial objects.

First, let us consider the case where the observer O is moving with a speed of v_O and the sound source S is stationary, as in Figure 13.23. For simplicity, we shall assume that the air is stationary and that the observer moves directly toward the source (considered as a point source). We shall take the frequency of the source to be f, the wavelength to be λ, and the speed of sound to be v. A stationary observer would detect f vibrations each second, where $f = v/\lambda$. (That is, when the source and observer are both at rest, the observed frequency must equal the true frequency of the source.) However, the observer moving toward the source with the speed v_O receives an additional number of vibrations per second, equal to v_O/λ. Hence, the frequency the observer hears, f', is *increased:* $f' = v/\lambda + v_O/\lambda$. Because $\lambda = v/f$, we can express f' as

$$f' = f\left(\frac{v + v_O}{v}\right) \qquad \text{(Observer moving toward source)} \qquad \textbf{[13.25]}$$

Now consider the situation in which the source moves with a speed of v_S relative to the medium, and the observer is at rest. If the source moves directly toward observer A in Figure 13.24a, the wave fronts seen by the observer are closer to each other than they would be if the source were at rest. As a result, the wavelength λ' measured by observer A is shorter than the true wavelength λ of the source. During each vibration, with a duration of T (the period), the source moves a distance of $v_S T = v_S/f$. Therefore, the wavelength is *shortened* by this amount, and the observed wavelength has the shorter value $\lambda' = \lambda - v_S/f$. Because $\lambda = v/f$, the frequency heard by observer A is

$$f' = \frac{v}{\lambda'} = f\left(\frac{v}{v - v_S}\right) \qquad \text{(source moving toward observer)} \qquad \textbf{[13.26]}$$

That is, the frequency is *increased* when the source moves toward the observer. In a similar manner, if the source moves away from observer B at rest, the sign of v_S is reversed in Equation 13.26.

Finally, if both the source and the observer are in motion, one finds the following general formula for the observed frequency:

Frequency heard with •
observer and source in motion

$$f' = f\left(\frac{v \pm v_O}{v \mp v_S}\right) \qquad \textbf{[13.27]}$$

In this expression, the upper signs ($+ v_O$ and $- v_S$) refer to motion of the one *toward* the other, and the lower signs ($- v_O$ and $+ v_S$) refer to motion of one *away from* the other.

When working with any Doppler effect problem, a convenient rule to remember concerning signs is the following: The word *toward* is associated with an *increase* in the observed frequency. The words *away from* are associated with a *decrease* in the observed frequency.

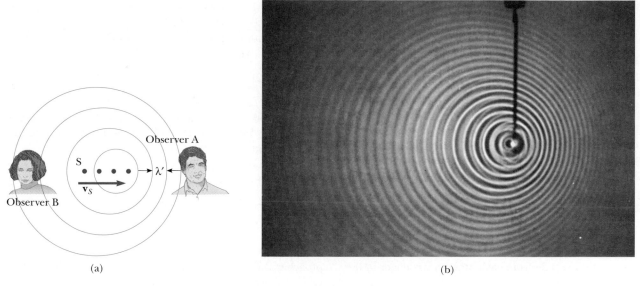

Figure 13.24 (a) A source S moving with a speed v_S toward a stationary observer A and away from a stationary observer B. Observer A hears an increased frequency, and observer B hears a decreased frequency. (b) The Doppler effect in water observed in a ripple tank. *(Courtesy Educational Development Center, Newton, Mass.)*

Thinking Physics 6

Suppose you place your stereo speakers far apart and run past them from right to left, or left to right. If you run rapidly enough and have excellent pitch discrimination, you may notice that the music that is playing seems to be out of tune when you are between the speakers. Why?

Reasoning When you are between the speakers, you are running away from one of them and toward the other. Thus, there is a Doppler shift downward for the sound from the speaker behind you and a Doppler shift upward for the sound from the speaker ahead of you. As a result, the sound from the two speakers will not be in tune. A calculation shows that a world-class sprint runner could run fast enough to generate about a semitone difference in the sound from the two speakers.

CONCEPTUAL PROBLEM 4

Suppose the wind blows. Does this cause a Doppler effect for an observer listening to a sound moving through the air? Is it like a moving source or a moving observer?

CONCEPTUAL PROBLEM 5

You are driving toward a cliff and you honk your horn. Is there a Doppler shift of the sound when you hear the echo? Is it like a moving source or moving observer? What if the reflection occurs not from a cliff but from the forward edge of a huge alien spacecraft which is moving toward you as you drive?

Example 13.6   The Noisy Siren

An ambulance travels down a highway at a speed of 33.5 m/s (75 mi/h). Its siren emits sound at a frequency of 400 Hz. What is the frequency heard by a passenger in a car traveling at 24.6 m/s (55 mi/h) in the opposite direction as the car approaches the ambulance and as the car moves away from the ambulance?

Solution   Let us take the velocity of sound in air to be $v = 343$ m/s. We can use Equation 13.27 in both cases. As the ambulance and car approach each other, the observed apparent frequency is

$$f' = f\left(\frac{v + v_O}{v - v_S}\right) = (400 \text{ Hz}) \left(\frac{343 \text{ m/s} + 24.6 \text{ m/s}}{343 \text{ m/s} - 33.5 \text{ m/s}}\right)$$

$$= \boxed{475 \text{ Hz}}$$

Likewise, as they recede from each other, a passenger in the car hears a frequency

$$f' = f\left(\frac{v - v_O}{v + v_S}\right) = (400 \text{ Hz}) \left(\frac{343 \text{ m/s} - 24.6 \text{ m/s}}{343 \text{ m/s} + 33.5 \text{ m/s}}\right)$$

$$= \boxed{338 \text{ Hz}}$$

The *change* in frequency as detected by the passenger in the car is $475 - 338 = 137$ Hz, which is more than 30% of the actual frequency emitted.

EXERCISE 7   Suppose that the passenger car is parked on the side of the highway as the ambulance travels down the highway at the speed of 33.5 m/s. What frequency will the passenger in the car hear as the ambulance (a) approaches the parked car and (b) recedes from the parked car?   Answer   (a) 443 Hz   (b) 364 Hz

EXERCISE 8   A band is playing on a moving truck. The band strikes the note middle C (262 Hz), but it is heard by spectators ahead of the truck as C# (277 Hz). How fast is the truck moving?   Answer   18.6 m/s

SUMMARY

A **transverse wave** is a wave in which the particles of the medium move in a direction perpendicular to the direction of the wave velocity. An example is a wave on a stretched string.

Longitudinal waves are waves in which the particles of the medium move parallel to the direction of the wave speed. Sound waves in air are longitudinal.

Any one-dimensional wave traveling with a speed of v in the positive x direction can be represented by a wave function of the form $y = f(x - vt)$. Likewise, the wave function for a wave traveling in the negative x direction has the form $y = f(x + vt)$. The shape of the wave at any instant (a snapshot of the wave) is obtained by holding t constant.

The **superposition principle** says that when two or more waves move through a medium, the resultant wave function equals the algebraic sum of the individual wave functions. Waves that obey this principle are said to be *linear*. When two waves combine in space, they interfere to produce a resultant wave.

The wave function for a one-dimensional harmonic wave traveling to the right can be expressed as

$$y = A \sin\left[\frac{2\pi}{\lambda}(x - vt)\right] = A \sin(kx - \omega t) \qquad \textbf{[13.6, 13.12]}$$

where A is the amplitude, λ is the wavelength, k is the angular wave number, and ω is the angular frequency. If T is the period and f is the frequency, then v, k, and ω can be written

$$v = \frac{\lambda}{T} = f\lambda \qquad\qquad \textbf{[13.7, 13.14]}$$

$$k \equiv \frac{2\pi}{\lambda} \qquad\qquad \textbf{[13.10]}$$

$$\omega \equiv \frac{2\pi}{T} = 2\pi f \qquad\qquad \textbf{[13.11]}$$

The speed of a wave traveling on a stretched string of mass per unit length μ and tension F is

$$v = \sqrt{\frac{F}{\mu}} \qquad\qquad \textbf{[13.19]}$$

When a pulse traveling on a string meets a fixed end, the pulse is reflected and inverted. If the pulse reaches a free end, it is reflected but not inverted.

The **power** transmitted by a harmonic wave on a stretched string is

$$P = \tfrac{1}{2}\mu\omega^2 A^2 v \qquad\qquad \textbf{[13.21]}$$

The change in frequency heard by an observer whenever there is relative motion between a wave source and the observer is called the **Doppler effect.** When the source and observer are moving toward each other, the observer hears a frequency that is higher than the true frequency of the source. When the source and observer are moving away from each other, the observer hears a frequency that is lower than the true frequency of the source.

CONCEPTUAL QUESTIONS

1. How would you set up a longitudinal wave in a stretched spring? Would it be possible to set up a transverse wave in a spring?

2. By what factor would you have to increase the tension in a taut string in order to double the wave speed?

3. When traveling on a taut string, does a wave pulse always invert on reflection? Explain.

4. Does the vertical speed of a segment of a horizontal taut string through which a wave is traveling depend on the wave speed?

5. A vibrating source generates a sinusoidal wave on a string under constant tension. If the power delivered to the string is doubled, by what factor does the amplitude change? Does the wave speed change under these circumstances?

6. Consider a wave traveling on a taut rope. What is the difference, if any, between the speed of the wave and the speed of a small section of the rope?

7. What happens to the wavelength of a wave on a string when the frequency is doubled? What happens to its speed? Assume the tension in the string remains the same.

8. When all the strings on a guitar are stretched to the same tension, will the speed of a wave along the more massive bass strings be faster or slower than the speed of a wave on the lighter strings?

9. If you stretch a rubber hose and pluck it, you can observe a pulse traveling up and down the hose. What happens to the speed if you stretch the hose tighter? If you fill the hose with water?

10. In a longitudinal wave in a spring, the coils move back and forth in the direction of wave motion. Does the speed of the wave depend on the maximum speed of each coil?

11. When two waves interfere, can the amplitude of the resultant wave be larger than either of the two original waves? Under what conditions?

12. A solid can transport both longitudinal waves and transverse waves, but a fluid can transport only longitudinal waves. Why?

13. In an earthquake both S (transverse) and P (longitudinal) waves are sent out. The S waves travel through the Earth more slowly than the P waves (4.5 km/s versus 7.8 km/s). By detecting the time of arrival of the waves, how can one determine how far away the epicenter of the quake was? How many detection centers are necessary to pinpoint the location of the epicenter?

PROBLEMS

Section 13.3   One-Dimensional Transverse
Traveling Waves

1. At $t = 0$, a transverse wave pulse in a wire is described by the function

$$y = \frac{6}{x^2 + 3}$$

where x and y are in meters. Write the function $y(x, t)$ that describes this wave if it is traveling in the positive x direction with a speed of 4.50 m/s.

2. Ocean waves with a crest-to-crest distance of 10.0 m can be described by

$$y(x, t) = (0.800 \text{ m})\sin [0.628(x - vt)]$$

where $v = 1.20$ m/s. (a) Sketch $y(x, t)$ at $t = 0$. (b) Sketch $y(x, t)$ at $t = 2.00$ s. Note how the entire waveform has shifted 2.40 m in the positive x direction in this time interval.

3. A wave moving along the x axis is described by

$$y(x, t) = 5.0e^{-(x+5.0t)^2}$$

where x is in meters and t is in seconds. Determine (a) the direction of the wave motion and (b) the speed of the wave.

4. Two wave pulses A and B are moving in opposite directions along a taut string with a speed of 2.00 cm/s. The amplitude of A is twice the amplitude of B. The pulses are shown in Figure P13.4 at $t = 0$. Sketch the shape of the string at $t = 1, 1.5, 2, 2.5,$ and 3 s.

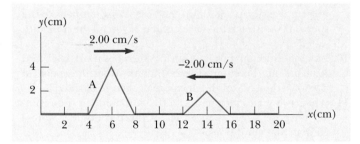

Figure P13.4

Section 13.4   Sinusoidal Traveling Waves

5. A sinusoidal wave is traveling along a rope. The oscillator that generates the wave completes 40.0 vibrations in 30.0 s. Also, a given peak travels 425 cm along the rope in 10.0 s. What is the wavelength?

6. A sinusoidal wave of wavelength 2.00 m and amplitude 0.100 m travels with a speed of 1.00 m/s on a string. Initially, the left end of the string is at the origin and the wave moves from left to right. Find (a) the frequency and angular frequency, (b) the angular wave number, and (c) the wave function for this wave. Determine the equation of motion for (d) the left end of the string, and (e) the point on the string at $x = 1.50$ m to the right of the left end. (f) What is the maximum speed of any point on the string?

7. (a) Write the expression for the function $y(x, t)$ for a sinusoidal wave traveling along a rope in the *negative x* direction with the following characteristics: $A = 8.00$ cm, $\lambda = 80.0$ cm, $f = 3.00$ Hz, and $y(0, 0) = 0$. (b) Write the expression for y as a function of x and t for the wave in part (a) assuming that $y(10 \text{ cm}, 0) = 0$.

8. A transverse wave on a string is described by

$$y = (0.120 \text{ m})\sin \pi(x/8 + 4t)$$

(a) Determine the transverse speed and acceleration of the string at $t = 0.200$ s for the point on the string located at $x = 1.60$ m. (b) What are the wavelength, period, and speed of propagation of this wave?

9. A wave is described by $y = (2.00 \text{ cm}) \sin(kx - \omega t)$, where $k = 2.11$ rad/m, $\omega = 3.62$ rad/s, x is in meters, and t is in seconds. Determine the amplitude, wavelength, frequency, and speed of the wave.

10. A sinusoidal wave traveling in the $-x$ direction (to the left) has an amplitude of 20.0 cm, a wavelength of 35.0 cm, and a frequency of 12.0 Hz. The displacement of the wave at $t = 0$, $x = 0$ is $y = -3.00$ cm and has a positive velocity here. (a) Sketch the wave at $t = 0$. (b) Find the angular wave number, period, angular frequency, and phase velocity of the wave. (c) Write an expression for the wave function $y(x, t)$.

11. A transverse sinusoidal wave on a string has a period $T = 25.0$ ms and travels in the negative x direction with a speed of 30.0 m/s. At $t = 0$, a particle on the string at $x = 0$ has a displacement of 2.00 cm and is moving in the negative y direction with a speed of 2.00 m/s. (a) What is the amplitude of the wave? (b) What is the initial phase angle? (c) What is the maximum transverse speed of the string? (d) Write the wave function for the wave.

Section 13.5   Superposition and Interference of Waves

12. Two pulses traveling on the same string are described by

$$y_1 = \frac{5}{(3x - 4t)^2 + 2} \quad \text{and} \quad y_2 = \frac{-5}{(3x + 4t - 6)^2 + 2}$$

(a) In which direction does each pulse travel? (b) At what time do the two cancel? (c) At what point do the two waves always cancel?

13. Two sinusoidal waves in a string are defined by the functions

$$y_1 = (2.00 \text{ cm})\sin(20.0x - 30.0t)$$

$$y_2 = (2.00 \text{ cm})\sin(25.0x - 40.0t)$$

where y and x are in centimeters and t is in seconds. (a) What is the phase difference between these two waves at the point $x = 5.00$ cm at $t = 2.00$ s? (b) What is the positive x value closest to the origin for which the two phases differ by π at $t = 2.00$ s? (This is where the two waves add to zero.)

Section 13.6 The Speed of Transverse Waves on Strings

14. A piano string of mass-per-length 5.00×10^{-3} kg/m is under a tension of 1350 N. Find the speed with which a wave travels on this string.

15. A phone cord is 4.00 m long. The cord has a mass of 0.200 kg. If a transverse wave pulse travels from the receiver to the phone box in 0.100 s, what is the tension in the cord?

16. A 30.0-m steel wire and a 20.0-m copper wire, both with 1.00-mm diameters, are connected end to end and stretched to a tension of 150 N. How long does it take a transverse wave to travel the entire length of the two wires?

17. Transverse pulses travel with a speed of 200 m/s along a taut copper wire the diameter of which is 1.50 mm. What is the tension in the wire? (The density of copper is 8.92 g/cm^3.)

18. A light string of mass-per-unit length 8.00 g/m has its ends tied to two walls separated by a distance equal to three fourths the length of the string (Fig. P13.18). A mass m is suspended from the center of the string, putting a tension in the string. (a) Find an expression for the transverse wave speed in the string as a function of the hanging mass.

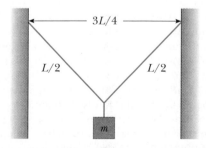

Figure P13.18

(b) How much mass should be suspended from the string to have a wave speed of 60.0 m/s?

19. An astronaut on the Moon wishes to measure the local value of the free-fall acceleration g_{Moon} by timing pulses traveling down a wire that has a large mass suspended from it. Assume a wire of mass 4.00 g is 1.60 m long and has a 3.00 kg mass suspended from it. A pulse requires 36.1 ms to traverse the length of the wire. Calculate g_{Moon} from these data. (You may neglect the mass of the wire when calculating the tension in it.)

Section 13.8 Energy Transmitted by Sinusoidal Waves on Strings

20. A taut rope has a mass of 0.180 kg and a length of 3.60 m. What power must be supplied to the rope in order to generate sinusoidal waves having an amplitude of 0.100 m and a wavelength of 0.500 m and traveling with a speed of 30.0 m/s?

21. Sinusoidal waves 5.00 cm in amplitude are to be transmitted along a string of linear density 4.00×10^{-2} kg/m. If the maximum power delivered by the source is 300 W and the string is under a tension of 100 N, what is the highest vibrational frequency at which the source can operate?

22. It is found that a 6.00-m segment of a long string contains four complete waves and has a mass of 180 g. The string is vibrating sinusoidally with a frequency of 50.0 Hz and a peak-to-valley displacement of 15.0 cm. (*Peak-to-valley* means the vertical distance from the farthest positive displacement to the farthest negative displacement.) (a) Write down the function that describes this wave traveling in the positive x direction. (b) Determine the power being supplied to the string.

23. A horizontal string can transmit a maximum power of P_0 (without breaking) if a wave with amplitude A and angular frequency ω is traveling along it. In order to increase this maximum power, a student folds the string and uses this double string as a transmitter. Determine the maximum power that can be transmitted along the double string, assuming that the tension remains the same.

24. A two-dimensional water wave spreads in circular wave fronts. Show that the amplitude A at a distance r from the initial disturbance is proportional to $1/\sqrt{r}$. (*Hint:* Consider the energy concentrated in the outward-moving ripple.)

Section 13.9 Sound Waves

(In this section, use the following values as needed unless otherwise specified: the equilibrium density of air, $\rho = 1.20$ kg/m^3; the speed of sound in air, $v = 343$ m/s. Also, pressure variations, ΔP, are measured relative to atmospheric pressure.)

25. A flowerpot is knocked off a balcony 20.0 m above the sidewalk heading for a man of height 1.75 m standing below. How high from the ground can the flowerpot be after which it would be too late for a shouted warning to reach the man in time? Assume that the man below requires 0.300 s to respond to the warning.

26. A rescue plane flies horizontally at a constant speed searching for a disabled boat. When the plane is directly above the boat, the boat's crew blows a loud horn. By the time the plane's sound detector perceives the horn's sound, the plane has traveled a distance equal to one-half its altitude above the ocean. If it takes the sound 2.00 s to reach the plane, determine (a) the speed of the plane, and (b) its altitude. Take the speed of sound to be 343 m/s.

27. A sound wave in a cylinder is described by Equations 13.22 through 13.24. Show that $\Delta P = \pm \rho v \omega \sqrt{s_{max}^2 - s^2}$.

28. Calculate the pressure amplitude of a 2.00-kHz sound wave in air if the displacement amplitude is 2.00×10^{-8} m.

29. An experimenter wishes to generate in air a sound wave that has a displacement amplitude of 5.50×10^{-6} m. The pressure amplitude is to be limited to 8.40×10^{-1} N/m². What is the minimum wavelength the sound wave can have?

30. A sound wave in air has a pressure amplitude of 4.00 N/m² and a frequency of 5.00 kHz. Take $\Delta P = 0$ at the point $x = 0$ when $t = 0$. (a) What is ΔP at $x = 0$ when $t = 2.00 \times 10^{-4}$ s, and (b) what is ΔP at $x = 0.0200$ m when $t = 0$?

the tuning fork when waves of frequency 485 Hz reach the release point? Take the speed of sound in air to be 340 m/s.

35. A block with a speaker bolted to it is connected to a spring having spring constant $k = 20.0$ N/m as in Figure P13.35. The total mass of the block and speaker is 5.00 kg, and the amplitude of this unit's motion is 0.500 m. If the speaker emits sound waves of frequency 440 Hz, determine the range in frequencies heard by the person to the right of the speaker.

Figure P13.35

Section 13.10   The Doppler Effect

31. A commuter train passes a passenger platform at a constant speed of 40.0 m/s. The train horn is sounded at its characteristic frequency of 320 Hz. (a) What total change in frequency is detected by a person on the platform as the train goes from approaching to receding? (b) What wavelength is detected by a person on the platform as the train approaches?

32. A driver traveling northbound on a highway is driving at a speed of 25.0 m/s. A police car driving southbound at a speed of 40.0 m/s approaches with its siren sounding at a base frequency of 2500 Hz. (a) What frequency is observed by the driver as the police car approaches? (b) What frequency is detected by the driver after the police car passes him? (c) Repeat parts (a) and (b) for the case when the police car is traveling northbound.

33. Standing at a crosswalk, you hear a frequency of 560 Hz from the siren of an approaching police car. After the police car passes, the observed frequency of the siren is 480 Hz. Determine the car's speed from these observations.

34. A tuning fork vibrating at 512 Hz falls from rest and accelerates at 9.80 m/s². How far below the point of release is

Additional Problems

36. Unoccupied by spectators, a large set of football bleachers has solid seats and risers. You stand on the field in front of it and sharply clap two wooden boards together once. You hear from the bleachers a sound with definite pitch, which may remind you of a short toot on a trumpet. Account for this sound. Compute order-of-magnitude estimates for its frequency, wavelength, and duration on the basis of data you specify.

37. A 2.00-kg block hangs from a rubber string, being supported so that the string is not stretched. The unstretched length of the string is 0.500 m and its mass is 5.00 g. The spring constant for the string is 100 N/m. The block is released and stops at the lowest point. (a) Determine the tension in the string when the block is at this lowest point. (b) What is the length of the string in this "stretched" position? (c) Find the speed of a transverse wave in the string if the block is held in this lowest position.

38. A block of mass M hangs from a rubber string, being supported so that the string is not stretched. The unstretched length of the string is L_0 and its mass is m, much less than M. The spring constant for the string is k. The block is released and stops at the lowest point. (a) Determine the

tension in the string when the block is at this lowest point. (b) What is the length of the string in this "stretched" position? (c) Find the speed of a transverse wave in the string if the block is held in this lowest position.

39. The wave function for a traveling wave on a taut string is (in SI units)

$$y(x, t) = (0.350 \text{ m})\sin(10\pi t - 3\pi x + \pi/4)$$

(a) What is the speed and direction of travel of the wave? (b) What is the vertical displacement of the string at $t = 0$, $x = 0.100$ m? (c) What are the wavelength and frequency of the wave? (d) What is the maximum magnitude of the transverse speed of the string?

40. A stone is dropped into a deep canyon and is heard to strike the bottom 10.2 s after release. The speed of sound waves in air is 343 m/s. How deep is the canyon? What would be the percentage error in the depth if the time required for the sound to reach the canyon rim were ignored?

41. The ocean floor is underlain by a layer of basalt that constitutes the crust, or uppermost layer of the Earth, in this region. Below this crust is found denser peridotite rock, which forms the Earth's mantle. The boundary between these two layers is called the *Mohorovičić discontinuity* ("Moho" for short). If an explosive charge is set off at the surface of the basalt, it generates a seismic wave that is reflected back at the Moho. If the speed of this wave in basalt is 6.50 km/s, and the two-way travel time is 1.85 s, what is the thickness of this oceanic crust?

42. A worker strikes a steel pipeline with a hammer, generating both longitudinal and transverse waves. Reflected waves return 2.40 s apart. How far away is the reflection point? (For steel, $v_{long} = 6.20$ km/s and $v_{trans} = 3.20$ km/s.)

43. A rope of total mass m and length L is suspended vertically. Show that a transverse wave pulse will travel the length of the rope in a time $t = 2\sqrt{L/g}$. (*Hint:* First find an expression for the wave speed at any point a distance x from the lower end by considering the tension in the rope as resulting from the weight of the segment below that point.)

44. A train whistle ($f = 400$ Hz) sounds higher or lower in pitch depending on whether it approaches or recedes. (a) Prove that the difference in frequency between the approaching and receding train whistle is

$$\Delta f = \frac{2f\left(\dfrac{u}{v}\right)}{1 - \dfrac{u^2}{v^2}} \qquad \begin{array}{l} u = \text{speed of train} \\ v = \text{speed of sound} \end{array}$$

(b) Calculate this difference for a train moving at a speed of 130 km/h. Take the speed of sound in air to be 340 m/s.

45. In order to be able to determine her speed, a skydiver carries a tone generator. A friend on the ground at the landing site has equipment for receiving and analyzing sound waves. While the skydiver is falling at terminal speed, her tone generator emits a steady tone of 1800 Hz. (Assume that the air is calm and that the sound speed is 343 m/s, independent of altitude.) (a) If her friend on the ground (directly beneath the skydiver) receives waves of frequency 2150 Hz, what is the skydiver's speed of descent? (b) If the skydiver were also carrying sound-receiving equipment sensitive enough to detect waves reflected from the ground, what frequency would she receive?

46. A wave pulse traveling along a string of linear mass density is described by the relationship

$$y = [A_0 e^{-bx}]\sin(kx - \omega t)$$

where the factors in brackets before the sine are said to be the amplitude. (a) What is the power $P(x)$ carried by this wave at a point x? (b) What is the power $P(0)$ carried by this wave at the origin? (c) Compute the ratio $P(x)/P(0)$.

47. A bat, moving at 5.00 m/s, is chasing a flying insect. If the bat emits a 40.0-kHz chirp and receives back an echo at 40.4 kHz, at what speed is the insect moving toward or away from the bat? (Take the speed of sound in air to be $v = 340$ m/s.)

48. An earthquake on the ocean floor in the Gulf of Alaska induces a *tsunami* (sometimes called a *tidal wave*) that reaches Hilo, Hawaii, 4450 km distant, in a time of 9 h 30 min. Tsunamis have enormous wavelengths (100 km to 200 km), and for such waves the propagation speed is $v \approx \sqrt{g\overline{d}}$, where $\overline{d}$ is the average depth of the water. From the information given, find the average wave speed and the average ocean depth between Alaska and Hawaii. (This method was used in 1856 to estimate the average depth of the Pacific Ocean long before soundings were made to give a direct determination.)

Spreadsheet Problem

S1. Two transverse wave pulses traveling in opposite directions along the x axis are represented by the following wave functions:

$$y_1(x, t) = \frac{6}{(x - 3t)^2} \qquad y_2(x, t) = -\frac{3}{(x + 3t)^2}$$

where x and y are measured in centimeters and t is in seconds. Write a spreadsheet or program to add the two pulses and obtain the shape of the composite waveform $y_{tot} = y_1 + y_2$ as a function of time. Plot y_{tot} versus x. Make separate plots for $t = 0, 0.5, 1, 1.5, 2, 2.5,$ and 3.0 s.

ANSWERS TO CONCEPTUAL PROBLEMS

1. A pulse in a long line of people is longitudinal, since the movement of people is parallel to the direction of propagation of the pulse. The speed is determined by the reaction time of the people and the speed with which they can move, once a space opens up. There is also a psychological factor, in that people will not want to fill a space that opens up in front of them too quickly, so as not to intimidate the person in front of them. The "wave" at a stadium is transverse, since the fans stand up vertically, as the wave sweeps past them horizontally. The speed of this pulse depends on the limits of the fans' abilities to rise and sit rapidly, and on psychological factors associated with the anticipation of seeing the pulse approach the observer's location.

2. Both the radio wave and the sound wave have the same wavelength. The speed of propagation for electromagnetic waves, however, is much higher than the speed of sound, so the frequency of the radio wave is much higher than that of the sound wave. Audible sound frequencies range from about 15 Hz to 15 000 Hz, while radio frequencies are in kilohertz for AM radio and megahertz for FM.

3. A wave on a massless string would have an infinite speed of propagation because its linear mass density is zero.

4. Wind can change a Doppler shift. Both v_O and v_S in our equation must be interpreted as speeds of observer and source relative to air. If source and observer are moving relative to each other, the observer will hear one shifted frequency in still air and a different shifted frequency if wind is blowing. However, if source and observer are at rest relative to each other, there is no shift when wind blows past.

5. The echo *is* Doppler shifted, and the shift is like both a moving source and a moving observer. The sound which leaves your horn in the forward direction is Doppler shifted to a higher frequency, because it is coming from a moving source. As the sound reflects back and comes toward you, you are a moving observer, so there is a second Doppler shift to an even higher frequency. If the sound reflects from the spacecraft moving toward you, there is another moving source shift to an even higher frequency. The reflecting surface of the spacecraft acts as a moving source.

14

Superposition and Standing Waves

An important aspect of waves is their combined effect when two or more travel in the same medium. For instance, what happens to a string when a wave traveling toward a fixed end is reflected back on itself?

In a linear medium—that is, one in which the restoring force of the medium is proportional to the displacement of the medium—one can apply the **principle of superposition** to obtain the resultant disturbance. This principle can be applied to many types of waves under certain conditions, including waves on strings, sound waves, surface water waves, and electromagnetic waves. The superposition principle states that the net displacement of any part of the disturbed medium equals the vector sum of the displacements caused by the individual waves. In Chapter 13, the term **interference** was used to describe the effect produced by combining two waves that are moving simultaneously through a medium.

This chapter is concerned with the superposition principle as it applies to sinusoidal waves. If the sinusoidal waves that com-

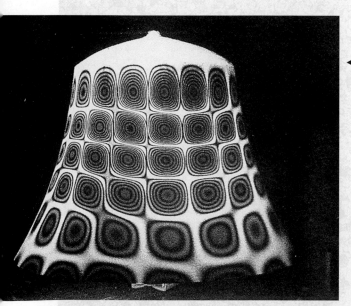

◀ **Photograph of standing waves on a vibrating handbell. The photograph was taken using a technique called** *time-average holographic interferometry.* **Such a hologram consists of millions of superimposed holograms, but this hologram emphasizes the two positions of maximum deflection of the vibrating bell. The pattern recorded here occurred at a frequency of 2684 Hz, but other patterns are observed at lower and higher frequencies.**
(Courtesy of Professor Thomas D. Rossing, Northern Illinois University)

389

bine in a given medium have the same frequency and wavelength, one finds that a stationary pattern, called a **standing wave,** can be produced at certain frequencies under certain circumstances. For example, a stretched string fixed at both ends has a discrete set of standing wave patterns, called **modes of vibration,** that depend on the tension and the mass-per-unit-length of the string. These modes of vibration are found in the strings of stringed musical instruments. Other musical instruments, such as the organ and flute, make use of the natural frequencies of sound waves in hollow pipes. Such frequencies depend on the length and shape of the pipe and on whether the ends are open or closed.

In this chapter we also consider the superposition and interference of waves with different frequencies and wavelengths. When two sound waves with nearly the same frequency interfere, one hears variations in loudness called *beats*. The beat frequency corresponds to the rate of alternation between constructive and destructive interference. Finally, we describe how any complex periodic waveform can, in general, be described by a sum of sine and cosine functions.

14.1 • SUPERPOSITION AND INTERFERENCE OF SINUSOIDAL WAVES

Let us apply the superposition principle to two sinusoidal waves traveling in the same direction in a medium. If the two waves are traveling to the right and have the same frequency, wavelength, and amplitude but differ in phase, we can express their individual wave functions as

$$y_1 = A \sin(kx - \omega t) \qquad \text{and} \qquad y_2 = A \sin(kx - \omega t - \phi)$$

Hence, the resultant wave function, y, is

$$y = y_1 + y_2 = A[\sin(kx - \omega t) + \sin(kx - \omega t - \phi)]$$

In order to simplify this expression, it is convenient to use the trigonometric identity

$$\sin a + \sin b = 2 \cos\left(\frac{a - b}{2}\right) \sin\left(\frac{a + b}{2}\right)$$

If we let $a = kx - \omega t$ and $b = kx - \omega t - \phi$, the resultant wave, y, reduces to

Resultant of two traveling •
sinusoidal waves

$$y = \left(2A \cos \frac{\phi}{2}\right) \sin\left(kx - \omega t - \frac{\phi}{2}\right) \qquad \text{[14.1]}$$

This result has several important features. The resultant wave function, y, is also harmonic and has the *same* frequency and wavelength as the individual waves. The amplitude of the resultant wave is $2A \cos(\phi/2)$ and the phase is $\phi/2$. If the phase constant, ϕ, equals 0, then $\cos(\phi/2) = \cos 0 = 1$ and the amplitude of the resultant wave is $2A$. In other words, the amplitude of the resultant wave is twice the amplitude of either individual wave. In this case, the waves are said to be everywhere *in phase* and to thus **interfere constructively.** That is, the crests and troughs of the individual waves occur at the same positions, as is shown by the blue lines in Figure 14.1a. In general, constructive interference occurs when $\cos(\phi/2) = \pm 1$, or when $\phi = 0, 2\pi, 4\pi, \ldots$. However, if ϕ is equal to π radians or to any *odd* multiple of π then $\cos(\phi/2) = \cos(\pi/2) = 0$ and the resultant wave has *zero* amplitude everywhere. In this case, the two waves **interfere destructively.** That is, the crest of one

Constructive interference •

Destructive interference •

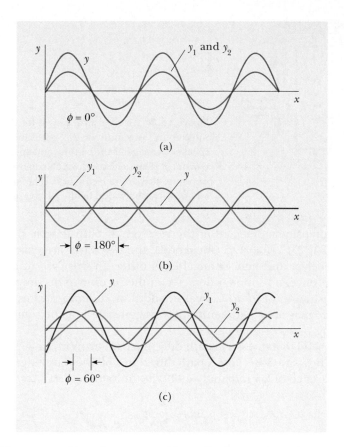

Figure 14.1   The superposition of two waves with amplitudes y_1 and y_2, where $y_1 = y_2$.
(a) When the two waves are in phase, the result is constructive interference. (b) When the two
waves are 180° out of phase, the result is destructive interference. (c) When the phase angle lies
in the range $0 < \phi < 180°$, the resultant y falls somewhere between that shown in (a) and that
shown in (b).

wave coincides with the trough of the second (Fig. 14.1b) and their displacements
cancel at every point. Finally, when the phase constant has an arbitrary value between 0 and π, as in Figure 14.1c, the resultant wave has an amplitude the value of
which is somewhere between 0 and $2A$.

Interference of Waves

One simple device for demonstrating interference of sound waves is illustrated in
Figure 14.2. Sound from speaker S is sent into a tube at P, at which point there is
a T-shaped junction. Half the sound power travels in one direction, and half in the
opposite direction. Thus, the sound waves that reach receiver R at the other side
can travel along either of two paths. The total distance from speaker to receiver is
called the **path length,** r. The length of the lower path is fixed at r_1. The upper
path length, r_2, can be varied by sliding the U-shaped tube (similar to that on a slide
trombone). When the difference in the path lengths, $\Delta r = |r_2 - r_1|$, is either zero
or some integral multiple of the wavelength λ, the two waves reaching the receiver
are in phase and interfere constructively, as in Figure 14.1a. In this case, a maximum

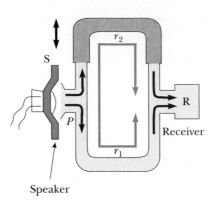

Figure 14.2 An acoustical system for demonstrating interference of sound waves. Sound from the speaker propagates into a tube and splits into two parts at P. The two waves, which superimpose at the opposite side, are detected at R. The upper path length r_2 can be varied by the sliding section.

in the sound intensity is detected at the receiver. If path length r_2 is adjusted so that Δr is $\lambda/2$, $3\lambda/2$, ... $n\lambda/2$ (for n odd), the two waves are exactly 180° out of phase at the receiver and hence cancel each other. In this case of completely destructive interference, no sound is detected at the receiver. This simple experiment is a striking illustration of interference. In addition, it demonstrates the fact that a phase difference may arise between two waves generated by the same source when they travel along paths of unequal lengths.

It is often useful to express the path difference in terms of the phase difference, ϕ, between the two waves. Because a path difference of one wavelength corresponds to a phase difference of 2π radians, we obtain the ratio $\lambda/2\pi = \Delta r/\phi$, or

Relationship between path •
difference and phase angle

$$\Delta r = \frac{\lambda}{2\pi} \phi \qquad\qquad [14.2]$$

Nature provides many other examples of interference phenomena. Later in the book we shall describe several interesting interference effects involving light waves.

Thinking Physics 1

If stereo speakers are connected to the amplifier "out of phase," one speaker is moving outward when the other is moving inward. This results in a weakness in the bass notes, which can be corrected by reversing the wires on one of the speaker connections. Why is there only an effect on the bass notes in this case and not the treble notes?

Reasoning Imagine that you are sitting in front of the speakers, midway between them. Then, the sound from each speaker travels the same distance to you, and there is no phase difference in the sound that is due to a path difference. Because the speakers are connected out of phase, the sound waves are half a wavelength out of phase on leaving the speaker and, consequently, on arriving at your ear. As a result, there will be cancellation of the sound for all frequencies in the ideal case of a zero-size head located exactly on the midpoint between the speakers. If the ideal head were moved off the centerline, there will be an additional phase difference introduced by the path-length difference for the sound from the two speakers. In the case of low frequency, long wavelength bass notes, the path-length differences will be a small fraction of a wavelength, so there will still be significant cancellation. For the high-frequency, short-wavelength treble notes, a small movement of the ideal head will result in a much larger fraction of a wavelength in path length difference or even multiple wavelengths. Thus, the treble notes could be in phase with this head movement. If we now add the fact that the head is not of zero size and the fact that there are two ears, we can see that the complete cancellation is not possible, and, with even small movements of the head, one or both ears will be at or near maxima for the treble notes. The size of the head is much smaller than bass wavelengths, however, so there is significant weakening of the bass over much of the region in front of the speakers.

CONCEPTUAL PROBLEM 1

As oppositely moving pulses of the same shape (one upward, one downward) on a string pass through each other, there is one instant at which the string shows no displacement from the equilibrium position at any point. Has the energy carried by the pulses disappeared at this instant in time? If not, where is it?

Example 14.1 Two Speakers Driven by the Same Source

A pair of speakers placed 3.00 m apart are driven by the same oscillator (Fig. 14.3). A listener is originally at point O, which is located 8.00 m from the center of the line connecting the two speakers. The listener then walks to point P, which is a perpendicular distance 0.350 m from O before reaching the *first minimum* in sound intensity. What is the frequency of the oscillator?

Solution The first minimum occurs when the two waves reaching the listener at P are 180° out of phase—in other words, when their path difference equals $\lambda/2$. In order to calculate the path difference, we must first find the path lengths r_1 and r_2. Making use of the two shaded triangles in Figure 14.3, we find the path lengths to be

$$r_1 = \sqrt{(8.00 \text{ m})^2 + (1.15 \text{ m})^2} = 8.08 \text{ m}$$

$$r_2 = \sqrt{(8.00 \text{ m})^2 + (1.85 \text{ m})^2} = 8.21 \text{ m}$$

Hence, the path difference is $r_2 - r_1 = 0.13$ m. Because we require that this path difference be equal to $\lambda/2$ for the first minimum, we find that $\lambda = 0.26$ m.

To obtain the oscillator frequency, we can use $v = \lambda f$, where v is the speed of sound in air, 343 m/s:

$$f = \frac{v}{\lambda} = \frac{343 \text{ m/s}}{0.26 \text{ m}} = 1.3 \text{ kHz}$$

EXERCISE 1 If the oscillator frequency is adjusted such that the listener hears the first minimum at a distance of 0.75 m from O, what is the new frequency? Answer 0.63 kHz

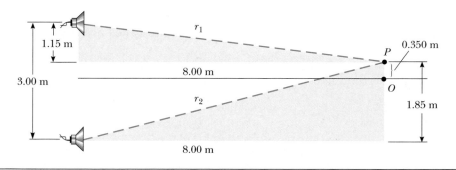

Figure 14.3 (Example 14.1)

14.2 • STANDING WAVES

If a stretched string is clamped at both ends, waves traveling in both directions are reflected from the ends. The incident and reflected waves combine according to the superposition principle.

Consider two sinusoidal waves, in the same medium, that have the same amplitude, frequency, and wavelength but are traveling in *opposite* directions. If we neglect any phase angle ϕ for simplicity, their wave functions can be written

$$y_1 = A \sin(kx - \omega t) \qquad \text{and} \qquad y_2 = A \sin(kx + \omega t)$$

where y_1 represents a wave traveling to the right and y_2 represents a wave traveling to the left. Adding these two functions gives the resultant wave function, y:

$$y = y_1 + y_2 = A \sin(kx - \omega t) + A \sin(kx + \omega t)$$

where $k = 2\pi/\lambda$ and $\omega = 2\pi f$, as usual. Using the trigonometric identity $\sin(a \pm b) = \sin a \cos b \pm \cos a \sin b$, this reduces to

Wave function for a standing • wave

$$y = (2A \sin kx)\cos \omega t \qquad\qquad [14.3]$$

This expression represents the wave function of a **standing wave.** From this result, we see that a standing wave has an angular frequency of ω and an amplitude of $2A \sin kx$ (the quantity in the parentheses of Eq. 14.3). That is, every particle of the string vibrates in simple harmonic motion with the same frequency. However, the amplitude of motion of a given particle depends on x. This is in contrast to the situation involving a traveling sinusoidal wave, in which all particles oscillate with the same amplitude as well as the same frequency.

Because the amplitude of the standing wave at any value of x is equal to $2A \sin kx$, we see that the *maximum* amplitude has the value $2A$. This occurs when the coordinate x satisfies the condition $\sin kx = 1$, or when

$$kx = \frac{\pi}{2}, \frac{3\pi}{2}, \frac{5\pi}{2}, \ldots$$

Because $k = 2\pi/\lambda$, the positions of maximum amplitude, called **antinodes,** are

Position of antinodes •

$$x = \frac{\lambda}{4}, \frac{3\lambda}{4}, \frac{5\lambda}{4}, \ldots = \frac{n\lambda}{4} \qquad\qquad [14.4]$$

where $n = 1, 3, 5, \ldots$. Note that **adjacent antinodes are separated by a distance of $\lambda/2$.**

In a similar way, the standing wave has a *minimum* amplitude of zero when x satisfies the condition $\sin kx = 0$ or when

$$kx = \pi, 2\pi, 3\pi, \ldots$$

giving

Position of nodes •

$$x = \frac{\lambda}{2}, \lambda, \frac{3\lambda}{2}, \ldots = \frac{n\lambda}{2} \qquad\qquad [14.5]$$

where $n = 1, 2, 3, \ldots$. **These points of zero amplitude, called nodes, are also spaced by $\lambda/2$.** The distance between a node and an adjacent antinode is $\lambda/4$.

The standing wave patterns produced at various times by two waves traveling in opposite directions are depicted graphically in Figure 14.4. The upper part of each figure represents the individual traveling waves, and the lower part represents the standing wave patterns. The nodes of the standing wave are labeled N, and the antinodes are labeled A. At $t = 0$ (Fig. 14.4a), the two waves are spatially identical,

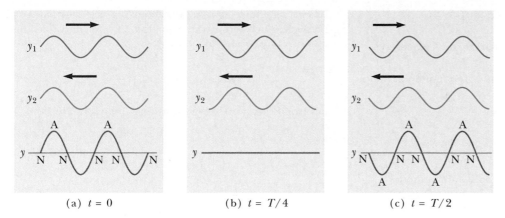

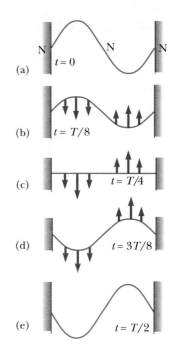

Figure 14.4 Standing wave patterns at various times produced by two waves of equal amplitude traveling in opposite directions. For the resultant wave y, the nodes (N) are points of zero displacement and the antinodes (A) are points of maximum displacement.

Figure 14.5 A standing wave pattern in a taut string showing snapshots during one half-cycle. (a) At $t = 0$, the string is momentarily at rest, and so $K = 0$ and all of the energy is potential energy U associated with the vertical displacements of the string segments. (b) At $t = T/8$, the string is in motion, and the energy is half kinetic and half potential. (c) At $t = T/4$, the string is horizontal (undeformed) and, therefore, $U = 0$; all of the energy is kinetic. The motion continues as indicated in (d) and (e), and ultimately the initial configuration in part (a) is repeated.

giving a standing wave of maximum amplitude, $2A$. One quarter of a period later, at $t = T/4$ (Fig. 14.4b), the individual waves have moved one quarter of a wavelength (one to the right and the other to the left). At this time, the individual displacements are equal and opposite for all values of x, and hence the resultant wave has zero displacement everywhere. At $t = T/2$ (Fig. 14.4c), the individual waves are again identical spatially, producing a standing wave pattern that is inverted relative to the $t = 0$ pattern.

It is instructive to describe the energy associated with the motion of a standing wave. To illustrate this point, consider a standing wave formed on a stretched string that is fixed at each end, as in Figure 14.5. Except for the nodes, which are stationary, all points on the string oscillate vertically with the same frequency. Furthermore, different points have different amplitudes of motion. Figure 14.5 represents snapshots of the standing wave at various times over one half-cycle. The nodes represent points on the string that are always at rest, and the locations of the maxima and minima never change. Each point on the string executes simple harmonic motion in the vertical direction. That is, one can view the standing wave as a large number of oscillators vibrating parallel to each other. The energy of the vibrating string continually alternates between elastic potential energy, at which time the string is momentarily stationary (Fig. 14.5a), and kinetic energy, at which time the string is horizontal and the particles have their maximum speed (Fig. 14.5c). The string particles have both potential energy and kinetic energy at intermediate times (Figs. 14.5b and 14.5d).

Example 14.2   Formation of a Standing Wave

Two waves traveling in opposite directions produce a standing wave. The individual wave functions are

$$y_1 = (4.0 \text{ cm}) \sin(3.0x - 2.0t)$$
$$y_2 = (4.0 \text{ cm}) \sin(3.0x + 2.0t)$$

where x and y are in centimeters. (a) Find the maximum displacement of the motion at $x = 2.3$ cm.

Solution When the two waves are summed, the result is a standing wave the function of which is given by Equation 14.3, with $A_0 = 4.0$ cm and $k = 3.0$ rad/cm:

$$y = (2A_0 \sin kx)\cos \omega t = [(8.0 \text{ cm})\sin 3.0x]\cos \omega t$$

Thus, the maximum displacement of the motion at the position $x = 2.3$ cm is

$$y_{\text{max}} = (8.0 \text{ cm})\sin 3.0x]_{x=2.3}$$
$$= (8.0 \text{ cm})\sin(6.9 \text{ rad}) = 4.6 \text{ cm}$$

(b) Find the positions of the nodes and antinodes.

Solution   Because $k = 2\pi/\lambda = 3$ rad/cm, we see that $\lambda = 2\pi/3$ cm. Therefore, from Equation 14.4 we find that the antinodes are located at

$$x = n\left(\frac{\pi}{6}\right) \text{ cm} \qquad (n = 1, 3, 5, \ldots)$$

and from Equation 14.5 we find that the nodes are located at

$$x = n\frac{\lambda}{2} = n\left(\frac{\pi}{3}\right) \text{ cm} \qquad (n = 1, 2, 3, \ldots)$$

14.3 • NATURAL FREQUENCIES IN A STRETCHED STRING

Consider a string that has a length of L and is fixed at both ends, as in Figure 14.6. Standing waves are set up in the string by a continuous superposition of waves incident on and reflected from the ends. The string has a number of natural patterns of vibration, called **normal modes.** Each of these modes has a characteristic frequency; the frequencies are easily calculated.

First, note that the ends of the string must be nodes, because these points are fixed. If the string is displaced at its midpoint and released, all modes that do not have a node at the center point are excited. Consider the normal mode in which the center of the string is an antinode (Fig. 14.6b). For this normal mode, the length of the string equals $\lambda/2$ (the distance between nodes):

$$L = \lambda_1/2 \qquad \text{or} \qquad \lambda_1 = 2L$$

The next normal mode, of wavelength λ_2 (Fig. 14.6c), occurs when the length of the string equals one wavelength, that is, when $\lambda_2 = L$. The third normal mode (Fig. 14.6d) corresponds to the case in which the length equals $3\lambda/2$; therefore,

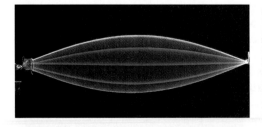
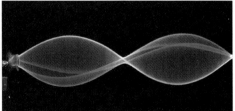
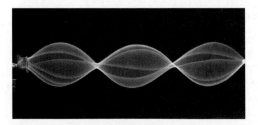

Multiflash photographs of standing wave patterns in a cord driven by a vibrator at the left end. The single loop pattern at the top left represents the fundamental ($n = 1$), the two-loop pattern at the right represents the second harmonic ($n = 2$), and the three-loop pattern at the lower left represents the third harmonic ($n = 3$). *(Richard Megna, Fundamental Photographs, NYC)*

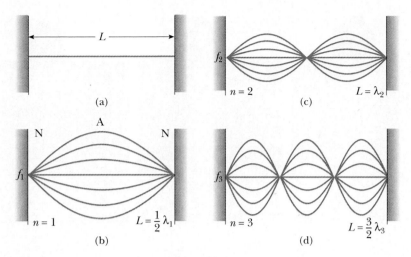

Figure 14.6 (a) A string of length L fixed at both ends. The normal modes of vibration form a harmonic series: (b) the fundamental frequency, or first harmonic; (c) the second harmonic; and (d) the third harmonic.

$\lambda_3 = 2L/3$. In general, the wavelengths of the various normal modes can be conveniently expressed as

$$\lambda_n = \frac{2L}{n} \qquad (n = 1, 2, 3, \ldots)$$ [14.6]

• *Wavelengths of normal modes*

where the index n refers to the nth mode of vibration. The natural frequencies associated with these modes are obtained from the relationship $f = v/\lambda$, where **the wave speed, v, is the same for all frequencies.** Using Equation 14.6, we find that the frequencies of the normal modes are

$$f_n = \frac{v}{\lambda_n} = \frac{n}{2L}v \qquad (n = 1, 2, 3, \ldots)$$ [14.7]

• *Frequencies of normal modes as functions of wave speed and length of string*

Because $v = \sqrt{F/\mu}$ (Eq. 13.19), where F is the tension in the string and μ is its mass per unit length, we can express the natural frequencies of a stretched string (sometimes called *harmonics*) as

$$f_n = \frac{n}{2L}\sqrt{\frac{F}{\mu}} \qquad (n = 1, 2, 3, \ldots)$$ [14.8]

• *Frequencies of normal modes as functions of string tension and linear mass density*

The lowest frequency, corresponding to $n = 1$, is called the **fundamental frequency, f_1,** and is

$$f_1 = \frac{1}{2L}\sqrt{\frac{F}{\mu}}$$ [14.9]

• *Fundamental frequency of a taut string*

It is clear that the frequencies of the remaining modes are integral multiples of the fundamental frequency—that is, $2f_1$, $3f_1$, $4f_1$, and so on. These higher natural frequencies, together with the fundamental frequency, form a **harmonic series.** The fundamental, f_1, is the first harmonic; the frequency $f_2 = 2f_1$ is the second harmonic; the frequency f_n is the nth harmonic.

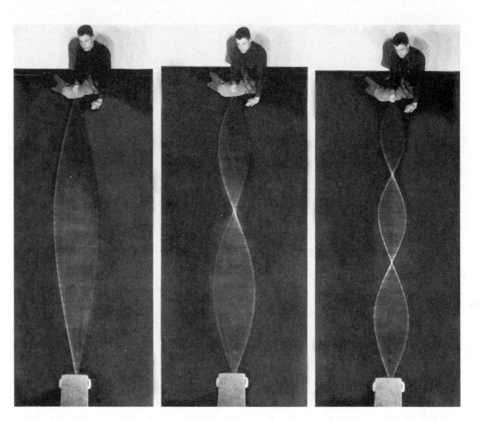

Photographs of standing waves. As one end of the rope is moved from side to side with increasing frequency, patterns with more and more loops are formed; only certain definite frequencies will produce fixed patterns.   *(Photos, Education Development Center, Newton, Mass.)*

We can obtain the foregoing results in an alternative manner. Because we require that the string be fixed at $x = 0$ and $x = L$, the wave function $y(x, t)$ given by Equation 14.3 must be *zero* at these points for *all* times. That is, the boundary conditions require that $y(0, t) = 0$ and $y(L, t) = 0$ for all values of t. Because $y = (2A \sin kx) \cos \omega t$, the first condition, $y(0, t) = 0$, is automatically satisfied because $\sin kx = 0$ at $x = 0$. To meet the second condition, $y(L, t) = 0$, we require that $\sin kL = 0$. This condition is satisfied when the angle kL equals an integral multiple of π (180°). Therefore, the allowed values of k are[1]

$$k_n L = n\pi \qquad (n = 1, 2, 3, \ldots) \qquad \textbf{[14.10]}$$

Because $k_n = 2\pi/\lambda_n$, we find that

$$\left(\frac{2\pi}{\lambda_n}\right) L = n\pi \qquad \text{or} \qquad \lambda_n = \frac{2L}{n}$$

which is identical to Equation 14.6.

When a stretched string is distorted to a shape that corresponds to any one of its harmonics, after being released it will vibrate at the frequency of that harmonic. However, if the string is plucked or bowed, the resulting vibration will include frequencies of various harmonics, including the fundamental. In effect, the string

---

[1] We exclude $n = 0$, because this corresponds to the trivial case in which no wave exists ($k = 0$).

"selects" the normal-mode frequencies when disturbed by a nonharmonic disturbance (for example, a finger plucking it).

Figure 14.7 shows a stretched string vibrating with its first and second harmonics simultaneously. In this figure, the combined vibration is the superposition of the two vibrations shown in Figures 14.6b and 14.6c. The larger loop corresponds to the fundamental frequency of vibration, f_1, and the smaller loops correspond to the second harmonic, f_2. In general, the resulting motion, or displacement, of the string can be described by a superposition of the various harmonic wave functions, with different frequencies and amplitudes. Hence, the sound that one hears corresponds to a complex waveform associated with these modes of vibration. (We shall return to this point in Section 14.6.)

The frequency of a stringed instrument can be changed either by varying the strings' tension, F, or by changing their length, L. For example, the tension in the strings of guitars and violins is adjusted by a screw mechanism or by turning pegs on the neck of the instrument. As the tension increases, the frequencies of the normal modes increase according to Equation 14.8. Once the instrument is "tuned," the player varies the frequency by moving his or her fingers along the neck, thereby changing the length of the vibrating portion of the string. As this length is reduced, the frequency increases, because the normal-mode frequencies are inversely proportional to (vibrating) string length.

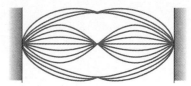

Figure 14.7   Multiple exposures of a string vibrating simultaneously in its first and second harmonics.

CONCEPTUAL PROBLEM 2

Guitarists sometimes play a "harmonic," by lightly touching a string at the exact center and plucking the string. The result is a clear note one octave higher than the fundamental of the string, even though the string is not pressed to the fingerboard. Why does this happen?

CONCEPTUAL PROBLEM 3

An archer shoots an arrow from a bow. Does the string of the bow exhibit standing waves after the arrow leaves? If so, and if the bow is perfectly symmetric so that the arrow leaves from the center of the string, what harmonics are excited?

Example 14.3   Give Me a C Note

A middle C string of the C-major scale on a piano has a fundamental frequency of 262 Hz, and the A note has a fundamental frequency of 440 Hz. (a) Calculate the frequencies of the next two harmonics of the C string.

Solution   Because $f_1 = 262$ Hz, we can use Equations 14.8 and 14.9 to find the frequencies f_2 and f_3:

$$f_2 = 2f_1 = \boxed{524 \text{ Hz}}$$

$$f_3 = 3f_1 = \boxed{786 \text{ Hz}}$$

(b) If the strings for the A and C notes are assumed to have the same mass per unit length and the same length, determine the ratio of tensions in the two strings.

Solution   Using Equation 14.8 for the two strings vibrating at their fundamental frequencies gives

$$f_{1A} = \frac{1}{2L}\sqrt{F_A/\mu} \quad \text{and} \quad f_{1C} = \frac{1}{2L}\sqrt{F_C/\mu}$$

$$f_{1A}/f_{1C} = \sqrt{F_A/F_C}$$

$$F_A/F_C = (f_{1A}/f_{1C})^2 = (440/262)^2 = \boxed{2.82}$$

(c) In a real piano, the assumption we made in (b) is only half true. The string densities are equal, but the A string is 64% as long as the C string. What is the ratio of their tensions?

$$f_{1A}/f_{1C} = (L_C/L_A)\sqrt{F_A/F_C} = (100/64)\sqrt{F_A/F_C}$$

$$F_A/F_C = (0.64)^2(440/262)^2 = \boxed{1.16}$$

EXERCISE 2 A stretched string is 160 cm long and has a linear density of 0.0150 g/cm. What tension in the string will result in a second harmonic of 460 Hz?   Answer   813 N

EXERCISE 3 A string of linear density 1.0×10^{-3} kg/m and length 3.0 m is stretched between two points. One end is vibrated transversely at 200 Hz. What tension in the string will establish a standing-wave pattern with three loops along the string's length?   Answer   160 N

14.4 · STANDING WAVES IN AIR COLUMNS

Standing longitudinal waves can be set up in a tube of air, such as an organ pipe, as the result of interference between longitudinal waves traveling in opposite directions. The phase relationship between the incident wave and the wave reflected from one end depends on whether that end is open or closed. This is analogous to the phase relationships between incident and reflected transverse waves at the ends of a string. **The closed end of an air column is a displacement node,** just as the fixed end of a vibrating string is a displacement node. As a result, the reflected wave at a closed end of a tube of air is 180° out of phase with the incident wave. Furthermore, because the pressure wave is 90° out of phase with the displacement wave (Section 13.9), **the closed end of an air column corresponds to a pressure antinode** (that is, a point of maximum pressure variation).

If the end of an air column is open to the atmosphere, the air molecules have complete freedom of motion. The wave reflected from an open end is nearly in phase with the incident wave when the tube's diameter is small relative to the wavelength of the sound. As a consequence, **the open end of an air column is approximately a displacement antinode and a pressure node.**

Strictly speaking, the open end of an air column is not exactly an antinode. A condensation does not reach full expansion until it passes somewhat beyond an open end. For a thin-walled tube of circular cross-section, this correction that must be added for each open end is about $0.6R$, where R is the tube's radius. Hence, the effective length of the tube is somewhat greater than the true length, L.

The first three modes of vibration of a pipe that is open at both ends are shown in Figure 14.8a. Note that in this figure, the standing longitudinal waves are drawn as transverse waves because it is difficult to draw longitudinal displacements. When air is directed into the pipe from the left, longitudinal standing waves are formed and the pipe resonates at its natural frequencies. All modes of vibration are excited simultaneously (although not with the same amplitude). Note that the ends are displacement antinodes (approximately). In the fundamental mode, the wavelength is twice the length of the pipe, and hence the frequency of the fundamental, f_1, is $v/2L$. In a similar way, the frequencies of the higher harmonics are $2f_1$, $3f_1$, ... Thus, in a pipe that is open at both ends, the natural frequencies of vibration form a harmonic series; that is, the higher harmonics are integral multiples of the fundamental frequency. Because all harmonics are present, we can express the natural frequencies of vibration as

Natural frequencies of a pipe •
open at both ends

$$f_n = n\,\frac{v}{2L} \qquad (n = 1, 2, 3, \ . \ . \ . \ ) \qquad \textbf{[14.11]}$$

where v is the speed of sound in air.

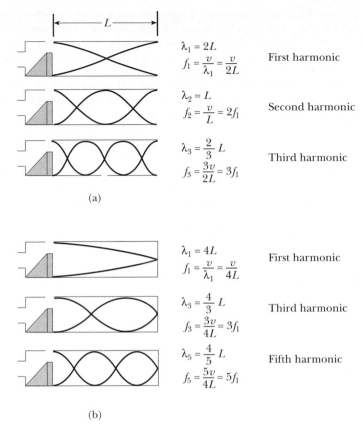

$\lambda_1 = 2L$
$f_1 = \dfrac{v}{\lambda_1} = \dfrac{v}{2L}$ First harmonic

$\lambda_2 = L$
$f_2 = \dfrac{v}{L} = 2f_1$ Second harmonic

$\lambda_3 = \dfrac{2}{3}L$
$f_3 = \dfrac{3v}{2L} = 3f_1$ Third harmonic

(a)

$\lambda_1 = 4L$
$f_1 = \dfrac{v}{\lambda_1} = \dfrac{v}{4L}$ First harmonic

$\lambda_3 = \dfrac{4}{3}L$
$f_3 = \dfrac{3v}{4L} = 3f_1$ Third harmonic

$\lambda_5 = \dfrac{4}{5}L$
$f_5 = \dfrac{5v}{4L} = 5f_1$ Fifth harmonic

(b)

Figure 14.8   (a) Standing longitudinal waves in an organ pipe open at both ends. The natural frequencies that form a harmonic series are f_1, $2f_1$, $3f_1$, ... (b) Standing longitudinal waves in an organ pipe closed at one end. Only the *odd* harmonics are present, and so the natural frequencies are f_1, $3f_1$, $5f_1$, ...

If a pipe is closed at one end and open at the other, the closed end is a displacement node (Fig. 14.8b). In this case, the wavelength for the fundamental mode is four times the length of the tube. Hence, the fundamental, f_1, is equal to $v/4L$, and the frequencies of the higher harmonics are equal to $3f_1$, $5f_1$, ... That is, **in a pipe that is closed at one end, only odd harmonics are present,** and these are

$$f_n = n\,\frac{v}{4L} \qquad (n = 1, 3, 5, \ldots)$$ [14.12]

• *Natural frequencies of a pipe closed at one end and open at the other*

Standing waves in air columns are the primary sources of the sounds produced by wind instruments.

Thinking Physics 2

Passing ocean waves sometimes cause the water in a harbor to undergo very large oscillations, called *seiches*. Why would this happen?

Reasoning   Water in a harbor is enclosed and possesses a natural frequency based on the size of the harbor. This is similar to the natural frequency of the enclosed air in a bottle, which can be excited by blowing across the edge of the opening. Ocean waves pass by the opening of the harbor at a certain frequency. If this frequency matches that of the enclosed harbor, then a large standing wave can be set up in the water by resonance. This can be simulated by carrying a fish tank with water. If your walking frequency matches the natural frequency of the water as it sloshes back and forth, a large standing wave in the fish tank can be established.

Thinking Physics 3

A bugle has no valves, keys, slides, or finger holes—how can it play a song?

Reasoning   Songs for the bugle are limited to harmonics of the fundamental frequency, because there is no control over frequencies without valves, keys, slides, or finger holes. The player obtains different notes by changing the tension in the lips as the bugle is played, in order to excite different harmonics. The normal playing range of a bugle is among the third, fourth, fifth, and sixth harmonics of the fundamental. For example, "Reveille" is played with just the three notes G, C, and E, and "Taps" is played with these three notes and the G one octave above the lower G.

Thinking Physics 4

If an orchestra doesn't warm up before a performance, the strings go flat and the wind instruments go sharp during the performance. Why?

Reasoning   Without warming up, all the instruments will be at room temperature at the beginning of the concert. As the wind instruments are played, they fill with warm air from the player's exhalation. The increase in temperature of the air in the instrument causes an increase in the speed of sound, which raises the resonance frequencies of the air columns. As a result, the instruments go sharp. The strings on the stringed instruments also increase in temperature due to the friction of rubbing with the bow. This results in thermal expansion, which causes a decrease in the tension in the strings. With a decrease in tension, the wave speed on the strings drops, and the fundamental frequencies decrease. Thus, the stringed instruments go flat.

CONCEPTUAL PROBLEM 4

In Balboa Park in San Diego, California, there is an outdoor organ. Does the fundamental frequency of a particular pipe on this organ change on hot and cold days? How about on days with high and low atmospheric pressure?

CONCEPTUAL PROBLEM 5

If you have a series of identical glass bottles, with varying amounts of water in them, you can play musical notes by either striking the bottles with a spoon, or blowing across the open tops of the bottles. When hitting the bottles, the frequency of the note decreases as the water level rises. When blowing on the bottles, the frequency of the note increases as the water level rises. Why is the behavior of the frequency different in these two cases?

Example 14.4   Resonance in a Pipe

A pipe has a length of 1.23 m. (a) Determine the frequencies of the first three harmonics if the pipe is open at each end. Take $v = 343$ m/s as the speed of sound in air.

Solution   The first harmonic of a pipe open at both ends is

$$f_1 = \frac{v}{2L} = \frac{343 \text{ m/s}}{2(1.23 \text{ m})} = \boxed{139 \text{ Hz}}$$

Because all harmonics are present, the second and third harmonics are $f_2 = 2f_1 = 278$ Hz and $f_3 = 3f_1 = 417$ Hz.

(b) What are the three frequencies determined in part (a) if the pipe is closed at one end?

Solution   The fundamental frequency of a pipe closed at one end is

$$f_1 = \frac{v}{4L} = \frac{343 \text{ m/s}}{4(1.23 \text{ m})} = \boxed{69.7 \text{ Hz}}$$

In this case, only odd harmonics are present, and so the next two resonances have frequencies $f_3 = 3f_1 = 209$ Hz and $f_5 = 5f_1 = 349$ Hz.

(c) For the pipe open at both ends, how many harmonics are present in the normal human hearing range (20 to 20 000 Hz)?

Solution   Because all harmonics are present, $f_n = nf_1$. For $f_n = 20\ 000$ Hz, we have $n = 20\ 000/139 = 144$, so that 144 harmonics are present in the audible range. Actually, only the first few harmonics have sufficient amplitude to be heard.

Example 14.5   Measuring the Frequency of a Tuning Fork

A simple apparatus for demonstrating resonance in a tube is described in Figure 14.9a. A long, vertical tube open at both ends is partially submerged in a beaker of water, and a vibrating tuning fork of unknown frequency is placed near the top. The length of the air column, L, is adjusted by moving the tube vertically. The sound waves generated by the fork are reinforced when the length of the air column corresponds to one of the resonant frequencies of the tube.

For a certain tube, the smallest value of L for which a peak occurs in the sound intensity is 9.00 cm. From this measurement, determine the frequency of the tuning fork and the value of L for the next two resonant modes.

Reasoning   Although the tube is open at both ends to allow the water in, the water surface acts like a wall at one end of the length L. Therefore this setup represents a pipe closed at one end and the fundamental has a frequency of $v/4L$ (Fig. 14.9b).

Solution   Taking $v = 343$ m/s for the speed of sound in air and $L = 0.0900$ m, we get

$$f_1 = \frac{v}{4L} = \frac{343 \text{ m/s}}{4(0.0900 \text{ m})} = 953 \text{ Hz}$$

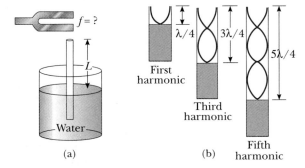

Figure 14.9   (Example 14.5) (a) Apparatus for demonstrating the resonance of sound waves in a tube closed at one end. The length L of the air column is varied by moving the tube vertically while it is partially submerged in water. (b) The first three normal modes of the system shown in part (a).

From this information about the fundamental mode, we see that the wavelength is $\lambda = 4L = 0.360$ m. Because the frequency of the source is constant, the next two resonance modes (Fig. 14.9b) correspond to lengths of $3\lambda/4 = 0.270$ m and $5\lambda/4 = 0.450$ m.

EXERCISE 4   The longest pipe on a certain organ is 4.88 m. What is the fundamental frequency (at 0.0°C) if the nondriven end of the pipe is (a) closed and (b) open?
Answer   (a) 17.0 Hz   (b) 34.0 Hz

EXERCISE 5   An air column 2.00 m in length is open at both ends. The frequency of a certain harmonic is 410 Hz, and the frequency of the next higher harmonic is 492 Hz. Determine the speed of sound in the air column.   Answer   328 m/s

14.5 • BEATS: INTERFERENCE IN TIME

The interference phenomena with which we have been dealing so far involve the superposition of two or more waves with the same frequency, traveling in opposite directions. Because the resultant waveform in this case depends on the coordinates of the disturbed medium, we can refer to the phenomenon as *spatial interference.* Standing waves in strings and pipes are common examples of spatial interference.

We now consider another type of interference effect, one that results from the superposition of two waves with slightly *different frequencies.* In this case, when the two waves are observed at a given point, they are periodically in and out of phase. That is, there occurs a temporal alternation between constructive and destructive interference. We refer to this phenomenon as **interference in time** or **temporal interference.**

For example, if two tuning forks of slightly different frequencies are struck, one hears a sound of pulsating intensity, called a **beat.**

Definition of beats •

> A **beat** is the periodic variation in intensity at a given point due to the superposition of two waves with slightly different frequencies.

The number of beats one hears per second, or **beat frequency,** equals the difference in frequency between the two sources. The maximum beat frequency that the human ear can detect is about 20 beats/s. When the beat frequency exceeds this value, it blends indistinguishably with the compound sounds producing the beats.

One can use beats to tune a stringed instrument, such as a piano, by beating a note against a reference tone of known frequency. The string can then be adjusted to equal the frequency of the reference by tightening or loosening it until the beats become too infrequent to notice.

Consider two waves with equal amplitudes, traveling through a medium with slightly different frequencies, f_1 and f_2. We can represent the displacement that each wave would produce at a point as

$$y_1 = A \cos 2\pi f_1 t \qquad \text{and} \qquad y_2 = A \cos 2\pi f_2 t$$

Using the superposition principle, we find that the resultant displacement at that point is given by

$$y = y_1 + y_2 = A(\cos 2\pi f_1 t + \cos 2\pi f_2 t)$$

It is convenient to write this in a form that uses the trigonometric identity

$$\cos a + \cos b = 2 \cos \left(\frac{a - b}{2} \right) \cos \left(\frac{a + b}{2} \right)$$

Letting $a = 2\pi f_1 t$ and $b = 2\pi f_2 t$, we find that

Resultant of two waves of •
different frequencies but equal
amplitude

$$y = 2A \cos 2\pi \left(\frac{f_1 - f_2}{2} \right) t \cos 2\pi \left(\frac{f_1 + f_2}{2} \right) t \qquad \textbf{[14.13]}$$

Graphs demonstrating the individual waveforms as well as the resultant wave are shown in Figure 14.10. From the factors in Equation 14.13, we see that the resultant

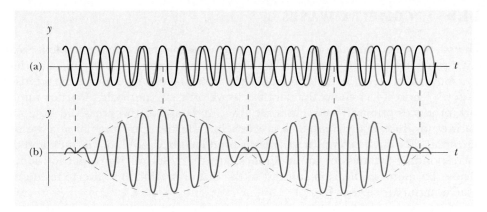

Figure 14.10 Beats are formed by the combination of two waves of slightly different frequencies traveling in the same direction. (a) The individual waves. (b) The combined wave has an amplitude (broken line) that oscillates in time.

vibration at a point has an effective frequency equal to the average frequency, $(f_1 + f_2)/2$, and an amplitude of

$$A = 2A \cos 2\pi \left(\frac{f_1 - f_2}{2} \right) t \qquad \textbf{[14.14]}$$

That is, the *amplitude varies in time* with a frequency of $(f_1 - f_2)/2$. When f_1 is close to f_2, this amplitude variation is slow, as illustrated by the envelope (broken line) of the resultant waveform in Figure 14.10b.

Note that a beat, or a maximum in amplitude, will be detected whenever

$$\cos 2\pi \left(\frac{f_1 - f_2}{2} \right) t = \pm 1$$

That is, there are *two* maxima in each cycle. Because the amplitude varies with frequency as $(f_1 - f_2)/2$, the number of beats per second, or the beat frequency, f_b, is twice this value:

$$f_b = f_1 - f_2 \qquad \textbf{[14.15]} \qquad \bullet \quad \textit{Beat frequency}$$

For instance, if two tuning forks vibrate individually at frequencies of 438 Hz and 442 Hz, respectively, the resultant sound wave of the combination has a frequency of 440 Hz (the musical note A) and a beat frequency of 4 Hz. That is, the listener hears the 440-Hz sound wave go through an intensity maximum four times every second.

CONCEPTUAL PROBLEM 6

You have a standard tuning fork whose frequency you know and a second tuning fork whose frequency you don't know. When you play them together, you hear a beat frequency of 4 Hz. You know that the frequency of the mystery tuning fork is different from that of the standard fork by 4 Hz, but you can't tell whether it is higher or lower. What could you do to determine this? (*Hint:* You are chewing gum at the time that you perform these measurements.)

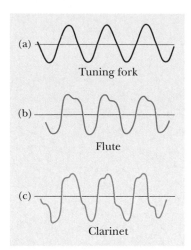

Figure 14.11 Waveform produced by (a) a tuning fork, (b) a flute, and (c) a clarinet, each at approximately the same frequency. *(Adapted from C. A. Culver, Musical Acoustics, 4th ed., New York, McGraw-Hill, 1956, p. 128.)*

14.6 • COMPLEX WAVES

The sound wave patterns produced by most instruments are very complex. Some characteristic waveforms produced by a tuning fork, a harmonic flute, and a clarinet are shown in Figure 14.11. Although each instrument has its own characteristic pattern, Figure 14.11 shows that all three waveforms are periodic. A struck tuning fork produces primarily one harmonic (the fundamental), whereas the flute and clarinet produce many frequencies, which include the fundamental and various harmonics. Thus, the complex waveforms produced by a violin or clarinet, and the corresponding richness of musical tones, are the result of the superposition of various harmonics. This is in contrast to the drum, in which the overtones do not form a harmonic series.

Analysis of complex waveforms appears at first sight to be a formidable task. However, if the waveform is periodic, it can be represented with arbitrary precision by the combination of a sufficiently large number of sinusoidal waves that form a harmonic series. In fact, one can represent any periodic function or any function over a finite interval as a series of sine and cosine terms by using a mathematical technique based on *Fourier's theorem*. The corresponding sum of terms that represents the periodic waveform is called a **Fourier series.**

Let $y(t)$ be any function that is periodic in time, with a period of T, so that $y(t + T) = y(t)$. **Fourier's theorem** states that this function can be written

$$y(t) = \sum_n (A_n \sin 2\pi f_n t + B_n \cos 2\pi f_n t) \qquad [14.16]$$

where the lowest frequency is $f_1 = 1/T$.

The higher frequencies are integral multiples of the fundamental, and so $f_n = nf_1$. The coefficients A_n and B_n represent the amplitudes of the various waves. The amplitude of the nth harmonic is proportional to $\sqrt{A_n{}^2 + B_n{}^2}$, and its intensity is proportional to $A_n{}^2 + B_n{}^2$.

Figure 14.12 represents a harmonic analysis of the waveforms shown in Figure 14.11. Note the variation of relative intensity with harmonic content for the flute

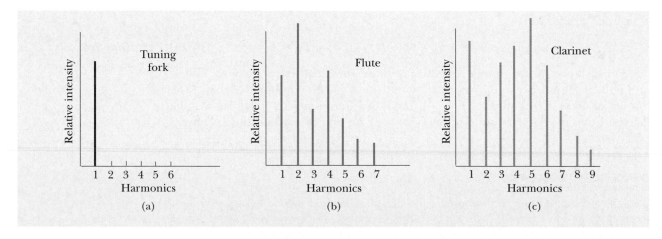

Figure 14.12 Harmonics of the waveforms shown in Figure 14.11. Note the variations in intensity of the various harmonics. *(Adapted from C. A. Culver, Musical Acoustics, 4th ed., New York, McGraw-Hill, 1956.)*

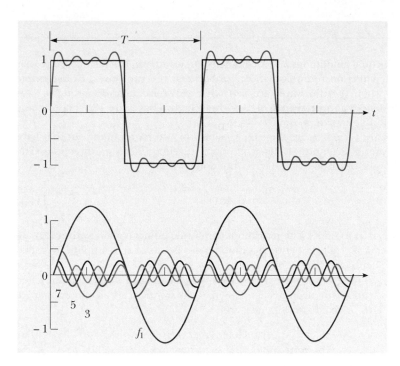

Figure 14.13 Harmonic synthesis of a square wave, which can be represented by the sum of odd harmonics of the fundamental. *(From M. L. Warren, Introductory Physics, New York, W. H. Freeman, 1979, p. 178; by permission of the publisher.)*

and clarinet. In general, any musical sound contains components that are members of a harmonic set with varying relative intensities.

As an example of Fourier synthesis, consider the periodic square wave shown in Figure 14.13. The square wave is synthesized by a series of *odd* harmonics of the fundamental. The series contains only sine functions (that is, $B_n = 0$ for all n). Only the first four odd harmonics and their respective amplitudes are shown. One obtains a better fit to the true waveform by adding more harmonics.

Using modern technology, one can produce a variety of musical tones by mixing harmonics with varying amplitudes. Electronic music synthesizers that do this are now widely used.

Thinking Physics 5

A professor performs a demonstration in which he breathes helium and then speaks with a comical voice. One student explains, "The velocity of sound in helium is higher than in air, so the fundamental frequency of the standing waves in the mouth is increased." Another student says, "No, the fundamental frequency is determined by the vocal folds and cannot be changed. Only the *quality* of the voice has changed." Which student is correct?

Reasoning The second student is correct. The fundamental frequency of the complex tone from the voice is determined by the vibration of the vocal folds and is not changed by substituting a different gas in the mouth. The introduction of the helium into the mouth results in higher harmonics being excited more than in the normal voice, but the fundamental frequency of the voice is the same—only the quality has changed. The unusual inclusion of the higher frequency harmonics results in a common description of this effect as a "high-pitched" voice, but this is incorrect.

SUMMARY

When two waves with equal amplitudes and frequencies superimpose, the resultant wave has an amplitude that depends on the phase angle, ϕ, between the two waves. **Constructive interference** occurs when the two waves are *in phase* everywhere, corresponding to $\phi = 0, 2\pi, 4\pi, \ldots$. **Destructive interference** occurs when the two waves are 180° out of phase everywhere, corresponding to $\phi = \pi, 3\pi, 5\pi, \ldots$.

Standing waves are formed from the superposition of two harmonic waves that have the same frequency, amplitude, and wavelength but are traveling in *opposite* directions. The resultant standing wave is described by the wave function

$$y = (2A \sin kx)\cos \omega t \qquad\qquad \textbf{[14.3]}$$

Hence, its amplitude varies as $\sin kx$. The maximum amplitude points (called **antinodes**) are separated by a distance $\lambda/2$. Halfway between antinodes are points of zero amplitude (called **nodes**). The distance between adjacent nodes is $\lambda/2$.

One can set up standing waves with specific frequencies in such systems as stretched strings, hollow pipes, rods, and drumheads. The natural frequencies of vibration of a stretched string of length, L, fixed at both ends, are

$$f_n = \frac{n}{2L}\sqrt{\frac{F}{\mu}} \qquad (n = 1, 2, 3, \ldots) \qquad\qquad \textbf{[14.8]}$$

where F is the tension in the string and μ is its mass-per-unit length. The natural frequencies of vibration form a **harmonic series**—that is, $f_1, 2f_1, 3f_1, \ldots$.

The standing wave patterns for longitudinal waves in a hollow pipe depend on whether the ends of the pipe are open or closed. If the pipe is open at both ends, the natural frequencies of vibration form a harmonic series. If one end is closed, only odd harmonics of the fundamental are present.

The phenomenon of **beats** occurs as a result of the superposition of two traveling waves of slightly different frequencies. For sound waves at a given point, one hears an alternation in sound intensity with time. Thus, beats correspond to *interference as time passes*.

Any periodic waveform can be represented by the combination of the sinusoidal waves that form a harmonic series. The process is called *Fourier synthesis* and is based on *Fourier's theorem*.

CONCEPTUAL QUESTIONS

1. When two waves interfere constructively or destructively, is there any gain or loss in energy? Explain.

2. Does the phenomenon of wave interference apply only to sinusoidal waves?

3. Some singers claim to be able to shatter a wine glass by maintaining a certain vocal pitch over a period of several seconds (Fig. Q14.3). What mechanism causes the glass to break? (The glass must be very clean in order to break.)

4. What limits the amplitude of motion of a real vibrating system that is driven at one of its natural frequencies?

5. Explain why all harmonics are present in an organ pipe open at both ends, but only the odd harmonics are present in a pipe closed at one end.

6. An airplane mechanic notices that the sound from a twin-engine aircraft rapidly varies in loudness when both engines are running. What could be causing this variation from loud to soft?

7. Why does a vibrating guitar string sound louder when placed on the instrument than it would if allowed to vibrate in the air while off the instrument?

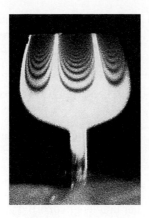

Figure Q14.3 (Question 3) *(Left)* Standing wave pattern in a vibrating wine glass. The wine glass will shatter if the amplitude of vibration becomes too large. *(Courtesy of Professor Thomas D. Rossing, Northern Illinois University)* *(Right)* A wine glass shattered by the amplified sound of a human voice. *(© Ben Rose 1992/The IMAGE Bank)*

8. When the base of a vibrating tuning fork is placed against a chalkboard, the sound becomes louder. How does this affect the length of time for which the fork vibrates? Does this agree with conservation of energy?

9. To keep animals away from their cars, some people mount short, thin pipes on the fenders. The pipes give out a high-pitched wail when the cars are moving. How do they create the sound?

10. If you wet your fingers and lightly run them around the rim of a fine wine glass, a high-pitched sound is heard. Why? How could you produce various musical notes with a set of wine glasses?

11. When a bell is rung, standing waves are set up around the bell's circumference. What boundary conditions must be satisfied by the resonant wavelengths? How does a crack in the bell, such as in the Liberty Bell, affect the satisfying of the boundary conditions and the sound emanating from the bell?

12. Explain why your voice seems to sound better than usual when you sing in the shower.

13. What is the purpose of the slide on a trombone or the valves on a trumpet?

14. Despite a reasonably steady hand, a certain person often spills his coffee when carrying it to his seat. Discuss resonance as a possible cause of this difficulty and devise a means for solving the problem.

PROBLEMS

Section 14.1 Superposition and Interference of Sinusoidal Waves

1. Two harmonic waves are described by

$$y_1 = (5.00 \text{ m}) \sin[\pi(4.00x - 1200t)]$$

$$y_2 = (5.00 \text{ m}) \sin[\pi(4.00x - 1200t - 0.250)]$$

where x, y_1, and y_2 are in meters and t is in seconds. (a) What is the amplitude of the resultant wave? (b) What is the frequency of the resultant wave?

2. A sinusoidal wave is described by

$$y_1 = (0.0800 \text{ m}) \sin[2\pi(0.100x - 80.0t)]$$

where y_1 and x are in meters and t is in seconds. Write an expression for a wave that has the same frequency, amplitude, and wavelength as y_1 but when added to y_1 gives a resultant with an amplitude of $8\sqrt{3}$ cm.

3. Two waves are traveling in the same direction along a stretched string. Each has an amplitude of 4.00 cm and they are 90.0° out of phase. Find the amplitude of the resultant wave.

4. Two identical sinusoidal waves with wavelengths of 3.00 m travel in the same direction at a speed of 2.00 m/s. The second wave originates from the same point as the first but at a later time. Determine the minimum possible time interval between the starting moments of the two waves if the amplitude of the resultant wave is the same as that of the two initial waves.

5. Two speakers are driven by a common oscillator at 800 Hz and face each other at a distance of 1.25 m. Locate the points along a line joining the two speakers at which relative minima would be expected. (Use $v = 343$ m/s.)

6. A tuning fork generates sound waves with a frequency of 246 Hz. The waves travel in opposite directions along a hall-

way, are reflected by walls, and return. What is the phase difference between the reflected waves when they meet? The corridor is 47.0 m long and the tuning fork is located 14.0 m from one end. The speed of sound in air is 343 m/s.

Section 14.2   Standing Waves

7. Use the trigonometric identity

$$\sin(a \pm b) = \sin a \cos b \pm \cos a \sin b$$

to show that the resultant of two wave functions each of amplitude A, angular frequency ω, and propagation number k and traveling in opposite directions can be written

$$y = (2A \sin kx) \cos \omega t$$

For simplicity, set all phase angles ϕ to zero.

8. Two sinusoidal waves traveling in opposite directions interfere to produce a standing wave described by

$$y = (1.50 \text{ m}) \sin(0.400x) \cos(200t)$$

where x is in meters and t is in seconds. Determine the wavelength, frequency, and speed of the interfering waves.

9. Two waves in a long string are given by

$$y_1 = (0.0150 \text{ m}) \cos\left(\frac{x}{2} - 40t\right)$$

$$y_2 = (0.0150 \text{ m}) \cos\left(\frac{x}{2} + 40t\right)$$

where the y's and x are in meters and t is in seconds. (a) Determine the positions of the nodes of the resulting standing wave. (b) What is the maximum displacement at the position $x = 0.400$ m?

10. Two waves that set up a standing wave in a long string are given by

$$y_1 = A \sin(kx - \omega t + \phi)$$

$$y_2 = A \sin(kx + \omega t)$$

Show (a) that the addition of the arbitrary phase angle will change only the position of the nodes and (b) that the distance between nodes remains constant.

Section 14.3   Natural Frequencies in a Stretched String

11. A student wants to establish a standing wave on a wire that is 1.80 m long and clamped at both ends. The wave speed is 540 m/s. What is the minimum frequency the student should apply to set up standing waves?

12. A standing wave is established in a 120–cm-long string fixed at both ends. The string vibrates in four segments when driven at 120 Hz. (a) Determine the wavelength. (b) What is the fundamental frequency?

13. A cello A-string vibrates in its fundamental mode with a frequency of 220 vibrations/s. The vibrating segment is 70.0 cm long and has a mass of 1.20 g. (a) Find the tension in the string. (b) Determine the frequency of the harmonic that causes the string to vibrate in three segments.

14. A string of length L, mass-per-unit length μ, and tension F is vibrating at its fundamental frequency. What effect will the following have on the fundamental frequency? (a) The length of the string is doubled, with all other factors held constant. (b) The mass per unit length is doubled, with all other factors held constant. (c) The tension is doubled, with all other factors held constant.

15. A 60.0-cm guitar string under a tension of 50.0 N has a mass per unit length of 0.100 g/cm. What is the highest resonant frequency that can be heard by a person capable of hearing frequencies up to 20 000 Hz?

16. A stretched wire vibrates in its fundamental mode at a frequency of 400 vibrations/s. What would be the fundamental frequency if the wire were half as long, with twice the diameter and with four times the tension?

17. A 2.00–m-long wire having a mass of 0.100 kg is fixed at both ends. The tension in the wire is maintained at 20.0 N. What are the frequencies of the first three allowed modes of vibration? If a node is observed at a point 0.400 m from one end, in what mode and with what frequency is it vibrating?

18. A violin string has a length of 0.350 m and is tuned to concert G, having a frequency $f_G = 392$ Hz. Where must the violinist place her finger to play concert A, having a frequency $f_A = 440$ Hz? If this position is to remain correct to one half the width of a finger (i.e., to within 0.600 cm), the string tension cannot be allowed to slip by more than what fraction?

19. Find the fundamental frequency and the next three frequencies that could cause a standing wave pattern on a string that is 30.0 m long, has a mass per unit length 9.00×10^{-3} kg/m, and is stretched to a tension of 20.0 N.

20. Standing-wave vibrations are set up in a crystal goblet with two nodes and two antinodes equally spaced around the 20.0-cm circumference of its rim. If transverse waves move around the glass at 900 m/s, an opera singer would have to produce a high harmonic with what frequency in order to shatter the glass with a resonant vibration?

Section 14.4   Standing Waves in Air Columns

(In this section, unless otherwise indicated, assume that the speed of sound in air is 343 m/s.)

21. Calculate the length for a pipe that has a fundamental frequency of 240 Hz if the pipe is (a) closed at one end and (b) open at both ends.

22. A glass tube (open at both ends) of length L is positioned near an audio speaker of frequency $f = 0.680$ kHz. For what values of L will the tube resonate with the speaker?

23. The overall length of a piccolo is 32.0 cm. The resonating air column vibrates as a pipe open at both ends. (a) Find the frequency of the lowest note a piccolo can play, assuming the speed of sound in air is 340 m/s. (b) Opening holes in the side effectively shortens the length of the resonant column. If the highest note a piccolo can sound is 4000 Hz, find the distance between adjacent nodes for this mode of vibration.

24. The fundamental frequency of an open organ pipe corresponds to middle C (261.6 Hz on the chromatic musical scale). The third resonance of a closed organ pipe has the same frequency. What are the lengths of the two pipes?

25. A tuning fork the frequency of which is f is used to set up a resonance condition in a pipe. Write an expression for the length that will cause the pipe to resonate in its nth mode if it is (a) open at both ends; (b) closed at one end. (Assume that the speed of sound is v.)

26. Do not stick anything into your ear! Estimate the length of your ear canal, from its opening at the external ear in to the eardrum. If you regard the canal as a tube that is open at one end and closed at the other, at approximately what fundamental frequency would you expect your hearing to be most sensitive? Explain why you can hear especially soft sounds just around this frequency.

27. A shower stall measures 86.0 cm × 86.0 cm × 210 cm. When you sing in the shower, which frequencies will sound the richest (resonate), assuming the shower acts as a pipe closed at both ends (nodes at opposite sides)? Assume also that the human voice ranges from 130 Hz to 2000 Hz (not necessarily one person's voice, however). Let the speed of sound in the hot shower stall be 355 m/s.

28. An open pipe 0.400 m in length is placed vertically in a cylindrical bucket, nearly touching the bottom of the bucket, which has area 0.100 m². Water is slowly poured into the bucket until a sounding tuning fork of frequency 440 Hz, held over the pipe, produces resonance. Find the mass of water in the bucket at this moment.

29. Water is pumped into a long cylinder at a rate of 18.0 cm³/s. The radius of the cylinder is 4.00 cm, and at the open top of the cylinder there is a tuning fork vibrating with a frequency of 200 Hz. As the water rises, how much time elapses between successive resonances?

30. Water is pumped into a long cylinder at a volume flow rate R. The radius of the cylinder is r (cm), and at the open top of the cylinder there is a tuning fork vibrating with a frequency f. As the water rises, how much time elapses between successive resonances?

31. A piece of metal pipe is just the right length so that when it is cut into two pieces, their lowest resonance frequencies are 256 Hz for one and 440 Hz for the other. (a) What resonant frequency would have been produced by the original length of pipe, and (b) how long was the original piece?

32. A piece of cardboard tubing, closed at one end, is just the right length so that when it is cut into two (unequal) pieces, their lowest resonant frequencies are 256 Hz for the piece with the closed end and 440 Hz for the piece with both ends open. (a) What resonant frequency would have been produced by the original cardboard tubing, and (b) how long was the original piece?

Section 14.5 Beats: Interference in Time (Optional)

33. In certain ranges of a piano keyboard, more than one string is tuned to the same note to provide extra loudness. For example, the note at 110 Hz has two strings at this pitch. If one string slips from its normal tension of 600 N to 540 N, what beat frequency will be heard when the two strings are struck simultaneously?

34. A student holds a tuning fork oscillating at 256 Hz. He walks toward a wall at a constant speed of 1.33 m/s. (a) What beat frequency does he observe between the tuning fork and its echo? (b) How fast must he walk away from the wall to observe a beat frequency of 5.00 Hz?

35. A flute is designed so that it plays a frequency of 261.6 Hz, middle C, when all the holes are covered and the temperature is 20.0°C. (a) Consider the flute as a pipe that is open at both ends; find the length of the flute, assuming that middle C is the fundamental. (b) A second player, nearby in a colder room, also attempts to play middle C on an identical flute. A beat frequency of 3.00 Hz is heard. What is the temperature of the room? The speed of sound in air is described by

$$v = (331 \text{ m/s}) \sqrt{1 + \frac{T}{273°}}$$

where T is the Celsius temperature.

Additional Problems

36. Two loudspeakers are placed on a wall, 2.00 m apart. A listener stands directly in front of one of the speakers, 3.00 m from the wall. The speakers are being driven by a single oscillator at a frequency of 300 Hz. (a) What is the phase difference between the two waves when they reach the observer? (b) What is the frequency closest to 300 Hz to which the oscillator may be adjusted such that the observer will hear minimal sound?

37. On a marimba, the wooden bar that sounds a tone when struck vibrates as a transverse standing wave with three antinodes and two nodes. The lowest frequency note is 87.0 Hz, produced by a bar 40.0 cm long. (a) Find the speed of transverse waves on the bar. (b) The loudness and duration of the emitted sound are enhanced by a resonant pipe suspended vertically below the center of the bar. If the pipe is open at the top end only and the speed of sound in air is 340 m/s, what is the length of the pipe required to resonate with the bar in part (a)?

(Problem 37) Marimba players in Mexico City.   *(Murray Greenberg)*

38. A speaker at the front of a room and an identical speaker at the rear of the room are being driven by the same oscillator at 456 Hz. A student walks at a uniform rate of 1.50 m/s along the length of the room. How many beats does the student hear per second?

39. Two train whistles have identical frequencies of 180 Hz. When one train is at rest in the station sounding its whistle, a beat frequency of 2.00 Hz is heard from a moving train. What two possible speeds and directions can the moving train have?

40. Jane waits on a railroad platform, while two trains approach from the same direction at equal speeds of 8.00 m/s. Both trains are blowing their whistles (which have the same frequency), and one train is some distance behind the other. After the first train passes Jane, but before the second train passes her, she hears beats of frequency 4.00 Hz. What is the frequency of the train whistles?

41. A string (mass = 4.80 g, length = 2.00 m, and tension = 48.0 N), fixed at both ends, vibrates in its second ($n = 2$) natural mode. What is the wavelength in air of the sound generated by this vibrating string?

42. A pipe that is open at both ends has a fundamental frequency of 300 Hz when the speed of sound in air is 333 m/s. (a) What is the length of the pipe? (b) What is the frequency of the second harmonic when the temperature of the air is increased so that the speed of sound in the pipe is 344 m/s?

43. A string with a mass of 8.00 g and a length of 5.00 m has one end attached to a wall; the other end is draped over a pulley and attached to a hanging mass of 4.00 kg. If the string is plucked, what is the fundamental frequency of vibration?

44. In a major chord on the physical-pitch musical scale, the frequencies are in the ratios 4:5:6:8. A set of pipes, closed at one end, are to be cut so that when sounded in their fundamental mode, they will sound out a major chord. (a) What is the ratio of the lengths of the pipes? (b) What length pipes are needed if the lowest frequency of the chord is 256 Hz? (c) What are the frequencies of this chord?

45. Two wires are welded together. The wires are of the same material, but one is twice the diameter of the other one. They are subjected to a tension of 4.60 N. The thin wire has a length of 40.0 cm and a linear mass density of 2.00 g/m. The combination is fixed at both ends and vibrated in such a way that two antinodes are present with the node between them being right at the weld. (a) What is the frequency of vibration? (b) How long is the thick wire?

46. Two identical strings, each fixed at both ends, are arranged near each other. If string A starts oscillating in its fundamental mode, it is observed that string B will begin vibrating in its third ($n = 3$) natural mode. Determine the ratio of the tension of string B to the tension of string A.

47. A standing wave is set up in a string of variable length and tension by a vibrator of variable frequency. When the vibrator has a frequency f, using a string of length L and tension F, there are n antinodes set up in the string. (a) If the length of the string is doubled, by what factor should the frequency be changed to get the same number of antinodes? (b) If the frequency and length are held constant, what tension will produce $n + 1$ antinodes? (c) If the frequency is tripled and the length halved, by what factor should the tension be changed to get twice as many antinodes?

48. Radar detects the speed of a car, using the Doppler shift of microwaves that are reflected off the moving car, by beating the received wave with the transmitted wave and measuring the difference. The Doppler shift for microwaves is

$$f = f_0 \sqrt{\frac{c + v}{c - v}}$$

where f_0 is the transmitted frequency, c is the speed of light (3×10^8 m/s), and v is the relative speed of the two objects. (a) Show that the wave that reflects back to the source has a frequency

$$f = f_0 \frac{(c + v)}{(c - v)}$$

(b) Show that the expression for the beat frequency of the microwaves may be written as $f_b = 2v/\lambda$. (Because the beat frequency is much smaller than the transmitted frequency, use the approximation $f + f_0 = 2f_0$.) (c) What beat frequency is measured for a speed of 30.0 m/s (67 mph) if the microwaves have a frequency of 10.0 GHz? (1 GHz = 10^9 Hz.) (d) If the beat frequency measurement is accurate to ± 5 Hz, how accurate is the velocity measurement?

49. If two adjacent natural frequencies of an organ pipe are determined to be 0.550 kHz and 0.650 kHz, calculate the fundamental frequency and length of this pipe. (Use $v = 340$ m/s.)

50. A 0.0100 kg and 2.00–m-long wire is fixed at both ends and vibrates in its simplest mode under tension 200 N. When a tuning fork is placed near the wire, a beat frequency of 5.00 Hz is heard. (a) What is the frequency (frequencies) of the tuning fork? (b) What should the tension in the wire be to make the beats disappear?

51. The wave function for a standing wave is given in Equation 14.3 as $y = 2A \sin kx \cos \omega t$. (a) Rewrite this wave function in terms of the wavelength λ and wavespeed v of the wave. (b) Write the wave function of the simplest standing-wave vibration of a stretched string of length L. (c) Write the wave function for the second harmonic. (d) Generalize these results and write the wave function for the nth resonance vibration.

▪ Spreadsheet Problems

S1. Spreadsheet 14.1 adds two traveling waves at some fixed time t. The resultant wave function is

$$y = y_1 + y_2 = A_1 \sin(k_1 x - \omega_1 t + \phi_1)$$
$$+ A_2 \sin(k_2 x - \omega_2 t + \phi_2)$$

Add two waves traveling in opposite directions having the same wavelengths, the same phases, and the same speeds. (a) Use $A_1 = A_2 = 0.10$ m, $\omega_1 = -\omega_2 = 3.0$ rad/s, $\phi_1 = \phi_2 = 0$, and $k_1 = k_2 = 2.0$ rad/m. View the associated graph. Choose different values of t to see the time evolution of the resultant wave function. Do you get standing waves? (b) Repeat part (a) using $A_1 = 0.10$ m, $A_2 = 0.20$ m.

S2. Use Spreadsheet 14.1 to add two traveling waves that differ only in phase. (a) Choose $A_1 = A_2 = 0.10$ m, $\omega_1 = \omega_2 = 2.5$ rad/s, $k_1 = k_2 = 1.0$ rad/m, $\phi_1 = 0$, and $\phi_2 = 0, \pi/8, \pi/4, \pi/2$, and π. View the associated graphs. For which values of ϕ_2 do you get constructive interference? Destructive interference? (b) Repeat part (a) using $A_2 = 0.20$ m.

S3. Use Spreadsheet 14.1 to add two traveling waves with different wavelengths. (a) Choose $A_1 = A_2 = 0.10$ m,

$\omega_1 = \omega_2 = 3.0$ rad/s, $k_1 = 2.0$ rad/m, $k_2 = 1.0$ rad/m, and $\phi_1 = \phi_2 = 0$. View the associated graph. (b) Repeat part (a) with $k_2 = 3, 4, 5, 6, 7, 8, 9,$ and 10. Examine the graph for each case. Explain the appearance of the graphs.

S4. Three waves with the same frequency, wavelength, and amplitude are traveling in the same direction. Each differs in phase such that $(\phi_2 - \phi_1) = (\phi_3 - \phi_2) = \Delta\phi$. Modify Spreadsheet 14.1 to add the three waves at some fixed time t. Calculate the resultant wave function at $t = 0.0, 1.0$ s, and 2.0 s. Use $A_1 = A_2 = A_3 = 0.050$ m, $\omega_1 = \omega_2 = \omega_3 = 2.5$ rad/s, and $k_1 = k_2 = k_3 = 1.0$ rad/m. Choose $\Delta\phi = \pi/6$. View the associated graph. Repeat for different values of $\Delta\phi$.

S5. Write a spreadsheet or computer program to add two traveling waves of different frequencies:

$$y_1 = y_0 \cos 2\pi f_1 t \qquad y_2 = y_0 \cos 2\pi f_2 t$$

If the two frequencies are close to each other, beats will be produced. Does your program show the beats? From your numerical results, how does the beat frequency relate to the frequencies of the two waves?

S6. The Fourier theorem states that any periodic wave of frequency f, no matter how complicated, can be expressed as a sum of even or odd harmonic functions. That is,

$$y(t) = \sum_{n=0}^{\infty} A_n \sin(n\omega t + \phi_n)$$

where $\omega = 2\pi f$ is the fundamental angular frequency. (a) Use $A_1 = 1, A_2 = 1/2, A_3 = 1/3$, and so on. All the ϕ_n's are zero. (b) Use $A_1 = 1, A_3 = 1/3, A_5 = 1/5$, and so on. All the even A_n's and all the ϕ_n's are zero. Write a spreadsheet or computer program that will calculate this sum. You should design your spreadsheet or program so that the number of terms in the sum can be easily specified. Start with just the first term ($n = 1$). Then rerun the calculation with more terms in the sum ($n = 2, 3, 4, \ldots 20$). Plot the sum so that one or two periods are displayed. Note how the waveform builds up as more terms are added.

ANSWERS TO CONCEPTUAL PROBLEMS

1. At the instant at which there is no displacement of the string, the string is still moving. Thus, the energy is stored at that instant completely as kinetic energy of the string.

2. At the center of the string, there is a node for the second harmonic, as well as for every even-numbered harmonic. By placing the finger at the center and plucking, the guitarist is eliminating any harmonic that does not have a node at that point, which is all of the odd harmonics. The even harmonics can vibrate relatively freely with the finger at the center because they exhibit no displacement at that point. The result is a sound with a mixture of frequencies that are in-

teger multiples of the second harmonic, which is one octave higher than the fundamental.

3. The bow string is pulled away from equilibrium and released, similar to the way that a guitar string is pulled and released when it is plucked. Thus, standing waves will be excited in the bow string. If the arrow leaves from the exact center of the string, then a series of odd harmonics will be excited. Even harmonics will not be excited because they have a node at the point where the string exhibits its maximum displacement.

4. A change in temperature will result in a change in the speed

of sound and, therefore, in the fundamental frequency. The speed of sound for audible frequencies in the open atmosphere is only a function of temperature, and does not depend on the pressure. Thus, there will be no effect on the fundamental frequency of variations in atmospheric pressure.

5. When the bottles are struck, standing wave vibrations are established in the glass material of the bottles. The frequencies of these vibrations are determined by the tension in the glass and the mass loading of the glass material. As the water level rises, there is more mass loading, since the glass is in contact with the water. This increased mass loading decreases the frequency. On the other hand, blowing into a bottle establishes a standing wave vibration in the air cavity above the water. As the water level rises, the length of this cavity decreases, and the frequency rises.

6. If you take the gum out of your mouth and stick it to one of the tines of the mystery tuning fork, it will lower its frequency, due to the mass loading of the gum. Now, when the two forks are played together, the beat frequency will be different. If the beat frequency is lower, then the forks are closer together in frequency, so the mystery fork must be higher in frequency than the standard. If the beat frequency is higher, the frequencies are farther apart, and the mystery fork must have had a lower frequency. This argument assumes that the piece of gum is small enough that the frequency shift for the fork is less than 4 Hz.

15

Fluid Mechanics

atter is normally classified as being in one of three states: solid, liquid, or gaseous. Everyday experience tells us that a solid has a definite volume and shape. A brick maintains its familiar shape and size day in and day out. We also know that a liquid has a definite volume but no definite shape. Finally, an unconfined gas has neither definite volume nor definite shape. These definitions help us to picture the states of matter, but they are somewhat artificial. For example, asphalt and plastics are normally considered solids, but over long periods of time they tend to flow like liquids. Likewise, most substances can be a solid, liquid, or gas (or combinations of these), depending on the temperature and pressure. In general, the time it takes a particular substance to change its shape in response to an external force determines whether we treat the substance as a solid, liquid, or gas.

A **fluid** is a collection of molecules that are randomly arranged and held together by weak cohesive forces and forces exerted by the walls of a container. Both liquids and gases are fluids. In our treatment of the mechanics of fluids, we shall see that no new physical principles are needed to

◀ A scuba diver plays with an octopus in Poor Knights Island, New Zealand. As the diver descends to greater depths, the water pressure increases above atmospheric pressure, and the internal pressure of the body also increases accordingly to maintain equilibrium. Scuba divers have been able to swim at depths of more than 1000 ft. *(Darryl Torckler/Tony Stone Images)*

415

explain such effects as the buoyant force on a submerged object and the dynamic lift on an airplane wing. First, we consider a fluid at rest and derive an expression for the pressure exerted by the fluid as a function of its density and depth. We then treat fluids in motion, an area of study called fluid dynamics. A fluid in motion can be described by a model in which certain simplifying assumptions are made. We use this model to analyze some situations of practical importance. An underlying principle known as the **Bernoulli principle** enables us to determine relationships between the pressure, density, and velocity at every point in a fluid.

Figure 15.1 The force of the fluid on a submerged object at any point is perpendicular to the surface of the object. The force of the fluid on the walls of the container is perpendicular to the walls at all points.

15.1 • PRESSURE

The study of fluid mechanics involves the density of a substance, defined as its mass per unit volume. For this reason, Table 15.1 lists the densities of various substances. These values vary slightly with temperature, because the volume of a substance is temperature dependent (as we shall see in Chapter 16). Note that under standard conditions (0°C and atmospheric pressure) the densities of gases are about 1/1000 the densities of solids and liquids. This difference implies that the average molecular spacing in a gas under these conditions is about ten times greater in each dimension than in a solid or liquid.

Fluids do not sustain shearing stresses, and thus the only stress that can exist on an object submerged in a fluid is one that tends to compress the object. The force exerted by the fluid on the object is always perpendicular to the surfaces of the object, as shown in Figure 15.1.

The pressure at a specific point in a fluid can be measured with the device pictured in Figure 15.2. The device consists of an evacuated cylinder enclosing a light piston connected to a spring. As the device is submerged in a fluid, the fluid presses down on the top of the piston and compresses the spring until the inward force of the fluid is balanced by the outward force of the spring. The fluid pressure can be measured directly if the spring is calibrated in advance. This is accomplished by applying a known force to the spring to compress it a given distance.

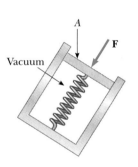

Figure 15.2 A simple device for measuring pressure in a fluid.

TABLE 15.1   Densities of Some Common Substances

Substance	ρ (kg/m^3)[a]	Substance	ρ (kg/m^3)[a]
Ice	0.917×10^3	Water	1.00×10^3
Aluminum	2.70×10^3	Sea water	1.03×10^3
Iron	7.86×10^3	Ethyl alcohol	0.806×10^3
Copper	8.92×10^3	Benzene	0.879×10^3
Silver	10.5×10^3	Mercury	13.6×10^3
Lead	11.3×10^3	Air	1.29
Gold	19.3×10^3	Oxygen	1.43
Platinum	21.4×10^3	Hydrogen	8.99×10^{-2}
Glycerine	1.26×10^3	Helium	1.79×10^{-1}

[a] All values are at standard atmospheric pressure and temperature (STP)—that is, atmospheric pressure and 0°C. To convert to grams per cubic centimeter, multiply by 10^{-3}.

If F is the magnitude of the normal force on the piston and A is the surface area of the piston, then the pressure, P, of the fluid at the level to which the device has been submerged is defined as the ratio of force to area:

$$P \equiv \frac{F}{A}$$ [15.1]

• Definition of pressure

To define the pressure at a specific point, consider a fluid acting on the device shown in Figure 15.2. If the normal force exerted by the fluid is F over a surface element of area δA that contains the point in question, then the pressure at that point is

$$P = \lim_{\delta A \to 0} \frac{F}{\delta A} = \frac{dF}{dA}$$ [15.2]

As we see in the next section, the pressure in a fluid varies with depth. Therefore, to get the total force on a flat wall of a container, we have to integrate Equation 15.2 over the surface.

Because pressure is force per unit area, it has units of N/m² in the SI system. Another name for the SI unit of pressure is **pascal** (Pa).

$$1 \text{ Pa} \equiv 1 \text{ N/m}^2$$ [15.3]

Snowshoes prevent the person from sinking into the soft snow because the person's weight is spread over a larger area, which reduces the pressure on the snow's surface. *(Earl Young/FPG)*

Thinking Physics 1

The daring physics professor, after a long lecture, stretches out for a nap on a bed of nails, as in the photograph. How is this possible?

Reasoning If you try to support your entire weight on a single nail, the pressure on your body is your weight divided by the very small area of the nail. This pressure is sufficiently large to penetrate the skin. However, if you distribute your weight over several hundred nails, as the professor is doing, the pressure is considerably reduced because the area that supports your weight is the total area of all nails in contact with your body. (Note that lying on a bed of nails is much more comfortable than sitting on a bed of nails. Standing on the bed of nails without shoes is not recommended.)

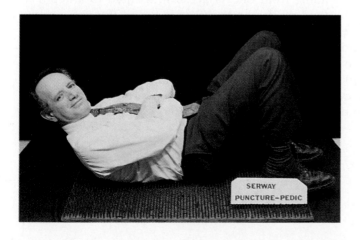

(Thinking Physics 1)   *(Jim Lehman)*

> ### *Thinking Physics 2*
>
> Suction cups can be used to hold objects onto surfaces. Why don't astronauts use suction cups to hold onto the outside surface of the space shuttle?
>
> **Reasoning**   The action of a suction cup depends on the fact that air is pushed out from under the cup when it is pressed against a surface. When released, it tends to spring back a bit, causing the trapped gas under the cup to expand and have a reduced pressure. Thus, the difference between the atmospheric pressure on the outside of the cup and the reduced pressure inside provides a net force pushing the cup against the surface. For astronauts in orbit around the Earth, there is no air outside the surface of the spacecraft. Thus, if a suction cup were to be pressed against the outside surface of the spacecraft, the pressure differential could not occur.

CONCEPTUAL PROBLEM 1

A woman wearing high-heeled shoes is invited into a home in which the kitchen has vinyl floor covering. Why should the homeowner be concerned?

EXERCISE 1   Estimate the density of the *nucleus* of an atom. What does this result suggest concerning the structure of matter? (Use the fact that the mass of a proton is equal to 1.67×10^{-27} kg and its radius is approximately 10^{-15} m.)   Answer 4.0×10^{17} kg/m^3; matter is mostly empty space.

15.2 • VARIATION OF PRESSURE WITH DEPTH

As divers know well, the pressure in the sea or a lake increases as they dive to greater depths. Likewise, atmospheric pressure decreases with increasing altitude. For this reason, aircraft flying at high altitudes must have pressurized cabins.

We now show how the pressure in a liquid increases linearly with depth. Consider a liquid of density ρ at rest and open to the atmosphere, as in Figure 15.3. Let us select a sample of the liquid contained within an imaginary cylinder of cross-sectional area A extending from the surface of the liquid to a depth h. The pressure exerted by the fluid on the bottom face is P, and the pressure on the top face of the cylinder is atmospheric pressure, P_0. Therefore, the upward force exerted by the liquid on the bottom of the cylinder is PA, and the downward force exerted by the atmosphere on the top is P_0A. Because the mass of liquid in the cylinder is $\rho V = \rho Ah$, the weight of the fluid in the cylinder is $w = \rho gV = \rho gAh$. Because the cylinder is in equilibrium, the upward force at the bottom must be greater than the downward force at the top of the sample to support its weight:

$$PA - P_0A = \rho gAh$$

or

$$P = P_0 + \rho gh \qquad\qquad [15.4]$$

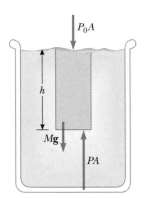

Figure 15.3   The variation of pressure with depth in a fluid. The net force on the volume of water within the darker region must be zero.

Variation of pressure with depth •

That is, the absolute pressure P at a depth h below the surface of a liquid open to the atmosphere is *greater* than atmospheric pressure by an amount ρgh.

In our calculations and end-of-chapter problems, we usually take atmospheric pressure to be

$$P_0 = 1.00 \text{ atm} \approx 1.013 \times 10^5 \text{ Pa}$$

Equation 15.4 verifies that the pressure is the same at all points having the same depth, independent of the shape of the container.

In view of the fact that the pressure in a liquid depends only on depth, any increase in pressure at the surface must be transmitted to every point in the fluid. This was first recognized by the French scientist Blaise Pascal (1623–1662) and is called **Pascal's law: A change in the pressure applied to an enclosed liquid is transmitted undiminished to every point of the liquid and to the walls of the container.**

• *Pascal's law*

An important application of Pascal's law is the hydraulic press, illustrated by Figure 15.4. A force F_1 is applied to a small piston of area A_1. The pressure is transmitted through a liquid to a larger piston of area A_2. Because the pressure is the same on both sides, we see that $P = F_1/A_1 = F_2/A_2$. Therefore, the force F_2 is larger than F_1 by the multiplying factor A_2/A_1. Hydraulic brakes, car lifts, hydraulic jacks, and forklifts all make use of this principle.

Thinking Physics 3

Blood pressure is normally measured with the cuff of the sphygmomanometer around the arm. Suppose that the blood pressure were measured with the cuff around the calf of the leg of a standing person. Would the reading of the blood pressure be the same here as it was for the arm?

Reasoning   The blood pressure measured at the calf would be larger than that measured at the arm. If we imagine the vascular system of the body to be a vessel containing a liquid (the blood), the pressure in the liquid will increase with depth. The blood at the calf is deeper in the liquid than that at the arm and is at a higher pressure.

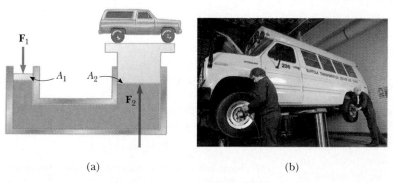

(a)   (b)

Figure 15.4   (a) Diagram of a hydraulic press. Because the increase in pressure is the same at the left and right sides, a small force $\mathbf{F}_1$ at the left produces a much larger force $\mathbf{F}_2$ at the right. (b) A bus under repair is supported by a hydraulic lift in a garage.   *(Superstock)*

(Conceptual Problem 2)   *(Henry Leap)*

CONCEPTUAL PROBLEM 2

A typical silo on a farm has many bands wrapped around its perimeter as shown in the photograph. Why is the spacing between successive bands smaller at the lower portions of the silo?

Example 15.1   The Car Lift

In a car lift used in a service station, compressed air exerts a force on a small piston of circular cross-section having a radius of 5.00 cm. This pressure is transmitted by a liquid to a second piston of radius 15.0 cm. What force must the compressed air exert in order to lift a car weighing 13 300 N? What air pressure will produce this force?

Solution   Because the pressure exerted by the compressed air is transmitted undiminished throughout the fluid, we have

$$F_1 = \left(\frac{A_1}{A_2}\right)F_2 = \frac{\pi(5.00 \times 10^{-2}\ \text{m})^2}{\pi(15.0 \times 10^{-2}\ \text{m})^2}\ (1.33 \times 10^4\ \text{N})$$

$$= 1.48 \times 10^3\ \text{N}$$

The air pressure that will produce this force is

$$P = \frac{F_1}{A_1} = \frac{1.48 \times 10^3\ \text{N}}{\pi(5.00 \times 10^{-2}\ \text{m})^2} = 1.88 \times 10^5\ \text{Pa}$$

This pressure is approximately twice atmospheric pressure.

The input work (the work done by $\mathbf{F}_1$) is equal to the output work (the work done by $\mathbf{F}_2$), so that energy is conserved.

Example 15.2   The Force on a Dam

Water is filled to a height H behind a dam of width w (Fig. 15.5). Determine the resultant force on the dam.

Reasoning   We cannot calculate the force on the dam by simply multiplying the area times the pressure, because the pressure varies with depth. The problem can be solved by finding the force dF on a narrow horizontal strip at depth h and then integrating the expression to find the total force on the dam.

Solution   The pressure at the depth h beneath the surface at the shaded portion is

$$P = \rho g h = \rho g (H - y)$$

(We have left out atmospheric pressure because it acts on both sides of the dam.) Using Equation 15.2, we find the force on the shaded strip of area $dA = w\,dy$ to be

$$dF = P\,dA = \rho g(H - y)w\,dy$$

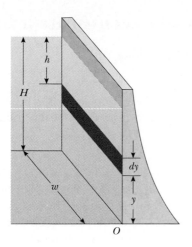

Figure 15.5   (Example 15.2) The total force on a dam must be obtained from the expression $F = \int P \, dA$, where dA is the area of the dark strip.

Therefore, the total force on the dam is

$$F = \int P \, dA = \int_0^H \rho g (H - y) w \, dy = \tfrac{1}{2} \rho g w H^2$$

Note that because the pressure increases with depth, the dam is designed so that its thickness increases with depth, as in Figure 15.5.

EXERCISE 2   Use the fact that the pressure increases linearly with depth to find the average pressure on the dam. Answer $\tfrac{1}{2}\rho g H$

EXERCISE 3   What is the hydrostatic force on the back of Grand Coulee Dam if the water in the reservoir is 150 m deep and the width of the dam is 1200 m? Answer 1.32×10^{11} N

EXERCISE 4   In some places, the Greenland ice sheet is 1.0 km thick. Estimate the pressure on the ground underneath the ice ($\rho_{ice} = 920$ kg/m^3).   Answer 9×10^6 Pa

15.3 • PRESSURE MEASUREMENTS

One simple device for measuring pressure is the open-tube manometer illustrated in Figure 15.6a. One end of a U-shaped tube containing a liquid is open to the atmosphere, and the other end is connected to a system of unknown pressure P. The difference in pressure $P - P_0$ is equal to $\rho g h$. Therefore, we see that $P = P_0 + \rho g h$. The pressure P is called the **absolute pressure,** and the difference $P - P_0$ is called the **gauge pressure.** For example, the pressure you measure in your bicycle tire is gauge pressure.

Another instrument used to measure pressure is the common barometer, invented by Evangelista Torricelli (1608–1647). A long tube closed at one end is filled with mercury and then inverted into a dish of mercury (Fig. 15.6b). The closed end of the tube is nearly a vacuum, so its pressure can be taken as zero. Therefore, it follows that $P_0 = \rho_{Hg} g h$, where ρ_{Hg} is the density of the mercury and h is the height of the mercury column. One atmosphere ($P_0 = 1$ atm) of pressure is defined to be the pressure equivalent of a column of mercury that is exactly 0.7600 m in height at 0°C, with $g = 9.80665$ m/s^2. At this temperature, mercury has a density of 13.595×10^3 kg/s^3; therefore

$$P_0 = \rho_{Hg} g h = (13.595 \times 10^3 \text{ kg/m}^3)(9.80665 \text{ m/s}^2)(0.7600 \text{ m})$$
$$= 1.013 \times 10^5 \text{ Pa}$$

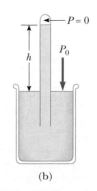
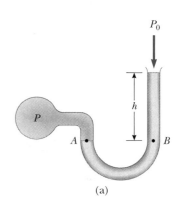

Figure 15.6   Two devices for measuring pressure: (a) an open-tube manometer and (b) a mercury barometer.

15.4 · BUOYANT FORCES AND ARCHIMEDES' PRINCIPLE

Archimedes' principle •

Figure 15.7 The external forces on the cube of water are the force of gravity **w** and the buoyant force **B**. Under equilibrium conditions, $B = w = mg$.

Archimedes (287–212 B.C.)

A Greek mathematician, physicist, and engineer, Archimedes was perhaps the greatest scientist of antiquity. According to legend, Archimedes was asked by King Hieron to determine whether the king's crown was made of pure gold or had been alloyed with some other metal. The task was to be performed without damaging the crown. Archimedes presumably arrived at a solution while taking a bath, noting a partial loss of weight after submerging his arms and legs in the water. As the story goes, he was so excited about his great discovery that he ran through the streets of Syracuse naked shouting, "Eureka!" which is Greek for "I have found it."

Archimedes' principle can be stated as follows:

> Any body completely or partially submerged in a fluid is buoyed up by a force the magnitude of which is equal to the weight of the fluid displaced by the body.

Everyone has experienced Archimedes' principle. Recall that it is relatively easy to lift someone if the person is in a swimming pool, whereas lifting that same individual on dry land is much harder. Evidently, water provides partial support to any object placed in it. The upward force that the fluid exerts on an object submerged in it is called the **buoyant force.** According to Archimedes' principle, **the magnitude of the buoyant force always equals the weight of the fluid displaced by the object.**

Archimedes' principle can be verified in the following manner. Suppose we focus our attention on the indicated cube of fluid in the container of Figure 15.7. This cube of fluid is in equilibrium under the action of the forces on it. One of these forces is the force of gravity. What cancels this downward force? Apparently, the rest of the fluid inside the container is holding it in equilibrium. Thus, the magnitude of the buoyant force is exactly equal to the weight of the fluid inside the cube:

$$B = w$$

Now, imagine that the cube of fluid is replaced by a cube of steel of the same dimensions. What is the buoyant force on the steel? The fluid surrounding a cube behaves in the same way whether a cube of fluid or a cube of steel is being buoyed up; therefore, **the buoyant force acting on the steel is the same as the buoyant force acting on a cube of fluid of the same dimensions.** This result applies for a submerged object of any shape, size, or density.

Let us show explicitly that the magnitude of the buoyant force is equal to the weight of the displaced fluid. The pressure at the bottom of the cube in Figure 15.7 is greater than the pressure at the top by an amount $\rho_f g h$, where ρ_f is the density of the fluid and h is the height of the cube. The pressure difference, ΔP, is equal to the buoyant force per unit area—that is, $\Delta P = B/A$. Thus, we see that $B = (\Delta P)A = (\rho_f g h)A = \rho_f g V$, where V is the volume of the cube. Because the mass of the fluid in the cube is $M = \rho_f V$, we see that

$$B = w = \rho_f V g = Mg \qquad [15.5]$$

where w is the weight of the displaced fluid.

Before proceeding with a few examples, it is instructive to compare the forces acting on a totally submerged object with those acting on a floating object.

Case I: A Totally Submerged Object   When an object is totally submerged in a fluid of density ρ_f, the magnitude of the upward buoyant force is $B = \rho_f V_0 g$, where V_0 is the volume of the object. If the object has a density ρ_0, its weight is $w = Mg = \rho_0 V_0 g$, and the net force on it is $B - w = (\rho_f - \rho_0) V_0 g$. Hence, if the

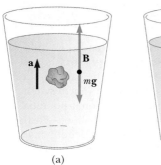
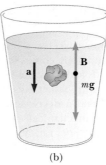

(a)   (b)

Figure 15.8   (a) A totally submerged object that is less dense than the fluid in which it is submerged will experience a net upward force. (b) A totally submerged object that is denser than the fluid sinks.

density of the object is less than the density of the fluid, as in Figure 15.8a, the unsupported object will accelerate upward. If the density of the object is greater than the density of the fluid as in Figure 15.8b, the unsupported object will sink.

Case II: A Floating Object   Now consider an object in static equilibrium floating on a fluid—that is, an object that is only partially submerged. In this case, the upward buoyant force is balanced by the downward force of gravity exerted on the object. If V is the volume of the fluid displaced by the object (which corresponds to that volume of the object beneath the fluid level), then the buoyant force has a magnitude $B = \rho_f V g$. Because the weight of the object is $w = Mg = \rho_0 V_0 g$, and $w = B$, we see that $\rho_f V g = \rho_0 V_0 g$, or

$$\frac{\rho_0}{\rho_f} = \frac{V}{V_0}$$ [15.6]

Under normal conditions, the average density of a fish is slightly greater than the density of water. This being the case, a fish would sink if it did not have some mechanism for adjusting its density. The fish accomplishes this by internally regulating the size of its swim bladder. In this manner, fish are able to swim to various depths.

Hot-air balloons over Albuquerque, New Mexico. Because hot air is less dense than cold air, there is a net upward buoyant force on the balloons.   *(William Moriarity/Rainbow)*

Thinking Physics 4

Suppose an office party is taking place on the top floor of a tall building. Carrying an iced soft drink, you step on the elevator, which begins to accelerate downward. What happens to the ice in the drink? Does it rise farther out of the liquid? Sink deeper into the liquid? Or is it unaffected by the motion?

Reasoning   The level of the ice in the liquid is unaffected by the motion. The acceleration of the elevator is equivalent to a change in the gravitational field, according to the principle of equivalence. If the elevator accelerates downward, one might be tempted to say that the effect is the same as if gravity decreases—the weight of the ice cube decreases, causing it to float higher in the liquid. Recall, however, that the magnitude of the buoyant force is equal to the weight of the liquid displaced by the ice cube. The weight of the liquid also decreases with the effectively decreased gravity. With both the weight of the ice cube and the buoyant force decreasing by the same factor, the level of the ice cube in the liquid is unaffected.

Thinking Physics 5

A florist delivery person is delivering a flower basket to a home. The basket includes an attached helium-filled balloon, which suddenly comes loose from the basket and begins to accelerate upward toward the sky. Startled by the release of the balloon, the delivery person drops the flower basket. As the basket falls, the basket–Earth system experiences an increase in kinetic energy and a decrease in gravitational potential energy, consistent with the conservation of mechanical energy. The balloon–Earth system, however, experiences an increase in *both* gravitational potential energy and kinetic energy. Is this inconsistent with the conservation of mechanical energy principle? If not, from where is the extra energy coming?

Reasoning   In the case of the system of the flower basket and the Earth, a good approximation to the motion of the basket can be made by ignoring the effects of the air. Thus, the conservation of mechanical energy principle is obeyed, with the exception of a small amount of transformation to internal energy of the air and the basket due to air friction. For the balloon–Earth system, we cannot ignore the effects of the air—it is the buoyant force of the air that causes the balloon to rise.

CONCEPTUAL PROBLEM 3

Atmospheric pressure varies from day to day. Does a ship float higher in the water on a high pressure day compared to a low pressure day?

CONCEPTUAL PROBLEM 4

Suppose a damaged ship just barely floats in the ocean after a hole in its hull has been plugged. It is pulled toward shore and into a river, heading toward a dry dock for repair. As it is pulled up the river, it sinks. Why?

CONCEPTUAL PROBLEM 5

A pound of styrofoam and a pound of lead have the same weight. If they are placed on an equal arm balance, will it balance?

CONCEPTUAL PROBLEM 6

A person in a boat floating in a small pond throws an anchor overboard. Does the level of the pond rise, fall, or remain the same?

Example 15.3   A Submerged Object

A piece of aluminum is suspended from a string and then completely immersed in a container of water (Fig. 15.9). The mass of the aluminum is 1.0 kg, and its density is 2.7×10^3 kg/m^3. Calculate the tension in the string before and after the aluminum is immersed.

Solution   When the aluminum is suspended in air, as in Figure 15.9a, the tension in the string, T_1 (the reading on the scale), is equal to the weight, Mg, of the aluminum, assuming that the buoyant force of air can be neglected:

$$T_1 = Mg = (1.0 \text{ kg})(9.80 \text{ m/s}^2) = \boxed{9.8 \text{ N}}$$

When immersed in water, the aluminum experiences an upward buoyant force **B**, as in Figure 15.9b, which reduces the tension in the string. Because the system is in equilibrium,

$$T_2 + B - Mg = 0$$
$$T_2 = Mg - B = 9.8 \text{ N} - B$$

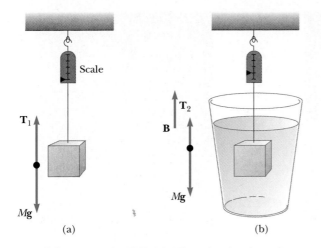

In order to calculate B, we must first calculate the volume of the aluminum:

$$V_{Al} = \frac{M}{\rho_{Al}} = \frac{1.0 \text{ kg}}{2.7 \times 10^3 \text{ kg/m}^3} = 3.7 \times 10^{-4} \text{ m}^3$$

Because the buoyant force equals the weight of the water displaced, we have

$$B = M_w g = \rho_w V_{Al} g$$
$$= (1.0 \times 10^3 \text{ kg/m}^3)(3.7 \times 10^{-4} \text{ m}^3)(9.80 \text{ m/s}^2)$$
$$= 3.6 \text{ N}$$

Therefore,

$$T_2 = 9.8 \text{ N} - B = 9.8 \text{ N} - 3.6 \text{ N} = \boxed{6.2 \text{ N}}$$

Figure 15.9 (Example 15.3) (a) When the aluminum is suspended in air, the scale reads the weight, Mg (neglecting the buoyancy of air). (b) When the aluminum is immersed in water, the buoyant force **B** reduces the scale reading to $T_2 = Mg - B$.

EXERCISE 5 What is the true weight of one cubic meter of a light plastic that has a specific gravity of 0.15? Note the true weight of an object is its weight in vacuum.
Answer   1.47 kN

15.5 · FLUID DYNAMICS

Thus far, our study of fluids has been restricted to fluids at rest. We now turn our attention to fluid dynamics—that is, fluids in motion. Instead of trying to study the motion of each particle of the fluid as a function of time, we describe the properties of the fluid at each point as a function of time.

Flow Characteristics

When fluid is in motion, its flow can be characterized as being one of two main types. The flow is said to be **steady** or **laminar** if each particle of the fluid follows a smooth path, so that the paths of different particles never cross each other, as in Figure 15.10. Thus, in steady flow, the velocity of the fluid at any point remains constant in time.

Above a certain critical speed, fluid flow becomes **turbulent.** Turbulent flow is an irregular flow characterized by small whirlpool-like regions, as in Figure 15.11. As an example, the flow of water in a stream becomes turbulent in regions in which rocks and other obstructions are encountered, often forming white-water rapids.

The term **viscosity** is commonly used in fluid flow to characterize the degree of internal friction in the fluid. This internal friction or viscous force is associated with the resistance of two adjacent layers of the fluid against moving relative to each other. Because of viscosity, part of the kinetic energy of a fluid is converted to thermal energy. This is similar to the mechanism by which an object sliding on a rough horizontal surface loses kinetic energy.

Figure 15.10   An illustration of streamline flow around an automobile in a test wind tunnel. The streamlines in the airflow are made visible by smoke particles.   *(Andy Sacks/Tony Stone Images)*

Figure 15.11   Turbulent flow: The tip of a rotating blade (the dark region at the top) forms a vortex in air that is being heated by an alcohol lamp (the wick is at the bottom). Note the air turbulence on both sides of the rotating blade.   *(Kim Vandiver and Harold Edgerton, Palm Press, Inc.)*

Because the motion of a real fluid is complex and not yet fully understood, we make some simplifying assumptions in our approach. As we shall see, many features of real fluids in motion can be understood by considering the behavior of an ideal fluid. In our model, we make the following four assumptions:

1. **Nonviscous fluid.** In a nonviscous fluid, internal friction is neglected. An object moving through the fluid experiences no viscous force.
2. **Incompressible fluid.** The density of an incompressible fluid is assumed to remain constant because it does not change due to a change in pressure on the fluid.
3. **Steady flow.** In steady flow, we assume that the velocity of the fluid at each point remains constant in time.
4. **Irrotational flow.** Fluid flow is irrotational if there is no angular momentum of the fluid about any point. If a small wheel placed anywhere in the fluid does not rotate about the wheel's center of mass, the flow is irrotational. (If the wheel were to rotate, as it would if turbulence were present, the flow would be rotational.)

Properties of an ideal fluid •

The first two assumptions are properties of our ideal fluid. The last two are descriptions of the way that the fluid flows.

Figure 15.12   This diagram represents a set of streamlines (blue lines). A particle at *P* follows one of these streamlines, and its velocity is tangent to the streamline at each point along its path.

15.6 • STREAMLINES AND THE EQUATION OF CONTINUITY

The path taken by a fluid particle under steady flow is called a **streamline.** The velocity of the fluid particle is always tangent to the streamline, as shown in Figure 15.12. No two streamlines can cross each other, for if they did, a fluid particle could move either way at the crossover point and then the flow would not be steady. A set of streamlines, as shown in Figure 15.12, forms what is called a *tube of flow*. Note that fluid particles cannot flow into or out of the sides of this tube, because if they did, the streamlines would be crossing each other.

Consider an ideal fluid flowing through a pipe of nonuniform size, as in Figure 15.13. The particles in the fluid move along the streamlines in steady flow. At all points the velocity of any particle is tangent to the streamline along which it moves.

In a small time interval, Δt, the fluid at the bottom end of the pipe moves a distance $\Delta x_1 = v_1 \Delta t$. If A_1 is the cross-sectional area in this region, then the mass contained in the shaded region is $\Delta m_1 = \rho A_1 \Delta x_1 = \rho A_1 v_1 \Delta t$. In a similar way, the fluid that moves through the upper end of the pipe in the time Δt has a mass $\Delta m_2 = \rho A_2 v_2 \Delta t$. However, because *mass is conserved* and because the flow is steady, the mass that crosses A_1 in a time Δt must equal the mass that crosses A_2 in a time Δt. That is, $\Delta m_1 = \Delta m_2$, or $\rho A_1 v_1 = \rho A_2 v_2$. Because the density is common to both sides of this expression, we get

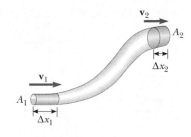

Figure 15.13 A fluid moving with streamline flow through a pipe of varying cross-sectional area. The volume of fluid flowing through A_1 in a time interval Δt must equal the volume flowing through A_2 in the same time interval. Therefore, $A_1 v_1 = A_2 v_2$.

$$A_1 v_1 = A_2 v_2 = \text{constant} \qquad [15.7]$$

This expression, called the **equation of continuity,** says that **the product of the area and the fluid speed at all points along the pipe is a constant.** Therefore, the speed is high where the tube is constricted and low where the tube is wide. The product Av, which has the dimensions of volume/time, is called the **volume flow rate.** The condition $Av =$ constant is equivalent to the fact that **the amount of fluid that enters one end of the tube in a given time interval equals the amount leaving in the same time interval, assuming no leaks.**

• *Equation of continuity*

Example 15.4   Filling a Water Bucket

A water hose 2.00 cm in diameter is used to fill a 20.0-liter bucket. If it takes 1.00 min to fill the bucket, what is the speed v at which the water leaves the hose? ($1\ L = 10^3\ cm^3$)

Solution   The cross-sectional area of the hose is

$$A = \pi r^2 = \pi \frac{d^2}{4} = \pi \left(\frac{2.00^2}{4} \right) cm^2 = \pi\ cm^2$$

According to the data given, the flow rate is equal to 20.0 liters/min. Equating this to the product Av gives

$$Av = 20.0\ \frac{L}{min} = \frac{20.0 \times 10^3\ cm^3}{60.0\ s}$$

$$v = \frac{20.0 \times 10^3\ cm^3}{(\pi\ cm^2)(60.0\ s)} = 106\ cm/s$$

EXERCISE 6   If the diameter of the hose is reduced to 1.00 cm, what will the speed of the water be as it leaves the hose, assuming the same flow rate?   Answer   424 cm/s

15.7 • BERNOULLI'S PRINCIPLE

As a fluid moves through a pipe of varying cross-section and elevation, the pressure changes along the pipe. In 1738 the Swiss physicist Daniel Bernoulli first derived an expression that relates the pressure to fluid speed and elevation.

Consider the flow of an ideal fluid through a nonuniform pipe in a time, Δt, as illustrated in Figure 15.14. The force on the lower end of the fluid is $P_1 A_1$, where P_1 is the pressure in section 1. The work done by this force is $W_1 = F_1 \Delta x_1 = P_1 A_1 \Delta x_1 = P_1 \Delta V$, where ΔV is the volume of section 1. In a similar manner, the

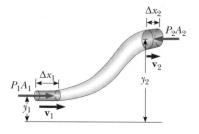

Figure 15.14 A fluid flowing through a constricted pipe with streamline flow. The fluid in the section of length Δx_1 moves to the section of length Δx_2. The volumes of fluid in the two sections are equal.

Daniel Bernoulli (1700–1782)

Bernoulli was a Swiss physicist and mathematician who made important discoveries in hydrodynamics. His most famous work, *Hydrodynamica*, published in 1738, is both a theoretical and a practical study of equilibrium, pressure, and velocity of fluids. In this publication, Bernoulli also attempted the first explanation of the behavior of gases with changing pressure and temperature; this was the beginning of the kinetic theory of gases. *(Corbis-Bettmann)*

work done on the fluid at the upper end in the time Δt is $W_2 = -P_2 A_2\,\Delta x_2 = -P_2\,\Delta V$. (The volume that passes through section 1 in a time Δt equals the volume that passes through section 2 in the same time interval.) This work is negative because the fluid force opposes the displacement. Thus the net work done by these forces in the time Δt is

$$W_F = (P_1 - P_2)\Delta V$$

If a fluid element of mass Δm enters the tube with speed v_1 and leaves with speed v_2, then the change in its kinetic energy is

$$\Delta K = \tfrac{1}{2}(\Delta m)v_2{}^2 - \tfrac{1}{2}(\Delta m)v_1{}^2$$

As the fluid element rises a vertical distance $y_2 - y_1$, the work done by gravity is negative and given by

$$W_g = -\Delta mg(y_2 - y_1)$$

Because the net work done on the fluid is $W_F + W_g$, we can apply the work-kinetic energy theorem to this volume of fluid to give

$$(P_1 - P_2)\Delta V - \Delta mg(y_2 - y_1) = \tfrac{1}{2}(\Delta m)v_2{}^2 - \tfrac{1}{2}(\Delta m)v_1{}^2$$

If we divide each term by ΔV, noting that $\rho = \Delta m/\Delta V$, and rearrange terms, the expression reduces to

$$P_1 + \tfrac{1}{2}\rho v_1{}^2 + \rho g y_1 = P_2 + \tfrac{1}{2}\rho v_2{}^2 + \rho g y_2 \qquad \textbf{[15.8]}$$

This is **Bernoulli's equation** as applied to an ideal fluid. It is often expressed as

$$P + \tfrac{1}{2}\rho v^2 + \rho g y = \text{constant} \qquad \textbf{[15.9]}$$

> Bernoulli's equation says that the sum of the pressure, (P), the kinetic energy per unit volume $(\tfrac{1}{2}\rho v^2)$, and gravitational potential energy per unit volume $(\rho g y)$ has the same value at all points along a streamline.

Note the Bernoulli's equation is *not* a sum of energy density terms because P is pressure, not energy density. When the fluid is at rest, $v_1 = v_2 = 0$ and Equation 15.8 becomes

$$P_1 - P_2 = \rho g(y_2 - y_1) = \rho g h$$

which agrees with Equation 15.4.

Thinking Physics 6

In tornadoes, it sometimes occurs that the windows of a house explode outward. Sometimes, the roof of a house is lifted off the house. What causes these events to happen? How does opening the windows in a tornado (which may seem counterintuitive!) help in theory?

Reasoning The wind speed in a tornado is very high. According to Bernoulli's principle, such high wind speeds will result in low pressures. When these winds pass by a window, the pressure outside the window is reduced well below that of the pressure on the inside surface, which is due to the still air inside the house. As a result, the net force outward can burst the window. When these high-speed winds pass over the roof of a house, the pressure above the roof is much lower than the pressure below the roof, again from the still air inside the house. The pressure difference can be enough to lift the roof off the house. By opening the windows, the pressure differential between the outside and inside air can be reduced by allowing a flow of air through the open windows. This will reduce the possibility of exploding windows or a flying roof.

CONCEPTUAL PROBLEM 7

When you are driving a small car on the freeway and a truck passes you at a high speed, you feel pulled *toward* the truck. Why?

Example 15.5   The Venturi Tube

The horizontal constricted pipe illustrated in Figure 15.15, known as a *Venturi tube,* can be used to measure flow speeds such as air speed in flight. Let us determine the flow speed at point 2 if the pressure difference $P_1 - P_2$ is known.

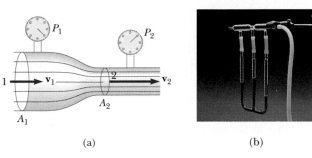

(a)   **(b)**

Figure 15.15   (a) (Example 15.5) The pressure P_1 is greater than the pressure P_2, because $v_1 < v_2$. This device can be used to measure the speed of fluid flow. (b) A Venturi tube.   *(Courtesy of Central Scientific Co.)*

Solution   Because the pipe is horizontal, $y_1 = y_2$ and Equation 15.8 applied to points 1 and 2 gives

$$P_1 + \tfrac{1}{2}\rho v_1{}^2 = P_2 + \tfrac{1}{2}\rho v_2{}^2$$

From the equation of continuity (Eq. 15.7), we see that $A_1 v_1 = A_2 v_2$, or

$$v_1 = \frac{A_2}{A_1} v_2$$

Substituting this expression into the previous equation gives

$$P_1 + \tfrac{1}{2}\rho \left(\frac{A_2}{A_1}\right)^2 v_2{}^2 = P_2 + \tfrac{1}{2}\rho v_2{}^2$$

$$v_2 = A_1 \sqrt{\frac{2(P_1 - P_2)}{\rho(A_1{}^2 - A_2{}^2)}}$$

We can also obtain an expression for v_1 using this result and the continuity equation. Note that because $A_2 < A_1$, it follows that $P_1 > P_2$. In other words, the pressure is reduced in the constricted part of the pipe. This result is somewhat analogous to the following situation: Consider a very crowded room in which people are squeezed together. As soon as a door is opened and people begin to exit, the squeezing (pressure) is least near the door where the motion (flow) is greatest.

EXERCISE 7   The Garfield Thomas water tunnel at Pennsylvania State University has a circular cross-section that constricts from a diameter of 3.6 m to the test section, which is 1.2 m in diameter. If the speed of flow is 3.0 m/s in the larger-diameter pipe, determine the speed of flow in the test section.   Answer   27 m/s

15.8 • OTHER APPLICATIONS OF BERNOULLI'S PRINCIPLE   O P T I O N A L

Consider the streamlines that flow around an airplane wing as shown in Figure 15.16. Let us assume that the airstream approaches the wing horizontally from the right with a velocity $\mathbf{v}_1$. The tilt of the wing causes the airstream to be deflected downward with a velocity $\mathbf{v}_2$. Because the airstream is deflected by the wing, the wing must exert a force on the airstream. According to Newton's third law, the airstream must exert an equal and opposite force $\mathbf{F}$ on the wing. This force has a

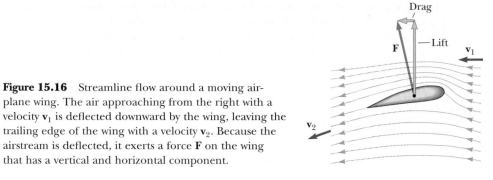

Figure 15.16 Streamline flow around a moving airplane wing. The air approaching from the right with a velocity $\mathbf{v}_1$ is deflected downward by the wing, leaving the trailing edge of the wing with a velocity $\mathbf{v}_2$. Because the airstream is deflected, it exerts a force $\mathbf{F}$ on the wing that has a vertical and horizontal component.

Figure 15.17 A stream of air passing over a tube dipped into a liquid will cause the liquid to rise in the tube as shown.

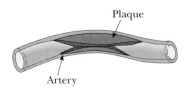

Figure 15.18 Blood must travel faster than normal through a constricted artery.

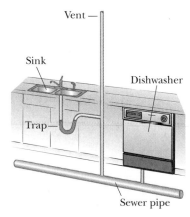

Figure 15.19   (Thinking Physics 7)

vertical component called the **lift** (or aerodynamic lift) and a horizontal component called **drag.** The lift depends on several factors, such as the speed of the airplane, the area of the wing, its curvature, and the angle between the wing and the horizontal. As this angle increases, turbulent flow can set in above the wing to reduce the lift.

The lift on the wing is consistent with Bernoulli's equation. The speed of the airstream is greater above the wing, hence the air pressure above the wing is less than the pressure below the wing, which results in a net upward force.

In general, an object experiences lift by any effect that causes the fluid to change its direction as it flows past the object. Some factors that influence the lift are the shape of the object, its orientation with respect to the fluid flow, spinning motion (for example, a spinning baseball), and the texture of the object's surface.

A number of devices operate in the manner described in Figure 15.17. A stream of air passing over an open tube reduces the pressure above the tube. This reduction in pressure causes the liquid to rise into the air stream. The liquid is then dispersed into a fine spray of droplets. You might recognize that this so-called atomizer is used in perfume bottles and paint sprayers.

Bernoulli's equation explains one symptom called vascular flutter in a person with advanced arteriosclerosis. The artery is constricted as a result of an accumulation of plaque on its inner walls (Fig. 15.18). If the blood speed is sufficiently high in the constricted region, the artery may collapse under external pressure, causing a momentary interruption in blood flow. At this point, Bernoulli's equation does not apply, and the vessel reopens under arterial pressure. As the blood rushes through the constricted artery, the internal pressure drops and again the artery closes. Such variations in blood flow can be heard with a stethoscope. If the plaque becomes dislodged and ends up in a smaller vessel that delivers blood to the heart, the person can suffer a heart attack.

Thinking Physics 7

Consider the portion of a home plumbing system shown in Figure 15.19. The water trap in the pipe below the sink captures a plug of water that prevents sewer gas from finding its way from the sewer pipe, up the sink drain, and into the home. Suppose the dishwasher is draining, so that water is moving to the left in the sewer pipe. What is the purpose of the vent, which is open to the air above the roof of the house? In which direction is air moving at the opening of the vent, upward or downward?

Reasoning Let us imagine that the vent is not present, so that the drain pipe for the sink is simply connected through the trap to the sewer pipe. As water from the dishwasher moves to the left in the sewer pipe, the pressure in the sewer pipe is reduced below atmospheric pressure, according to Bernoulli's principle. The pressure at the drain in the sink is still at atmospheric pressure. Thus, this pressure differential can push the plug of water in the water trap of the sink down the drain pipe into the sewer pipe, removing it as a barrier to sewer gas. With the addition of the vent to the roof, the reduced pressure of the dishwasher water will result in air entering the vent pipe at the roof. This will keep the pressure in the vent pipe and the right-hand side of the sink drain pipe at a pressure close to atmospheric, so that the plug of water in the water trap will remain.

SUMMARY

The **pressure,** P, in a fluid is the force per unit area that the fluid exerts on any surface:

$$P \equiv \frac{F}{A} \qquad \text{[15.1]}$$

In the SI system, pressure has units of N/m^2, and $1 \ N/m^2 = 1$ pascal (Pa).

The pressure in a fluid varies with depth h according to the expression

$$P = P_0 + \rho g h \qquad \text{[15.4]}$$

where P_0 is atmospheric pressure ($=1.013 \times 10^5 \ N/m^2$) and ρ is the density of the fluid, assumed uniform.

Pascal's law states that when pressure is applied to an enclosed fluid, the pressure is transmitted undiminished to every point in the fluid and to every point on the walls of the container.

When an object is partially or fully submerged in a fluid, the fluid exerts an upward force on the object called the **buoyant force.** According to **Archimedes' principle,** the buoyant force is equal to the weight of the fluid displaced by the body.

Various aspects of fluid dynamics can be understood by assuming that the fluid is nonviscous and incompressible and that the fluid motion is a steady flow with no turbulence.

Using these assumptions, two important results regarding fluid flowing through a pipe of nonuniform size can be obtained:

- The flow rate through the pipe is a constant, which is equivalent to stating that the product of the cross-sectional area, A, and the speed, v, at any point is a constant. This gives the **equation of continuity:**

$$A_1 v_1 = A_2 v_2 = \text{constant} \qquad \text{[15.7]}$$

- The sum of the pressure, kinetic energy per unit volume, and gravitational potential energy per unit volume has the same value at all points along a streamline. This corresponds to **Bernoulli's equation:**

$$P + \tfrac{1}{2}\rho v^2 + \rho g y = \text{constant} \qquad \text{[15.9]}$$

CONCEPTUAL QUESTIONS

1. Figure Q15.1 shows aerial views from directly above two dams. Both dams are equally long (the vertical dimension in the diagram) and equally deep (into the page in the diagram). The dam on the left holds back a very large lake, and the dam on the right holds back a narrow river. Which dam has to be built more strongly?

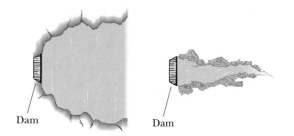

Dam   Dam

Figure Q15.1 (Conceptual Question 1)

2. Some physics students attach a long tube to the opening of a balloon. Leaving the balloon on the ground, the other end of the tube is hoisted to the roof of a multistory campus building. Students at the top of the building start pouring water into the tube. The students on the ground watch the balloon fill with water. On the roof, the students are surprised to see that the tube never seems to fill up—more and more water can continue to be poured into the tube. On the ground, the balloon bursts and drenches the students there. Why did the balloon-tube system never fill up, and why did the balloon burst?

3. A barge is carrying a load of gravel along a river. It approaches a low bridge, and the captain realizes that the top of the pile of gravel is not going to make it under the bridge. The captain orders the crew to quickly shovel gravel from the pile into the water. Is this a good decision?

4. Two drinking glasses having equal weights but different shapes and different cross-sectional areas are filled to the same level with water. According to the expression $P = P_0 + \rho gh$, the pressure is the same at the bottom of both glasses. In view of this, why does one weigh more than the other?

5. Pascal used a barometer with water as the working fluid. Why is it impractical to use water for a typical barometer?

6. A helium-filled balloon rises until its density becomes the same as that of the air. If a sealed submarine begins to sink, will it go all the way to the bottom of the ocean or will it stop when its density becomes the same as that of the surrounding water?

7. A fish rests on the bottom of a bucket of water while the bucket is being weighed. When the fish begins to swim around, does the weight change?

8. Lead has a greater density than iron, and both are denser than water. Is the buoyant force on a lead object greater than, less than, or equal to the buoyant force on an iron object of the same volume?

9. An ice cube is placed in a glass of water. What happens to the level of the water as the ice melts?

10. The water supply for a city is often provided from reservoirs built on high ground. Water flows from the reservoir, through pipes, and into your home when you turn the tap on your faucet. Why is the water flow more rapid out of a faucet on the first floor of a building than in an apartment on a higher floor?

11. If air from a hair dryer is blown over the top of a Ping-Pong ball, the ball can be suspended in air. Explain.

12. When ski-jumpers are airborne, why do they bend their bodies forward and keep their hands at their sides?

13. Explain why a sealed bottle partially filled with a liquid can float.

14. When is the buoyant force on a swimmer greater: after exhaling or after inhaling?

15. A piece of unpainted wood is partially submerged in a container filled with water. If the container is sealed and pressurized above atmospheric pressure, does the wood rise, fall, or remain at the same level? (*Hint:* Wood is porous.)

16. Because atmospheric pressure is about 10^5 N/m^2 and the area of a person's chest is about 0.13 m^2, the force of the atmosphere on one's chest is approximately 13 000 N. In view of this enormous force, why don't our bodies collapse?

17. If you release a ball while inside a freely falling elevator, the ball remains in front of you rather than falling to the floor, because the ball, the elevator, and you all experience the same downward acceleration, **g**. What happens if you repeat this experiment with a helium-filled balloon? (This one is tricky.)

18. Two identical ships set out to sea. One is loaded with a cargo of Styrofoam, and the other is empty. Which ship is more submerged?

19. A small piece of steel is tied to a block of wood. When the wood is placed in a tub of water with the steel on top, half of the block is submerged. If the block is inverted so that the steel is under water, does the amount of the block submerged increase, decrease, or remain the same? What happens to the water level in the tub when the block is inverted?

20. Prairie dogs ventilate their burrows by building a mound over one entrance, which is open to a stream of air. A second entrance at ground level is open to almost stagnant air. How does this construction create an air flow through the burrow?

21. An unopened can of diet cola floats when placed in a tank of water, whereas a can of regular cola of the same brand sinks in the tank. What do you suppose could explain this behavior?

PROBLEMS

Section 15.1 Pressure

1. A 50.0-kg woman balances on one heel of a high-heel shoe. If the heel is circular with radius 0.500 cm, what pressure does she exert on the floor?

2. A king orders a gold crown having a mass of 0.500 kg. When it arrives from the metalsmith, the volume of the crown is found to be 185 cm^3. Is the crown made of solid gold?

3. What is the total mass of the earth's atmosphere? (The radius of the Earth is 6.37×10^6 m, and atmospheric pressure at the surface is 1.013×10^5 N/m^2.)

Section 15.2 Variation of Pressure with Depth

4. Determine the absolute pressure at the bottom of a lake that is 30.0 m deep.

5. The spring of the pressure gauge shown in Figure 15.2 has a force constant of 1000 N/m, and the piston has a diameter of 2.00 cm. Find the depth in water for which the spring compresses by 0.500 cm.

6. The small piston of a hydraulic lift has a cross-sectional area of 3.00 cm^2, and the large piston has an area of 200 cm^2 (Fig. 15.4). What force must be applied to the small piston to raise a load of 15.0 kN? (In service stations this is usually accomplished with compressed air.)

7. What must be the contact area between a suction cup (completely exhausted) and a ceiling in order to support the weight of an 80.0-kg student?

8. A swimming pool has dimensions 30.0 m × 10.0 m and a flat bottom. When the pool is filled to a depth of 2.00 m with fresh water, what is the total force due to the water on the bottom? On each end? On each side?

9. Normal atmospheric pressure is 1.013×10^5 Pa. The approach of a storm causes the height of a mercury barometer to drop by 20.0 mm from the normal height. What is the atmospheric pressure? (The density of mercury is 13.59 g/cm^3.)

10. Mercury is poured into a U-shaped tube, as in Figure P15.10a. The left arm of the tube has a cross-sectional area of $A_1 = 10.0$ cm^2, and the right arm has a cross-sectional area of $A_2 = 5.00$ cm^2. One hundred grams of water are then poured into the right arm, as in Figure P15.10b. (a) Determine the length of the water column in the right arm of the U-shaped tube. (b) Given that the density of mercury is 13.6 g/cm^3, what distance, h, does the mercury rise in the left arm?

11. Blaise Pascal duplicated Torricelli's barometer using (as a Frenchman would) a red Bordeaux wine as the working liquid (Fig. P15.11). The density of the wine he used was 984 kg/m^3. What was the height h of the wine column for normal atmospheric pressure? Would you expect the vacuum above the column to be as good as for mercury?

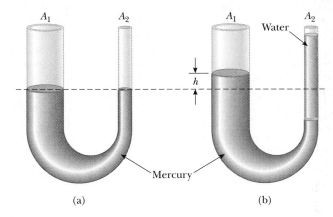

Figure P15.10

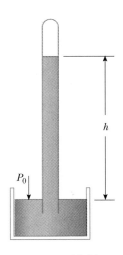

Figure P15.11

Section 15.4 Buoyant Forces and Archimedes' Principle

12. A Styrofoam slab has a thickness of 10.0 cm and a density of 300 kg/m^3. What is the area of the slab if it floats just awash in fresh water when a 75.0-kg swimmer is aboard?

13. A Styrofoam slab has thickness h and density ρ_s. What is the area of the slab if it floats with its top surface just awash in fresh water when a swimmer of mass m is on top?

14. A balloon is filled with 400 m^3 of helium. How big a payload can the balloon lift? (The density of air is 1.29 kg/m^3; the density of helium is 0.180 kg/m^3.)

15. A cube of wood 20.0 cm on a side and having a density of 650 kg/m^3 floats on water. (a) What is the distance from the top of the cube to the water level? (b) How much lead

weight has to be placed on top of the cube so that its top is just level with the water?

16. About how many helium-filled toy balloons would be required to lift you? Because helium is an irreplaceable resource, develop a theoretical answer rather than an experimental answer. In your solution state what physical quantities you take as data and the values you measure or estimate for them.

17. A plastic sphere floats in water with 50.0 percent of its volume submerged. This same sphere floats in oil with 40.0 percent of its volume submerged. Determine the densities of the oil and the sphere.

18. A frog in a hemispherical pod finds that he just floats without sinking into a sea of blue-green ooze with density 1.35 g/cm³ (Fig. P15.18). If the pod has radius 6.00 cm and negligible mass, what is the mass of the frog?

Figure P15.18

19. How many cubic meters of helium are required to lift a balloon with a 400-kg payload to a height of 8000 m? ($\rho_{He} = 0.18$ kg/m³.) Assume the balloon maintains a constant volume and that the density of air decreases with altitude z according to the expression $\rho_{air} = \rho_0 e^{-z/8000}$, where z is in meters, and ρ_0 ($=1.25$ kg/m³) is the density of air at sea level.

20. A 10.0-kg block of metal measuring 15.0 cm $\times$ 10.0 cm $\times$ 10.0 cm is suspended from a scale and immersed in water, as in Figure 15.9b. The 15.0-cm dimension is vertical, and the top of the block is 5.00 cm from the surface of the water. (a) What are the forces on the top and bottom of the block? (Take $P_0 = 1.0130 \times 10^5$ N/m².) (b) What is the reading of the spring scale? (c) Show that the buoyant force equals the difference between the forces at the top and bottom of the block.

Section 15.5   Fluid Dynamics

Section 15.6   Streamlines and the Equation of Continuity

Section 15.7   Bernoulli's Principle

21. The volume flow rate of water through a horizontal pipe is 2.00 m³/min. Determine the speed of flow at a point

at which the diameter of the pipe is (a) 10.0 cm, (b) 5.00 cm.

22. The legendary Dutch boy who saved Holland by placing his finger in the hole of a dike had a finger 1.20 cm in diameter. Assuming the hole was 2.00-m below the surface of the sea (density 1030 kg/m³), (a) what was the force on his finger? (b) If he removed his finger from the hole, how long would it take the released water to fill 1 acre of land to a depth of 1 foot, assuming the hole remained constant in size? (A typical U.S. family of four uses 1 acre-foot of water, 1234 m³, in 1 year.)

23. A large storage tank, open at the top and filled with water, develops a small hole in its side at a point 16.0 m below the water level. If the rate of flow from the leak is 2.50×10^{-3} m³/min, determine (a) the speed at which the water leaves the hole and (b) the diameter of the hole.

24. A horizontal pipe 10.0 cm in diameter has a smooth reduction to a pipe 5.00 cm in diameter. If the pressure of the water in the larger pipe is 8.00×10^4 Pa and the pressure in the smaller pipe is 6.00×10^4 Pa, at what rate does water flow through the pipes?

25. Water is pumped from the Colorado River up to Grand Canyon Village through a 15.0-cm diameter pipe. The river is at 564 m elevation and the village is at 2096 m. (a) What is the minimum pressure the water must be pumped at to arrive at the village? (b) If 4500 m³ are pumped per day, what is the speed of the water in the pipe? (c) What additional pressure is necessary to deliver this flow? *Note:* You may assume that the acceleration due to gravity and the density of air are constant over this range of elevations.

26. Old Faithful geyser in Yellowstone Park erupts at approximately 1-hour intervals, and the height of the fountain reaches 40.0 m. (a) With what speed does the water leave the ground? (b) What is the pressure (above atmospheric) in the heated underground chamber?

Section 15.8   Other Applications of Bernoulli's Principle (Optional)

27. An airplane has a mass of 1.60×10^4 kg, and each wing has an area of 40.0 m². During level flight, the pressure on the lower wing surface is 7.00×10^4 Pa. Determine the pressure on the upper wing surface.

28. A Venturi tube may be used as a fluid flow meter (Fig. 15.15). If the difference in pressure $P_1 - P_2 = 21.0$ kPa, find the fluid flow rate in m³/s given that the radius of the outlet tube is 1.00 cm, the radius of the inlet tube is 2.00 cm, and the fluid is gasoline ($\rho = 700$ kg/m³).

29. A Pitot tube can be used to determine the velocity of air flow by measuring the difference between the total pressure and the static pressure (Fig. P15.29). If the fluid in the tube is mercury, density $\rho_{Hg} = 13\,600$ kg/m³, and $\Delta h = 5.00$ cm, find the speed of air flow. (Assume that the air is stagnant at point A and take $\rho_{air} = 1.25$ kg/m³.)

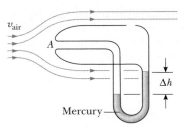

Figure P15.29

30. A siphon is used to drain water from a tank, as indicated in Figure P15.30. The siphon has a uniform diameter. Assume steady flow. (a) If the distance $h = 1.00$ m, find the speed of outflow at the end of the siphon. (b) What is the limitation on the height of the top of the siphon above the water surface? (In order to have a continuous flow of liquid, the pressure in the liquid cannot go below the vapor pressure of the liquid.)

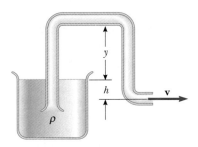

Figure P15.30

31. A large storage tank is filled to a height h_0. The tank is punctured at a height h from the bottom of the tank (Fig. P15.31). (a) Prove that the speed at which the water comes out is $\sqrt{2g(h_0 - h)}$ if the flow is steady and frictionless. (b) How far from the tank will the stream land?

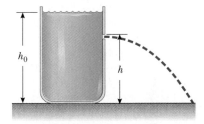

Figure P15.31

32. A hole is punched at height h in the side of a container of height h_0, full of water as in Figure P15.31. If the water is to shoot as far as possible horizontally, (a) how far from the bottom of the container should the hole be punched?

(b) Neglecting friction losses, how far (initially) from the side of the container will the water land?

Additional Problems

33. A Ping-Pong ball has a diameter of 3.80 cm and average density of 0.0840 g/cm^3. What force would be required to hold it completely submerged under water?

34. Figure P15.34 shows a water tank with a valve at the bottom. If this valve is opened, what is the maximum height attained by the water stream coming out of the right side of the tank? Assume that $h = 10.0$ m, $L = 2.00$ m, and $\theta = 30.0°$ and that the cross-sectional area at A is very large compared with that at B.

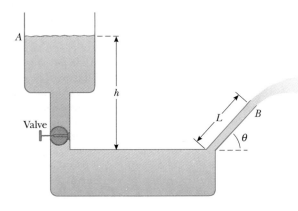

Figure P15.34

35. A helium-filled balloon is tied to a 2.00–m-long, 0.0500-kg uniform string. The balloon is spherical with a radius of 0.400 m. When released, it lifts a length, h, of string and then remains in equilibrium, as in Figure P15.35. Determine the value of h. The envelope of the balloon has mass 0.250 kg.

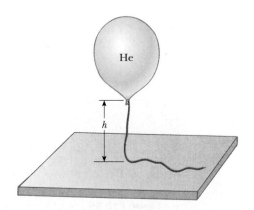

Figure P15.35

36. Water is forced out of a fire extinguisher by air pressure, as shown in Figure P15.36. How much gauge air pressure (above atmospheric) is required for a water jet to have a speed of 30.0 m/s when the water level is 0.500 m below the nozzle?

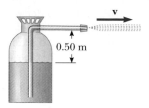

Figure P15.36

37. Torricelli was the first to realize that we live at the bottom of an ocean of air. He correctly surmised that the pressure of our atmosphere is due to the weight of the air. The density of air at 0°C at the Earth's surface is 1.29 kg/m³. The density actually decreases with increasing altitude as the atmosphere thins out. But *if we assume* that the density is constant (1.29 kg/m³) up to some altitude *h* and zero above, then *h* would represent the thickness of our atmosphere. Use this model to determine the value of *h* that gives a pressure of 1.00 atm at the surface of the Earth. Would the peak of Mt. Everest rise above the surface of such an atmosphere?

38. The true weight of a body is its weight when measured in a vacuum in which there are no buoyant forces. A body of volume *V* is weighed in air on a balance using weights of density ρ. If the density of air is ρ_a and the balance reads *w'*, show that the true weight *w* is

$$w = w' + \left(V - \frac{w'}{\rho g} \right) \rho_a g$$

39. A wooden dowel has a diameter of 1.20 cm. It floats in water with 0.400 cm of its diameter above water (Fig. P15.39). Determine the density of the dowel.

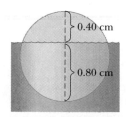

Figure P15.39

40. A 1.00-kg beaker containing 2.00 kg of oil (density = 916.0 kg/m³) rests on a scale. A 2.00-kg block of iron is suspended from a spring scale and completely submerged in the oil, as in Figure P15.40. Determine the equilibrium readings of both scales.

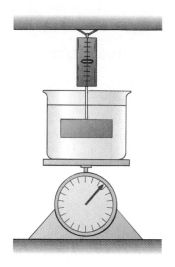

Figure P15.40

41. If a 1-megaton nuclear weapon is exploded at ground level, the peak overpressure (that is, the pressure increase above normal atmospheric pressure) will be 0.200 atm at a distance of 6.00 km. What force due to such an explosion will be exerted on the side of a house with dimensions 4.50 m × 22.0 m?

42. A light spring of constant *k* = 90.0 N/m rests vertically on a table (Fig. P15.42a). A 2.00-g balloon is filled with helium (density = 0.180 kg/m³) to a volume of 5.00 m³ and connected to the spring, causing it to expand as in Figure P15.42b. Determine the expansion length *L* when the balloon is in equilibrium.

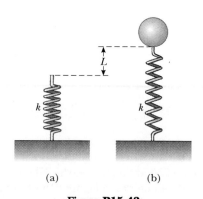

(a)   (b)

Figure P15.42

43. With reference to Figure 15.5, show that the total torque exerted by the water behind the dam about an axis through O is $\frac{1}{6}\rho g w H^3$. Show that the effective line of action of the total force exerted by the water is at a distance $\frac{1}{3}H$ above O.

44. About 1654 Otto von Guericke, inventor of the air pump, evacuated a sphere made of two brass hemispheres. Two teams of eight horses each could pull the hemispheres apart only on some trials, and then "with greatest difficulty," making a sound like a cannon firing (Fig. P15.44). (a) Show that the force F required to pull the evacuated hemispheres apart is $\pi R^2(P_0 - P)$, where R is the radius of the hemispheres and P is the pressure inside the hemispheres, which is much less than P_0. (b) Determine the force if $P = 0.100P_0$ and $R = 0.300$ m.

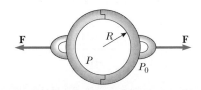

Figure P15.44 *(Henry Leap and Jim Lehman)*

45. In 1983, the United States began coining the cent piece out of copper-clad zinc rather than pure copper. If the mass of the old copper cent is 3.083 g and that of the new cent is 2.517 g, calculate the percent of zinc (by volume) in the new cent. The density of copper is 8.960 g/cm³ and that of zinc is 7.133 g/cm³. The new and old coins have the same volume.

46. A thin spherical shell of mass 4.00 kg and diameter 0.200 m is filled with helium (density = 0.180 kg/m³). It is then released from rest on the bottom of a pool of water that is 4.00 m deep. (a) Neglecting frictional effects, show that the shell rises with constant acceleration, and determine the value of that acceleration. (b) How long will it take for the top of the shell to reach the water surface?

47. An incompressible, nonviscous fluid initially rests in the vertical portion of the pipe shown in Figure P15.47a, where $L = 2.00$ m. When the valve is opened, the fluid flows into the horizontal section of the pipe. What is the speed of the

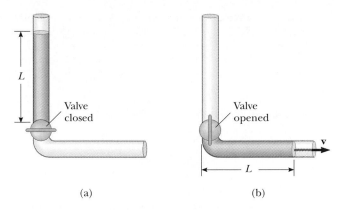

Figure P15.47

fluid when it is entirely in the horizontal section as in Figure P15.47b? Assume the cross-sectional area of the entire pipe is constant.

48. Water falls over a dam of height h meters at a rate of R kg/s. (a) Show that the power available from the water is

$$P = Rgh$$

where g is the acceleration of gravity. (b) Each hydroelectric unit at the Grand Coulee Dam discharges water at a rate of 8.50×10^5 kg/s from a height of 87.0 m. The power developed by the falling water is converted to electric power with an efficiency of 85.0%. How much electric power is produced by each hydroelectric unit?

49. Show that the variation of atmospheric pressure with altitude is given by $P = P_0 e^{-\alpha h}$, where $\alpha = \rho_0 g/P_0$, P_0 is atmospheric pressure at some reference level, and ρ_0 is the atmospheric density at this level. Assume that the decrease in atmospheric pressure with increasing altitude is given by Equation 15.4, so that $dP/dh = -\rho g$, and that the density of air is proportional to the pressure.

50. A cube of ice the edge of which is 20.0 mm is floating in a glass of ice-cold water with one of its faces parallel to the water surface. (a) How far below the water surface is the bottom face of the block? (b) Ice-cold ethyl alcohol is gently poured onto the water surface to form a layer 5.00 mm thick above the water. When the ice cube attains hydrostatic equilibrium again, what will be the distance from the top of the water to the bottom face of the block? (c) Additional cold ethyl alcohol is poured onto the water surface until the top surface of the alcohol coincides with the top surface of the ice cube (in hydrostatic equilibrium). How thick is the required layer of ethyl alcohol?

51. A light balloon filled with helium of density 0.180 kg/m³ is tied to a light string of length $L = 3.00$ m. The string is tied to the ground, forming an "inverted" simple pendulum as in Figure P15.51a. If the balloon is displaced slightly from equilibrium, as in Figure P15.51b, (a) show that the motion

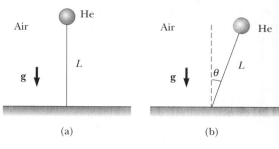

Figure P15.51

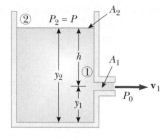

Figure P15.52

is simple harmonic, and (b) determine the period of the motion. Take the density of air to be 1.29 kg/m³, and ignore any energy lost due to air friction.

52. *Torricelli's Law:* A tank, with a cover, containing a liquid of density ρ has a hole in its side at a distance y_1 from the bottom (Fig. P15.52). The diameter of the hole is small compared to the diameter of the tank. The air above the liquid

is maintained at a pressure P. Assume steady, frictionless flow. Show that the speed at which the fluid leaves the hole when the liquid level is a distance h above the hole is

$$v_1 = \sqrt{\frac{2(P - P_0)}{\rho} + 2gh}$$

ANSWERS TO CONCEPTUAL PROBLEMS

1. She can exert enough pressure on the floor to dent or puncture the floor covering. The large pressure is caused by the fact that her weight is distributed over the very small cross-sectional area of her high heels. If you are the homeowner, you might want to suggest that she remove her high heels and put on some slippers.

2. If you think of the grain stored in the silo as a fluid, the pressure the grain exerts on the walls of the silo increases with increasing depth just as water pressure in a lake increases with increasing depth. Thus, the spacing between bands is made smaller at the lower portions to overcome the larger outward forces on the walls in these regions.

3. The level of floating of a ship in the water is unaffected by the atmospheric pressure. Air exerts negligible buoyant force compared to water. The buoyant force results from the pressure *differential* in the fluid. On a high-pressure day, the pressure at all points in the water is higher than on a low-pressure day. Since water is almost incompressible, however, the rate of change of pressure with depth is the same, resulting in no change in the buoyant force.

4. In the ocean, the ship floats due to the buoyant force from *salt water*. Salt water is denser than fresh water. As the ship is pulled up the river, the buoyant force from the fresh water in the river is not sufficient to support the weight of the ship, and it sinks.

5. The balance will not be in equilibrium—the lead side will be lower. Despite the fact that the weights on both sides of the balance are the same, the styrofoam, due to its larger volume, will experience a larger buoyant force from the surrounding air. Thus, the net force of the weight and buoyant force is larger, in the downward direction, for the lead than for the styrofoam.

6. The level of the pond falls. This is because the anchor displaces more water while in the boat. A floating object displaces a volume of water whose weight is equal to the weight of the object. A submerged object displaces a volume of water equal to the volume of the object. Because the density of the anchor is greater than that of water, a volume of water that weighs the same as the anchor will be greater than the volume of the anchor.

7. As the truck passes, the air between your car and the truck is compressed into the channel between you and the truck and moves at a higher speed than when your car is in the open. According to Bernoulli's principle, this high-speed air has a lower pressure than the air on the outer side of your car. The difference in pressure provides a net force on your car toward the truck.

APPENDIX A

TABLE A.1 Conversion Factors

Length					
	m	**cm**	**km**	**in.**	**ft**	**mi**
1 meter	1	10^2	10^{-3}	39.37	3.281	6.214×10^{-4}
1 centimeter	10^{-2}	1	10^{-5}	0.3937	3.281×10^{-2}	6.214×10^{-6}
1 kilometer	10^3	10^5	1	3.937×10^4	3.281×10^3	0.6214
1 inch	2.540×10^{-2}	2.540	2.540×10^{-5}	1	8.333×10^{-2}	1.578×10^{-5}
1 foot	0.3048	30.48	3.048×10^{-4}	12	1	1.894×10^{-4}
1 mile	1609	1.609×10^5	1.609	6.336×10^4	5280	1

Mass				
	kg	**g**	**slug**	**u**
1 kilogram	1	10^3	6.852×10^{-2}	6.024×10^{26}
1 gram	10^{-3}	1	6.852×10^{-5}	6.024×10^{23}
1 slug	14.59	1.459×10^4	1	8.789×10^{27}
1 atomic mass unit	1.660×10^{-27}	1.660×10^{-24}	1.137×10^{-28}	1

Time				
	s	**min**	**h**	**day**	**year**
1 second	1	1.667×10^{-2}	2.778×10^{-4}	1.157×10^{-5}	3.169×10^{-8}
1 minute	60	1	1.667×10^{-2}	6.994×10^{-4}	1.901×10^{-6}
1 hour	3600	60	1	4.167×10^{-2}	1.141×10^{-4}
1 day	8.640×10^4	1440	24	1	2.738×10^{-3}
1 year	3.156×10^7	5.259×10^5	8.766×10^3	365.2	1

Speed			
	m/s	**cm/s**	**ft/s**	**mi/h**
1 meter/second	1	10^2	3.281	2.237
1 centimeter/second	10^{-2}	1	3.281×10^{-2}	2.237×10^{-2}
1 foot/second	0.3048	30.48	1	0.6818
1 mile/hour	0.4470	44.70	1.467	1

Note: 1 mi/min = 60 mi/h = 88 ft/s

	Force		
	N	**dyn**	**lb**
1 newton	1	10^5	0.2248
1 dyne	10^{-5}	1	2.248×10^{-6}
1 pound	4.448	4.448×10^5	1

	Work, Energy, Heat		
	J	**erg**	**ft·lb**
1 joule	1	10^7	0.7376
1 erg	10^{-7}	1	7.376×10^{-8}
1 ft·lb	1.356	1.356×10^7	1
1 eV	1.602×10^{-19}	1.602×10^{-12}	1.182×10^{-19}
1 cal	4.186	4.186×10^7	3.087
1 Btu	1.055×10^{-3}	1.055×10^{10}	7.779×10^2
1 kWh	3.600×10^6	3.600×10^{13}	2.655×10^6

	eV	**cal**	**Btu**	**kWh**
1 joule	6.242×10^{18}	0.2389	9.481×10^{-4}	2.778×10^{-7}
1 erg	6.242×10^{11}	2.389×10^{-8}	9.481×10^{-11}	2.778×10^{-14}
1 ft·lb	8.464×10^{18}	0.3239	1.285×10^{-3}	3.766×10^{-7}
1 eV	1	3.827×10^{-20}	1.519×10^{-22}	4.450×10^{-26}
1 cal	2.613×10^{19}	1	3.968×10^{-3}	1.163×10^{-6}
1 Btu	6.585×10^{21}	2.520×10^2	1	2.930×10^{-4}
1 kWh	2.247×10^{25}	8.601×10^5	3.413×10^2	1

	Pressure		
	Pa	**dyn/cm^2**	**atm**
1 pascal = 1 newton/meter^2	1	10	9.869×10^{-6}
1 dyne/centimeter^2	10^{-1}	1	9.869×10^{-7}
1 atmosphere	1.013×10^5	1.013×10^6	1
1 centimeter mercury*	1.333×10^3	1.333×10^4	1.316×10^{-2}
1 pound/inch^2	6.895×10^3	6.895×10^4	6.805×10^{-2}
1 pound/foot^2	47.88	4.788×10^2	4.725×10^{-4}

	cm Hg	**lb/in.^{2}**	**lb/ft^2**
1 pascal = newton/meter^2	7.501×10^{-4}	1.450×10^{-4}	2.089×10^{-2}
1 dyne/centimeter^2	7.501×10^{-5}	1.450×10^{-5}	2.089×10^{-3}
1 atmosphere	76	14.70	2.116×10^3
1 centimeter mercury*	1	0.1943	27.85
1 pound/inch^2	5.171	1	144
1 pound/foot^2	3.591×10^{-2}	6.944×10^{-3}	1

* At 0°C and at a location at which the free-fall acceleration has its "standard" value, 9.80665 m/s^2.

TABLE A.2 Symbols, Dimensions, and Units of Physical Quantities

Quantity	Common Symbol	Unit*	Dimensions†	Unit in Terms of Base SI Units
Acceleration	**a**	m/s^2	L/T^2	m/s^2
Amount of substance	n	mole		mol
Angle	θ, ϕ	radian (rad)	1	
Angular acceleration	$\boldsymbol{\alpha}$	rad/s^2	T^{-2}	s^{-2}
Angular frequency	ω	rad/s	T^{-1}	s^{-1}
Angular momentum	**L**	$kg \cdot m^2/s$	ML^2/T	$kg \cdot m^2/s$
Angular velocity	$\boldsymbol{\omega}$	rad/s	T^{-1}	s^{-1}
Area	A	m^2	L^2	m^2
Atomic number	Z			
Capacitance	C	farad (F) (=C/V)	Q^2T^2/ML^2	$A^2 \cdot s^4/kg \cdot m^2$
Charge	q, Q, e	coulomb (C)	Q	$A \cdot s$
Charge density				
Line	λ	C/m	Q/L	$A \cdot s/m$
Surface	σ	C/m^2	Q/L^2	$A \cdot s/m^2$
Volume	ρ	C/m^3	Q/L^3	$A \cdot s/m^3$
Conductivity	σ	$1/\Omega \cdot m$	Q^2T/ML^3	$A^2 \cdot s^3/kg \cdot m^3$
Current	I	AMPERE	Q/T	A
Current density	**J**	A/m^2	Q/T^2	A/m^2
Density	ρ	kg/m^3	M/L^3	kg/m^3
Dielectric constant	κ			
Displacement	**s**	METER	L	m
Distance	d, h			
Length	ℓ, L			
Electric dipole moment	**p**	$C \cdot m$	QL	$A \cdot s \cdot m$
Electric field	**E**	V/m (=N/C)	ML/QT^2	$kg \cdot m/A \cdot s^3$
Electric flux	Φ	$V \cdot m$	ML^3/QT^2	$kg \cdot m^3/A \cdot s^3$
Electromotive force	$\mathcal{E}$	volt (V)	ML^2/QT^2	$kg \cdot m^2/A \cdot s^3$
Energy	E, U, K	joule (J)	ML^2/T^2	$kg \cdot m^2/s^2$
Entropy	S	J/K	$ML^2/T^2 \cdot K$	$kg \cdot m^2/s^2 \cdot K$
Force	**F**	newton (N)	ML/T^2	$kg \cdot m/s^2$
Frequency	f	hertz (Hz)	T^{-1}	s^{-1}
Heat	Q	joule (J)	ML^2/T^2	$kg \cdot m^2/s^2$
Inductance	L	henry (H)	ML^2/Q^2	$kg \cdot m^2/A^2 \cdot s^2$
Magnetic dipole moment	$\boldsymbol{\mu}$	$N \cdot m/T$	QL^2/T	$A \cdot m^2$
Magnetic field	**B**	tesla (T) (=Wb/m²)	M/QT	$kg/A \cdot s^2$
Magnetic flux	Φ_B	weber (Wb)	ML^2/QT	$kg \cdot m^2/A \cdot s^2$
Mass	m, M	KILOGRAM	M	kg
Molar specific heat	C	$J/Mol \cdot K$		$kg \cdot m^2/s^2 \cdot mol \cdot K$
Moment of inertia	I	$kg \cdot m^2$	ML^2	$kg \cdot m^2$
Momentum	**p**	$kg \cdot m/s$	ML/T	$kg \cdot m/s$
Period	T	s	T	s
Permeability of space	μ_0	$N/A^2 (=H/m)$	ML/Q^2T	$kg \cdot m/A^2 \cdot s^2$
Permittivity of space	ε_0	$C^2/N \cdot m^2 (=F/m)$	Q^2T^2/ML^3	$A^2 \cdot s^4/kg \cdot m^3$
Potential (voltage)	V	volt (V) (=J/C)	ML^2/QT^2	$kg \cdot m^2/A \cdot s^3$
Power	P	watt (W) (=J/s)	ML^2/T^3	$kg \cdot m^2/s^3$
Pressure	P, p	pascal (Pa) = (N/m²)	M/LT^2	$kg/m \cdot s^2$

continued

* The base SI units are given in uppercase letters.

† The symbols M, L, T, and Q denote mass, length, time, and charge, respectively.

TABLE A.2   *(Continued)*

Quantity	Common Symbol	Unit*	Dimensions[†]	Unit in Terms of Base SI Units
Resistance	R	ohm $(\Omega)\,(=V/A)$	ML^2/Q^2T	$kg\cdot m^2/A^2\cdot s^3$
Specific heat	c	$J/kg\cdot K$	$L^2/T^2\cdot K$	$m^2/s^2\cdot K$
Temperature	T	KELVIN	K	K
Time	t	SECOND	T	s
Torque	τ	$N\cdot m$	ML^2/T^2	$kg\cdot m^2/s^2$
Speed	v	m/s	L/T	m/s
Volume	V	m^3	L^3	m^3
Wavelength	λ	m	L	m
Work	W	joule $(J)\,(=N\cdot m)$	ML^2/T^2	$kg\cdot m^2/s^2$

* The base SI units are given in uppercase letters.

[†] The symbols M, L, T, and Q denote mass, length, time, and charge, respectively.

TABLE A.3   **Table of Selected Atomic Masses**[a]

Z	Element	Symbol	Chemical Atomic Mass (u)	Mass Number (* Indicates Radioactive) A	Atomic Mass (u)	Percentage Abundance	Half-Life (if Radioactive) $T_{1/2}$
0	(Neutron)	n		1*	1.008 665		10.4 m
1	Hydrogen	H	1.0079	1	1.007 825	99.985	
	Deuterium	D		2	2.014 102	0.015	
	Tritium	T		3*	3.016 049		12.33 y
2	Helium	He	4.00260	3	3.016 029	0.00014	
				4	4.002 602	99.99986	
3	Lithium	Li	6.941	6	6.015 121	7.5	
				7	7.016 003	92.5	
4	Beryllium	Be	9.0122	7*	7.016 928		53.3 d
				9	9.012 174	100	
5	Boron	B	10.81	10	10.012 936	19.9	
				11	11.009 305	80.1	
6	Carbon	C	12.011	11*	11.011 433		20.4 m
				12	12.000 000	98.90	
				13	13.003 355	1.10	
				14*	14.003 242		5730 y
7	Nitrogen	N	14.0067	13*	13.005 738		9.96 m
				14	14.003 074	99.63	
				15	15.000 108	0.37	
8	Oxygen	O	15.9994	15*	15.003 065		122
				16	15.994 915	99.761	
				18	17.999 160	0.20	
9	Fluorine	F	18.99840	19	18.998 404	100	
10	Neon	Ne	20.180	20	19.992 435	90.48	
				22	21.991 383	9.25	
11	Sodium	Na	22.98987	22*	21.994 434		2.61 y
				23	22.989 770	100	
				24*	23.990 961		14.96 h

TABLE A.3 (*Continued*)

Z	Element	Symbol	Chemical Atomic Mass (u)	Mass Number (* Indicates Radioactive) A	Atomic Mass (u)	Percentage Abundance	Half-Life (if Radioactive) $T_{1/2}$
12	Magnesium	Mg	24.305	24	23.985 042	78.99	
				25	24.985 838	10.00	
				26	25.982 594	11.01	
13	Aluminum	Al	26.98154	27	26.981 538	100	
14	Silicon	Si	28.086	28	27.976 927	92.23	
				29	28.976 495	4.67	
				30	29.973 770	3.10	
15	Phosphorus	P	30.97376	31	30.973 762	100	
				32*	31.973 908		14.26 d
16	Sulfur	S	32.066	32	31.972 071	95.02	
				33	32.971 459	0.75	
				34	33.967 867	4.21	
				35*	34.969 033		87.5 d
17	Chlorine	Cl	35.453	35	34.968 853	75.77	
				37	36.965 903	24.23	
18	Argon	Ar	39.948	36	35.967 547	0.337	
				40	39.962 384	99.600	
19	Potassium	K	39.0983	39	38.963 708	93.2581	
				40*	39.964 000	0.0117	1.28×10^9 y
				41	40.961 827	6.7302	
20	Calcium	Ca	40.08	40	39.962 591	96.941	
				44	43.955 481	2.086	
21	Scandium	Sc	44.9559	45	44.955 911	100	
22	Titanium	Ti	47.88	46	45.952 630	8.0	
				47	46.951 765	7.3	
				48	47.947 947	73.8	
				49	48.947 871	5.5	
				50	49.944 792	5.4	
23	Vanadium	V	50.9415	50*	49.947 161	0.25	1.5×10^{17} y
				51	50.943 962	99.75	
24	Chromium	Cr	51.996	50	49.946 047	4.345	
				52	51.940 511	83.79	
				53	52.940 652	9.50	
				54	53.938 883	2.365	
25	Manganese	Mn	54.93805	55	54.938 048	100	
26	Iron	Fe	55.847	54	53.939 613	5.9	
				56	55.934 940	91.72	
				57	56.935 396	2.1	
				58	57.933 278	0.28	

continued

---

[a] The masses in the sixth column are atomic masses, which include the mass of Z electrons. Data are from the National Nuclear Data Center, Brookhaven National Laboratory, prepared by Jagdish K. Tuli, July 1990. The data are based on experimental results reported in *Nuclear Data Sheets* and *Nuclear Physics* and also from *Chart of the Nuclides*, 14th ed. Atomic masses are based on those by A. H. Wapstra, G. Audi, and R. Hoekstra. Isotopic abundances are based on those by N. E. Holden.

TABLE A.3   *(Continued)*

Z	Element	Symbol	Chemical Atomic Mass (u)	Mass Number (* Indicates Radioactive) A	Atomic Mass (u)	Percentage Abundance	Half-Life (if Radioactive) $T_{1/2}$
27	Cobalt	Co	58.93320	59	58.933 198	100	
				60*	59.933 820		5.27 y
28	Nickel	Ni	58.693	58	57.935 346	68.077	
				60	59.930 789	26.223	
				62	61.928 346	3.634	
29	Copper	Cu	63.54	63	62.929 599	69.17	
				65	64.927 791	30.83	
30	Zinc	Zn	65.39	64	63.929 144	48.6	
				66	65.926 035	27.9	
				67	66.927 129	4.1	
				68	67.924 845	18.8	
31	Gallium	Ga	69.723	69	68.925 580	60.108	
				71	70.924 703	39.892	
32	Germanium	Ge	72.61	70	69.924 250	21.23	
				72	71.922 079	27.66	
				73	72.923 462	7.73	
				74	73.921 177	35.94	
				76	75.921 402	7.44	
33	Arsenic	As	74.9216	75	74.921 594	100	
34	Selenium	Se	78.96	74	73.922 474	0.89	
				76	75.919 212	9.36	
				77	76.919 913	7.63	
				78	77.917 307	23.78	
				80	79.916 519	49.61	
				82*	81.916 697	8.73	1.4×10^{20} y
35	Bromine	Br	79.904	79	78.918 336	50.69	
				81	80.916 287	49.31	
36	Krypton	Kr	83.80	78	77.920 400	0.35	
				80	79.916 377	2.25	
				82	81.913 481	11.6	
				83	82.914 136	11.5	
				84	83.911 508	57.0	
				86	85.910 615	17.3	
37	Rubidium	Rb	85.468	85	84.911 793	72.17	
				87*	86.909 186	27.83	4.75×10^{10} y
38	Strontium	Sr	87.62	86	85.909 266	9.86	
				87	86.908 883	7.00	
				88	87.905 618	82.58	
				90*	89.907 737		29.1 y
39	Yttrium	Y	88.9058	89	88.905 847	100	
40	Zirconium	Zr	91.224	90	89.904 702	51.45	
				91	90.905 643	11.22	
				92	91.905 038	17.15	
				94	93.906 314	17.38	
				96	95.908 274	2.80	
41	Niobium	Nb	92.9064	93	92.906 376	100	

TABLE A.3   *(Continued)*

Z	Element	Symbol	Chemical Atomic Mass (u)	Mass Number (* Indicates Radioactive) A	Atomic Mass (u)	Percentage Abundance	Half-Life (if Radioactive) $T_{1/2}$
42	Molybdenum	Mo	95.94	92	91.906 807	14.84	
				94	93.905 085	9.25	
				95	94.905 841	15.92	
				96	95.904 678	16.68	
				97	96.906 020	9.55	
				98	97.905 407	24.13	
				100	99.907 476	9.63	
43	Technetium	Tc		98*	97.907 215		4.2×10^6 y
				99*	98.906 254		2.1×10^5 y
44	Ruthenium	Ru	101.07	96	95.907 597	5.54	
				99	98.905 939	12.7	
				100	99.904 219	12.6	
				101	100.905 558	17.1	
				102	101.904 348	31.6	
				104	103.905 428	18.6	
45	Rhodium	Rh	102.9055	103	102.905 502	100	
46	Palladium	Pd	106.42	104	103.904 033	11.14	
				105	104.905 082	22.33	
				106	105.903 481	27.33	
				108	107.903 893	26.46	
				110	109.905 158	11.72	
47	Silver	Ag	107.868	107	106.905 091	51.84	
				109	108.904 754	48.16	
48	Cadmium	Cd	112.41	110	109.903 004	12.49	
				111	110.904 182	12.80	
				112	111.902 760	24.13	
				113*	112.904 401	12.22	9.3×10^{15} y
				114	113.903 359	28.73	
				116	115.904 755	7.49	
49	Indium	In	114.82	113	112.904 060	4.3	
				115*	114.903 876	95.7	4.4×10^{14} y
50	Tin	Sn	118.71	116	115.901 743	14.53	
				117	116.902 953	7.68	
				118	117.901 605	24.22	
				119	118.903 308	8.58	
				120	119.902 197	32.59	
				122	121.903 439	4.63	
				124	123.905 274	5.79	
51	Antimony	Sb	121.76	121	120.903 820	57.36	
				123	122.904 215	42.64	
52	Tellurium	Te	127.60	122	121.903 052	2.59	
				124	123.902 817	4.79	
				125	124.904 429	7.12	
				126	125.903 309	18.93	
				128*	127.904 463	31.70	$>8 \times 10^{24}$ y
				130*	129.906 228	33.87	$\leq 1.25 \times 10^{21}$ y

continued

TABLE A.3 *(Continued)*

Z	Element	Symbol	Chemical Atomic Mass (u)	Mass Number (* Indicates Radioactive) A	Atomic Mass (u)	Percentage Abundance	Half-Life (if Radioactive) $T_{1/2}$
53	Iodine	I	126.9045	127	126.904 474	100	
				129*	128.904 984		1.6×10^7 y
54	Xenon	Xe	131.29	129	128.904 779	26.4	
				130	129.903 509	4.1	
				131	130.905 069	21.2	
				132	131.904 141	26.9	
				134	133.905 394	10.4	
				136*	135.907 215	8.9	$\geq 2.36 \times 10^{21}$ y
55	Cesium	Cs	132.9054	133	132.905 436	100	
56	Barium	Ba	137.33	134	133.904 492	2.42	
				135	134.905 671	6.593	
				136	135.904 559	7.85	
				137	136.905 816	11.23	
				138	137.905 236	71.70	
57	Lanthanum	La	138.905	139	138.906 346	99.9098	
58	Cerium	Ce	140.12	140	139.905 434	88.43	
				142*	141.909 241	11.13	$>5 \times 10^{16}$ y
59	Praseodymium	Pr	140.9076	141	140.907 647	100	
60	Neodymium	Nd	144.24	142	141.907 718	27.13	
				143	142.909 809	12.18	
				144*	143.910 082	23.80	2.3×10^{15} y
				145	144.912 568	8.30	
				146	145.913 113	17.19	
				148	147.916 888	5.76	
				150*	149.920 887	5.64	$>1 \times 10^{18}$ y
61	Promethium	Pm		145*	144.912 745		17.7 y
62	Samarium	Sm	150.36	144	143.911 996	3.1	
				147*	146.914 894	15.0	1.06×10^{11} y
				148*	147.914 819	11.3	7×10^{15} y
				149*	148.917 180	13.8	$>2 \times 10^{15}$ y
				150	149.917 273	7.4	
				152	151.919 728	26.7	
				154	153.922 206	22.7	
63	Europium	Eu	151.96	151	150.919 846	47.8	
				153	152.921 226	52.2	
64	Gadolinium	Gd	157.25	155	154.922 618	14.80	
				156	155.922 119	20.47	
				157	156.923 957	15.65	
				158	157.924 099	24.84	
				160	159.927 050	21.86	
65	Terbium	Tb	158.9253	159	158.925 345	100	
66	Dysprosium	Dy	162.50	161	160.926 930	18.9	
				162	161.926 796	25.5	
				163	162.928 729	24.9	
				164	163.929 172	28.2	
67	Holmium	Ho	164.9303	165	164.930 316	100	

TABLE A.3 *(Continued)*

Z	Element	Symbol	Chemical Atomic Mass (u)	Mass Number (* Indicates Radioactive) A	Atomic Mass (u)	Percentage Abundance	Half-Life (if Radioactive) $T_{1/2}$
68	Erbium	Er	167.26	166	165.930 292	33.6	
				167	166.932 047	22.95	
				168	167.932 369	27.8	
				170	169.935 462	14.9	
69	Thulium	Tm	168.9342	169	168.934 213	100	
70	Ytterbium	Yb	173.04	170	169.934 761	3.05	
				171	170.936 324	14.3	
				172	171.936 380	21.9	
				173	172.938 209	16.12	
				174	173.938 861	31.8	
				176	175.942 564	12.7	
71	Lutecium	Lu	174.967	175	174.940 772	97.41	
				176*	175.942 679	2.59	3.78×10^{10} y
72	Hafnium	Hf	178.49	176	175.941 404	5.206	
				177	176.943 218	18.606	
				178	177.943 697	27.297	
				179	178.945 813	13.629	
				180	179.946 547	35.100	
73	Tantalum	Ta	180.9479	181	180.947 993	99.988	
74	Tungsten (Wolfram)	W	183.85	182	181.948 202	26.3	
				183	182.950 221	14.28	
				184	183.950 929	30.7	
				186	185.954 358	28.6	
75	Rhenium	Re	186.207	185	184.952 951	37.40	
				187*	186.955 746	62.60	4.4×10^{10} y
76	Osmium	Os	190.2	188	187.955 832	13.3	
				189	188.958 139	16.1	
				190	189.958 439	26.4	
				192	191.961 468	41.0	
77	Iridium	Ir	192.2	191	190.960 585	37.3	
				193	192.962 916	62.7	
78	Platinum	Pt	195.08	194	193.962 655	32.9	
				195	194.964 765	33.8	
				196	195.964 926	25.3	
				198	197.967 867	7.2	
79	Gold	Au	196.9665	197	196.966 543	100	
80	Mercury	Hg	200.59	198	197.966 743	9.97	
				199	198.968 253	16.87	
				200	199.968 299	23.10	
				201	200.970 276	13.10	
				202	201.970 617	29.86	
				204	203.973 466	6.87	
81	Thallium	Tl	204.383	203	202.972 320	29.524	
				205	204.974 400	70.476	
		(Th C″)		208*	207.981 992		3.053 m

continued

TABLE A.3   *(Continued)*

Z	Element	Symbol	Chemical Atomic Mass (u)	Mass Number (* Indicates Radioactive) A	Atomic Mass (u)	Percentage Abundance	Half-Life (if Radioactive) $T_{1/2}$
82	Lead	Pb	207.2	204*	203.973 020	1.4	$\geqslant 1.4 \times 10^{17}$ y
				206	205.974 440	24.1	
				207	206.975 871	22.1	
				208	207.976 627	52.4	
		(Ra D)		210*	209.984 163		22.3 y
		(Ac B)		211*	210.988 734		36.1 m
		(Th B)		212*	211.991 872		10.64 h
		(Ra B)		214*	213.999 798		26.8 m
83	Bismuth	Bi	208.9803	209	208.980 374	100	
		(Th C)		211*	210.987 254		2.14 m
84	Polonium	Po					
		(Ra F)		210*	209.982 848		138.38 d
		(Ra C′)		214*	213.995 177		164 μs
85	Astatine	At		218*	218.008 685		1.6 s
86	Radon	Rn		222*	222.017 571		3.823 d
87	Francium	Fr					
		(Ac K)		223*	223.019 733		22 m
88	Radium	Ra		226*	226.025 402		1600 y
		(Ms Th$_1$)		228*	228.031 064		5.75 y
89	Actinium	Ac		227*	227.027 749		21.77 y
90	Thorium	Th	232.0381				
		(Rd Th)		228*	228.028 716		1.913 y
				232*	232.038 051	100	1.40×10^{10} y
91	Protactinium	Pa		231*	231.035 880		32.760 y
92	Uranium	U	238.0289	232*	232.037 131		69 y
				233*	233.039 630		1.59×10^5 y
		(Ac U)		235*	235.043 924	0.720	7.04×10^8 y
				236*	236.045 562		2.34×10^7 y
		(UI)		238*	238.050 784	99.2745	4.47×10^9 y
93	Neptunium	Np		237*	237.048 168		2.14×10^6 y
94	Plutonium	Pu		239*	239.052 157		24,120 y
				242*	242.058 737		3.73×10^5 y
				244*	244.064 200		8.1×10^7 y

Mathematics Review

These mathematics appendices are intended as a brief review of operations and methods. Early in your course, you should be facile with basic algebraic techniques, analytic geometry, and trigonometry. The appendices on differential and integral calculus are more detailed and are intended for those students who have difficulty applying calculus concepts to physical situations. Mathematical symbols used in the text are given in the following table.

Mathematical Symbols Used in the Text and Their Meanings

Symbol	Meaning
$=$	is equal to
$\equiv$	is defined as
$\neq$	is not equal to
$\propto$	is proportional to
$>$	is greater than
$<$	is less than
$\gg (\ll)$	is much greater (less) than
$\approx$	is approximately equal to
Δx	the change in x
$\displaystyle\sum_{i=1}^{N} x_i$	the sum of all quantities x_i from $i = 1$ to $i = N$
$\lvert x \rvert$	the magnitude of x (always a positive quantity)
$\Delta x \to 0$	Δx approaches zero
$\dfrac{dx}{dt}$	the derivative of x with respect to t
$\dfrac{\partial x}{\partial t}$	the partial derivative of x with respect to t
$\displaystyle\int$	integral

B.1 · SCIENTIFIC NOTATION

Many quantities that scientists deal with often have very large or very small values. For example, the speed of light is about 300 000 000 m/s, and the ink required to make the dot over an i in this textbook has a mass of about 0.000 000 001 kg.

Obviously, it is very cumbersome to read, write, and keep track of numbers such as these. We avoid this problem by using the powers of the number 10:

$$10^0 = 1$$

$$10^1 = 10$$

$$10^2 = 10 \times 10 = 100$$

$$10^3 = 10 \times 10 \times 10 = 1000$$

$$10^4 = 10 \times 10 \times 10 \times 10 = 10\,000$$

$$10^5 = 10 \times 10 \times 10 \times 10 \times 10 = 100\,000$$

and so on. The number of zeros corresponds to the power to which 10 is raised, called the exponent of 10. For example, the speed of light, $300\,000\,000$ m/s, can be expressed as 3×10^8 m/s.

For numbers less than one, we note the following:

$$10^{-1} = \frac{1}{10} = 0.1$$

$$10^{-2} = \frac{1}{10 \times 10} = 0.01$$

$$10^{-3} = \frac{1}{10 \times 10 \times 10} = 0.001$$

$$10^{-4} = \frac{1}{10 \times 10 \times 10 \times 10} = 0.0001$$

$$10^{-5} = \frac{1}{10 \times 10 \times 10 \times 10 \times 10} = 0.00001$$

In these cases, the number of places that the decimal point lies to the left of the digit 1 equals the value of the (negative) exponent. Numbers that are expressed as some power of 10 multiplied by another number between 1 and 10 are said to be in **scientific notation.** For example, the scientific notation for $5\,943\,000\,000$ is 5.943×10^9, and that for 0.0000832 is 8.32×10^{-5}.

When numbers expressed in scientific notation are being multiplied, the following general rule is very useful:

$$10^n \times 10^m = 10^{n+m} \qquad\qquad \textbf{[B.1]}$$

where n and m can be *any* numbers (not necessarily integers). For example, $10^2 \times 10^5 = 10^7$. The rule also applies if one of the exponents is negative. For example, $10^3 \times 10^{-8} = 10^{-5}$.

When dividing numbers expressed in scientific notation, note that

$$\frac{10^n}{10^m} = 10^n \times 10^{-m} = 10^{n-m} \qquad\qquad \textbf{[B.2]}$$

EXERCISES   With help from these rules, verify the answers to the following:

1. $86\,400 = 8.64 \times 10^4$
2. $9\,816\,762.5 = 9.8167625 \times 10^6$

3. $0.0000000398 = 3.98 \times 10^{-8}$
4. $(4 \times 10^8)(9 \times 10^9) = 3.6 \times 10^{18}$
5. $(3 \times 10^7)(6 \times 10^{-12}) = 1.8 \times 10^{-4}$
6. $\dfrac{75 \times 10^{-11}}{5 \times 10^{-3}} = 1.5 \times 10^{-7}$
7. $\dfrac{(3 \times 10^6)(8 \times 10^{-2})}{(2 \times 10^{17})(6 \times 10^5)} = 2 \times 10^{-18}$

B.2 · ALGEBRA

Some Basic Rules

When algebraic operations are performed, the laws of arithmetic apply. Symbols such as x, y, and z are usually used to represent quantities that are not specified—the **unknowns.**

First, consider the equation

$$8x = 32$$

If we wish to solve for x, we can divide (or multiply) each side of the equation by the same factor without destroying the equality. In this case, if we divide both sides by 8, we have

$$\frac{8x}{8} = \frac{32}{8}$$

$$x = 4$$

Next consider the equation

$$x + 2 = 8$$

In this type of expression, we can add or subtract the same quantity from each side. If we subtract 2 from each side, we get

$$x + 2 - 2 = 8 - 2$$

$$x = 6$$

In general, if $x + a = b$, then $x = b - a$.

Now consider the equation

$$\frac{x}{5} = 9$$

If we multiply each side by 5, we are left with x on the left by itself and 45 on the right:

$$\left(\frac{x}{5}\right)(5) = 9 \times 5$$

$$x = 45$$

In all cases, **whatever operation is performed on the left side of the equality must also be performed on the right side.**

The following rules for multiplying, dividing, adding, and subtracting fractions should be recalled, where a, b, and c are three numbers:

	Rule	**Example**
Multiplying	$\left(\dfrac{a}{b}\right)\left(\dfrac{c}{d}\right) = \dfrac{ac}{bd}$	$\left(\dfrac{2}{3}\right)\left(\dfrac{4}{5}\right) = \dfrac{8}{15}$
Dividing	$\dfrac{(a/b)}{(c/d)} = \dfrac{ad}{cb}$	$\dfrac{2/3}{4/5} = \dfrac{(2)(5)}{(4)(3)} = \dfrac{10}{12}$
Adding	$\dfrac{a}{b} \pm \dfrac{c}{d} = \dfrac{ad \pm bc}{bd}$	$\dfrac{2}{3} - \dfrac{4}{5} = \dfrac{(2)(5) - (4)(3)}{(3)(5)} = -\dfrac{2}{15}$

EXERCISES   In the following exercises, solve for x:

Answers

1. $a = \dfrac{1}{1 + x}$ $x = \dfrac{1 - a}{a}$

2. $3x - 5 = 13$ $x = 6$

3. $ax - 5 = bx + 2$ $x = \dfrac{7}{a - b}$

4. $\dfrac{5}{2x + 6} = \dfrac{3}{4x + 8}$ $x = -\dfrac{11}{7}$

Powers

When powers of a given quantity x are multiplied, the following rule applies:

$$x^n x^m = x^{n+m} \qquad \text{[B.3]}$$

For example, $x^2 x^4 = x^{2+4} = x^6$.

When dividing the powers of a given quantity, the rule is

$$\frac{x^n}{x^m} = x^{n-m} \qquad \text{[B.4]}$$

For example, $x^8 / x^2 = x^{8-2} = x^6$.

A power that is a fraction, such as $\frac{1}{3}$, corresponds to a root as follows:

$$x^{1/n} = \sqrt[n]{x} \qquad \text{[B.5]}$$

For example, $4^{1/3} = \sqrt[3]{4} = 1.5874$. (A scientific calculator is useful for such calculations.)

Finally, any quantity x^n that is raised to the mth power is

$$(x^n)^m = x^{nm} \qquad \text{[B.6]}$$

Table B.1 summarizes the rules of exponents.

TABLE B.1   Rules of Exponents

$x^0 = 1$
$x^1 = x$
$x^n x^m = x^{n+m}$
$x^n / x^m = x^{n-m}$
$x^{1/n} = \sqrt[n]{x}$
$(x^n)^m = x^{nm}$

EXERCISES   Verify the following:

1. $3^2 \times 3^3 = 243$
2. $x^5 x^{-8} = x^{-3}$
3. $x^{10}/x^{-5} = x^{15}$
4. $5^{1/3} = 1.709975$ (Use your calculator.)
5. $60^{1/4} = 2.783158$ (Use your calculator.)
6. $(x^4)^3 = x^{12}$

Factoring

Some useful formulas for factoring an equation are

$$ax + ay + az = a(x + y + z) \qquad \text{Common factor}$$

$$a^2 + 2ab + b^2 = (a + b)^2 \qquad \text{Perfect square}$$

$$a^2 - b^2 = (a + b)(a - b) \qquad \text{Differences of squares}$$

Quadratic Equations

The general form of a quadratic equation is

$$ax^2 + bx + c = 0 \qquad \text{[B.7]}$$

where x is the unknown quantity and a, b, and c are numerical factors referred to as **coefficients** of the equation. This equation has two roots, given by

$$x = \frac{-b \pm \sqrt{b^2 - 4ac}}{2a} \qquad \text{[B.8]}$$

If $b^2 \geq 4ac$, the roots are real.

Example 1

The equation $x^2 + 5x + 4 = 0$ has the following roots, corresponding to the two signs of the square-root term:

$$x = \frac{-5 \pm \sqrt{5^2 - (4)(1)(4)}}{2(1)} = \frac{-5 \pm \sqrt{9}}{2} = \frac{-5 \pm 3}{2}$$

That is,

$$x_+ = \frac{-5 + 3}{2} = -1 \qquad x_- = \frac{-5 - 3}{2} = -4$$

where x_+ denotes the root corresponding to the positive sign and x_- denotes the root corresponding to the negative sign.

EXERCISES   Solve the following quadratic equations:

Answers

1. $x^2 + 2x - 3 = 0$ $x_+ = 1$ $x_- = -3$
2. $2x^2 - 5x + 2 = 0$ $x_+ = 2$ $x_- = \frac{1}{2}$
3. $2x^2 - 4x - 9 = 0$ $x_+ = 1 + \sqrt{22}/2$ $x_- = 1 - \sqrt{22}/2$

Linear Equations

A linear equation has the general form

$$y = ax + b \qquad \text{[B.9]}$$

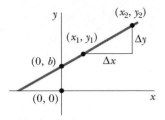

Figure B.1

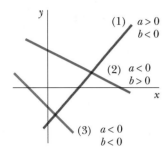

Figure B.2

where a and b are constants. This equation is referred to as linear because the graph of y versus x is a straight line, as shown in Figure B.1. The constant b, called the **intercept,** represents the value of y at which the straight line intersects the y axis. The constant a is equal to the slope of the straight line. If any two points on the straight line are specified by the coordinates (x_1, y_1) and (x_2, y_2), as in Figure B.1, then the **slope** of the straight line can be expressed as

$$\text{Slope} = \frac{y_2 - y_1}{x_2 - x_1} = \frac{\Delta y}{\Delta x} \qquad \textbf{[B.10]}$$

Note that a and b can have either positive or negative values. If $a > 0$, the straight line has a *positive* slope, as in Figure B.1. If $a < 0$, the straight line has a *negative* slope. In Figure B.1, both a and b are positive. Three other possible situations are shown in Figure B.2: $a > 0$, $b < 0$; $a < 0$, $b > 0$; and $a < 0$, $b < 0$.

EXERCISES

1. Draw graphs of the following straight lines:
   (a) $y = 5x + 3$ (b) $y = -2x + 4$ (c) $y = -3x - 6$.
2. Find the slopes of the straight lines described in Exercise 1.
   Answers   (a) 5   (b) -2 (c) -3
3. Find the slopes of the straight lines that pass through the following sets of point:
   (a) $(0, -4)$ and $(4, 2)$;   (b) $(0, 0)$ and $(2, -5)$;   (c) $(-5, 2)$ and $(4, -2)$.
   Answers   (a) 3/2   (b) $-5/2$ (c) $-4/9$

Solving Simultaneous Linear Equations

Consider an equation such as $3x + 5y = 15$, which has two unknowns, x and y. Such an equation does not have a unique solution. That is, $(x = 0, y = 3)$, $(x = 5, y = 0)$, and $(x = 2, y = 9/5)$ are all solutions to this equation.

If a problem has two unknowns, a unique solution is possible only if there are *two* independent equations. In general, if a problem has n unknowns, its solution requires n equations. In order to solve two simultaneous equations involving two unknowns, x and y, we solve one of the equations for x in terms of y and substitute this expression into the other equation.

Example 2

Solve the following two simultaneous equations:

$$(1) \ 5x + y = -8$$
$$(2) \ 2x - 2y = 4$$

Solution   From (2), $x = y + 2$. Substitution of this into (1) gives

$$5(y + 2) + y = -8$$
$$6y = -18$$
$$y = -3$$
$$x = y + 2 = \boxed{-1}$$

Alternative Solution   Multiply each term in (1) by the factor 2 and add the result to (2):

$$10x + 2y = -16$$
$$\underline{2x - 2y = 4}$$
$$12x = -12$$
$$x = -1$$
$$y = x - 2 = \boxed{-3}$$

Two linear equations with two unknowns can also be solved by a graphical method. If the straight lines corresponding to the two equations are plotted in a conventional coordinate system, the intersection of the two lines represents the solution. For example, consider the two equations

$$x - y = 2$$
$$x - 2y = -1$$

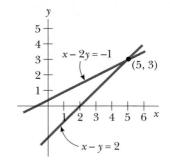

Figure B.3

These are plotted in Figure B.3. The intersection of the two lines has the coordinates $x = 5$, $y = 3$. This represents the solution to the equations. You should check this solution by the analytical technique discussed earlier.

EXERCISES Solve the following pairs of simultaneous equations involving two unknowns:

Answers

1. $x + y = 8$ $x = 5, y = 3$
    $x - y = 2$
2. $98 - T = 10a$ $T = 65, a = 3.27$
    $T - 49 = 5a$
3. $6x + 2y = 6$ $x = 2, y = -3$
    $8x - 4y = 28$

Logarithms

Suppose that a quantity, x, is expressed as a power of some quantity a:

$$x = a^y \qquad \text{[B.11]}$$

The number a is called the **base** number. The **logarithm** of x with respect to the base a is equal to the exponent to which the base must be raised in order to satisfy the expression $x = a^y$:

$$y = \log_a x \qquad \text{[B.12]}$$

Conversely, the **antilogarithm** of y is the number x:

$$x = \text{antilog}_a y \qquad \text{[B.13]}$$

In practice, the two bases most often used are base 10, called the *common* logarithm base, and base $e = 2.718 \ldots$, called the *natural* logarithm base. When common logarithms are used,

$$y = \log_{10} x \qquad (\text{or } x = 10^y) \qquad \text{[B.14]}$$

When natural logarithms are used,

$$y = \log_e x = \ln x \qquad (\text{or } x = e^y) \qquad \text{[B.15]}$$

For example, $\log_{10} 52 = 1.716$, and so $\text{antilog}_{10} 1.716 = 10^{1.716} = 52$. Likewise, $\ln 52 = 3.951$, so $\text{antiln } 3.951 = e^{3.951} = 52$.

In general, note that you can convert between base 10 and base e with the equality

$$\ln x = (2.302585) \log_{10} x \qquad \text{[B.16]}$$

Finally, some useful properties of logarithms are as follows:

$$\log(ab) = \log a + \log b$$

$$\log(a/b) = \log a - \log b$$

$$\log(a^n) = n \log a$$

$$\ln e = 1$$

$$\ln e^a = a$$

$$\ln\left(\frac{1}{a}\right) = -\ln a$$

B.3 · GEOMETRY

The **distance,** d, between two points the coordinates of which are (x_1, y_1) and (x_2, y_2) is

$$d = \sqrt{(x_2 - x_1)^2 + (y_2 - y_1)^2} \qquad \text{[B.17]}$$

The arc length, s, of a circular arc (Fig. B.4) is proportional to the radius, r, for a fixed value of θ (in radians) — the **radian measure:**

$$s = r\theta$$

$$\theta = \frac{s}{r} \qquad \text{[B.18]}$$

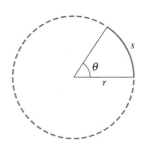

Figure B.4

Table B.2 gives the areas and volumes for several geometric shapes used throughout this book:

TABLE B.2   Useful Information for Geometry

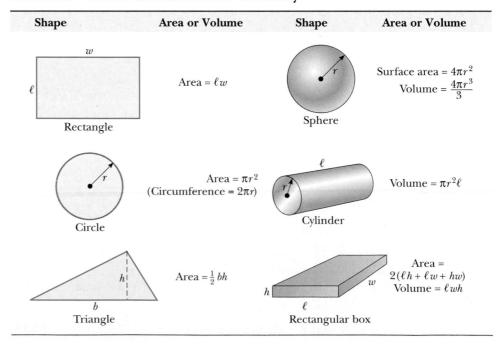

Shape	Area or Volume	Shape	Area or Volume
Rectangle	Area $= \ell w$	Sphere	Surface area $= 4\pi r^2$ Volume $= \dfrac{4\pi r^3}{3}$
Circle	Area $= \pi r^2$ (Circumference $= 2\pi r$)	Cylinder	Volume $= \pi r^2 \ell$
Triangle	Area $= \frac{1}{2} bh$	Rectangular box	Area $=$ $2(\ell h + \ell w + hw)$ Volume $= \ell w h$

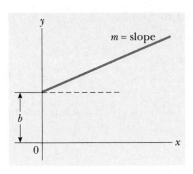

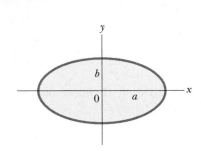

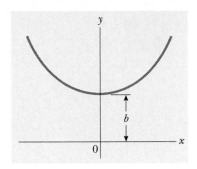

Figure B.5   **Figure B.6**   **Figure B.7**

The equation of a **straight line** (Fig. B.5) is

$$y = mx + b \qquad\qquad \textbf{[B.19]}$$

where b is the y intercept and m is the slope of the line.

The equation of a **circle** of radius R centered at the origin is

$$x^2 + y^2 = R^2 \qquad\qquad \textbf{[B.20]}$$

The equation of an **ellipse** with the origin at its center (Fig. B.6) is

$$\frac{x^2}{a^2} + \frac{y^2}{b^2} = 1 \qquad\qquad \textbf{[B.21]}$$

where a is the length of the semimajor axis and b is the length of the semiminor axis.

The equation of a **parabola** the vertex of which is at $y = b$ (Fig. B.7) is

$$y = ax^2 + b \qquad\qquad \textbf{[B.22]}$$

The equation of a **rectangular hyperbola** (Fig. B.8) is

$$xy = \text{constant} \qquad\qquad \textbf{[B.23]}$$

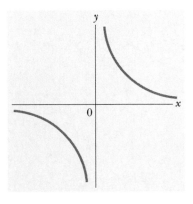

Figure B.8

B.4 · TRIGONOMETRY

That portion of mathematics based on the special properties of the right triangle is called trigonometry. By definition, a right triangle is one containing a 90° angle. Consider the right triangle in Figure B.9, where side a is opposite the angle θ, side b is adjacent to the angle θ, and side c is the hypotenuse of the triangle, The three basic trigonometric functions defined by such a triangle are the sine (sin), cosine (cos), tangent (tan) functions. In terms of the angle θ, these functions are defined by

$$\sin\theta \equiv \frac{\text{side opposite }\theta}{\text{hypotenuse}} = \frac{a}{c} \qquad\qquad \textbf{[B.24]}$$

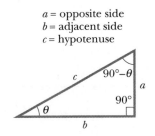

a = opposite side
b = adjacent side
c = hypotenuse

Figure B.9

$$\cos \theta \equiv \frac{\text{side adjacent to } \theta}{\text{hypotenuse}} = \frac{b}{c} \qquad [\textbf{B.25}]$$

$$\tan \theta \equiv \frac{\text{side opposite } \theta}{\text{side adjacent to } \theta} = \frac{a}{b} \qquad [\textbf{B.26}]$$

The Pythagorean theorem provides the following relationship between the sides of a right triangle:

$$c^2 = a^2 + b^2 \qquad [\textbf{B.27}]$$

From the preceding definitions and the Pythagorean theorem, it follows that

$$\sin^2 \theta + \cos^2 \theta = 1$$

$$\tan \theta = \frac{\sin \theta}{\cos \theta}$$

The cosecant, secant, and cotangent functions are defined by

$$\csc \theta \equiv \frac{1}{\sin \theta} \qquad \sec \theta \equiv \frac{1}{\cos \theta} \qquad \cot \theta \equiv \frac{1}{\tan \theta}$$

The following relations come directly from the right triangle shown in Figure B.9:

$$\begin{cases} \sin \theta = \cos(90° - \theta) \\ \cos \theta = \sin(90° - \theta) \\ \cot \theta = \tan(90° - \theta) \end{cases}$$

Some properties of trigonometric functions are as follows:

$$\begin{cases} \sin(-\theta) = -\sin \theta \\ \cos(-\theta) = \cos \theta \\ \tan(-\theta) = -\tan \theta \end{cases}$$

The following relations apply to *any* triangle, as shown in Figure B.10:

$$\alpha + \beta + \gamma = 180°$$

$$\text{Law of cosines} \quad \begin{cases} a^2 = b^2 + c^2 - 2bc \cos \alpha \\ b^2 = a^2 + c^2 - 2ac \cos \beta \\ c^2 = a^2 + b^2 - 2ab \cos \gamma \end{cases}$$

$$\text{Law of sines} \quad \begin{cases} \dfrac{a}{\sin \alpha} = \dfrac{b}{\sin \beta} = \dfrac{c}{\sin \gamma} \end{cases}$$

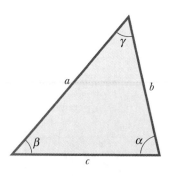

Figure B.10

Table B.3 lists a number of useful trigonometric identities.

TABLE B.3 Some Trigonometric Identities

$$\sin^2 \theta + \cos^2 \theta = 1 \qquad \csc^2 \theta = 1 + \cot^2 \theta$$

$$\sec^2 \theta = 1 + \tan^2 \theta \qquad \sin^2 \frac{\theta}{2} = \tfrac{1}{2}(1 - \cos \theta)$$

$$\sin 2\theta = 2 \sin \theta \cos \theta \qquad \cos^2 \frac{\theta}{2} = \tfrac{1}{2}(1 + \cos \theta)$$

$$\cos 2\theta = \cos^2 \theta - \sin^2 \theta \qquad 1 - \cos \theta = 2 \sin^2 \frac{\theta}{2}$$

$$\tan 2\theta = \frac{2 \tan \theta}{1 - \tan^2 \theta} \qquad \tan \frac{\theta}{2} = \sqrt{\frac{1 - \cos \theta}{1 + \cos \theta}}$$

$$\sin(A \pm B) = \sin A \cos B \pm \cos A \sin B$$

$$\cos(A \pm B) = \cos A \cos B \mp \sin A \sin B$$

$$\sin A \pm \sin B = 2 \sin[\tfrac{1}{2}(A \pm B)] \cos[\tfrac{1}{2}(A \mp B)]$$

$$\cos A + \cos B = 2 \cos[\tfrac{1}{2}(A + B)] \cos[\tfrac{1}{2}(A - B)]$$

$$\cos A - \cos B = 2 \sin[\tfrac{1}{2}(A + B)] \sin[\tfrac{1}{2}(B - A)]$$

Example 3

Consider the right triangle in Figure B.11, in which $a = 2$, $b = 5$, and c is unknown. From the Pythagorean theorem, we have

$$c^2 = a^2 + b^2 = 2^2 + 5^2 = 4 + 25 = 29$$

$$c = \sqrt{29} = \boxed{5.39}$$

To find the angle θ, note that

$$\tan \theta = \frac{a}{b} = \frac{2}{5} = 0.400$$

From a table of functions or from a calculator, we have

$$\theta = \tan^{-1}(0.400) = \boxed{21.8°}$$

where $\tan^{-1}(0.400)$ is the notation for "angle the tangent of which is 0.400," sometimes written as arctan (0.400).

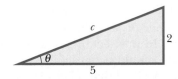

Figure B.11

EXERCISES

1. In Figure B.12, find (a) the side opposite θ, (b) the side adjacent to ϕ, (c) cos θ, (d) sin ϕ, and (e) tan ϕ.
   Answers (a) 3, (b) 3, (c) $\frac{4}{5}$, (d) $\frac{4}{5}$, (e) $\frac{4}{3}$
2. In a certain right triangle, the two sides that are perpendicular to each other are 5 m and 7 m long. What is the length of the third side of the triangle?
   Answer 8.60 m
3. A right triangle has a hypotenuse of length 3 m, and one of its angles is 30°. What are the lengths of (a) the side opposite the 30° angle and (b) the side adjacent to the 30° angle?
   Answers (a) 1.5 m, (b) 2.60 m

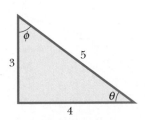

Figure B.12

B.5 • SERIES EXPANSIONS

$$(a + b)^n = a^n + \frac{n}{1!} a^{n-1} b + \frac{n(n-1)}{2!} a^{n-2} b^2 + \cdots$$

$$(1 + x)^n = 1 + nx + \frac{n(n-1)}{2!} x^2 + \cdots$$

$$e^x = 1 + x + \frac{x^2}{2!} + \frac{x^3}{3!} + \cdots$$

$$\ln(1 \pm x) = \pm x - \tfrac{1}{2} x^2 \pm \tfrac{1}{3} x^3 - \cdots$$

$$\left. \begin{aligned} \sin x &= x - \frac{x^3}{3!} + \frac{x^5}{5!} - \cdots \\[4pt] \cos x &= 1 - \frac{x^2}{2!} + \frac{x^4}{4!} - \cdots \\[4pt] \tan x &= x + \frac{x^3}{3} + \frac{2x^5}{15} + \cdots \quad |x| < \pi/2 \end{aligned} \right\} \quad x \text{ in radians}$$

For $x \ll 1$, the following approximations can be used:

$$(1 + x)^n \approx 1 + nx \qquad \sin x \approx x$$

$$e^x \approx 1 + x \qquad \cos x \approx 1$$

$$\ln(1 \pm x) \approx \pm x \qquad \tan x \approx x$$

B.6 • DIFFERENTIAL CALCULUS

In various branches of science, it is sometimes necessary to use the basic tools of calculus, first invented by Newton, to describe physical phenomena. The use of calculus is fundamental in the treatment of a variety of problems in newtonian mechanics, electricity, and magnetism. In this section, we simply state some basic properties and "rules of thumb," which should be a useful review.

First, a **function relation** must be specified that relates one variable to another (such as a coordinate as a function of time). Suppose one of the variables is called y (the dependent variable), and the other, x (the independent variable). We might have a function relation such as

$$y(x) = ax^3 + bx^2 + cx + d$$

If a, b, c, and d are specified constants, then y can be calculated for any value of x. We usually deal with continuous functions, that is, those for which y varies "smoothly" with x.

The **derivative** of y with respect to x is defined as the limit of the slopes of chords drawn between two points on the y versus x curve as Δx approaches zero. Mathematically, we write this definition as

$$\frac{dy}{dx} = \lim_{\Delta x \to 0} \frac{\Delta y}{\Delta x} = \lim_{\Delta x \to 0} \frac{y(x + \Delta x) - y(x)}{\Delta x} \qquad \textbf{[B.28]}$$

where Δy and Δx are defined as $\Delta x = x_2 - x_1$ and $\Delta y = y_2 - y_1$ (see Fig. B.13).

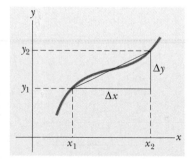

Figure B.13

A useful expression to remember when $y(x) = ax^n$, where a is a *constant* and n is *any* positive or negative number (integer or fraction), is

$$\frac{dy}{dx} = nax^{n-1} \qquad \text{[B.29]}$$

If $y(x)$ is a polynomial or algebraic function of x, we apply Equation B.29 to *each* term in the polynomial and take $da/dx = 0$. It is important to note that dy/dx *does not* mean dy divided by dx but is simply a notation of the limiting process of the derivative, as defined by Equation B.28. In Examples 4 through 7, we evaluate the derivatives of several well-behaved functions.

Example 4

Suppose $y(x)$ (that is, y as a function of x) is given by

$$y(x) = ax^3 + bx + c$$

where a and b are constants. Then it follows that

$$y(x + \Delta x) = a(x + \Delta x)^3 + b(x + \Delta x) + c$$
$$= a(x^3 + 3x^2\,\Delta x + 3x\,\Delta x^2 + \Delta x^3)$$
$$+ b(x + \Delta x) + c$$

so

$$\Delta y = y(x + \Delta x) - y(x) = a(3x^2\,\Delta x + 3x\,\Delta x^2 + \Delta x^3) + b\,\Delta x$$

Substituting this into Equation B.28 gives

$$\frac{dy}{dx} = \lim_{\Delta x \to 0} \frac{\Delta y}{\Delta x} = \lim_{\Delta x \to 0} [3ax^2 + 3x\,\Delta x + \Delta x^2] + b$$

$$\frac{dy}{dx} = 3ax^2 + b$$

Example 5

$$y(x) = 8x^5 + 4x^3 + 2x + 7$$

Solution Applying Equation B.29 to each term independently and remembering that d/dx (constant) $= 0$, we have

$$\frac{dy}{dx} = 8(5)x^4 + 4(3)x^2 + 2(1)x^0 + 0$$

$$\frac{dy}{dx} = 40x^4 + 12x^2 + 2$$

Special Properties of the Derivative

A. Derivative of the Product of Two Functions. If a function, y, is given by the product of two functions—say, $g(x)$ and $h(x)$—then the derivative of y is defined as

$$\frac{d}{dx} f(x) = \frac{d}{dx} [g(x)h(x)] = g\frac{dh}{dx} + h\frac{dg}{dx} \qquad \text{[B.30]}$$

B. Derivative of the Sum of Two Functions. If a function, y, is equal to the sum of two functions, then the derivative of the sum is equal to the sum of the derivatives:

$$\frac{d}{dx} f(x) = \frac{d}{dx} [g(x) + h(x)] = \frac{dg}{dx} + \frac{dh}{dx} \qquad \text{[B.31]}$$

C. Chain Rule of Differential Calculus. If $y = f(x)$ and x is a function of some other variable, z, then dy/dx can be written as the product of two derivatives:

$$\frac{dy}{dx} = \frac{dy}{dz}\frac{dz}{dx} \qquad\qquad \text{[B.32]}$$

D. The Second Derivative. The second derivative of y with respect to x is defined as the derivative of the function dy/dx (that is, the derivative of the derivative). It is usually written

$$\frac{d^2y}{dx^2} = \frac{d}{dx}\left(\frac{dy}{dx}\right) \qquad\qquad \text{[B.33]}$$

Example 6

Find the first derivative of $y(x) = x^3/(x + 1)^2$ with respect to x.

Solution We can rewrite this function in the alternate form $y(x) = x^3(x + 1)^{-2}$ and apply Equation B.30 directly:

$$\frac{dy}{dx} = (x + 1)^{-2}\frac{d}{dx}(x^3) + x^3\frac{d}{dx}(x + 1)^{-2}$$

$$= (x + 1)^{-2}\,3x^2 + x^3(-2)(x + 1)^{-3}$$

$$\frac{dy}{dx} = \frac{3x^2}{(x + 1)^2} - \frac{2x^3}{(x + 1)^3}$$

Example 7

A useful formula that follows from Equation B.30 is the derivative of the quotient of two functions. Show that the expression is given by

$$\frac{d}{dx}\left[\frac{g(x)}{h(x)}\right] = \frac{h\dfrac{dg}{dx} - g\dfrac{dh}{dx}}{h^2}$$

Solution We can write the quotient as gh^{-1} and then apply Equations B.29 and B.30:

$$\frac{d}{dx}\left(\frac{g}{h}\right) = \frac{d}{dx}(gh^{-1}) = g\frac{d}{dx}(h^{-1}) + h^{-1}\frac{d}{dx}(g)$$

$$= -gh^{-2}\frac{dh}{dx} + h^{-1}\frac{dg}{dx}$$

$$= \frac{h\dfrac{dg}{dx} - g\dfrac{dh}{dx}}{h^2}$$

TABLE B.4   Derivatives for Several Functions

$\dfrac{d}{dx}(a) = 0$	$\dfrac{d}{dx}(\cos ax) = -a\sin ax$	$\dfrac{d}{dx}(\sec x) = \tan x\sec x$
$\dfrac{d}{dx}(ax^n) = nax^{n-1}$	$\dfrac{d}{dx}(\tan ax) = a\sec^2 ax$	$\dfrac{d}{dx}(\csc x) = -\cot x\csc x$
$\dfrac{d}{dx}(e^{ax}) = ae^{ax}$	$\dfrac{d}{dx}(\cot ax) = -a\csc^2 ax$	$\dfrac{d}{dx}(\ln ax) = \dfrac{1}{x}$
$\dfrac{d}{dx}(\sin ax) = a\cos ax$		

Note: The letters a and n are constants.

Some of the most commonly used derivatives of functions are listed in Table B.4.

B.7 · INTEGRAL CALCULUS

We think of integration as the inverse of differentiation. As an example, consider the expression

$$f(x) = \frac{dy}{dx} = 3ax^2 + b$$

which was the result of differentiating the function

$$y(x) = ax^3 + bx + c$$

in Example 4. We can write the first expression, $dy = f(x)\ dx = (3ax^2 + b)\ dx$, and obtain $y(x)$ by "summing" over all values of x. Mathematically, we write this inverse operation as

$$y(x) = \int f(x)\ dx$$

For the function $f(x)$ already given,

$$y(x) = \int (3ax^2 + b)\ dx = ax^3 + bx + c$$

where c is a constant of the integration. This type of integral is called an *indefinite integral* because its value depends on the choice of the constant c.

A general **indefinite integral,** $I(x)$, is defined as

$$I(x) = \int f(x)\ dx \qquad\qquad\qquad \textbf{[B.34]}$$

where $f(x)$ is called the *integrand* and $f(x) = \dfrac{dI(x)}{dx}$.

For a *general continuous* function $f(x)$, the integral can be described as the area under the curve bounded by $f(x)$ and the x axis, between two specified values of x—say, x_1 and x_2—as in Figure B.14.

The area of the shaded element is approximately $f_i\ \Delta x_i$. If we sum all these area elements from x_1 to x_2 and take the limit of this sum as $\Delta x_i \rightarrow 0$, we obtain the *true* area under the curve bounded by $f(x)$ and x, between the limits x_1 and x_2:

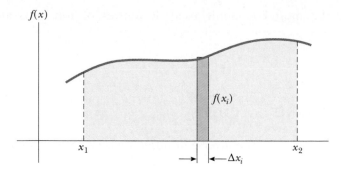

Figure B.14

$$\text{Area} = \lim_{\Delta x_i \to 0} \sum_i f(x_i)\,\Delta x_i = \int_{x_1}^{x_2} f(x)\,dx \qquad \textbf{[B.35]}$$

Integrals of the type defined by Equation B.35 are called **definite integrals.**

One of the common types of integrals that arise in practical situations has the form

$$\int x^n\,dx = \frac{x^{n+1}}{n+1} + c \qquad (n \neq -1) \qquad \textbf{[B.36]}$$

This result is obvious because differentiation of the right-hand side with respect to x gives $f(x) = x^n$ directly. If the limits of the integration are known, this integral becomes a *definite integral* and is written

$$\int_{x_1}^{x_2} x^n\,dx = \frac{x_2^{\,n+1} - x_1^{\,n+1}}{n+1} \qquad (n \neq -1) \qquad \textbf{[B.37]}$$

Examples

1. $\displaystyle\int_0^a x^2\,dx = \frac{x^3}{3}\bigg]_0^a = \frac{a^3}{3}$

2. $\displaystyle\int_0^b x^{3/2}\,dx = \frac{x^{5/2}}{5/2}\bigg]_0^b = \frac{2}{5}\,b^{5/2}$

3. $\displaystyle\int_3^5 x\,dx = \frac{x^2}{2}\bigg]_3^5 = \frac{5^2 - 3^2}{2} = 8$

Partial Integration

Sometimes it is useful to apply the method of *partial integration* to evaluate certain integrals. The method uses the property that

$$\int u \, dv = uv - \int v \, du \qquad\qquad \text{[B.38]}$$

where u and v are *carefully* chosen in order to reduce a complex integral to a simpler one. In many cases, several reductions have to be made. Consider the example

$$I(x) = \int x^2 e^x \, dx$$

This can be evaluated by integrating by parts twice. First, if we choose $u = x^2$, $v = e^x$, we get

$$\int x^2 e^x \, dx = \int x^2 \, d(e^x) = x^2 e^x - 2 \int e^x x \, dx + c_1$$

Now, in the second term, choose $u = x$, $v = e^x$, which gives

$$\int x^2 e^x \, dx = x^2 e^x - 2xe^x + 2 \int e^x \, dx + c_1$$

or

$$\int x^2 e^x \, dx = x^2 e^x - 2xe^x + 2e^x + c_2$$

The Perfect Differential

Another useful method to remember is the use of the *perfect differential*. That is, we should sometimes look for a change of variable such that the differential of the function is the differential of the independent variable appearing in the integrand. For example, consider the integral

$$I(x) = \int \cos^2 x \sin x \, dx$$

This becomes easy to evaluate if we rewrite the differential as $d(\cos x) = -\sin x \, dx$. The integral then becomes

$$\int \cos^2 x \sin x \, dx = - \int \cos^2 x \, d(\cos x)$$

If we now change variables, letting $y = \cos x$, we get

$$\int \cos^2 x \sin x \, dx = - \int y^2 \, dy = -\frac{y^3}{3} + c = -\frac{\cos^3 x}{3} + c$$

TABLE B.5   **Some Indefinite Integrals (an arbitrary constant should be added to each of these integrals)**

$$\int x^n\, dx = \frac{x^{n+1}}{n+1} \quad \text{(provided } n \neq -1)$$

$$\int xe^{ax}\, dx = \frac{e^{ax}}{a^2}(ax-1)$$

$$\int \frac{dx}{x} = \int x^{-1}\, dx = \ln x$$

$$\int \frac{dx}{a+be^{cx}} = \frac{x}{a} - \frac{1}{ac}\ln(a+be^{cx})$$

$$\int \frac{dx}{a+bx} = \frac{1}{b}\ln(a+bx)$$

$$\int \sin ax\, dx = -\frac{1}{a}\cos ax$$

$$\int \frac{dx}{(a+bx)^2} = -\frac{1}{b(a+bx)}$$

$$\int \cos ax\, dx = \frac{1}{a}\sin ax$$

$$\int \frac{dx}{a^2+x^2} = \frac{1}{a}\tan^{-1}\frac{x}{a}$$

$$\int \tan ax\, dx = -\frac{1}{a}\ln(\cos ax) = \frac{1}{a}\ln(\sec ax)$$

$$\int \frac{dx}{a^2-x^2} = \frac{1}{2a}\ln\frac{a+x}{a-x} \quad (a^2-x^2>0)$$

$$\int \cot ax\, dx = \frac{1}{a}\ln(\sin ax)$$

$$\int \frac{dx}{x^2-a^2} = \frac{1}{2a}\ln\frac{x-a}{x+a} \quad (x^2-a^2>0)$$

$$\int \sec ax\, dx = \frac{1}{a}\ln(\sec ax + \tan ax) = \frac{1}{a}\ln\left[\tan\left(\frac{ax}{2}+\frac{\pi}{4}\right)\right]$$

$$\int \frac{x\, dx}{a^2 \pm x^2} = \pm\tfrac{1}{2}\ln(a^2 \pm x^2)$$

$$\int \csc ax\, dx = \frac{1}{a}\ln(\csc ax - \cot ax) = \frac{1}{a}\ln\left(\tan\frac{ax}{2}\right)$$

$$\int \frac{dx}{\sqrt{a^2-x^2}} = \sin^{-1}\frac{x}{a} = -\cos^{-1}\frac{x}{a} \quad (a^2-x^2>0)$$

$$\int \sin^2 ax\, dx = \frac{x}{2} - \frac{\sin 2ax}{4a}$$

$$\int \frac{dx}{\sqrt{x^2 \pm a^2}} = \ln(x+\sqrt{x^2 \pm a^2})$$

$$\int \cos^2 ax\, dx = \frac{x}{2} + \frac{\sin 2ax}{4a}$$

$$\int \frac{x\, dx}{\sqrt{a^2-x^2}} = \sqrt{a^2-x^2}$$

$$\int \frac{dx}{\sin^2 ax} = -\frac{1}{a}\cot ax$$

$$\int \frac{x\, dx}{\sqrt{x^2 \pm a^2}} = \sqrt{x^2 \pm a^2}$$

$$\int \frac{dx}{\cos^2 ax} = \frac{1}{a}\tan ax$$

$$\int \sqrt{a^2-x^2}\, dx = \tfrac{1}{2}\left(x\sqrt{a^2-x^2} + a^2\sin^{-1}\frac{x}{a}\right)$$

$$\int \tan^2 ax\, dx = \frac{1}{a}(\tan ax) - x$$

$$\int x\sqrt{a^2-x^2}\, dx = -\tfrac{1}{3}(a^2-x^2)^{3/2}$$

$$\int \cot^2 ax\, dx = -\frac{1}{a}(\cot ax) - x$$

$$\int \sqrt{x^2 \pm a^2}\, dx = \tfrac{1}{2}[x\sqrt{x^2 \pm a^2} \pm a^2\ln(x+\sqrt{x^2 \pm a^2})]$$

$$\int \sin^{-1} ax\, dx = x(\sin^{-1} ax) + \frac{\sqrt{1-a^2x^2}}{a}$$

$$\int x(\sqrt{x^2 \pm a^2})\, dx = \tfrac{1}{3}(x^2 \pm a^2)^{3/2}$$

$$\int \cos^{-1} ax\, dx = x(\cos^{-1} ax) - \frac{\sqrt{1-a^2x^2}}{a}$$

$$\int e^{ax}\, dx = \frac{1}{a}e^{ax}$$

$$\int \frac{dx}{(x^2+a^2)^{3/2}} = \frac{x}{a^2\sqrt{x^2+a^2}}$$

$$\int \ln ax\, dx = (x\ln ax) - x$$

$$\int \frac{x\, dx}{(x^2+a^2)^{3/2}} = -\frac{1}{\sqrt{x^2+a^2}}$$

Table B.5 lists some useful indefinite integrals. Table B.6 gives Gauss's probability integral and other definite integrals. A more complete list can be found in various handbooks, such as *The Handbook of Chemistry and Physics,* CRC Press.

TABLE B.6   Gauss's Probability Integral and Related Integrals

$$I_0 = \int_0^\infty e^{-ax^2}\, dx = \tfrac{1}{2}\sqrt{\frac{\pi}{\alpha}} \qquad \text{(Gauss's probability integral)}$$

$$I_1 = \int_0^\infty xe^{-ax^2}\, dx = \frac{1}{2\alpha}$$

$$I_2 = \int_0^\infty x^2 e^{-ax^2}\, dx = -\frac{dI_0}{d\alpha} = \frac{1}{4}\sqrt{\frac{\pi}{\alpha^3}}$$

$$I_3 = \int_0^\infty x^3 e^{-ax^2}\, dx = -\frac{dI_1}{d\alpha} = \frac{1}{2\alpha^2}$$

$$I_4 = \int_0^\infty x^4 e^{-ax^2}\, dx = \frac{d^2 I_0}{d\alpha^2} = \frac{3}{8}\sqrt{\frac{\pi}{\alpha^5}}$$

$$I_5 = \int_0^\infty x^5 e^{-ax^2}\, dx = \frac{d^2 I_1}{d\alpha^2} = \frac{1}{\alpha^3}$$

$$\cdot$$
$$\cdot$$
$$\cdot$$

$$I_{2n} = (-1)^n \frac{d^n}{d\alpha^n} I_0$$

$$I_{2n+1} = (-1)^n \frac{d^n}{d\alpha^n} I_1$$

APPENDIX C

Periodic Table of the Elements

Group I	Group II	Transition elements							
H 1 1.0080 $1s^1$									
Li 3 6.94 $2s^1$	**Be** 4 9.012 $2s^2$								
Na 11 22.99 $3s^1$	**Mg** 12 24.31 $3s^2$								
K 19 39.102 $4s^1$	**Ca** 20 40.08 $4s^2$	**Sc** 21 44.96 $3d^14s^2$	**Ti** 22 47.90 $3d^24s^2$	**V** 23 50.94 $3d^34s^2$	**Cr** 24 51.996 $3d^54s^1$	**Mn** 25 54.94 $3d^54s^2$	**Fe** 26 55.85 $3d^64s^2$	**Co** 27 58.93 $3d^74s^2$
Rb 37 85.47 $5s^1$	**Sr** 38 87.62 $5s^2$	**Y** 39 88.906 $4d^15s^2$	**Zr** 40 91.22 $4d^25s^2$	**Nb** 41 92.91 $4d^45s^1$	**Mo** 42 95.94 $4d^55s^1$	**Tc** 43 (99) $4d^55s^2$	**Ru** 44 101.1 $4d^75s^1$	**Rh** 45 102.91 $4d^85s^1$
Cs 55 132.91 $6s^1$	**Ba** 56 137.34 $6s^2$	57-71*	**Hf** 72 178.49 $5d^26s^2$	**Ta** 73 180.95 $5d^36s^2$	**W** 74 183.85 $5d^46s^2$	**Re** 75 186.2 $5d^56s^2$	**Os** 76 190.2 $5d^66s^2$	**Ir** 77 192.2 $5d^76s^2$
Fr 87 (223) $7s^1$	**Ra** 88 (226) $7s^2$	89-103**	**Rf‡** 104 (261) $6d^27s^2$	**Ha** 105 (262) $6d^37s^2$	**Sg** 106 (263)	**Ns** 107 (262)	**Hs** 108 (265)	**Mt** 109 (266)

Symbol — **Ca** 20 — Atomic number
Atomic mass † — 40.08
$4s^2$ — Electron configuration

*Lanthanide series

La 57 138.91 $5d^16s^2$	**Ce** 58 140.12 $5d^14f^16s^2$	**Pr** 59 140.91 $4f^36s^2$	**Nd** 60 144.24 $4f^46s^2$	**Pm** 61 (147) $4f^56s^2$	**Sm** 62 150.4 $4f^66s^2$
Ac 89 (227) $6d^17s^2$	**Th** 90 (232) $6d^27s^2$	**Pa** 91 (231) $5f^26d^17s^2$	**U** 92 (238) $5f^36d^17s^2$	**Np** 93 (239) $5f^46d^17s^2$	**Pu** 94 (239) $5f^66d^07s^2$

**Actinide series

† Atomic mass values given are averaged over isotopes in the percentages in which they exist in nature.
  For an unstable element, mass number of the most stable known isotope is given in parentheses.
‡ The names for elements 104–109 are controversial. The names given here are those recommended
  by the discoverers.

				Group III	Group IV	Group V	Group VI	Group VII	Group 0
								H 1 1.0080 $1s^1$	**He** 2 4.0026 $1s^2$
				B 5 10.81 $2p^1$	**C** 6 12.011 $2p^2$	**N** 7 14.007 $2p^3$	**O** 8 15.999 $2p^4$	**F** 9 18.998 $2p^5$	**Ne** 10 20.18 $2p^6$
				Al 13 26.98 $3p^1$	**Si** 14 28.09 $3p^2$	**P** 15 30.97 $3p^3$	**S** 16 32.06 $3p^4$	**Cl** 17 35.453 $3p^5$	**Ar** 18 39.948 $3p^6$
Ni 28 58.71 $3d^84s^2$	**Cu** 29 63.54 $3d^{10}4s^2$	**Zn** 30 65.37 $3d^{10}4s^2$	**Ga** 31 69.72 $4p^1$	**Ge** 32 72.59 $4p^2$	**As** 33 74.92 $4p^3$	**Se** 34 78.96 $4p^4$	**Br** 35 79.91 $4p^5$	**Kr** 36 83.80 $4p^6$
Pd 46 106.4 $4d^{10}$	**Ag** 47 107.87 $4d^{10}5s^1$	**Cd** 48 112.40 $4d^{10}5s^2$	**In** 49 114.82 $5p^1$	**Sn** 50 118.69 $5p^2$	**Sb** 51 121.75 $5p^3$	**Te** 52 127.60 $5p^4$	**I** 53 126.90 $5p^5$	**Xe** 54 131.30 $5p^6$
Pt 78 195.09 $5d^96s^1$	**Au** 79 196.97 $5d^{10}6s^1$	**Hg** 80 200.59 $5d^{10}6s^2$	**Tl** 81 204.37 $6p^1$	**Pb** 82 207.2 $6p^2$	**Bi** 83 208.98 $6p^3$	**Po** 84 (210) $6p^4$	**At** 85 (218) $6p^5$	**Rn** 86 (222) $6p^6$
110 Discovered Nov. 1994	111 Discovered Dec. 1994							

Eu 63 152.0 $4f^76s^2$	**Gd** 64 157.25 $5d^14f^76s^2$	**Tb** 65 158.92 $5d^14f^86s^2$	**Dy** 66 162.50 $4f^{10}6s^2$	**Ho** 67 164.93 $4f^{11}6s^2$	**Er** 68 167.26 $4f^{12}6s^2$	**Tm** 69 168.93 $4f^{13}6s^2$	**Yb** 70 173.04 $4f^{14}6s^2$	**Lu** 71 174.97 $5d^14f^{14}6s^2$
Am 95 (243) $5f^76d^07s^2$	**Cm** 96 (245) $5f^76d^17s^2$	**Bk** 97 (247) $5f^86d^17s^2$	**Cf** 98 (249) $5f^{10}6d^07s^2$	**Es** 99 (254) $5f^{11}6d^07s^2$	**Fm** 100 (253) $5f^{12}6d^07s^2$	**Md** 101 (255) $5f^{13}6d^07s^2$	**No** 102 (255) $6d^07s^2$	**Lr** 103 (257) $6d^17s^2$

APPENDIX D

SI Units

TABLE D.1 SI Base Units

Base Quantity	SI Base Unit	
	Name	Symbol
Length	meter	m
Mass	kilogram	kg
Time	second	s
Electric current	ampere	A
Temperature	kelvin	K
Amount of substance	mole	mol
Luminous intensity	candela	cd

TABLE D.2 Some Derived SI Units

Quantity	Name	Symbol	Expression in Terms of Base Units	Expression in Terms of Other SI Units
Plane angle	radian	rad	m/m	
Frequency	hertz	Hz	s^{-1}	
Force	newton	N	$kg \cdot m/s^2$	J/m
Pressure	pascal	Pa	$kg/m \cdot s^2$	N/m^2
Energy:work	joule	J	$kg \cdot m^2/s^2$	$N \cdot m$
Power	watt	W	$kg \cdot m^2/s^3$	J/s
Electric charge	coulomb	C	$A \cdot s$	
Electric potential (emf)	volt	V	$kg \cdot m^2/A \cdot s^3$	W/A
Capacitance	farad	F	$A^2 \cdot s^4/kg \cdot m^2$	C/V
Electric resistance	ohm	Ω	$kg \cdot m^2/A^2 \cdot s^2$	V/A
Magnetic flux	weber	Wb	$kg \cdot m^2/A \cdot s^2$	$V \cdot s$
Magnetic field intensity	tesla	T	$kg/A \cdot s^2$	Wb/m^2
Inductance	henry	H	$kg \cdot m^2/A^2 \cdot s^2$	Wb/A

APPENDIX E

Spreadsheet Problems

Overview

Students come to introductory physics courses with a wide variety of computing experience. Many are already accomplished programmers in one or more programming languages (BASIC, Pascal, FORTRAN, and so forth). Others have never even turned on a computer. To further complicate matters, a wide variety of hardware environments exists, although the vast majority can be classified as IBM/compatible (MS-DOS) or Macintosh environments. We have designed the spreadsheet problems and *Spreadsheet Investigations in Physics* to be usable by and useful to students in all these diverse situations. Our goal is to enable students to investigate a range of physical phenomena and obtain a feel for the physics. Merely "getting the right answer" by plugging numbers into a formula and comparing the result to the answer in the back of the book is discouraged.

Spreadsheets are particularly valuable in exploratory investigations. Once you have constructed a spreadsheet, you can simply vary the parameters and see instantly how things change. Even more important is the ease with which you can construct accurate graphs of relations between physical variables. When you change a parameter, you can view the effects of the change upon the graphs simply by pressing a key or clicking a mouse button. "What if" questions can be easily addressed and depicted graphically.

How to Use the Templates

The computer spreadsheet problems are arranged by difficulty level. The least difficult problems are coded in black. For most of these problems, spreadsheets are provided on disk. Only the input parameters need to be changed. Problems of moderate difficulty are coded in blue. These require additional analysis and the provided spreadsheets will need to be modified to solve them. The most challenging problems are coded in magenta. For most of these, you must develop your own spreadsheets. The emphasis should be on understanding what the results mean rather than just getting an answer. For example, one spreadsheet problem explores the strategies for fuel burn rates in rocket propulsion. How long and at what rate should the rocket fuel be burned to obtain the highest velocity or the greatest distance?

Software Requirements

The spreadsheet template files are provided on a high-density (1.44 MB) 3.5″ diskette, which can be obtained in either of two forms. The disk for the Macintosh environment will contain files in the file format used by the program Microsoft Excel 5.0 for the Macintosh.

Since there are many spreadsheet programs used in the IBM/compatible environment, the disk for Windows and DOS users will contain three different versions of the spreadsheet template files. These versions are chosen to allow the diskette to be usable by as many students as possible. The file ***123WIN.EXE*** contains templates in the format suitable for use in the program Lotus 1-2-3 for Windows, Release 4 or 5. Files in this format have filenames in the form *.WK4, where * stands for a filename of eight letters or less. The file ***EXCELWIN.EXE*** contains templates in the format used by Microsoft Excel for Windows, Version 5.0; files in this format have the form *.XLS. Finally, for students using an earlier version of Lotus 1-2-3 or Excel or using other popular spreadsheet programs (including non-Windows programs), a third file, ***123DOS.EXE,*** contains templates in the format introduced with the versions 2.x of DOS-based Lotus 1-2-3; files in this format have the form *.WK1. The Lotus WK1 format can be read directly by almost all spreadsheet programs, including Lotus 1-2-3 (versions 2.0 and later), Excel, Microsoft Works, and Quattro Pro. The f(g) Scholar program can also import WK1 spreadsheets; however, some minor format changes of the templates are needed.

To fit these versions of the template files onto one diskette, the diskette is supplied with the files in compressed form. These files are self-extracting; they can be decompressed by a simple procedure without using any external programs. Choose the version you want, copy the compressed file to a subdirectory on a hard-disk, and execute the file. For example, to install the Lotus for Windows templates in the subdirectory C:\INVEST, do the following steps:

A. Make sure that the subdirectory C:\INVEST exists. If it does not exist, make the subdirectory.
B. Copy the file 123WIN.EXE from the distribution disk to the subdirectory C:\INVEST.
C. Run or launch the file 123WIN.EXE. The templates will be extracted and are then ready to be used.

The templates on the Macintosh diskette are for Excel 5.0 for the Macintosh. The templates are not compressed. Copy these templates to a folder; they are ready to be used just as they are.

Hardware Requirements

You will need a microcomputer that can run one of the spreadsheet programs in one of their many versions. Your computer should be connected to a printer that can print text and graphics. Older versions of the software will run on a 8086/8088 MS-DOS system with just a single floppy disk drive or on a 512K Macintosh with two floppy drives. Newer versions require a more powerful computer to run effectively. For example, to run Excel For Windows, Version 5.0, you must have a hard disk with about 15 megabytes of disk memory available and 4 to 8 megabytes of RAM memory. Your software manuals will tell you exactly what you need to run the particular version of your spreadsheet program. However, all the problems require only a minimal computer system.

There are many different software versions and many different computer configurations. You might have one floppy drive, two floppy drives, a hard disk, a local area network, and so on. The combinations are almost endless. Our best suggestion is to read your software manual and ask your instructor or computer laboratory personnel how to start your spreadsheet program.

Spreadsheet Tutorial

Some students will have the needed computer and spreadsheet skills to start work-ing with the templates immediately. Other students, who have not used a spread-sheet program before, will need some instruction. We have written *Spreadsheet In-vestigations in Physics* for both these groups of students. The first part contains a spreadsheet tutorial that the novice student can use *independently* to gain the re-quired spreadsheet skills. A two- or three-hour initial session with the tutorial and the computer is all that most students need to get started. Once the students have mastered the basic operations of the spreadsheet program, they should try one or two of the easier problems.

Because very few introductory physics students have studied numerical meth-ods, we have also included a brief introduction to numerical methods in *Spreadsheet Investigations in Physics*. This section covers numerical interpolation, differentiation, integration, and the solution of simple differential equations. The student should not try to master all of this material at one time; only study the sections that are needed to solve the currently assigned problems.

The templates supplied on the distribution diskettes with *Spreadsheet Investiga-tions in Physics* constitute an outline. You must enter the appropriate data and pa-rameters. The parameters must be adjusted to fit the needs of your problem. Feel free to change any parameters, to expand or decrease the number of rows of output, and to change the size of increments (for example, in time or distance). Most templates have graphs associated with them. You may have to adjust the ranges of variables plotted and the scales for the axes. See the tutorials for how to adjust the appearance of your graphs.

Answers to Odd-Numbered Problems

Chapter 1

1. (a) $4.00 \text{ u} = 6.64 \times 10^{-27} \text{ kg}$
   (b) $55.9 \text{ u} = 9.28 \times 10^{-26} \text{ kg}$
   (c) $207 \text{ u} = 3.44 \times 10^{-25} \text{ kg}$
3. 2.86 cm
5. (a) 623 kg/m^3
   (b) Yes, because this is less than 1000 kg/m^3
7. $0.579 \, t \, \text{ft}^3/\text{s} + 1.19 \times 10^{-9} t^2 \text{ ft}^3/\text{s}^2$
11. (a) $6.31 \times 10^4 \text{ AU}$ (b) $1.33 \times 10^{11} \text{ AU}$
13. $151 \, \mu\text{m}$
15. (a) $1.609 \text{ km/h} = 1 \text{ mi/h}$ (b) 88 km/h   (c) 16 km/h
17. 1.19×10^{57}
19. $\sim 10^2$ tuners
21. (a) 797   (b) 11   (c) 18
23. (a) $346 \text{ m}^2 \pm 13 \text{ m}^2$ (b) $(66.0 \pm 1.3) \text{ m}$
25. $(-2.75, -4.76) \text{ m}$
27. (a) $(2.17 \text{ m}, 1.25 \text{ m})$ and $(-1.90 \text{ m}, 3.29 \text{ m})$
    (b) 4.55 m
29. (a) 10.0 m   (b) 15.7 m   (c) 0
31. (a) 5.2 m at 60°   (b) 3.0 m at 330°
    (c) 3.0 m at 150°   (d) 5.2 m at 300°
33. 47.2 units at 122°
35. 227 paces at 165°
37. (a) $R_x = 49.5$ units; $R_y = 27.1$ units
    (b) 56.4 units at 28.7°
39. 240 m at 237°
41. 196 cm at $-14.7°$
43. 0.449%
45. $5 \times 10^9 \text{ gal}$
47. $6 \times 10^{20} \text{ kg}$
49. 2.29 km
51. $\sim 10^{11}$
S1. (a) A plot of log T versus log L for the given data shows an approximate linear relationship. Applying the least-squares method to the data yields 0.503 for the slope and 0.302 for the intercept of the straight line that best fits the data. The data points and the best-fit line, log $T = 0.503$ log $L + 0.302$ are shown in the figure at the top of the second column.

Chapter 2

1. (a) 2.30 m/s   (b) 16.1 m/s   (c) 11.5 m/s
3. (a) 5 m/s   (b) 1.2 m/s   (c) -2.5 m/s
   (d) -3.3 m/s   (e) 0

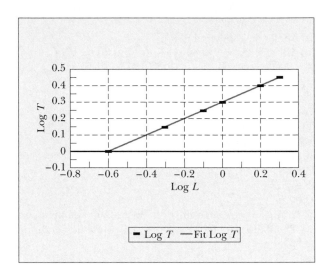

5. (a) $2v_1 v_2/(v_1 + v_2)$ (b) 0
7. (a) -2.4 m/s   (b) -3.2 m/s   (c) 4 s
9. (a) 5 m/s   (b) -2.5 m/s   (c) 0
   (d) 5 m/s
11. $1.34 \times 10^4 \text{ m/s}^2$
13. (a) 2.0 m   (b) -3.0 m/s   (c) -2.0 m/s^2
15. (a) 1.3 m/s^2 (b) 2 m/s^2 at 3 s
    (c) at $t = 6$ s and for $t > 10$ s   (d) -1.5 m/s^2 at 8 s
17. -16.0 cm/s^2
19. 160 ft
21. (a) 20.0 s   (b) no
23. 3.10 m/s
25. (a) -202 m/s^2 (b) 198 m
27. (a) 10.0 m/s up   (b) 4.68 m/s down
29. (a) 29.4 m/s   (b) 44.1 m
31. (a) 7.82 m   (b) 0.782 s
33. (a) 3.00 m/s   (b) 6.00 s   (c) -0.300 m/s^2
    (d) 2.05 m/s
35. Yes, at 11.4 s and 212 m
37. 0.509 s
39. (a) 5.43 m/s^2 and 3.83 m/s^2
    (b) 10.9 m/s and 11.5 m/s   (c) Maggie by 2.62 m
41. $\sim 10^3 \text{ m/s}^2$
43. (a) 2.99 s   (b) -15.4 m/s
    (c) 31.3 m/s down and 34.9 m/s down
45. (a) 26.4 m   (b) 6.88%

47.   (a) 5.46 s   (b) 73.0 m   (c) 26.7 m/s and 22.6 m/s
49.   (c) $v = 0$ and $a = v_0^2/h$ (d) $v = v_0$ and $a = 0$
51. $v/\sqrt{3}$
S3.   (a) At approximately $t = 37.0$ s after the police car starts,
      or 42.0 s after the first police officer sees the speeder.
      (b) 74.0 m/s   (c) 1370 m

Chapter 3

1.   (a) 4.87 km at 209° from east   (b) 23.3 m/s
      (c) 13.5 m/s at 209°
3.   (a) $18.0t\,\mathbf{i} + (4.00t - 4.90t^2)\mathbf{j}$
      (b) $18.0\mathbf{i} + (4.00 - 9.80t)\mathbf{j}$ (c) -9.80 m/s² $\mathbf{j}$
      (d) $(54.0\mathbf{i} - 32.1\mathbf{j})$ m   (e) $(18.0\mathbf{i} - 25.4\mathbf{j})$ m/s
      (f) -9.80 m/s² $\mathbf{j}$
5.   (a) $(2.00\mathbf{i} + 3.00\mathbf{j})$ m/s²
      (b) $(3.00t + t^2)\mathbf{i}$ m + $(1.50t^2 - 2.00t)\mathbf{j}$ m
7.   (a) $a_x = 0.800$ m/s²; $a_y = -0.300$ m/s²   (b) 339°
      (c) $(360\mathbf{i} - 72.7\mathbf{j})$ m, $-15.2°$
9.   (a) 3.34 m/s $\mathbf{i}$ (b) $-50.9°$
11.   74.5 m/s
13.   48.6 m/s
15.   0.600 m/s²
17.   (a) 2.67 s   (b) $29.9\mathbf{i}$ m/s   (c) $(29.9\mathbf{i} - 26.2\mathbf{j})$ m/s
19.   (a) The ball clears by 0.889 m   (b) while descending
21.   67.8°
23.   (a) 1.03 s   (b) $(8.67\mathbf{i} + 4.20\mathbf{j})$ m/s   (c) 25.8°
25.   377 m/s²
27.   (a) 6.00 rev/s   (b) 1.52 km/s²   (c) 1.28 km/s²
29.   (a) 7.90 km/s   (b) 5.07 ks = 1.41 h
31.   1.48 m/s²
33.   (a) 13.0 m/s²   (b) 5.70 m/s   (c) 7.50 m/s²
35.   (a) $4.00\mathbf{i}$ m/s   (b) $(4.00\mathbf{i} + 6.00\mathbf{j})$ m
37.   (a) 41.7 m/s   (b) 3.81 s
      (c) $v_x = 34.1$ m/s; $v_y = -13.4$ m/s; 36.6 m/s
39.   8.94 m/s at $-63.4°$
41.   20.0 m
43.   (a) 6.80 km
      (b) 3.00 km vertically above the impact point
      (c) 66.2°
45.   (a) 46.5 m/s   (b) $-77.6°$ (c) 6.34 s
47.   (a) 20.0 m/s, 5.00 s   (b) $(16.0\mathbf{i} - 27.1\mathbf{j})$ m/s
      (c) 6.54 s   (d) 24.6 m$\mathbf{i}$
49.   (a) 43.2 m   (b) $v_x = 9.66$ m/s; $v_y = -25.6$ m/s
51.   Less than 265 m and more than 3.48 km
S1.   There are an infinite number of solutions to the prob-
      lem; however, there is one practical limitation. We can
      estimate that the maximum speed that the punter can
      give the ball is about 20 to 30 m/s. Use a speed in this
      range.
S3.   There are an infinite number of solutions. For example
      if $v_0 = 22.32$ m/s at 45°, then the ball clears the crossbar
      in 3.01 s. Or if $v_0 = 24.60$ m/s at 30°, then the ball clears
      the crossbar in 2.23 s. The time it takes the ball to clear

the crossbar is immaterial, since in all likelihood time
will run out before the clock is stopped.

Chapter 4

1.   (a) 1/3   (b) 0.750 m/s²
3. $(6.00\mathbf{i} + 15.0\mathbf{j})$ N; 16.2 N
5.   (a) 1.44 m   (b) $(50.8\mathbf{i} + 1.40\mathbf{j})$ N
7.   (a) $(2.50\mathbf{i} + 5.00\mathbf{j})$ N   (b) 5.59 N
9.   (a) 3.64×10^{-18} N
      (b) 8.93×10^{-30} N, which is 408 billion times smaller
11.   (a) 534 N down   (b) 54.4 kg
13.   (a) 5.00 m/s² at 36.9°   (b) 6.08 m/s² at 25.3°
15.   (a) $\sim 10^{-22}$ m/s²   (b) $\sim 10^{-23}$ m
17.   (a) 0.200 m/s² forward   (b) 10.0 m   (c) 2.00 m/s
19.   (a) 15.0 lb up   (b) 5.00 lb up   (c) 0
21.   (a) 31.5 N, 37.5 N, 49.0 N   (b) 113 N, 56.6 N, 98.0 N
23.   (b) 514 N, 557 N, 325 N
25.   (a) 5.10 kN   (b) 3620 kg
27.   8.66 N east
29.   3.73 m
31. $a = F/(m_1 + m_2)$; $T = Fm_1/(m_1 + m_2)$
33.   (a) $a_1 = 2a_2$
      (b) $T_1 = m_1m_2g/(2m_1 + \frac{1}{2}m_2)$ and
      $T_2 = m_1m_2g/(m_1 + \frac{1}{4}m_2)$
      (c) $a_1 = m_2g/(2m_1 + \frac{1}{2}m_2)$ and
      $a_2 = m_2g/(4m_1 + m_2)$
35.   (a) 706 N   (b) 814 N   (c) 706 N   (d) 648 N
37.   (a) 3.00 s   (b) $(18.0\mathbf{i} - 9.00\,\mathbf{j})$ m   (c) 20.1 m
39.   1.66 MN
41.   (a)

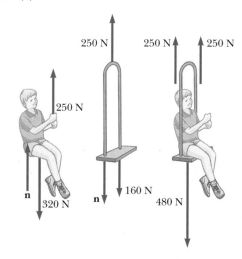

      (b) 0.408 m/s²   (c) 83.3 N
43.   1.18 kN
45.   (a) $Mg/2$, $Mg/2$, $Mg/2$, $3Mg/2$, Mg (b) $Mg/2$
47. $(M + m_1 + m_2)(m_2g/m_1)$
49.   (a) 30.7°   (b) 0.843 N

Chapter 5

1. $\mu_s = 0.306$; $\mu_k = 0.245$
3. (a) 256 m   (b) 42.7 m
5. (a) 1.78 m/s^2 (b) 0.368   (c) 9.37 N
   (d) 2.67 m/s
7. (a) 0.161   (b) 1.01 m/s^2
9. 37.8 N
11. (a) 0.931 m/s^2 (b) 6.10 cm
13. Any speed up to 8.08 m/s
15. (a) 9.13×10^{22} m/s^2 (b) 83.2 nN
17. (a) static friction   (b) 0.0850
19. (a) 68.6 N toward the center of the circle and 784 N up
    (b) 0.857 m/s^2
21. No. The jungle-lord needs a vine of minimal tensile strength 1.38 kN.
23. (a) 6670 N up   (b) 20.3 m/s
25. 3.13 m/s
27. (a) 32.8 s^{-1} (b) 9.80 m/s^2 down
    (c) 4.90 m/s^2 down
29. (a) 0.0347 s^{-1} (b) 2.50 m/s   (c) $a = -cv$
31. (a) 13.7 m/s down
    (b)

t(s)	x(m)	v(m/s)
0	0	0
0.2	0	-1.96
0.4	-0.392	-3.88
. . .	. . .	. . .
1.0	-3.77	-8.71
2.0	-14.4	-12.56
4.0	-41.0	-13.67

33. (a) 49.5 m/s down and 4.95 m/s down
    (b)

t(s)	y(m)	v(m/s)
0	1000	0
1.00	995	-9.70
2.00	980	-18.6
10.0	674	-47.7
10.1	671	-16.7
12.0	659	-4.95
145	0	-4.95

35. 2.97 nN
37. 0.612 m/s^2 toward the Earth
39. (a) 19.3°   (b) 4.21 N
41. 2.14 rev/min
43. (b) 732 N down at the Equator and 735 N down at the poles
45. 20.1°
47. (b) 2.54 s; 23.6 rev/min
49. 12.8 N
51. 1.26×10^{32} kg
53. $\sim 10^{-7}$ N toward you

Chapter 6

1. 15.0 MJ
3. (a) 32.8 mJ   (b) -32.8 mJ
5. 4.70 kJ

7. 14.0
9. (a) 16.0 J   (b) 36.9°
11. (a) 11.3°   (b) 156°   (c) 82.3°
13. (a) 7.50 J   (b) 15.0 J   (c) 7.50 J   (d) 30.0 J
15. 50.0 J
17. (a) 575 N/m   (b) 46.0 J
19. 12.0 J
21. (a) 1.20 J   (b) 5.00 m/s   (c) 6.30 J
23. (a) 60.0 J   (b) 60.0 J
25. (a) 79.4 N   (b) 1.49 kJ   (c) $W_f = -1.49$ kJ
27. 2.04 m
29. (a) -168 J   (b) 184 J   (c) 500 J
    (d) 148 J   (e) 5.64 m/s
31. 234 J
33. 875 W
35. 685 bundles
37. (a) 20.6 kJ   (b) 686 W
39. 90.0 J
43. (a) $(2.0 + 24t^2 + 72t^4)$ J   (b) $12t$ m/s^2; $48t$ N
    (c) $(48t + 288t^3)$ W   (d) 1250 J
45. 878 kN
47. (a) 4.12 m   (b) 3.35 m
49. 1.68 m/s
51. (a) A plot of F as the independent variable and L as the dependent variable shows that the last four points tend to vary the most from a straight line. However, because the first point has the largest percentage deviation from a straight line, we probably are not justified in throwing any of the data points out. The slope of the best straight line obtained from the least-squares fit is 8.654 545 5 mm/N.   (b) $k = 0.116$ N/mm $= 116$ N/m.
    (c) The least-squares fit we used was of the form $L = aF + b$; solving for F gives $F = L/a - b/a = 0.116L - 0.561$. Hence, for $L = 105$ mm, $F = 12.7$ mm.

Chapter 7

1. (a) 259 kJ, 0, 259 kJ   (b) 0, -259 kJ, 259 kJ
3. (a) 80.0 J   (b) 10.7 J   (c) 0
5. (a) 40.0 J   (b) -40.0 J   (c) 62.5 J
7. (a) 22.0 J, 40.0 J   (b) Yes.
9. $v = (3gR)^{1/2}$, 0.0980 N down
11. (a) 4.43 m/s   (b) 5.00 m
13. 1.84 m
15. (a) 18.5 km, 51.0 km   (b) 10.0 MJ
17. At $h = 2H/3$ or at $h = R$, whichever is smaller
19. 2.00 m/s, 2.79 m/s, 3.19 m/s
21. 3.74 m/s
23. (a) -160 J   (b) 73.5 J   (c) 28.8 N   (d) 0.679
25. (a) 24.5 m/s   (b) Yes   (c) 206 m
    (d) unrealistic
29. 289 m
31. (a) 1.84×10^9 kg/m^3 (b) 3.27 Mm/s^2
    (c) -2.08×10^{13} J

33. (a) -1.67×10^{-14} J   (b) At the center
35. (a) $+$ at B, $-$ at D, 0 at A, C, and E
    (b) C stable; A and E unstable
    (c)

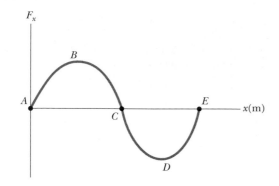

37. A/r^2
39. (a) 0.225 J   (b) $W_{nc} = -0.363$ J   (c) No; the normal force changes in a complicated way because of the curved slope.
41. 0.328
43. 0.115
45. 1.24 m/s
47. (a) 0.400 m   (b) 4.10 m/s   (c) The block stays on the track
49. 914 N/m
51. 2.06 m/s
S1. If the particle has an initial energy less than 2471 J, it will be trapped in the potential well; for example, if $E_T = 1000$ J, it will be confined approximately to -3.15 m $\leq x \leq 4.58$ m.
S3. $x = 0$ is a point of stable equilibrium; $x = 8.33$ m is a point of unstable equilibrium.

Chapter 8
1. $(9.00\mathbf{i} - 12.0\mathbf{j})$ kg·m/s; 15.0 kg·m/s
3. 6.25 cm/s west
5. (a) 6.00 m/s   (b) 8.40 J
7. (a) 13.5 N·s   (b) 9.00 kN   (c) 18.0 kN
9. 260 N normal to the wall
11. 87.5 N
13. 301 m/s
15. $(4M/m)(g\ell)^{1/2}$
17. (a) 20.9 m/s east   (b) 8.68 kJ into thermal energy
19. (a) 0.284   (b) 115 fJ and 45.4 fJ
21. 91.2 m/s
23. 0.556 m
25. (a) 2.88 m/s at 32.3° north of east   (b) 783 J into thermal energy
27. 2.50 m/s at $-60.0°$
29. (a) 1.07 m/s at $-29.7°$ (b) 0.318
31. (a) $(-9.33\mathbf{i} - 8.33\mathbf{j})$ Mm/s   (b) 439 fJ
33. $\mathbf{r}_{CM} = (0\mathbf{i} + 1.00\mathbf{j})$ m

35. $\mathbf{r}_{CM} = (11.7\mathbf{i} + 13.3\mathbf{j})$ cm
37. 0.00673 nm from the oxygen nucleus along the bisector of the angle
39. 0.700 m
41. (a) 39.0 MN   (b) 3.20 m/s² up
43. (a) 442 metric tons   (b) 19.2 metric tons
45. 1.39 km/s
47. (a) -0.667 m/s   (b) 0.952 m
49. 240 s
51. (a) 100 m/s   (b) 374 J
53. $2v_0$ and 0
55. $\sim 10^{-23}$ m/s to 10^{-22} m/s
S1. (a) The maximum acceleration is 100 m/s². It occurs at the end of the burn time of 80 s, when the rocket has its smallest mass. The maximum speed that the rocket reaches is 3.22 km/s.   (b) The speed reaches half its maximum after 55.5 s; if the acceleration were constant, it would reach half its maximum speed at 40 s (half the burn time), but the acceleration is always increasing during the burn time.
S3. The disadvantages are that the ship has to withstand twice the acceleration and that it has only traveled half as far when the fuel burns out. The advantage is that it takes half as long to reach its final speed; hence, it travels farther in 100 s.

Chapter 9
5. $0.866c$
7. 64.9/min; 10.6/min
9. 1.54 ns
11. $0.800c$
13. (a) 2.18 μs   (b) 649 m
15. $0.696c$
17. $0.960c$
19. 0.0880
21. (a) 2.73×10^{-24} kg·m/s   (b) 1.58×10^{-22} kg·m/s
    (c) 5.64×10^{-22} kg·m/s
23. (a) $0.850c$ (b) No change
25. $0.285c$
27. (a) 939 MeV   (b) 3.01 GeV   (c) 2.07 GeV
29. (a) 0.582 MeV   (b) 2.45 MeV
31. $\sim 10^{-15}$
33. 42.1 g/cm³
37. 0.687%
39. (a) a few hundred seconds   (b) $\sim 10^8$ km
41. (a) 83.3 ns   (b) 23.7 m
45. Yes, with 18.8 m to spare
47. (b) For v small compared to c, the relativistic agrees with the classical expression. As v approaches c, the acceleration approaches zero, so the object can never reach or surpass the speed of light.
    (c) Perform $\int (1 - v^2/c^2)^{-3/2}\, dv = (qE/m) \int dt$ to obtain $v = qEct(m^2c^2 + q^2E^2t^2)^{-1/2}$ and then

$\int dx = \int qEct(m^2c^2 + q^2E^2t^2)^{-1/2} \, dt$ to obtain
$x = (c/qE)[(m^2c^2 + q^2E^2t^2)^{1/2} - mc]$

S1. (a)

Z	0.2	0.5	1.0	2.0
v/c	0.180	0.385	0.600	0.800

    (b) For $Z = 3.8$, $v/c = 0.870$

S3. For $Z = 0.2$, $r = 2710$ Mly; $Z = 0.5$, $r = 5770$ Mly;
    $Z = 1.0$, $r = 9000$ Mly; and $Z = 2.0$, $r = 12\,000$ Mly and
    $Z = 3.8$, $r = 13\,800$ Mly

Chapter 10
1. (a) 1.99×10^{-7} rad/s   (b) 2.65×10^{-6} rad/s
3. (a) 5.24s   (b) 27.4 rad
5. (a) 822 rad/s^2 (b) 4.21×10^3 rad
7. 50.0 rev
9. (a) 25.0 rad/s   (b) 39.8 rad/s^2 (c) 0.628 s
11. (a) 126 rad/s   (b) 3.77 m/s   (c) 1.26 km/s^2
    (d) 20.1 m
13. 0.545
15. (a) 143 kg·m^2 (b) 2.57 kJ
17. -3.55 N·m
19. (a) $-17.0\mathbf{k}$ (b) 70.5°
21. No. The cross product must be perpendicular to each of
    its factors, having zero dot product with either of them.
23. 2.94 kN on each rear wheel and 4.41 kN on each front
    wheel
25. 1.46 kN; 1.33 kN (to the right) and 2.58 kN (upward)
27. (a) 24.0 N·m   (b) 0.0356 rad/s^2 (c) 1.07 m/s^2
29. $17.5\mathbf{k}$ kg·m^2/s
31. $60.0\mathbf{k}$ kg·m^2/s
33. $-m\ell gt \cos\theta\mathbf{k}$
35. (a) 0.360 rad/s counterclockwise   (b) 99.9 J
37. (a) 7.20×10^{-3} kg·m^2/s   (b) 9.47 rad/s
39. $\sim 10^0$ kg·m^2
41. (a) 500 J   (b) 250 J   (c) 750 J
43. $3g/2L$ and $3g/2$
45. (a) $2(Rg/3)^{1/2}$ (b) $4(Rg/3)^{1/2}$ (c) $(Rg)^{1/2}$
47. 1.21×10^{-4} kg·m^2
49. (a) 118 N and 156 N   (b) 1.19 kg·m^2
51. (a) $(2mgd \sin\theta + kd^2)^{1/2} (I + mR^2)^{-1/2}$
    (b) 1.74 rad/s
53. (a) 3750 kg·m^2/s   (b) 1.88 kJ   (c) 3750 kg·m^2/s
    (d) 10.0 m/s   (e) 7.50 kJ   (f) 5.62 kJ
55. $3w/8$
S1. The answer is not unique because the torque $\tau = FR$.

Chapter 11
1. The sun: 5.90 mN versus 33.3 μN
3. 2/3
5. 346 Mm
7. 1.27

9. 1.90×10^{27} kg
11. 8.92×10^7 m
13. 16.6 km/s
15. 469 MJ
19. 15.6 km/s
21. 11.8 km/s
23. (a) 2.19 Mm/s   (b) 2.18 aJ   (c) -4.36 aJ
25. (a) 0.212 nm   (b) 9.95×10^{-25} kg·m/s
    (c) 2.11×10^{-34} kg·m^2/s   (d) 3.40 eV
    (e) -6.80 eV   (f) -3.40 eV
29. 492 m/s
31. 0.0572 rad/s
33. 7.41×10^{-10} N
35. (a) $m_2(2G/d)^{1/2}(m_1 + m_2)^{-1/2}$ and
    $m_1(2G/d)^{1/2}(m_1 + m_2)^{-1/2}$ and $(2G/d)^{1/2}(m_1 + m_2)^{1/2}$
    (b) 1.07×10^{32} J and 2.67×10^{31} J
37. (a) 850 MJ   (b) 2.71 GJ
39. (a) 7.34×10^{22} kg   (b) 1.63 km/s
    (c) 1.32×10^{10} J
41. 119 km
43. (a) B   (b) A   (c) B and C
47. 1.89 Mm/s
S5. The kinetic energy is

$$K = \tfrac{1}{2}mv^2 = \tfrac{1}{2}m(v_x^2 + v_y^2)$$

The potential energy is

$$U = \frac{GM_E m}{R_E} - \frac{GM_E m}{r}$$

where G is the universal gravitation constant, M_E is the
mass of the Earth, R_E is the radius of the Earth, m is the
mass of the satellite, and r is the distance from the center
of the Earth to the satellite. The total energy, $K + U$,
is constant. There may be a small change in the numerical value of the total energy because of numerical errors
during the integration of the differential equations.

Chapter 12
1. (a) 1.50 Hz, 0.667 s   (b) 4.00 m   (c) π rad
    (d) 2.83 m
3. 18.8 m/s, 7.11 km/s^2
5. (b) 18.8 cm/s, 0.333 s   (c) 178 cm/s^2, 0.500 s
    (d) 12.0 cm
7. (a) 575 N/m   (b) 46.0 J
9. (a) 40.0 cm/s, 160 cm/s^2
    (b) 32.0 cm/s, -96.0 cm/s^2
    (c) 0.232 s
11. 0.628 m/s
13. 2.23 m/s
15. (a) 126 N/m   (b) 0.178 m
17. (a) 28.0 mJ   (b) 1.02 m/s   (c) 12.2 mJ
    (d) 15.8 mJ
19. (a) 1.55 m   (b) 6.06 s

21. (a) 0.820 m/s   (b) 2.57 rad/s^2 (c) 0.641 N
25. (a) 2.09 s   (b) 4.08% larger
27. 1.00×10^{-3} s^{-1}
31. (a) 1.00 s   (b) 5.09 cm
33. (a) 38.6 N/m   (b) 1.32 kg
35. $f = (2\pi L)^{-1} (gL + kh^2/M)^{1/2}$
37. 6.62 cm
39. (a) 3.00 s   (b) 14.3 J   (c) 25.5°
41. (a) $\frac{1}{2}(M + m/3)v^2$ (b) $2\pi(M + m/3)^{1/2} k^{-1/2}$
47. After 42.1 minutes if the hole is frictionless
S3. The periods increase as θ_0 increases. These periods are always greater than those calculated using $T_0 = 2\pi\sqrt{L/g}$. For example, if $\theta_0 = 45° = \pi/4$ rad, and $L = 1$ m, then $T_0 = 2.00\,709$ s, but the actual period is 2.08 s. (*Note:* Due to numerical errors in the integration of the differential equations, the amplitudes may tend to increase. If this occurs, try smaller time steps.)

Chapter 13

1. $y = 6 [(x - 4.5t)^2 + 3]^{-1}$
3. (a) left   (b) 5.00 m/s
5. 0.319 m
7. (a) $y = (8.00$ cm$)\sin(7.85x + 6.00\pi t)$
   (b) $y = (8.00$ cm$)\sin(7.85x + 6.00\pi t - 0.785)$
9. 2.00 cm, 2.98 m, 0.576 Hz, 1.72 m/s
11. (a) 2.15 cm   (b) 0.379 rad   (c) 541 cm/s
    (d) $y = (2.15$ cm$)\cos(80\pi t + 8.38x + 0.379)$
13. (a) $-73.5°$ (b) 0.858 cm
15. 80.0 N
17. 631 N
19. 1.64 m/s^2
21. 55.1 Hz
23. $(\sqrt{2})P_0$
25. 14.7 m
29. 5.81 m
31. (a) A drop by 75.7 Hz   (b) 0.948 m
33. 26.4 m/s
35. 439 Hz to 441 Hz
37. (a) 39.2 N   (b) 0.892 m   (c) 83.6 m/s
39. (a) 3.33**i** m/s   (b) -5.48 cm   (c) 0.667 m, 5.00 Hz
    (d) 11.0 m/s
41. 6.01 km
45. (a) 55.8 m/s   (b) 2500 Hz
47. Bat is gaining at 1.69 m/s

Chapter 14

1. (a) 9.24 m   (b) 600 Hz
3. 5.66 cm
5. At 0.0891 m, 0.303 m, 0.518 m, 0.732 m, 0.947 m, 1.16 m from one speaker
9. (a) $x = \pi, 3\pi, 5\pi, \ldots$ (b) 0.0294 m
11. 150 Hz

13. (a) 163 N   (b) 660 Hz
15. 19.976 kHz
17. 5.00 Hz, 10.0 Hz, 15.0 Hz; the fifth state, 25.0 Hz
19. 0.786 Hz, 1.57 Hz, 2.36 Hz, 3.14 Hz
21. (a) 0.357 m   (b) 0.715 m
23. (a) 531 Hz   (b) 42.5 mm
25. (a) $L = nv/2f$ where $n = 1, 2, 3, \ldots$
    (b) $L = (2n - 1)v/4f$
27. $n(206$ Hz$)$ and $n + 1(84.5$ Hz$)$ for $n = 1, 2, 3, \ldots$
29. 240 s
31. (a) 162 Hz   (b) 1.06 m
33. 5.64 beats/s
35. (a) 0.655 m   (b) 11.7°C
37. (a) 34.8 m/s   (b) 0.977 m
39. 3.85 m/s away from the station or 3.77 m/s toward the station
41. 4.86 m
43. 15.7 Hz
45. (a) 59.9 Hz   (b) 20.0 cm
47. (a) 1/2   (b) $[n/(n + 1)]^2 F$ (c) 9/16
49. 50.0 Hz, 1.70 m
51. (a) $2A \sin(2\pi x/\lambda)\cos(2\pi v t/\lambda)$
    (b) $2A \sin(\pi x/L)\cos(\pi v t/L)$
    (c) $2A \sin(2\pi x/L)\cos(2\pi v t/L)$
    (d) $2A \sin(n\pi x/L)\cos(n\pi v t/L)$
S4. The following steps can be used to modify the spreadsheet.

   1. MOVE the entire Y_T column—one column to the right.
   2. COPY the Y_2 column—one column to the right. EDIT the heading label to Y_3.
   3. COPY the second wave input data block to the right and EDIT the labels to reflect the third wave.
   4. EDIT the Y_3 column to reflect the third wave data block addresses.
   5. EDIT the Y_T column to include Y_1, Y_2, and Y_3 in the sum.

S6. Plot $Y(t)$ versus ωt. You may want to start with three terms in the series and then add additional terms and watch how $Y(t)$ changes.

Chapter 15

1. 6.24 MPa
3. 5.27×10^{18} kg
5. 1.62 m
7. 7.74×10^{-3} m^2
9. 98.6 kPa
11. 10.5 m; no
13. $m/[(\rho_w - \rho_s)h]$
15. (a) 7.00 cm   (b) 2.80 kg
17. 1250 kg/m^3 and 500 kg/m^3
19. 1430 m^3

21. (a) 4.24 m/s   (b) 17.0 m/s
23. (a) 17.7 m/s   (b) 1.73 mm
25. (a) 1 atm + 15.0 MPa   (b) 2.98 m/s
     (c) 4.45 kPa
27. 68.0 kPa
29. 103 m/s
31. (b) $2\sqrt{h(h_0 - h)}$
33. 0.258 N

35. 1.91 m
37. 8.01 km; yes
39. 709 kg/m^3
41. 2.00 MN
45. 90.04%
47. 4.43 m/s
51. (b) 1.40 s

Index

Page numbers in *italics* indicate illustrations; page numbers followed by "n" indicate footnotes; page numbers followed by "t" indicate tables.